INTRODUCTION TO FOURIER ANALYSIS

INSTRUCTOR'S MANUAL

INTRODUCTION TO FOURIER ANALYSIS

INSTRUCTOR'S MANUAL

Norman Morrison
University of Cape Town
South Africa

A Wiley-Interscience Publication
JOHN WILEY & SONS, INC.
New York • Chichester • Brisbane • Toronto • Singapore

Please address any comments regarding this manual to:

Norman Morrison
Electrical Engineering
University of Cape Town
P/Bag Rondebosch, South Africa 7700

Email: norman@uctvax.uct.ac.za

Note: This manual has been written in conjuction with Version 1.30 of the FFT disk. While the differences are minor, if your version number is lower than 1.30 you may wish to contact Wiley-Interscience for an update.

Library of Congress Cataloging in Publication Data:

ISBN 0-471-12848-1 (paper)

10 9 8 7 6 5 4 3 2 1

Contents

Chapter 2 Fourier Series for Periodic Functions

2.1 (1)

$$\begin{aligned} \sin(A+B) &= \sin(A)\cos(B) + \cos(A)\sin(B) \\ +\ \sin(A-B) &= \sin(A)\cos(B) - \cos(A)\sin(B) \end{aligned}$$

$$\sin(A+B) + \sin(A-B) = 2\sin(A)\cos(B)$$

$$\sin(A)\cos(B) = \tfrac{1}{2}[\sin(A+B) + \sin(A-B)] \qquad \text{QED}$$

(2)

$$\begin{aligned} \sin(A+B) &= \sin(A)\cos(B) + \cos(A)\sin(B) \\ -\ \sin(A-B) &= -\sin(A)\cos(B) + \cos(A)\sin(B) \end{aligned}$$

$$\sin(A+B) - \sin(A-B) = 2\cos(A)\sin(B)$$

$$\cos(A)\sin(B) = \tfrac{1}{2}[\sin(A+B) - \sin(A-B)] \qquad \text{QED}$$

(3)

$$\begin{aligned} \cos(A+B) &= \cos(A)\cos(B) - \sin(A)\sin(B) \\ +\ \cos(A-B) &= \cos(A)\cos(B) + \sin(A)\sin(B) \end{aligned}$$

$$\cos(A+B) + \cos(A-B) = 2\cos(A)\cos(B)$$

$$\cos(A)\cos(B) = \tfrac{1}{2}[\cos(A+B) + \cos(A-B)] \qquad \text{QED}$$

(4)

$$\begin{aligned} \cos(A-B) &= \cos(A)\cos(B) + \sin(A)\sin(B) \\ -\ \cos(A+B) &= -\cos(A)\cos(B) + \sin(A)\sin(B) \end{aligned}$$

$$\cos(A-B) - \cos(A+B) = 2\sin(A)\sin(B)$$

$$\sin(A)\sin(B) = \tfrac{1}{2}[\cos(A-B) - \cos(A+B)] \qquad \text{QED}$$

2.2

(a)

$$\int_{-1}^{1} \cos(t)\cos(2\omega_0 t)\,dt = 2\int_0^1 \cos(t)\cos(2\omega_0 t)\,dt$$

(using the fact that the integrand is even)

$$= \int_0^1 ([\cos(1+2\omega_0)t] + [\cos(1-2\omega_0)t])\,dt$$

(using the result of Exercise 2.1(c))

$$= \frac{\sin[(1+2\omega_0)t]}{1+2\omega_0}\Bigg|_0^1 + \frac{\sin[(1-2\omega_0)t]}{1-2\omega_0}\Bigg|_0^1$$

$$= \frac{\sin(1+2\omega_0)}{1+2\omega_0} + \frac{\sin(1-2\omega_0)}{1-2\omega_0}$$

(b)

$$\int_0^2 \sin(\pi t)\cos(n\pi t)\,dt = \tfrac{1}{2}\int_0^2 (\sin[(1+n)\pi t] + \sin[(1-n)\pi t])\,dt$$

$$= -\tfrac{1}{2}\frac{\cos[(1+n)\pi t}{(1+n)\pi}\Bigg|_0^2 - \tfrac{1}{2}\frac{\cos[(1-n)\pi t]}{(1-n)\pi}\Bigg|_0^2$$

$$= \tfrac{1}{2}\frac{1-\cos[(1+n)2\pi]}{(1+n)\pi} + \tfrac{1}{2}\frac{1-\cos[(1-n)2\pi]}{(1-n)\pi} = 0$$

since $\cos[(1+n)2\pi] = \cos(2\pi)\cos(n2\pi) - \sin(2\pi)\sin(n2\pi) = 1$

and $\cos[(1-n)2\pi] = \cos(2\pi)\cos(n2\pi) + \sin(2\pi)\sin(n2\pi) = 1$

(c)

$$\int_0^1 \cos(\pi t)\sin(n\pi t)\,dt = \tfrac{1}{2}\int_0^1 (\sin[(1+n)\pi t] - \sin[(1-n)\pi t]\,dt$$

$$= -\tfrac{1}{2}\frac{\cos[(1+n)\pi t]}{(1+n)\pi}\Bigg|_0^1 + \tfrac{1}{2}\frac{\cos[(1-n)\pi t]}{(1-n)\pi}\Bigg|_0^1$$

$$= \tfrac{1}{2}\frac{1-\cos[(1+n)\pi]}{(1+n)\pi} - \tfrac{1}{2}\frac{1-\cos[(1-n)\pi]}{(1-n)\pi}$$

$$= \tfrac{1}{2}\frac{1-\cos(\pi)\cos(n\pi)}{(1+n)\pi} - \tfrac{1}{2}\frac{1-\cos(\pi)\cos(n\pi)}{(1-n)\pi}$$

$$= \tfrac{1}{2}\frac{1+\cos(n\pi)}{(1+n)\pi} - \tfrac{1}{2}\frac{1+\cos(n\pi)}{(1-n)\pi}$$

$$= \frac{1+(-1)^n}{2\pi}\left[\frac{1}{1+n} - \frac{1}{1-n}\right]$$

$$= \frac{1+(-1)^n}{\pi}\left[\frac{n}{n^2-1}\right] = \begin{cases} (2/\pi)[n/(n^2-1)] & (n \text{ even}) \\ 0 & (n \text{ odd}) \end{cases}$$

$$\lim_{n\to 1} \frac{1+\cos(n\pi)}{n^2-1} = \lim_{n\to 1} \frac{-\pi\sin(n\pi)}{2n} = 0$$

(d) $$\int_{-2}^{2} \sin(t)\,\sin(n\pi t)\,dt = 2\int_{0}^{2} \sin(t)\,\sin(n\pi t)\,dt$$

$$= \int_0^2 (\cos[(1-n\pi)t] - \cos[(1+n\pi)t])\,dt = \left.\frac{\sin[(1-n\pi)t]}{1-n\pi}\right|_0^2 - \left.\frac{\sin[(1+n\pi)t]}{1+n\pi}\right|_0^2$$

$$= \frac{\sin[(1-n\pi)2]}{1-n\pi} - \frac{\sin[(1+n\pi)2]}{1+n\pi} = \frac{\sin(2)\cos(2n\pi)}{1-n\pi} - \frac{\sin(2)\cos(2n\pi)}{1+n\pi}$$

$$= \sin(2)\left[\frac{1}{1-n\pi} - \frac{1}{1+n\pi}\right] = \frac{2n\pi\sin(2)}{1-n^2\pi^2}$$

2.5 (a) $\cos(2\pi t)$ $\qquad \omega_0 = 2\pi,\ T_0 = 2\pi/\omega_0 = 1$

$= \tfrac{1}{2}\exp(-j2\pi t) + \tfrac{1}{2}\exp(j2\pi t)$

and so $F(-1) = \tfrac{1}{2} = F(1)$

(b) $1 + \sin(t) + \sin(2t)$ $\qquad \omega_0 = 1,\ T_0 = 2\pi$

$= 1 + (1/2j)\exp(jt) - (1/2j)\exp(-jt) + (1/2j)\exp(j2t) - (1/2j)\exp(-j2t)$

$F(-2) = j/2,\ F(-1) = j/2,\ F(0) = 1,\ F(1) = -j/2,\ F(2) = -j/2$

(c) $3 + 4\cos(\pi t) - 2\cos(2\pi t) + 6\cos(5\pi t)$ $\qquad \omega_0 = \pi,\ T_0 = 2$

$= 3 + 2\exp(-j\pi t) + 2\exp(j\pi t) - \exp(-j2\pi t) - \exp(j2\pi t)$

$+ 3\exp(-j5\pi t) + 3\exp(j5\pi t)$

$F(-5)=3,\ F(-2)=-1,\ F(-1)=2,\ F(0)=3,\ F(1)=2,\ F(2)=-1,\ F(5)=3$

(d) $5 - 3\sin(2t) + 2\sin(8t) + 5\sin(10t)$ $\qquad \omega_0 = 2,\ T_0 = \pi$

$= 5 - (3/2j)\exp(j2t) + (3/2j)\exp(-j2t) + (1/j)\exp(j8t)$

$- (1/j)\exp(-j8t) + (5/2j)\exp(j10t) - (5/2j)\exp(-j10t)$

$F(-5)=5j/2,\ F(-4)=j,\ F(-1)=-3j/2,\ F(0)=5,\ F(1)=3j/2,\ F(4)=-j,\ F(5)=-5j/2$

$= 4 - (1/j)\exp(j3t) + (1/j)\exp(-j3t) + (3/2)\exp(j2t)$

$+ (3/2)\exp(-j2t) + (2/j)\exp(j5t) - (2/j)\exp(-j5t)$

$F(-5)=2j,\ F(-3)=-j,\ F(-2)=3/2,\ F(0)=4,\ F(2)=3/2,\ F(3)=j,\ F(5)=-2j$

(f) $3\cos(100\pi t) + 5\sin(70\pi t)$ $\qquad \omega_0 = 10\pi,\ T_0 = 1/5$

$= (3/2)\exp(j100t) + (3/2)\exp(-j100t)$

$+ (5/2j)\exp(j70t) - (5/2j)\exp(-j70t)$

$F(-10) = 3/2,\ F(-7) = 5j/2,\ F(7) = -5j/2,\ F(10) = 3/2$

(g) $\cos(2\pi t/3) + 2\cos(5\pi t/3)$ $\qquad \omega_0 = \pi/3,\ T_0 = 6$

$= \tfrac{1}{2}\exp(j2\pi t/3) + \tfrac{1}{2}\exp(-j2\pi t/3) + \exp(j5\pi t/3) + \exp(-j5\pi t/3)$

$F(-5) = 1,\ F(-2) = \tfrac{1}{2},\ F(2) = \tfrac{1}{2},\ F(5) = 1$

(h) $\cos(t/2) - 2\sin(t/3)$ $\qquad \omega_0 = 1/6,\ T_0 = 12\pi$

$= \tfrac{1}{2}\exp(jt/2) + \tfrac{1}{2}\exp(-jt/2) - (1/j)\exp(jt/3) + (1/j)\exp(-jt/3)$

$F(-3) = \tfrac{1}{2},\ F(-2) = -j,\ F(2) = j,\ F(3) = \tfrac{1}{2}$

(i) $\cos(2t) + \sin(\pi t)$ $\qquad$ Not periodic, no Fourier series, no ω_0, no T_0

(j) $$f_p(t) = \tfrac{1}{4} + \sum_{n=1}^{4}\left[\frac{1}{n^2}\cos(nt/5) + \frac{(-1)^n}{(2n+1)^2}\sin(nt/5)\right] \qquad \omega_0=1/5,\ T_0=10\pi$$

$$f_p(t) = \tfrac{1}{4} + \sum_{n=1}^{4}\left[\frac{1}{2n^2}\exp(jnt/5) + \frac{1}{2n^2}\exp(-jnt/5)\right]$$

$$+ \sum_{n=1}^{4}\left[\frac{(-1)^n}{2j(2n+1)^2}\exp(jnt/5) - \frac{(-1)^n}{2j(2n+1)^2}\exp(-jnt/5)\right]$$

$$f_p(t) = \tfrac{1}{4} + \sum_{n=1}^{4}\left[\frac{1}{2n^2} - j\frac{(-1)^n}{2(2n+1)^2}\exp(jnt/5)\right]$$

$$+ \sum_{n=1}^{4}\left[\frac{1}{2n^2} + j\frac{(-1)^n}{2(2n+1)^2}\exp(-jnt/5)\right]$$

$$F(n) = \frac{1}{2n^2} - j\frac{(-1)^n}{2(2n+1)^2}, \quad F(-n) = \frac{1}{2n^2} + j\frac{(-1)^n}{2(2n+1)^2} \quad (n = 1,2,3,4) \quad F(0) = \tfrac{1}{4}$$

(k) $$f_p(t) = \tfrac{1}{2} + \frac{2}{\pi}\sum_{k=1}^{3} \frac{1}{2k-1}\sin[(2k-1)2\pi t] \quad \omega_0 = 2\pi, \quad T_0 = 1$$

$$f_p(t) = \tfrac{1}{2} + \frac{2}{\pi}\sum_{k=1}^{3}\left[\frac{1}{2j(2k-1)}\exp[j(2k-1)2\pi t] - \frac{1}{2j(2k-1)}\exp[-j(2k-1)2\pi t]\right]$$

$F(0) = \tfrac{1}{2}$, $F(k) = -j/\pi(2k-1)$, $F(-k) = j/\pi(2k-1)$, $(k = 1, 2, 3)$

2.6 (a) $f(t) = 2\sin(3t)$ $\qquad \omega_0 = 3, \quad T_0 = 2\pi/3$

$$= (1/j)\exp(j3t) - (1/j)\exp(-j3t)$$

$F(-1) = j$, $F(1) = -j$, $P(-1) = |F(-1)|^2 = 1$, $P(1) = 1$

$$P_{tot} = \sum_{n=-\infty}^{\infty} P(n) = 2$$

(b) $f(t) = 3\cos(2t)$ $\qquad \omega_0 = 2, \quad T_0 = \pi$

$$= (3/2)\exp(j2t) + (3/2)\exp(-j2t)$$

$F(-1) = 3/2$, $F(1) = 3/2$, $P(-1) = 9/4$, $P(1) = 9/4$, $P_{tot} = 4.5$

As a voltage across 100Ω, $P_{tot} = 0.045$, as a current through 100Ω, $P_{tot} = 450$

(c) $f(t) = 5\sin(5t) + 7\cos(6t) + 3$ $\qquad \omega_0 = 1, \quad T_0 = 2\pi$

$$= (5/2j)\exp(j5t) - (5/2j)\exp(-j5t) + (7/2)\exp(j6t) + (7/2)\exp(-j6t) + 3$$

$F(-6) = 7/2$, $F(-5) = 5j/2$, $F(0) = 3$, $F(5) = -5j/2$, $F(6) = 7/2$

$P(-6) = 49/4$, $P(-5) = 25/4$, $P(0) = 9$, $P(5) = 25/4$, $P(6) = 49/4$

$P_{tot} = 46$. As a voltage across 10KΩ, $P_{tot} = 4.6e-3$, as a current through 0.01Ω, $P_{tot} = 0.46$.

2.7 (a) $$f_p(t) = \sum_{n=-\infty}^{\infty} Sa(n\pi/2)\exp(jn\pi t/2) = \sum_{n=-\infty}^{\infty} F(n)\exp(jn\pi t/2)$$

$F(n) = Sa(n\pi/2)$, $\omega_0 = \pi/2$, $T = 4$

$$F(n)^* = Sa(n\pi/2), \quad F(-n) = \frac{\sin(-n\pi/2)}{-n\pi/2} = \frac{\sin(n\pi/2)}{n\pi/2} = Sa(n\pi/2)$$

and so $F(n)^* = F(-n)$. Thus $f_p(t)$ is real.

$A(n) = Sa(n\pi/2)$, $B(n) = 0$. Thus $f_p(t)$ is even.

$F(n)$ dies out like $1/n$. Thus $f_p(t)$ has discontinuities in its 0-th derivative. The average value of $f_p(t)$ is $F(0) = 1$.

(b) $$f_p(t) = \tfrac{1}{2} + \frac{j}{2\pi}\sum_{\substack{n=-\infty \\ n\neq 0}}^{\infty} \frac{(-1)^n - 1}{n}\exp(jn2\pi t)$$

$$F(n) = \begin{cases} \dfrac{j[(-1)^n - 1]}{2\pi n} & (n \neq 0) \\ \tfrac{1}{2} & (n = 0) \end{cases} \qquad \omega_0 = 2\pi, \quad T_0 = 1$$

For $n = 0$, $F(n)^* = F(-n)$, and for $n \neq 0$: $F(n)^* = \dfrac{-j[(-1)^n - 1]}{2\pi n}$,

$$F(-n) = \frac{j[(-1)^{-n} - 1]}{-2\pi n} = \frac{-j[(-1)^n - 1]}{2\pi n} = F(n)^* \quad \text{thus } f_p(t) \text{ is real.}$$

$A(n) = \tfrac{1}{2}$, $B(n) = 0$ for $n = 0$, and $A(n) = 0$, $B(n) \neq 0$ for $n \neq 0$

Thus $f_p(t)$ is neither odd nor even.

$F(n)$ dies out like $1/n$, and so $f_p(t)$ has discontinuities.

Average value of $f_p(t)$ is $F(0) = \tfrac{1}{2}$

(c) $$f_p(t) = \sum_{\substack{n=-\infty \\ n\neq 0}}^{\infty} \frac{\cos(n\pi/2) - j\sin(n\pi/2)}{n(n^2 - 4)}\exp(j4nt)$$

$$F(n) = \frac{\cos(n\pi/2) - j\sin(n\pi/2)}{n(n^2 - 4)} \quad (n \neq 0) \qquad \omega_0 = 4, \quad T_0 = \pi/2$$

$$F(n)^* = \frac{\cos(n\pi/2) + j\sin(n\pi/2)}{n(n^2 - 4)}$$

$$F(-n) = \frac{\cos(-n\pi/2) - j\sin(-n\pi/2)}{-n(n^2 - 4)} = \frac{-\cos(n\pi/2) - j\sin(n\pi/2)}{n(n^2 - 4)} \neq F(n)^*$$

(d) $$f_p(t) = \frac{2j}{\pi}\sum_{\substack{n=-\infty\\ n\neq 0}}^{\infty} \frac{1-(-1)^n}{n(n^2-4)}\exp(jnt)$$

$$F(n) = \frac{2j[1-(-1)^n]}{\pi n(n^2-4)} \qquad (n \neq 0) \qquad \omega_0 = 1, \quad T_0 = 2\pi$$

$$F(n)^* = \frac{-2j[1-(-1)^n]}{\pi n(n^2-4)}, \quad F(-n) = \frac{2j[1-(-1)^{-n}]}{-\pi n(n^2-4)} = \frac{-2j[1-(-1)^n]}{\pi n(n^2-4)} = F(n)^*$$

Thus $f_p(t)$ is real.

$$A(n) = 0, \quad B(n) = \frac{2j[1-(-1)^n]}{\pi n(n^2-4)}$$ and so $f_p(t)$ is odd.

$F(n)$ dies out like $1/n^3$. Thus $f_p(t)$ and $f'_p(t)$ are both everywhere continuous, and $f''_p(t)$ has discontinuities.

Average value of $f_p(t)$ is $F(0) = 0$.

(e) $$f_p(t) = \tfrac{1}{2}\sum_{n=-\infty}^{\infty} \frac{(-1)^n e - 1}{1 - jn\pi}\exp(jn\pi t)$$

$$F(n) = \tfrac{1}{2}\frac{(-1)^n e - 1}{1 - jn\pi}, \quad \omega_0 = \pi, \quad T_0 = 2.$$

$$F(n)^* = \tfrac{1}{2}\frac{(-1)^n e - 1}{1 + jn\pi} \qquad F(-n) = \tfrac{1}{2}\frac{(-1)^{-n}e - 1}{1 + jn\pi} = \tfrac{1}{2}\frac{(-1)^n e - 1}{1 + jn\pi} = F(n)^*$$

thus $f_p(t)$ is real.

$$F(n) = \tfrac{1}{2}\frac{(-1)^n e - 1}{1 - jn\pi} = \tfrac{1}{2}[(-1)^n e - 1]\frac{1 + jn\pi}{1 + n^2\pi^2}$$

and so $$A(n) = \frac{\tfrac{1}{2}[(-1)^n e - 1]}{1 + n^2\pi^2} \neq 0$$

$$B(n) = \frac{\tfrac{1}{2}[(-1)^n e - 1]jn\pi}{1 + n^2\pi^2} \neq 0$$

Thus $f_p(t)$ is neither odd nor even. $F(n)$ dies out like $1/n$, and so $f_p(t)$ is discontinuous. Average value of $f_p(t)$ is $F(0) = \tfrac{1}{2}[e - 1]$

2.8 $$f_p(t) = \sum_{n=-\infty}^{\infty} F(n)\exp(jn\omega_0 t), \quad \text{where} \quad \omega_0 = 2\pi/T_0$$

For each of these problems we are given $F(n)$ and T_0 and so we can assemble the Fourier series representation for $f_p(t)$. The results are given in the Answers.

To test if $f_p(t)$ is real or not we use: Is $F(n)^* = F(-n)$?

(a) $$F(n) = \begin{cases} \tfrac{1}{2} & (n = 0) \\ \dfrac{\cos(n\pi) - 1}{n^2\pi^2} & (n \neq 0) \end{cases} \qquad T_0 = 2$$

$F(n)^*$ is the same as $F(n)$.

$$F(-n) = \begin{cases} \tfrac{1}{2} & (n = 0) \\ \dfrac{\cos(-n\pi) - 1}{(-n)^2\pi^2} & (n \neq 0) \end{cases} = \begin{cases} \tfrac{1}{2} & (n = 0) \\ \dfrac{\cos(n\pi) - 1}{n^2\pi^2} & (n \neq 0) \end{cases}$$

which is also the same as $F(n)$. Thus $f_p(t)$ is real. Its average value is $F(0) = \tfrac{1}{2}$

(b) $$F(n) = j\frac{n\cos(n\pi/2) - \sin(n\pi/2)}{n^2} \qquad (n \neq 0) \qquad T_0 = \pi$$

$$F(n)^* = -j\frac{n\cos(n\pi/2) - \sin(n\pi/2)}{n^2} = -F(n)$$

$$F(-n) = j\frac{-n\cos(-n\pi/2) - \sin(-n\pi/2)}{(-n)^2} = j\frac{-n\cos(n\pi/2) + \sin(n\pi/2)}{n^2} = -F(n)$$

Thus $f_p(t)$ is real. Its average value is $F(0) = 0$.

(c) $$F(n) = \tfrac{1}{2}\frac{\sin(n\pi/2) - jn\cos(n\pi/2)}{n\pi/2} \qquad (\forall n) \qquad T_0 = 4$$

$$F(n)^* = \tfrac{1}{2}\frac{\sin(n\pi/2) + jn\cos(n\pi/2)}{n\pi/2}$$

$$F(-n) = \tfrac{1}{2}\frac{\sin(-n\pi/2) + jn\cos(-n\pi/2)}{-n\pi/2} = \tfrac{1}{2}\frac{\sin(n\pi/2) - jn\cos(n\pi/2)}{n\pi/2} \neq F(n)^*$$

and so $f_p(t)$ is complex. Its average value is $F(0)$ which we can find as follows:

$$F(0) = \tfrac{1}{2} \lim_{n \to 0} \frac{\sin(n\pi/2) - jn\cos(n\pi/2)}{n\pi/2}$$

$$= \tfrac{1}{2} \lim_{n \to 0} \frac{\sin(n\pi/2)}{n\pi/2} - \tfrac{1}{2} \lim_{n \to 0} \frac{jn\cos(n\pi/2)}{n\pi/2} = \tfrac{1}{2} - j/\pi$$

2.9 (a) Writing z and w in their polar form we have:

$z = |z| \exp[j\arg(z)]$ and $w = |w| \exp[j\arg(w)]$ and so

$$zw = |z|\exp[j\arg(z)]\ |w|\exp[j\arg(w)]$$
$$= |z||w| \exp[j\arg(z)]\exp[j\arg(w)]$$
$$= |z||w| \exp[j(\arg(z) + \arg(w))]$$

which is the polar form of zw. From this we see that

$|zw| = |z||w|$ and $\arg(zw) = (\arg(z) + \arg(w)$ QED

$$z/w = \frac{|z|\exp[j\arg(z)]}{|w|\exp[j\arg(w)]} = |z|/|w| \exp[j(\arg(z) - \arg(w))]$$

which is the polar form of z/w. From this we see that

$|z/w| = |z|/|w|$ and $\arg(zw) = (\arg(z) - \arg(w)$ QED

(b) (1) $F(n) = j\exp(-jn\pi/2)/(n\pi/2)$ $(n \neq 0)$

$$|F(n)| = |j\exp(-jn\pi/2)/(n\pi/2)|$$
$$= |j||\exp(-jn\pi/2)|/|(n\pi/2)|$$
$$= 1\times1/(|n|\pi/2) = 2/|n|\pi$$

$$\theta(n) = \arg(j\exp(-jn\pi/2)/(n\pi/2))$$
$$= \arg(j) + \arg[\exp(-jn\pi/2)] - \arg(n\pi/2)$$
$$= \tfrac{1}{2}\pi - \tfrac{1}{2}n\pi - \arg(n\pi/2) = \tfrac{1}{2}\pi(1-n) - \arg(n\pi/2)$$

For $n > 0$, $n\pi/2$ is a positive number, and so its argument is zero. For $n < 0$, $n\pi/2$ is a negative number and so its argument is π or $-\pi$. Thus

$\theta(n) = \tfrac{1}{2}\pi(1-n)$ $(n > 0)$ $\quad \theta(n) = \tfrac{1}{2}\pi(1-n) - \pi$

$= -\tfrac{1}{2}\pi(1+n)$ $(n < 0)$

(2) $$F(n) = \overbrace{\left[\frac{1 - \exp(-jn) - jn\exp(-jn)}{(jn)^2}\right]}^{F_1(n)} \exp(jn) \qquad (n \neq 0)$$

First we find the real and imaginary parts of $F_1(n)$ as follows:

$$F_1(n) = \frac{1 - \exp(-jn) - jn\exp(-jn)}{(jn)^2}$$
$$= \frac{1 - [\cos(n) - j\sin(n)] - jn[\cos(n) - j\sin(n)]}{-n^2}$$
$$= \frac{-1 + \cos(n) - j\sin(n) + jn\cos(n) + n\sin(n)}{n^2}$$
$$= \frac{\cos(n) + n\sin(n) - 1}{n^2} + j\frac{n\cos(n) - \sin(n)}{n^2}$$

Then

$$|F(n)| = |F_1(n)||\exp(jn)| = |F_1(n)|$$
$$= \left[[\cos(n) + n\sin(n) - 1]^2 + [n\cos(n) - \sin(n)]^2\right]^{\frac{1}{2}}/n^2$$

$$\arg[F(n)] = \arg[F_1(n)] + \arg[\exp(jn)]$$
$$= \tan^{-1}\left[\frac{n\cos(n) - \sin(n)}{\cos(n) + n\sin(n) - 1}\right] + n$$

(3) $$F(n) = \frac{(1 + jn)(2 - j3n)(4 + j5n)}{(6 + j7n)(8 + j9n)(10 + j11n)}$$

$$|F(n)| = \frac{|(1 + jn)||(2 - j3n)||(4 + j5n)|}{|(6 + j7n)||(8 + j9n)||(10 + j11n)|}$$
$$= \left[\frac{(1 + n^2)(4 + 9n^2)(16 + 25n^2)}{(36 + 49n^2)(64 + 81n^2)(100 + 121n^2)}\right]^{\frac{1}{2}}$$

$$\theta(n) = \arg(1 + jn) + \arg(2 - j3n) + \arg(4 + j5n) - \arg(6 + j7n)$$

$= \tan^{-1}(n/1) +$ [illegible]

$\quad - \tan^{-1}(7n/6) - \tan^{-1}(9n/8) + \tan^{-1}(11n/10) + n$

(c) (1) $|3+j| = [3^2+1^2]^{1/2} = 3.1623$
$\arg = \text{atn}(1/3) = 0.3218$ (first quadrant)

(2) $|1-3j| = [1^2+3^2]^{1/2} = 3.1623$
$\arg = \text{atn}(-3/1) = -1.249$ (fourth quadrant)

(3) $|-1-2j| = [1^2+2^2]^{1/2} = 2.2361$
$\arg = \text{atn}(-2/-1) = -2.0344$ (third quadrant)

(4) $|-3+4j| = [3^2+4^2]^{1/2} = 5$
$\arg = \text{atn}(4/-3) = 2.2143$ (second quadrant)

(5) $|-1| = 1 \quad \arg = \pi$ (-1 lies on the negative real line)

(6) $|\pi| = \pi \quad \arg = 0$ (π lies on the +ve real line)

(7) $|j\pi| = \pi \quad \arg = \pi/2$ ($j\pi$ lies on the +ve imag. axis)

(8) $|-j\pi/5| = \pi/5 \quad \arg = -\pi/2$ (it lies on the neg imag axis)

(9) $|1/j\pi| = 1/\pi \quad \arg(1/j\pi) = \arg(-j/\pi) = -\pi/2$ (it lies on the neg imag axis)

(10) $|-2/j\pi| = 2/\pi \quad \arg(-2/j\pi) = \arg(j2/\pi) = \pi/2$ (it lies on the +ve imag axis)

(11) $\exp(j\pi) = \cos(\pi) + j\sin(\pi) = -1 \quad$ magnit $= 1 \quad \arg = \pi$ (obvious by inspection)

(12) $|\exp(-j\pi/2)| = 1 \quad \arg = -\pi/2$

(13) $|-\exp(j\pi/2)| = 1 \quad \arg = -\pi/2$ (it lies on the neg imag axis)

(14) $|(1-2j)\exp(j\pi/2)| = |1-2j||\exp(j\pi/2)| = [1^2+2^2]^{1/2} = 2.2361$
$\arg = \arg(1-2j) + \arg[\exp(j\pi/2)] = \text{atn}(-2/1)+\pi/2 = 0.4636$

(15) $|-(1-j)/j\pi| = |-(1-j)|/|j\pi| = [1^2+1^2]^{1/2}/\pi = 0.4502$
$\arg[-(1-j)/j\pi] = \arg(-1+j) - \arg(j\pi) = \text{atn}(1/-1) - \pi/2$
$= (-0.7854+\pi) - \pi/2 = 0.7854$

(16) $|-(1-j)j\pi| = |-1+j||j\pi| = 1.4142\pi = 4.4429$
$\arg[(-1+j)j\pi] = \arg(-1+j) + \arg(j\pi) = (-0.7854+\pi) + \pi/2$
$= 3.9270$ which exceeds π and so we subtract 2π giving -2.3562

(17) $|(1+j)/(1-2j)| = |1+j|/|1-2j| = 2^{1/2}/5^{1/2} = 0.6325$
$\arg[(1+j)/(1-2j)] = \arg(1+j) - \arg(1-2j)$
$= \text{atn}(1/1) - \text{atn}(-2/1) = 0.7854 - 1.1071 = 1.8925$

(18) $|(-3+4j)/(-j4\pi-2)j\pi/3| = |-3+4j|/\Big[|-j4\pi-2||j\pi/3|\Big]$

$= (3^2+4^2)^{1/2}/[(4^2\pi^2+2^2)^{1/2}\pi/3] = 0.3752$

$= \text{atn}(4/-3) - \text{atn}(-4\pi/-2) - \pi/2$
$= [-1.249+\pi] - [1.4130-\pi] - \pi/2 = 2.3721$

(19) $|-j(1-3j)/(-2+j)e^{2j}| = \Big[|-j||(1-3j)|\Big]/\Big[|(-2+j)||e^{2j}|\Big]$

$= [1 \times 10^{1/2}]/[5^{1/2} \times 1] = 1.4142$

$\arg = \arg(-j) + \arg[(1-3j)] - \arg(-2+j) - \arg(e^{2j})$
$= -\pi/2 + \text{atn}(-3/1) - \text{atn}(1/-2) - 2$
$= -\pi/2 - 1.247 - [-0.4636+\pi] - 2$
$= -7.4978$ to which we add 2π giving -1.2146

(20) $|j[(-1)^3e^{-1} - j]/[1+j3\pi]| = |j||[(-1)^3e^{-1} - j]|/|1+j3\pi|$

$= 1 \times [e^{-2} + 1]^{1/2} / [1^2 + 9\pi^2]^{1/2} = 0.1124$

$\arg = \arg(j) + \arg(-e^{-1} - j) - \arg(1+3\pi j)$
$= \pi/2 + \text{atn}(-1/-e^{-1}) - \text{atn}(3\pi/1)$
$= \pi/2 + [1.2183 - \pi] - 1.4651 = -1.8176$

2.10

(a) $f_p(t) = \begin{cases} 1 & (0 < t < 1) \\ 0 & (1 < t < 2) \end{cases} \qquad f_p(t+2) = f_p(t)$

$T_0 = 2, \quad \omega_0 = 2\pi/T_0 = \pi$

(b) $n \neq 0$: $F(n) = 1/T_0 \int_0^{T_0} f_p(t)\exp(-jn\omega_0 t)\,dt = \frac{1}{2}\int_0^1 \exp(-jn\pi t)\,dt$

$$= \frac{1}{2}\,\frac{\exp(-jn\pi t)}{-jn\pi}\Bigg|_0^1 = \frac{1}{2}\,\frac{\exp(-jn\pi) - 1}{-jn\pi} = \frac{j[(-1)^n - 1]}{2n\pi}$$

$n = 0$: $F(n) = 1/T_0 \int_0^{T_0} f_p(t)\,dt = \frac{1}{2}\int_0^1 dt = \frac{1}{2}$

n	0	1	2	3	4	5
A(n)	½	0	0	0	0	0
B(n)	0	$-1/\pi$	0	$-1/3\pi$	0	$-1/5\pi$

(c) $f_p(t) = \ldots + (j/3\pi)\exp(-j3\pi t) + (j/\pi)\exp(-j\pi t) + \frac{1}{2}$

$\quad - (j/\pi)\exp(j\pi t) - (j/3\pi)\exp(3\pi t) + \ldots$

(d) Wherever t is an integer, $\exp(jn\pi t) = \exp(-jn\pi t)$ and so opposite pairs of these terms cancel each other for all values of n, leaving only the ½. This is the same result as predicted by Fourier's theorem in the text.

(e) The coefficients converge to zero like $1/n$.

(f) Steps for loading the waveform using N = 256, SAMPLED, T = 2, PERIODIC:

Main menu, Create Values, Continue, Create X, Real, Neither, Left, 2 intervals, 1, Continue, 1, 0, Go, Neither, Go. You are now back on the main menu with the waveform correctly loaded. Plotting **X** confirms that.

Run ANALYSIS, Show Numbers, F, Complex.

The FFT values were as shown in Table 2.6 in the text.

(g) The FFT values in Table 2.6 are very close to the values from the table in (b), and shown again below:

Formula values of F(n)		
n	A(n)	B(n)
0	½	0
1	0	-0.318309886
3	0	-0.106103295
5	0	-6.3661977e-2

(h) From the final column of Table 2.6 we see that the coefficients are dying out like $1/n$. In theory that is always true for n large but not necessarily for n small, but here we see it also for n small.

2.11

(a) $$f_p(t) = \begin{cases} -\tfrac{1}{2} & (-2 < t < -1) \\ 1 & (-1 < t < 1) \\ \tfrac{1}{2} & (1 < t < 2) \end{cases} \qquad f_p(t+4) = f_p(t) \qquad f_p(t) \text{ is even}$$

(b) $T_0 = 4, \quad \omega_0 = \pi/2$

$$n \neq 0: \ F(n) = 1/T_0 \int_{-T_0/2}^{T_0/2} f_p(t)\exp(-jn\omega_0 t)\,dt$$

$$= 2/T_0 \int_0^{T_0/2} f_p(t)\cos(jn\omega_0 t)\,dt \quad (\text{since } f_p(t) \text{ is even})$$

$$= \tfrac{1}{2}\int_0^1 \tfrac{1}{2}\cos(n\pi t/2)\,dt + \tfrac{1}{2}\int_1^2 (-\tfrac{1}{2})\cos(n\pi t/2)\,dt$$

$$= \tfrac{1}{4}\,\frac{\sin(n\pi t/2)}{n\pi/2}\Bigg|_0^1 - \tfrac{1}{4}\,\frac{\sin(n\pi t/2)}{n\pi/2}\Bigg|_1^2$$

$$= \tfrac{1}{4}\,\frac{\sin(n\pi/2)}{n\pi/2} - \tfrac{1}{4}\,\frac{\sin(n\pi) - \sin(n\pi/2)}{n\pi/2}$$

$$= \tfrac{1}{2}\,\frac{\sin(n\pi/2)}{n\pi/2} = \tfrac{1}{2}\,Sa(n\pi/2) \quad (n \neq 0)$$

$n = 0$: $F(0) = 0$, since the mean value of $f_p(t)$ is zero.

Observe that for n even, $\sin(n\pi/2) = 0$ and so only the odd values of n appear in the Fourier series. Thus

$$f_p(t) = \ldots + \tfrac{1}{2}\,\frac{\sin(-3\pi/2)}{-3\pi/2}\exp(-j3\pi t/2) + \tfrac{1}{2}\,\frac{\sin(-\pi/2)}{-\pi/2}\exp(-j\pi t/2)$$

$$+ \tfrac{1}{2}\,\frac{\sin(\pi/2)}{\pi/2}\exp(j\pi t/2) + \tfrac{1}{2}\,\frac{\sin(3\pi/2)}{3\pi/2}\exp(j3\pi t/2) + \ldots$$

(c) According to Fourier's theorem, the series should converge to zero at t = 1, 3, 5, ... In fact, at t = 1:

$$\frac{\sin(n\pi/2)}{n\pi/2}\exp(jn\pi t/2)\Bigg|_{t=1} + \frac{\sin(-n\pi/2)}{-n\pi/2}\exp(-jn\pi t/2)\Bigg|_{t=1}$$

$$= \frac{\sin(n\pi/2)}{n\pi/2}\exp(jn\pi/2) + \frac{\sin(n\pi/2)}{n\pi/2}\exp(-jn\pi/2)$$

and since n is odd, $\exp(jn\pi/2) = j \text{ or } -j$, $\exp(-jn\pi/2) = -j \text{ or } j$

and so the Fourier series sums to zero. The same is true for t any other odd integer.

(d) Coefficients converge to zero like $1/n$

(e) Steps for loading this waveform into the FFT system using N = 256, SAMPLED, T = 4, PERIODIC:

Main menu, Create Values, Continue, Create X, Real, Even, 2 intervals, 1, Continue, 1/2, -1/2, Go, Neither, Go. You are now back on the main menu with the waveform correctly loaded.

We obtain the following values:

F(1) = 0.31829391   F(3) = -0.10605536   F(5) = 6.358206e-2
F(7) = -4.53609330e-2   F(9) = 3.5223837e-2

(f) F(n) is real for all values of n because $f_p(t)$ is real and even.

2.12 (a) $f_p(t) = t \quad (0 < t < 1) \qquad f_p(t + 1) = f_p(t)$

$T_0 = 1, \quad \omega_0 = 2\pi$

$$n \neq 0:\ F(n) = 1/T_0 \int_0^{T_0} f_p(t)\ \exp(-jn\omega_0 t)\ dt = \int_0^1 t\ \exp(-jn2\pi t)\ dt$$

$$= \left.\frac{t\ \exp(-jn2\pi t)}{-jn2\pi}\right|_0^1 - \left.\frac{\exp(-jn2\pi t)}{(-jn2\pi t)^2}\right|_0^1$$

$$= \frac{\exp(-jn2\pi)}{-jn2\pi} - \frac{\exp(-jn2\pi) - 1}{(-jn2\pi t)^2} = j/2\pi n$$

$$n = 0:\ F(n) = 1/T_0 \int_0^{T_0} f_p(t)\ dt = \int_0^1 t\ dt = t^2/2 \Big|_0^1 = \tfrac{1}{2}$$

(b) To convert the coefficients to their real form we use (2.88). Thus

$a(0) = 2\ F(0) = 1 \qquad a(n) = 2\ A(n) = 0 \qquad b(n) = -2\ B(n) = -1/\pi n$

$$f_p(t) = a(0)/2 + \sum_{n=1}^{\infty} a(n)\ \cos(n\omega_0 t) + \sum_{n=1}^{\infty} b(n)\ \sin(n\omega_0 t)$$

$$= \tfrac{1}{2} - (1/\pi)\left[\frac{\sin(2\pi t)}{1} + \frac{\sin(4\pi t)}{2} + \frac{\sin(6\pi t)}{3} + \frac{\sin(8\pi t)}{4} + \ldots\right]$$

(c) At t any integer all of the sine terms are zero and so the series sums to ½ at all of the points of discontinuity.

(e) Steps for loading the waveform into the FFT system using N = 256, SAMPLED, T = 1, PERIODIC:

Main menu, Create Values, Continue, Create X, Real, Neither, Left, 1 interval, Continue, t, Go, Neither, Go. You are now back on the main menu with the waveform correctly loaded. Plotting **X** confirms that.

Run ANALYSIS, Show Numbers, F, Real.

We obtain the values appearing in Table 2.12a below where we also show the exact values computed from the formula as well as the relative errors in the FFT values.

Table 2.12a

	FFT estimates of a(n) and b(n)		Formula values of a(n) and b(n)		Percentage error in FFT estimate
n	a(n)	b(n)	a(n)	b(n)	
0	1	0	1	0	0
1	0	-0.31829391	0	-0.318309866	-5.019e-3
2	0	-0.15912298	0	-0.159154943	-2.008e-2
3	0	-0.10605535	0	-0.106103295	-4.519e-2
4	0	-7.9513545e-2	0	-7.9577471e-2	-8.033e-2
5	0	-6.3582062e-2	0	-6.3661977e-2	-1.255e-1

(f) Repeating the FFT run using N = 1024 we obtained the set of values appearing in Table 2.12b. Observe how much smaller the relative errors are for N = 1024 compared to N = 256.

Table 2.12b

	FFT estimates of a(n) and b(n)		Formula values of a(n) and b(n)		Percentage error in FFT estimate
n	a(n)	b(n)	a(n)	b(n)	
0	1	0	1	0	0
1	0	-0.31830889	0	-0.318309866	-3.129e-4
2	0	-0.15915295	0	-0.159154943	-1.252e-3
3	0	-0.10610030	0	-0.106103295	-2.823e-3
4	0	-7.9573477e-2	0	-7.9577471e-2	-5.020e-3
5	0	-6.3656984e-2	0	-6.3661977e-2	-7.843e-3

2.13 Consider first (2.100), and assume that $n \neq m$. Then

$$\int_{-T_0/2}^{T_0/2} \cos(n\omega_0 t)\cos(m\omega_0 t)\,dt = \tfrac{1}{2}\int_{-T_0/2}^{T_0/2}\Big[\cos[(n+m)\omega_0 t] + \cos[(n-m)\omega_0 t]\Big]\,dt$$

$$= \frac{\tfrac{1}{2}\sin[(n+m)\omega_0 t]}{(n+m)\omega_0}\Bigg|_{-T_0/2}^{T_0/2} + \frac{\tfrac{1}{2}\sin[(n-m)\omega_0 t]}{(n-m)\omega_0}\Bigg|_{-T_0/2}^{T_0/2}$$

$$= \frac{\sin[(n+m)\pi]}{(n+m)\omega_0} + \frac{\sin[(n-m)\pi]}{(n-m)\omega_0} = 0$$

Assume next that $n = m \neq 0$. Then

$$\int_{-T_0/2}^{T_0/2} \cos(n\omega_0 t)\cos(m\omega_0 t)\,dt = \int_{-T_0/2}^{T_0/2}\cos^2(n\omega_0 t)\,dt = \tfrac{1}{2}\int_{-T_0/2}^{T_0/2}[1 + \cos(2n\omega_0 t)]\,dt$$

$$= \tfrac{1}{2}t\Bigg|_{-T_0/2}^{T_0/2} + \tfrac{1}{2}\frac{\sin(2n\omega_0 t)}{2n\omega_0}\Bigg|_{-T_0/2}^{T_0/2} = T_0/2 + \frac{\sin(2n\pi)}{2n\omega_0} = T_0/2$$

Assume finally that $n = m = 0$. Then

$$\int_{-T_0/2}^{T_0/2} \cos(n\omega_0 t)\cos(m\omega_0 t)\,dt = \int_{-T_0/2}^{T_0/2} dt = T_0$$

This completes the proof of (2.100). Continuing with (2.101):

$$\int_{-T_0/2}^{T_0/2} \cos(n\omega_0 t)\sin(m\omega_0 t)\,dt$$

This integral is zero because the integrand is odd $\forall$ n and m.

Consider next (2.102), and assume that $n \neq m$. Then

$$\int_{-T_0/2}^{T_0/2} \sin(n\omega_0 t)\sin(m\omega_0 t)\,dt = \tfrac{1}{2}\int_{-T_0/2}^{T_0/2}\Big[\cos[(n-m)\omega_0 t] - \cos[(n+m)\omega_0 t]\Big]\,dt$$

$$= \frac{\tfrac{1}{2}\sin[(n-m)\omega_0 t]}{(n-m)\omega_0}\Bigg|_{-T_0/2}^{T_0/2} - \frac{\tfrac{1}{2}\sin[(n+m)\omega_0 t]}{(n+m)\omega_0}\Bigg|_{-T_0/2}^{T_0/2}$$

$$= \frac{\sin[(n+m)\pi]}{(n+m)\omega_0} + \frac{\sin[(n-m)\pi]}{(n-m)\omega_0} = 0$$

Assume next that $n = m \neq 0$. Then

$$\int_{-T_0/2}^{T_0/2} \sin(n\omega_0 t)\sin(m\omega_0 t)\,dt = \int_{-T_0/2}^{T_0/2}\sin^2(n\omega_0 t)\,dt = \tfrac{1}{2}\int_{-T_0/2}^{T_0/2}[1 - \cos(2n\omega_0 t)]\,dt$$

$$= \tfrac{1}{2}t\Bigg|_{-T_0/2}^{T_0/2} - \tfrac{1}{2}\frac{\sin(2n\omega_0 t)}{2n\omega_0}\Bigg|_{-T_0/2}^{T_0/2} = T_0/2 + \frac{\sin(2n\pi)}{2n\omega_0} = T_0/2$$

Assume finally that $n = m = 0$. Then

$$\int_{-T_0/2}^{T_0/2} \sin(n\omega_0 t)\sin(m\omega_0 t)\,dt = \int_{-T_0/2}^{T_0/2} 0\,dt = 0$$

This completes the proof of (2.102).

2.14 $f_p(t) = 3 - \cos(4t) + 2\sin(5t) - 7\cos(9t)$

(a) $\omega_0 = 1$ since that is the only integer which will divide into 4, 5, and 9. Then, $T_0 = 2\pi$.

(b)

$$P = (1/T_0)\int_{-T_0/2}^{T_0/2} |f_p(t)|^2\,dt$$

$$= (1/T_0)\int_{-T_0/2}^{T_0/2}\Big[3 - \cos(4t) + 2\sin(5t) - 7\cos(9t)\Big]^2\,dt$$

$$= (1/T_0)\int_{-T_0/2}^{T_0/2}\Big[9 + \cos^2(4t) + 4\sin^2(5t) + 49\cos^2(9t)\Big]\,dt$$

$$= (1/T_0) \int_{-T_0/2}^{T_0/2} \left[9 + \frac{1+\cos(8t)}{2} + 4\,\frac{1-\cos(5t)}{2} + 49\,\frac{1+\cos(18t)}{2} \right] dt$$

$$= (1/T_0)\ [9T_0 + \tfrac{1}{2}T_0 + 2T_0 + 49T_0/2] = 36$$

(c) Finding the complex Fourier coeffficients:

$$f_p(t) = 3 - \cos(4t) + 2\sin(5t) - 7\cos(9t)$$

$$= 3 - \tfrac{1}{2}\exp(j4t) - \tfrac{1}{2}\exp(-j4t) + (1/j)\exp(j5t) - (1/j)\exp(-j5t) - (7/2)\exp(j7t) - (7/2)\exp(-j7t)$$

and so the coefficients are:

$F(-9)=-7/2,\ F(-5)=j,\ F(-4)=-\tfrac{1}{2},\ F(0)=3,\ F(4)=-\tfrac{1}{2},\ F(5)=-j,\ F(9)=-7/2$

(d) Using Parseval's theorem, the power spectrum is

$P(-9)=49/4,\ P(-5)=1,\ P(-4)=\tfrac{1}{4},\ P(0)=9,\ P(4)=\tfrac{1}{4},\ P(5)=1,\ P(9)=49/4$

(e) The total average power from (d) is the sum of the terms, which equals 36. That is the same as the value obtained in (b).

2.15 (a) $\displaystyle\sum_{n=-\infty}^{\infty} F(n) \overset{?}{=} f_p(0)$

From the synthesis equation: $\displaystyle f_p(t) = \sum_{n=-\infty}^{\infty} F(n)\exp(jn\omega_0 t)$

Setting $t = 0$ gives $\displaystyle f_p(0) = \sum_{n=-\infty}^{\infty} F(n)$ QED

(b) $\displaystyle\int_{-T_0/2}^{T_0/2} f_p(t)\,dt \overset{?}{=} T_0F(0)$

From the analysis equation: $\displaystyle T_0F(n) = \int_{-T_0/2}^{T_0/2} f_p(t)\exp(-jn\omega_0 t)\,dt$

Setting $n = 0$ gives $\displaystyle\int_{-T_0/2}^{T_0/2} f_p(t)\,dt = T_0F(n)$ QED

$$F(n) = (1/T_0)\int_{-T_0/2}^{T_0/2} f_p(t)\exp(-jn\omega_0 t)\,dt = \int_0^{1/2}\exp(-jn2\pi t)\,dt = \left.\frac{\exp(-jn2\pi t)}{-jn2\pi}\right|_0^{1/2}$$

$$= \frac{\exp(-jn\pi) - 1}{-jn2\pi} = \frac{(-1)^n - 1}{-jn2\pi} = -j/n\pi \quad (n \text{ odd})$$

For $n = 0$: $\displaystyle F(0) = (1/T_0)\int_{-T_0/2}^{T_0/2} f_p(t)\,dt = \int_0^{1/2} dt = \tfrac{1}{2}$

The Fourier series is $\displaystyle f_p(t) = \tfrac{1}{2} - (j/\pi)\sum_{\substack{n=-\infty\\ n\text{ odd}}}^{\infty}(1/n)\exp(jn2\pi t)$

(b) For $t = 0$ the series becomes $\displaystyle f_p(0) = \tfrac{1}{2} - (j/\pi)\sum_{\substack{n=-\infty\\ n\text{ odd}}}^{\infty}(1/n)$

$$= \tfrac{1}{2} - (j/\pi)[\ldots -(1/5) -(1/3) -(1/1) +(1/1) +(1/3) +(1/5) + \ldots] = \tfrac{1}{2}$$

(because pairs of terms cancel each other out)

For $t = \tfrac{1}{2}$ the series becomes $\displaystyle f_p(\tfrac{1}{2}) = \tfrac{1}{2} - (j/\pi)\sum_{\substack{n=-\infty\\ n\text{ odd}}}^{\infty}(1/n)\exp(jn\pi)$

$$= \tfrac{1}{2} - (j/\pi)\sum_{\substack{n=-\infty\\ n\text{ odd}}}^{\infty}(1/n)(-1)^n = \tfrac{1}{2} - (j/\pi)[\ \ldots + (1/5) + (1/3) + (1/1) - (1/1) - (1/3) - (1/5) - \ldots\ ]$$

$= \tfrac{1}{2}$ (because pairs of terms again cancel each other out)

(c) For large n the Fourier coefficients converge like K/n^p where $p = 1$.

(d) The expression for the power spectrum is

$$P(n) = |F(n)|^2 = \begin{cases} \tfrac{1}{4} & (n = 0) \\ (1/n\pi)^2 & (n \text{ odd}) \end{cases}$$

(e) Magnitude, phase and power are as shown in the Answers in the text.

(f) The total average power in the waveform is

$$P_{tot} = (1/T_0)\int_0^{T_0} |f_p(t)|^2\,dt = \int_0^{\frac{1}{2}} 1\,dt = \tfrac{1}{2}$$

The sum of the power terms in the table in Answers is:
$P_{table} = 0.483263$ This is 96.65% of P_{tot}.

(g) Steps for loading the waveform using N = 1024, SAMPLED, T = 1, PERIODIC:

Main menu, Create Values, Continue, Create X, Real, Neither, Left, 2 intervals, 1/2, Continue, 1, 0, Go, Neither, Go. You are now back on the main menu with the waveform correctly loaded. Plotting **X** confirms that.

Run ANALYSIS, Show Numbers, F, Complex. The values displayed for $|F(n)|$, $\theta(n)$ and $P(n)$ confirm the theoretical values shown in the table in the Answers.

Run Postprocessors, F, Power, M = 512. The system gives: Total average power = 0.499512

For the sum of the terms in the table use M = 5. The system gives: Power for terms in the table = 0.483261 which is 96.75% of 0.499512

2.17

(a) $$f_p(t) = \begin{cases} 0, & (-1 < t < 0) \\ e^t & (0 < t < 1) \end{cases} \qquad f_p(t+2) = f_p(t)$$

$$F(n) = 1/T_0 \int_{-T_0/2}^{T_0/2} f_p(t)\exp(-jn\omega_0 t)\,dt \qquad T_0 = 2, \quad \omega_0 = \pi$$

$$= \tfrac{1}{2}\int_0^1 \exp(t)\exp(-jn\pi t)\,dt = \tfrac{1}{2}\int_0^1 \exp[(1-jn\pi)t]\,dt$$

$$= \tfrac{1}{2}\frac{\exp[(1-jn\pi)t]}{1-jn\pi}\Bigg|_0^1 = \tfrac{1}{2}\frac{\exp(1-jn\pi)-1}{1-jn\pi}$$

$$= \tfrac{1}{2}\frac{e[\cos(n\pi) - j\sin(n\pi)] - 1}{1-jn\pi} = \tfrac{1}{2}\frac{(-1)^n e - 1}{1-jn\pi}$$

"Introduction to Fourier Analysis", Instructor's Manual, Chapter 2

$$= \frac{(-1)^n e - 1}{2}\,\frac{1 + jn\pi}{1 + n^2\pi^2} \qquad \text{(for all n)}$$

The Fourier series is $$f_p(t) = \sum_{n=-\infty}^{\infty} \frac{(-1)^n e - 1}{2}\,\frac{1 + jn\pi}{1 + n^2\pi^2}\exp(jn\pi t)$$

(b) According to Fourier's theorem, the Fourier series converges as follows:

$t = -1$: $f_p(-1) = e/2$ $t = 0$: $f_p(0) = 1/2$ $t = 1$: $f_p(1) = e/2$

None of these is obvious from the series.

(c) See sketch in the Answers.

(d) For large n the Fourier coefficients converge like K/n^p where $p = 1$.

(e) Magnitude = $$|F(n)| = \frac{|(-1)^n e - 1|}{2(1 + n^2\pi^2)^{\frac{1}{2}}}$$

Using this formula the values shown in the Answers were computed.

Phase = $\theta(n) = \text{atn}[B(n)/A(n)]$ where

$$A(n) = \frac{(-1)^n e - 1}{2}\,\frac{1}{1 + n^2\pi^2} \qquad B(n) = \frac{(-1)^n e - 1}{2}\,\frac{n\pi}{1 + n^2\pi^2}$$

When n is odd then $(-1)^n e - 1$ is a negative number, and when n is even then it is a positive number. Thus:

$$\theta(n) = \begin{cases} \text{atn}(-n\pi/-1) & \text{(n odd)} \\ \text{atn}(n\pi/1) & \text{(n even)} \end{cases}$$

Using this formula, and assuming that $n > 0$, the values in the table in the Answers were computed. For $n < 0$ they were formed so that $\theta(n)$ was odd.

(f) The expression for the power spectrum is

$$P(n) = |F(n)|^2 = \frac{[(-1)^n e - 1]^2}{4(1 + n^2\pi^2)}$$

$$P_{tot} = (1/T_0)\int_{-T_0/2}^{T_0/2} |f_p(t)|^2\,dt = \tfrac{1}{2}\int_0^1 \exp(2t)\,dt = \frac{\exp(2) - 1}{4} = 1.597264$$

The power terms for $-5 <= n <= 5$ (see the table) total to [illegible]

(g) Steps for loading the waveform using $N = 1024$, SAMPLED, $T = 2$, PERIODIC:

Main menu, Create Values, Continue, Create X, Real, Neither, Left, 2 intervals, 1, Continue, EXP(t), 0, Go, Neither, Go. You are now back on the main menu with the waveform correctly loaded. Plotting **X** confirms that. Run ANALYSIS, Show Numbers, F, Complex. The values are the same as those in the Answers.

Run Postprocessors, F, Power, M = 512. The system gives: Total average power = 1.59522

For the sum of the terms in the table use M = 5. The system gives: Power for terms in the table = 1.52471 which is 95.58% of 1.59522

2.18 Using the results of Theorem 2.7 we can represent a given waveform by a series of the form

$$f_p(t) = a(0)/2 + \sum_{n=1}^{\infty} a(n)\cos(n\omega_0 t) + \sum_{n=1}^{\infty} b(n)\sin(n\omega_0 t) \qquad (A)$$

To find formulae for computing the coefficients $a(n)$ and $b(n)$ for any waveform $f_p(t)$ we use the same procedure as was used to derive (2.27) and (2.28) (first devised by Euler who used it rarely but now more properly associated with the name of Fourier).

Starting from (A) we multiply every term on both sides of the equation by *$\cos(m\omega_0 t)$ and then integrate over one complete* period, obtaining

$$\int_{-T_0/2}^{T_0/2} f_p(t)\cos(m\omega_0 t)\,dt$$

$$= \int_{-T_0/2}^{T_0/2} a(0)/2\cos(m\omega_0 t)\,dt$$

$$+ \int_{-T_0/2}^{T_0/2} \left[\sum_{n=1}^{\infty} a(n)\cos(n\omega_0 t)\right]\cos(m\omega_0 t)\,dt$$

$$+ \int_{-T_0/2}^{T_0/2} \left[\sum_{n=1}^{\infty} b(n)\sin(n\omega_0 t)\right]\cos(m\omega_0 t)\,dt$$

$$= a(0)/2\int_{-T_0/2}^{T_0/2}\cos(m\omega_0 t)\,dt \qquad (B)(1)$$

$$+ \sum_{n=1}^{\infty} a(n)\int_{-T_0/2}^{T_0/2}\cos(n\omega_0 t)\cos(m\omega_0 t)\,dt \qquad (B)(2)$$

$$+ \sum_{n=1}^{\infty} b(n)\int_{-T_0/2}^{T_0/2}\sin(n\omega_0 t)\cos(m\omega_0 t)\,dt \qquad (B)(3)$$

in which we again draw attention to the fact that we have quietly interchanged the order of the summation and integration operations. (This very important fact was discussed in the text.) We now examine (B)(1) through (B)(3) very carefully, assuming first that $m \neq 0$. The case $m = 0$ will be considered separately, further down.

The first term labelled (B)(1) is seen to be zero because it represents the area under a cosine over an interval that contains a number of complete periods. (It is also zero because of (2.100).)

The third term labelled (B)(3) is seen to be an infinite series of terms, every one of which is zero by the orthogonality property stated in (2.101), and so the entire sum must also be zero.

Finally we consider the second term labelled (B)(2). It too is an infinite series of terms, and by (2.100) we see that as long as $n \neq m$ all terms in that series are zero. However, **for one of those terms n is equal to m,** and by (2.100) that single term integrates to $T_0/2$. Thus the entire sum is equal to $a(m)\,T_0/2$, and so (B)(2) collapses down to the remarkably simple statement

$$\int_{-T_0/2}^{T_0/2} f_p(t)\cos(m\omega_0 t)\,dt = a(m)\,T_0/2 \qquad (C)$$

We now solve for $a(m)$ and then replace m with n, obtaining our desired formula as

$$a(n) = 2/T_0\int_{-T_0/2}^{T_0/2} f_p(t)\cos(n\omega_0 t)\,dt \qquad (n = 1, 2, 3 \ldots) \qquad (D)$$

This then is the desired expression for $a(n)$, which can then be evaluated for any given $f_p(t)$. By a similar argument the formula

for b(n) becomes

$$b(n) = 2/T_0 \int_{-T_0/2}^{T_0/2} f_p(t)\ \sin(n\omega_0 t)\ dt \qquad (n = 1, 2, 3 \ldots,) \qquad \text{(E)}$$

Returning now to the case where $m = 0$, only the term labelled (B)(1) is nonzero and using (2.100) we see that (B)(1) becomes

$$\int_{-T_0/2}^{T_0/2} f_p(t)\ \cos(0\omega_0 t)\ dt = a(0)/2 \int_{-T_0/2}^{T_0/2} dt = a(0)T_0/2 \qquad \text{(F)}$$

Solving for a(0) gives us

$$a(0) = 2/T_0 \int_{-T_0/2}^{T_0/2} f_p(t)\ dt \qquad \text{(G)}$$

which shows that (D) holds for $n = 0$ as well. (Had we used simply a(0) rather than a(0)/2 in (A) this latter fact would not have been true, and we would have been forced to make a separate statement for obtaining a(0).)

2.19
(a) Finding the real Fourier series for the periodic waveform shown in Figure 2.22:

$$f_p(t) = \begin{cases} 0 & (-2 < t < -1) \\ 1 + t & (-1 < t < 0) \\ 1 & (0 < t < 1) \\ 0 & (1 < t < 2) \end{cases} \qquad f_p(t + 4) = f_p(t)$$

$T_0 = 4$, $\omega_0 = \pi/2$. Then, for $n \neq 0$:

$$a(n) = (2/T_0) \int_{-T_0/2}^{T_0/2} f_p(t)\ \cos(n\omega_0 t)\ dt$$

$$= \tfrac{1}{2} \int_{-1}^{0} (1 + t)\ \cos(n\pi t/2)\ dt + \tfrac{1}{2} \int_{0}^{1} \cos(n\pi t/2)\ dt$$

$$= \frac{\tfrac{1}{2}(1 + t)\ \sin(n\pi t/2)}{n\pi/2} + \frac{\tfrac{1}{2}\cos(n\pi t/2)}{(n\pi/2)^2} \Bigg|_{-1}^{0} + \frac{\tfrac{1}{2}\sin(n\pi t/2)}{n\pi/2} \Bigg|_{0}^{1}$$

$$= \tfrac{1}{2} \frac{1 - \cos(n\pi/2)}{(n\pi/2)^2} + \tfrac{1}{2} \frac{\sin(n\pi/2)}{n\pi/2} \qquad (n \neq 0) \qquad \text{(A)}$$

For $n = 0$:

$$a(0) = (2/T_0) \int_{-T_0/2}^{T_0/2} f_p(t)\ dt = \tfrac{1}{2} \int_{-1}^{0} (1 + t)\ dt + \tfrac{1}{2} \int_{0}^{1} dt$$

$$= \tfrac{1}{4}(1 + t)^2 \Big|_{-1}^{0} + \tfrac{1}{2}t \Big|_{0}^{1} = .25 + .5 = .75 \qquad (n \neq 0) \qquad \text{(B)}$$

$$b(n) = (2/T_0) \int_{-T_0/2}^{T_0/2} f_p(t)\ \sin(n\omega_0 t)\ dt$$

$$= \tfrac{1}{2} \int_{-1}^{0} (1 + t)\ \sin(n\pi t/2)\ dt + \tfrac{1}{2} \int_{0}^{1} \sin(n\pi t/2)\ dt$$

$$= \frac{-\tfrac{1}{2}(1 + t)\ \cos(n\pi t/2)}{n\pi/2} - \frac{\tfrac{1}{2}\sin(n\pi t/2)}{(n\pi/2)^2} \Bigg|_{-1}^{0} - \frac{\tfrac{1}{2}\cos(n\pi t/2)}{n\pi/2} \Bigg|_{0}^{1}$$

$$= -\tfrac{1}{2} \frac{1}{n\pi/2} + \frac{\tfrac{1}{2}\sin(n\pi/2)}{(n\pi/2)^2} - \tfrac{1}{2} \frac{\cos(n\pi/2) - 1}{n\pi/2}$$

$$= \tfrac{1}{2} \frac{\sin(n\pi/2)}{(n\pi/2)^2} - \tfrac{1}{2} \frac{\cos(n\pi/2)}{n\pi/2} \qquad (n \neq 0) \qquad \text{(C)}$$

Using the expressions (A), (B), and (C) we can now set up the real Fourier series:

$$f_p(t) = a(0)/2 + \sum_{n=1}^{\infty} a(n)\ \cos(n\omega_0 t) + \sum_{n=1}^{\infty} b(n)\ \sin(n\omega_0 t) \qquad \text{(A)}$$

(b) Both a(n) and b(n) converge to zero like 1/n.

(c) According to Fourier's theorem the series converge to ½ at $t = 1$. However this result is not immediately obvious from

(d) Converting the series ... $A(n)$ and $B(n)$ from (2.88):

$$A(n) = a(n)/2 \qquad B(n) = -b(n)/2 \qquad \text{(D)}$$
$$A(0) = 3/8$$

$$A(n) = \tfrac{1}{4}\,\frac{1 - \cos(n\pi/2)}{(n\pi/2)^2} + \tfrac{1}{4}\,\frac{\sin(n\pi/2)}{n\pi/2} \qquad (n \neq 0) \qquad \text{(E)}$$

$$B(n) = -\tfrac{1}{4}\,\frac{\sin(n\pi/2)}{(n\pi/2)^2} + \tfrac{1}{4}\,\frac{\cos(n\pi/2)}{n\pi/2} \qquad (n \neq 0) \qquad \text{(F)}$$

The complex Fourier series is: $\quad f_p(t) = \sum_{n=-\infty}^{\infty} [A(n) + jB(n)] \exp(jn\pi t)$

(e) Finding the sum of all of the elements of the power spectrum:

$$P_{tot} = (1/T_0) \int_{-T_0/2}^{T_0/2} |f_p(t)|^2 \, dt$$

$$= \tfrac{1}{4} \int_{-1}^{0} (1+t)^2 \, dt + \tfrac{1}{4} \int_{0}^{1} dt = \frac{(1+t)^3}{12}\Bigg|_{-1}^{0} + 1/4 = 1/12 + 1/4 = 1/3$$

(f) Steps for loading the waveform using N = 256, SAMPLED, T = 4, PERIODIC: Main menu, Create Values, Continue, Create X, Real, Neither, Center, 4 intervals, -1, 0, 1, Continue, 0, 1+t, 1, 0, Go, Neither, Go. You are now back on the main menu with the waveform correctly loaded. Plotting X confirms that. Run ANALYSIS, Show Numbers, F, Real.

The following are some of the FFT values and formula values.

n	$a(n)_{FFT}$	$a(n)_{formula}$	$b(n)_{FFT}$	$b(n)_{formula}$
0	0.75	0.75	0	0
1	0.52094645	0.52995525	0.20265254	0.20264237
2	0.10134153	0.10132118	0.15912298	0.15915494
3	-8.352936e-2	-8.358748e-2	-2.252599e-2	-2.251582e-2

2.20 (a) If $f_p(t)$ is real and even then $A(n) \neq 0$, $B(n) = 0$

By (2.88): $\quad a(n) = 2\,A(n) \qquad b(n) = -2\,B(n)$

which means that $a(n) \neq 0$ and $b(n) = 0$, and so the real Fourier series consists only of cosines.

By (2.88): $\quad a(n) = 2\,A(n) \qquad b(n) = -2\,B(n)$

which means that $a(n) = 0$ and $b(n) \neq 0$, and so the real Fourier series consists only of sines.

2.21 (a) $f(t) = \cos^2(t) \qquad (0 < t < \pi)$

In order to obtain a series of cosines we must extend $f(t)$ so that it is an even function. We obtain

$$f_p(t) = \cos^2(t) \quad (0 < t < \pi) \qquad f_p(t+\pi) = f_p(t) \qquad T_0 = \pi, \quad \omega_0 = 2$$

For the sketch see the Answer section. Finding the values for $a(n)$ we have

$$\cos^2(t) = \tfrac{1}{2} + \tfrac{1}{2}\cos(2t)$$

and so $a(0) = 1$, $a(1) = \tfrac{1}{2}$. The real Fourier series is

$$f_p(t) = \tfrac{1}{2} + \tfrac{1}{2}\cos(2t)$$

(b) $f(t) = \sin(\pi t/2) \qquad (0 < t < 2)$

Extending $f(t)$ evenly in order to obtain a series of cosines we get

$$f_p(t) = \sin(\pi t/2) \quad (0 < t < 2) \qquad f_p(t+2) = f_p(t) \qquad T_0 = 2, \quad \omega_0 = \pi$$

For the sketch see the Answer section. Finding the values for $a(n)$ we have

$$a(n) = (2/T_0) \int_{-T_0/2}^{T_0/2} f_p(t)\cos(n\omega_0 t)\,dt = (4/T_0) \int_{0}^{T_0/2} f_p(t)\cos(n\omega_0 t)\,dt \quad \text{since } f_p(t) \text{ is even)}$$

$$= (4/2) \int_0^1 \overset{A}{\sin(\pi t/2)}\ \overset{B}{\cos(n\pi t)}\,dt = (4/4) \int_0^1 \Big[\sin[(\tfrac{1}{2}+n)\pi t] + \sin[(\tfrac{1}{2}-n)\pi t]\Big]\,dt$$

$$= -\left[\frac{\cos[(\tfrac{1}{2}+n)\pi t]}{(\tfrac{1}{2}+n)\pi}\Bigg|_0^1 + \frac{\cos[(\tfrac{1}{2}-n)\pi t]}{(\tfrac{1}{2}-n)\pi}\Bigg|_0^1\right]$$

$$= -\left[\frac{\cos[(\tfrac{1}{2}+n)\pi] - 1}{(\tfrac{1}{2}+n)\pi} + \frac{\cos[(\tfrac{1}{2}-n)\pi] - 1}{(\tfrac{1}{2}-n)\pi}\right]$$

But
$$\cos[(\tfrac{1}{2}+n)\pi] = \cos(\tfrac{1}{2}\pi)\cos(n\pi) - \sin(\tfrac{1}{2}\pi)\sin(n\pi) = 0$$
$$\cos[(\tfrac{1}{2}-n)\pi] = \cos(\tfrac{1}{2}\pi)\cos(n\pi) + \sin(\tfrac{1}{2}\pi)\sin(n\pi) = 0$$

$$= \frac{1}{(\frac{1}{2}+n)\pi} + \frac{1}{(\frac{1}{2}-n)\pi} = \frac{2}{\pi}\left[\frac{1}{1+2n} + \frac{1}{1-2n}\right] = \frac{4}{\pi}\,\frac{1}{1-4n^2} \qquad (n = 0,1,2,\ ..)$$

The real Fourier is as follows:

$$f_p(t) = a(0)/2 + \sum_{n=1}^{\infty} a(n)\cos(n\omega_0 t)\ dt = \frac{2}{\pi} + \frac{4}{\pi}\sum_{n=1}^{\infty}\frac{1}{1-4n^2}\cos(n\pi t)$$

$$= \frac{2}{\pi} - \frac{4}{\pi}\left[\frac{1}{3}\cos(\pi t) + \frac{1}{15}\cos(2\pi t) + \frac{1}{35}\cos(3\pi t) + \ldots\right]$$

(c) The coefficients are dying out like $1/n^2$ because the waveform is everywhere continuous but its first derivative has discontinuities.

(d) The average power is $\quad P = (1/T_0)\displaystyle\int_0^{T_0} |f_p(t)|^2\ dt$

$$= \tfrac{1}{2}\int_0^2 \sin^2(\pi t/2)\ dt = \tfrac{1}{4}\int_0^2 [1-\cos(\pi t)]\ dt = \tfrac{1}{4}\left[t - \frac{\sin(\pi t)}{\pi}\right]_0^2 = \tfrac{1}{2}$$

(e) Steps for loading the waveform in (b) using N = 256, SAMPLED, T = 2, PERIODIC:

Main menu, Create Values, Continue, Create X, Real, Even, 1 interval, Continue, SIN(pi*t/2), Go, Neither, Go. You are now back on the main menu with the waveform correctly loaded. Plotting **X** confirms that. Run ANALYSIS, Show Numbers, F, Real.

The following are some of the FFT values and formula values.

n	$a(n)_{FFT}$	$a(n)_{formula}$	$b(n)_{FFt}$	$b(n)_{formula}$
0	1.27322360	1.273239545	0	0
1	-0.42442916	-0.424413181	0	0
2	-8.4898617e-2	-8.4882636e-2	0	0
3	-3.6394256e-2	-3.6378272e-2	0	0

Run Postprocessors, F, Power, M = 128. The system gives: Total average power = 0.5

2.22 (a) $f(t) = \cos(\pi t/2) \qquad (0 < t < 2)$

Extending $f(t)$ oddly in order to obtain a series of sines we obtain

$f_p(t) = \cos(\pi t/2) \qquad (0 < t < 1)$

$f_p(t+2) = f_p(t) \quad f_p(t)$ is odd $\quad (T_0 = 2,\ \ \omega_0 = \pi)$

For the sketch see the Answer section. Finding the values for $b(n)$ we have

$$b(n) = (2/T_0)\int_{-T_0/2}^{T_0/2} f_p(t)\sin(n\omega_0 t)\ dt = (4/T_0)\int_0^{T_0/2} f_p(t)\sin(n\omega_0 t)\ dt \qquad (\text{since } f_p(t) \text{ is odd})$$

$$= (4/2)\int_0^1 \cos(\pi t/2)\sin(n\pi t)\ dt = 2\int_0^1\left[\frac{\sin[(n+\frac{1}{2})\pi t]}{2} + \frac{\sin[(n-\frac{1}{2})\pi t]}{2}\right] dt$$

$$= \left[\frac{-\cos[(n+\frac{1}{2})\pi t]}{(n+\frac{1}{2})\pi} - \frac{\cos[(n-\frac{1}{2})\pi t]}{(n-\frac{1}{2})\pi}\right]_0^1 = \frac{1-\cos[(n+\frac{1}{2})\pi]}{(n+\frac{1}{2})\pi} + \frac{1-\cos[(n-\frac{1}{2})\pi]}{(n-\frac{1}{2})\pi}$$

$$= \frac{2}{\pi}\left[\frac{1}{2n+1} + \frac{1}{2n-1}\right] = \frac{8n}{\pi(4n^2-1)} \qquad (n > 0)$$

$$f_p(t) = \sum_{n=1}^{\infty} b(n)\sin(n\omega_0 t) = \frac{8}{\pi}\sum_{n=1}^{\infty}\frac{n}{4n^2-1}\sin(n\pi t)$$

$$= \frac{8}{\pi}\left[\frac{1}{3}\sin(\pi t) + \frac{2}{15}\sin(2\pi t) + \frac{3}{35}\sin(3\pi t) + \ldots\right]$$

(b) $f(t) = \sin^2(t) \qquad (0 < t < \pi)$

Extending $f(t)$ oddly in order to obtain a series of sines we get

$f_p(t) = \sin^2(t) \qquad (0 < t < \pi)$

$f_p(t+2\pi) = f_p(t) \quad f_p(t)$ is odd $\quad (T_0 = 2\pi,\ \ \omega_0 = 1)$

$$b(n) = (2/T_0)\int_{-T_0/2}^{T_0/2} f_p(t)\,\sin(n\omega_0 t)\,dt = (4/T_0)\int_0^{T_0/2} f_p(t)\,\sin(n\omega_0 t)\,dt \quad \text{(since } f_p(t) \text{ is odd)}$$

$$= (4/2\pi)\int_0^{\pi} \sin^2(t)\,\sin(nt)\,dt = \frac{2}{\pi}\int_0^{\pi} \frac{1-\cos(2t)}{2}\,\sin(nt)\,dt$$

$$= \frac{1}{\pi}\int_0^{\pi}\sin(nt)\,dt - \frac{1}{\pi}\int_0^{\pi}\left[\frac{\sin[(n+2)t]}{2} + \frac{\sin[(n-2)t]}{2}\right]dt$$

$$= -\frac{1}{\pi}\frac{\cos(nt)}{n}\Big|_0^{\pi} + \frac{1}{2\pi}\left[\frac{\cos[(n+2)t]}{n+2} + \frac{\cos[(n-2)t]}{n-2}\right]_0^{\pi}$$

$$= \frac{1}{\pi}\frac{1-(-1)^n}{n} + \frac{1}{2\pi}\left[\frac{\cos[(n+2)\pi]-1}{n+2} + \frac{\cos[(n-2)\pi - 1]}{n-2}\right]$$

$$= \frac{1}{\pi}\frac{1-(-1)^n}{n} + \frac{1}{2\pi}\left[\frac{(-1)^n-1}{n+2} + \frac{(-1)^n - 1]}{n-2}\right]$$

$$= \frac{1}{\pi}\frac{1-(-1)^n}{n} - \frac{n}{\pi}\frac{1-(-1)^n}{n^2-4} = \frac{4}{\pi}\frac{(-1)^n-1}{n^3-4n} \qquad (n \neq 2)$$

For $n = 2$:

$$b(n) = \frac{1}{\pi}\int_0^{\pi}\sin(2t)\,dt - \frac{1}{\pi}\int_0^{\pi}\frac{\sin(4t)}{2}\,dt = 0$$

$$f_p(t) = \sum_{n=1}^{\infty} b(n)\,\sin(n\omega_0 t) = \frac{4}{\pi}\sum_{\substack{n=1\\ n\neq 2}}^{\infty}\frac{(-1)^n-1}{n^3-4n}\,\sin(nt)$$

$$= \frac{8}{\pi}\left[\frac{1}{3}\sin(t) - \frac{1}{15}\sin(3t) - \frac{1}{105}\sin(5t) - \dots\right]$$

first derivative are everywhere continuous but its second derivative is not.

(d) Steps for loading the waveform in (b) using $N = 1024$, SAMPLED, $T = 2\pi$, PERIODIC:

Main menu, Create Values, Continue, Create X, Real, Odd, 1 interval, Continue, SIN(t)^2, Go, Neither, Go. You are now back on the main menu with the waveform correctly loaded. Plotting **X** confirms that. Run ANALYSIS, Show Numbers, F, Real.

The following are some of the FFT values and formula values.

n	$b(n)_{FFT}$	$b(n)_{formula}$
0	0	0
1	0.84882636	0.848826363
3	-0.16976527	-0.169765273
5	-2.4252182e-2	-2.4252181e-2
7	-8.0840605e-3	-8.0840606e-3

The reason for this very close agreement between formula and FFT values lies in the fact that the coefficients are converging so rapidly. In Chapter 12 we discuss in detail the relationships between these two sets of values and it will become clear there why the rapid convergence leads to such good agreement.

Keep in mind that as we move away from $n = 0$ and approach $n = N/2$ the FFT values always pull away from the exact values to a greater or lesser extent, and the sort of agreement that is present in the above table begins to break down. In Chapter 12 we shall also see why this happens.

2.23 (a) For the sketch see the Answer section.

(b) The breakpoint is at $t = 1$.

At $t = 1$: $f_p(1^+) = t^2-4t+3\big|_{t=1} = 0$

$f_p(1^-) = 1-t^2\big|_{t=1} = 0$

At $t = 1$: $f'_p(1^+) = 2t-4\big|_{t=1} = -2$

$f'_p(1^-) = -2t\big|_{t=1} = -2$

From this we see that the two expressions result in a waveform that is everywhere continuous with a first derivative that is also everywhere continuous.

(c) We expect that the coefficients will die out like $1/n^3$.

(d) Steps for loading the waveform using N = 512, SAMPLED, T = 4, PERIODIC: Main menu, Create Values, Continue, Create X, Real, Even, 2 intervals, 1, Continue, 1 - t^2, t^2 - 4*t + 3, Go, Neither, Go. You are now back on the main menu with the waveform correctly loaded. Plotting **X** confirms that. Run ANALYSIS, Show Numbers, F, Real.

The following are some of the FFT values. From the final two columns we see that the FFT's estimates of the coefficients shown in the table are dying out approximately like $1/n^3$.

Table 2.8: FFT estimates of a(n)

n	a(n)	$\lvert a_1/a(n)\rvert$	n^3
1	1.03204910	1	1
3	-3.8224041e-2	26.9999998	27
5	8.2563923e-3	125.000008	125
7	-3.0088888e-3	343.000083	343

Note: The theorem states that the coefficients should die out like $1/n^3$ for LARGE n. In this case we see that they are also dying out at that rate for small n.

2.25 (a) Proof of **Bessel's theorem** for the average power in a real periodic waveform, namely:

$$P = 1/T_0 \int_{-T_0/2}^{T_0/2} |f_p(t)|^2\, dt = \frac{a(0)^2}{4} + \frac{1}{2}\sum_{n=1}^{\infty}\left[a(n)^2 + b(n)^2\right]$$

From Parseval's theorem:

$$P = 1/T_0 \int_{-T_0/2}^{T_0/2} |f_p(t)|^2\, dt = \sum_{n=-\infty}^{\infty} |F(n)|^2 = \sum_{n=-\infty}^{\infty} [A(n)^2 + B(n)^2] \qquad (1)$$

But $f_p(t)$ is real and so $A(n) = A(-n)$, and $B(n) = -B(-n)$. Moreover, from (2.88), $A(n) = a(n)/2$, $B(n) = -b(n)/2$. Thus (1) continues as

$$\ldots = A(0)^2 + 2\sum_{n=1}^{\infty}[A(n)^2 + B(n)^2] = a(0)^2/4 + 2\sum_{n=1}^{\infty}[a(n)^2/4 + b(n)^2/4]$$

$$= \frac{a(0)^2}{4} + \frac{1}{2}\sum_{n=1}^{\infty}\left[a(n)^2 + b(n)^2\right] \qquad \text{QED}$$

(b) Using Bessel's theorem we can now prove, for a real periodic

function, that the total power is equal to the sum of the power in the even and the odd parts, i.e.

$$P[f_p(t)] = P[f_{ev}(t)] + P[f_{od}(t)] \qquad (2)$$

By Bessel's theorem, if

$$f_p(t) = a(0)/2 + \sum_{n=1}^{\infty}[a(n)\cos(n\omega_0 t) + b(n)\sin(n\omega_0 t)]$$

then $$P = \frac{a(0)^2}{4} + \frac{1}{2}\sum_{n=1}^{\infty}[a(n)^2 + b(n)^2]$$

The power in the even part must thus be

$$P_{ev} = \frac{a(0)^2}{4} + \frac{1}{2}\sum_{n=1}^{\infty} a(n)^2$$

and the power in the odd part must thus be

$$P_{od} = \frac{1}{2}\sum_{n=1}^{\infty} b(n)^2$$

from which it is clear that the total power is the sum of the two, i.e.

$$P = P_{ev} + P_{od} \qquad \text{QED}$$

2.26(a) The analytical expression for the waveform shown in Fig 2.23 is

$$f_p(t) = \begin{cases} \tfrac{1}{2} & (0 < t < 1) \\ 1 & (1 < t < 2) \end{cases} \qquad f_p(t+4) = f_p(t)$$

Then $T_0 = 4$, $\omega_0 = \pi/2$. The complex Fourier coeffs will be given by

$$F(n) = 1/T_0 \int_0^{T_0} f_p(t)\exp(-jn\omega_0 t)\, dt$$

$$= \tfrac{1}{4}\int_0^1 \tfrac{1}{2}\exp(-jn\pi t/2)\, dt + \tfrac{1}{4}\int_1^2 \exp(-jn\pi t/2)\, dt$$

$$= 1/8\, \frac{\exp(-jn\pi t/2)}{}\Big|^1 + \tfrac{1}{4}\, \frac{\exp(-jn\pi t/2)}{}\Big|^2$$

$$= \frac{j}{4\pi n}\left[\exp(-jn\pi/2) - 1 + 2\exp(-jn\pi) - 2\exp(-jn\pi/2) \right]$$

$$= \frac{j}{4\pi n}\left[2(-1)^n - \exp(-jn\pi/2) - 1 \right] \qquad (n \neq 0)$$

For $n = 0$:

$$F(n) = 1/T_0 \int_0^{T_0} f_p(t)\, dt = \tfrac{1}{4}\int_0^1 \tfrac{1}{2}\, dt + \tfrac{1}{4}\int_1^2 dt = 1/8 + 1/4 = 3/8$$

Thus:

$$F(n) = \begin{cases} \dfrac{j}{4n\pi}\left[2(-1)^n - 1 - \exp(-jn\pi/2) \right] & (n \neq 0) \\ 3/8 & (n = 0) \end{cases}$$

(b) For sketches of the even and odd parts of this waveform see the Answer section. The expressions for their Fourier coefficients are obtained from $F(n)$ as follows:

$$F(n) = \begin{cases} \dfrac{j}{4n\pi}\left[2(-1)^n - 1 - \cos(n\pi/2) + j\sin(n\pi/2) \right] & (n \neq 0) \\ 3/8 & (n = 0) \end{cases}$$

$$= A(n) + jB(n)$$

and so the expression for the coefficients for the even part is

$$A(n) = \begin{cases} -\sin(n\pi/2)/4n\pi & (n \neq 0) \\ 3/8 & (n = 0) \end{cases}$$

and for the odd part it is

$$jB(n) = \frac{j}{4n\pi}\left[2(-1)^n - 1 - \cos(n\pi/2) \right] \qquad (n \neq 0)$$

(c) Steps for loading the waveform using N = 256, SAMPLED, T = 4, PERIODIC: Main menu, Create Values, Continue, Create X, Real, Neither, Left, 3 intervals, 1, 2, Continue, 1, 2, 0, Go, Neither, Go. You are now back on the main menu with the waveform correctly loaded. Plotting **X** confirms that.

F, [illegible] even part appears.

To create a plot of its odd part: Run Postprocessors, F, Copy F2 to F, Set FRE to zero, Quit, Run SYNTHESIS, Plot **Y**. The odd part appears.

(d) Steps for loading the expressions for A(n) and B(n) into the system using N = 256, SAMPLED, T = 4, PERIODIC:

Main menu, Create Values, Continue, Create F, Complex,

-SIN(n*pi/2)/(4*pi*n)

Alpha = 0, Division by zero, Go, A, 3/8, Go,

(2*(-1)^n - 1 - COS(n*pi/2))/(4*pi*n)

Go, Go, Run SYNTHESIS. Plot **Y**. The plot shows the original waveform but with bad ripple before and after each of the discontinuities.

To reduce the ripple: Main menu, Create Values, Continue, Use the Old-Problem, Accept, Alpha = 5, Division by zero, Go, A, 3/8, Go, Accept, Continue aliasing with Alpha = 5, Go, Go, Run SYNTHESIS, Plot **Y**,

The plot now shows the original waveform but with the ripple before and after each of the discontinuities **almost completely gone**. A larger value for Alpha would reduce it further.

2.27(a) The analytical expression for the periodic function shown in Figure 2.24 is

$f_p(t) = t \quad (0 < t < 1) \qquad f_p(t)$ is even $\qquad f_p(t + 2) = f_p(t)$

Then $T_0 = 2$, $\omega_0 = \pi$. The complex Fourier coefficients will be given by

$$F(n) = 1/T_0 \int_{-T_0/2}^{T_0/2} f_p(t) \exp(-jn\omega_0 t)\, dt = 2/T_0 \int_0^{T_0/2} f_p(t) \cos(-jn\omega_0 t)\, dt$$

(since $f_p(t)$ is even)

$$= 2/2 \int_0^1 t\cos(n\pi t)\, dt = \left. \frac{t\sin(n\pi t)}{n\pi} + \frac{\cos(n\pi t)}{(n\pi)^2} \right|_0^1$$

$$= \frac{\sin(n\pi)}{n\pi} + \frac{\cos(n\pi) - 1}{(n\pi)^2} = \frac{(-1)^n - 1}{n^2\pi^2} \qquad (n \neq 0)$$

For $n = 0$: $F(0) = 2/T_0 \int_0^{T_0/2} f_p(t)\, dt = 2/2 \int_0^1 t\, dt = ½$

Then the complex Fourier series is

$$f_p(t) = ½ + 1/\pi^2 \sum_{n=-\infty}^{\infty} \frac{(-1)^n - 1}{n^2} \exp(jn\pi t)$$

(b) For n large, $|F(n)|$ converges like K/n^2 because the waveform is everywhere continuous but its first derivative is not.

(c) Computing magnitude, phase and power for $-5 \le n \le 5$ we have

$$F(n) = \begin{cases} \dfrac{(-1)^n - 1}{n^2\pi^2} & (n \ne 0) \\ ½ & (n = 0) \end{cases}$$

from which
$$|F(n)| = \begin{cases} 2/n^2\pi^2 & (n \text{ odd}) \\ 0 & (n \text{ even and } \ne 0) \\ ½ & (n = 0) \end{cases}$$

For $\theta(n)$: $F(n)$ is
- negative for n odd
- zero for n even and $\ne 0$
- positive for $n = 0$

from which we have
$$\theta(n) = \begin{cases} \pi & (n \text{ odd and } > 0) \\ 0 & (n \text{ even}) \\ -\pi & (n \text{ odd an } < 0) \end{cases}$$
(selected to make $\theta(n)$ odd)

For power we square $|F(n)|$. This gives us the following table:

n	-5	-4	-3	-2	-1	0	1	2	3	4	5
$\|F(n)\|$	.008106	0	.022516	0	.202642	½	.202642	0	.022516	0	.008106
$\theta(n)$	$-\pi$	0	$-\pi$	0	$-\pi$	0	π	0	π	0	$-\pi$
P(n)	.000066	0	.000507	0	.041064	¼	.041064	0	.000507	0	.000066

(d) Computing total average power:

$$P = \frac{1}{T_0}\int_{-T_0/2}^{T_0/2} |f_p(t)|^2\, dt = \frac{2}{T_0}\int_0^{T_0/2} |f_p(t)|^2\, dt = \frac{2}{2}\int_0^1 t^2\, dt = 1/3$$

The three central power terms in the table add up to

$P(0) + P(1) + P(-1) = 0.332128$

which is 99.64% of the total power. Thus $n = 1$ will give a power sum that contains 99% of the total average power.

(e) Steps for loading the waveform using $N = 256$, SAMPLED, $T = 2$, PERIODIC:

Main menu, Create Values, Continue, Create X, Real, Even, 1 interval, Continue, t, Go, Neither, Go. You are now back on the main menu with the waveform correctly loaded. Plotting **X** confirms that.

Run ANALYSIS, Show Numbers, F, Complex. The values appearing in the above table were closely approximated by the FFT.

(f) Because the waveform has no discontinuities, the spectrum dies out like $1/n^2$, i.e. quite quickly. Thus to produce a good version of the FFT spectrum from the one that we obtained in (a) we will only have to alias a small amount.

Steps for loading the expression for A(n) into the system using N = 256, SAMPLED, T = 2, PERIODIC:

Main menu, Create Values, Continue, Create F, Real,

```
((-1)^n - 1)/(n*pi)^2
```

Note: True BASIC evaluates the operations with the following priorities: Highest ^, then /, then *, then + and -. The brackets shown above will ensure that A(n) is properly evaluated.

Alpha = 0, Division by zero, Go, A, 1/2, Go, Go, Run SYNTHESIS. Plot **Y**. The plot shows the original waveform almost perfectly.

To use Alpha = 5: Main menu, Create Values, Continue, Use the Old-Problem, Accept, Alpha = 5, Division by zero, Go, A, 1/2, Go, Go, Run SYNTHESIS, Plot **Y**,

The plot shows essentially the same version of the original waveform. With both level-0 and level-5 aliasing we observe that in fact no aliasing at all is required because of the rapid convergence of the spectrum.

2.28

$$F(n) = 1/T_0 \int_{-T_0/2}^{T_0/2} f_p(t)\ \exp(-jn\omega_0 t)\ dt$$

$$= 1/T_0 \int_{-T_0/2}^{T_0/2} [f_{ev}(t) + f_{od}(t)]\ [\cos(n\omega_0 t) - j\ \sin(n\omega_0 t)]dt$$

$$= 1/T_0 \int_{-T_0/2}^{T_0/2} \overset{A}{f_{ev}(t)\ \cos(n\omega_0 t)\ dt} + 1/T_0 \int_{-T_0/2}^{T_0/2} \overset{B}{f_{od}(t)\ \cos(n\omega_0 t)\ dt}$$

$$-j/T_0 \int_{-T_0/2}^{T_0/2} \overset{C}{f_{ev}(t)\ \sin(n\omega_0 t)\ dt} - j/T_0 \int_{-T_0/2}^{T_0/2} \overset{D}{f_{od}(t)\ \sin(n\omega_0 t)\ dt}$$

in which B and C are equal to zero because their integrands are odd, and so we have shown that

$$F(n) = 1/T_0 \int_{-T_0/2}^{T_0/2} f_p(t)\ \exp(-jn\omega_0 t)\ dt$$

$$= 1/T_0 \int_{-T_0/2}^{T_0/2} \overset{A}{f_{ev}(t)\ \cos(n\omega_0 t)\ dt} - j/T_0 \int_{-T_0/2}^{T_0/2} \overset{D}{f_{od}(t)\ \sin(n\omega_0 t)\ dt}$$

If $f_p(t)$ is even then $f_{od}(t)$ is zero and thus so D. Then

$$F(n) = 1/T_0 \int_{-T_0/2}^{T_0/2} \overset{A}{f_{ev}(t)\ \cos(n\omega_0 t)\ dt} = 2/T_0 \int_{0}^{T_0/2} f_{ev}(t)\ \cos(n\omega_0 t)\ dt$$

which means that (a) is false and (b) is true. Likewise if $f_p(t)$ is odd then A is zero and so

$$F(n) = -j/T_0 \int_{-T_0/2}^{T_0/2} \overset{D}{f_{od}(t)\ \sin(n\omega_0 t)\ dt} = -2j/T_0 \int_{0}^{T_0/2} f_{od}(t)\ \sin(n\omega_0 t)\ dt$$

which means that (c) is true.

2.29 (a)

$$f_p(t) = \begin{cases} t & (0 < t < 1) \\ 0 & (1 < t < 4) \end{cases} \qquad f_p(t + 4) = f_p(t)$$

For the sketch see the Answer section. $T_0 = 4$, $\omega_0 = \pi/2$.

$$F(n) = 1/T_0 \int_0^{T_0} f_p(t)\ \exp(-jn\omega_0 t)\ dt = \tfrac{1}{4} \int_0^1 t\ \exp(-jn\pi t/2)\ dt$$

$$= \tfrac{1}{4} \cdot \left[\frac{t\ \exp(-jn\pi t/2)}{-jn\pi/2} - \frac{\exp(-jn\pi t/2)}{(-jn\pi/2)^2} \right]_0^1$$

$$= \tfrac{1}{4} \left[\frac{\exp(-jn\pi/2)}{(-jn\pi/2)} - \frac{\exp(-jn\pi/2) - 1}{(jn\pi/2)^2} \right]$$

$$= \tfrac{1}{4} \left[\frac{-(jn\pi/2)\ \exp(-jn\pi/2) - \exp(-jn\pi/2) + 1}{(jn\pi/2)^2} \right]$$

$$= \tfrac{1}{4} \left[\frac{1 - \exp(-jn\pi/2)(1 + jn\pi/2)}{(jn\pi/2)^2} \right]$$

Splitting this into its real and imaginary parts we continue as

$$= \tfrac{1}{4} \left[\frac{1 - [\cos(n\pi/2) - j\ \sin(n\pi/2)](1 + jn\pi/2)}{-n^2\pi^2/4} \right]$$

$$= \left[\frac{\cos(n\pi/2) + (n\pi/2)\ \sin(n\pi/2) - 1}{n^2\pi^2} \right]$$

$$+ j \left[\frac{(n\pi/2)\ \cos(n\pi/2) - \sin(n\pi/2)}{n^2\pi^2} \right]$$

from which

$$A(n) = \left[\frac{\cos(n\pi/2) + (n\pi/2)\sin(n\pi/2) - 1}{n^2\pi^2} \right] \quad (n \neq 0)$$

$$B(n) = \left[\frac{(n\pi/2)\cos(n\pi/2) - \sin(n\pi/2)}{n^2\pi^2} \right] \quad (n \neq 0)$$

For $n = 0$, $$F(n) = 1/T_0 \int_0^{T_0} f_p(t)\, dt = \tfrac{1}{4} \int_0^1 t\, dt = 1/8$$

and so $A(0) = 1/8$, and $B(0) = 0$.

(b) For sketches of the even and odd parts see the Answer section. The expressions for their Fourier coefficients are

$f_{ev}(t)$ <===> $A(n)$ as defined above

$f_{od}(t)$ <===> $jB(n)$ as defined above

(c) Steps for loading the waveform using N = 256, SAMPLED, T = 4, PERIODIC:

Main menu, Create Values, Continue, Create X, Real, Neither, Left, 2 intervals, 1, Continue, t, Go, Neither, Go. You are now back on the main menu with the waveform correctly loaded. Plotting X confirms that. Run ANALYSIS.

To display the even part: Run Postprocessors, F, Copy F to F2, Set FIM to zero, Quit, Run SYNTHESIS, Plot Y. The even part appears.

To display the odd part: Run Postprocessors, F, Copy F2 to F, Set FRE to zero, Quit, Run SYNTHESIS, Plot Y. The odd part appears.

(d) Steps for loading the expressions for A(n) and B(n) into the system using N = 256, SAMPLED, T = 4, PERIODIC:

Main menu, Create Values, Continue, Create F, Complex,

```
( COS(n*pi/2) + (n*pi/2) * SIN(n*pi/2) - 1 )/(n*pi)^2
```

Note: True BASIC evaluates the operations with the following priorities: Highest ^, then /, then *, then + and -. The brackets shown above will ensure that A(n) is properly evaluated.

Alpha = 0, Division by zero


```
( (n*pi/2) * COS(n*pi/2) - SIN(n*pi/2) )/(n*pi)^2
```

Go, Go, Run SYNTHESIS, Plot Y. $f_p(t)$ will appear, with large oscillations before and after the discontinuities.

To eliminate the oscillations: Main menu, Create Values, Use the Old Problem, Accept A(n), Alpha = 5. (Use 20 if you have the patience or if you have a very fast machine.) Division by zero, Go, B, 0.125, Accept B(n), Continue aliasing with Alpha = 5, Go, Go, Run SYNTHESIS, Plot Y.

The oscillations are now much reduced. Using larger values for Alpha will make them as small as we please.

2.30 (a)

$$f_p(t) = \begin{cases} 0 & (-2 < t < -1) \\ t + 1 & (-1 < t < 0) \\ 1 & (0 < t < 1) \\ 0 & (1 < t < 2) \end{cases} \qquad f_p(t + 4) = f_p(t)$$

For the sketch, see the Answer section of the text. $T_0 = 4$, $\omega_0 = \pi/2$

$$F(n) = 1/T_0 \int_{-T_0/2}^{T_0/2} f_p(t) \exp(-jn\omega_0 t)\, dt$$

$$= \tfrac{1}{4} \int_{-1}^{0} (t + 1) \exp(-jn\pi t/2)\, dt + \tfrac{1}{4} \int_0^1 1 \exp(-jn\pi t/2)\, dt$$

$$= \tfrac{1}{4} \left[\frac{(t+1)\exp(-jn\pi t/2)}{-jn\pi/2} - \frac{\exp(-jn\pi t/2)}{(-jn\pi/2)^2} \right]_{-1}^{0} + \tfrac{1}{4} \left[\frac{\exp(-jn\pi t/2)}{-jn\pi/2} \right]_0^1$$

$$= \tfrac{1}{4} \left[\frac{1}{-jn\pi/2} - \frac{1 - \exp(jn\pi/2)}{(-jn\pi/2)^2} + \frac{\exp(-jn\pi t/2) - 1}{-jn\pi/2} \right]$$

$$= \tfrac{1}{4} \left[\frac{\exp(jn\pi/2) - 1}{\ldots} + \frac{\exp(-jn\pi t/2)}{\ldots} \right.$$

$$= \tfrac{1}{4}\left[\frac{\exp(jn\pi/2) - 1 - (jn\pi/2)\exp(-jn\pi/2)}{(-jn\pi/2)^2}\right]$$

$$F(n) = \tfrac{1}{4}\left[\frac{1 - \exp(jn\pi/2) + (jn\pi/2)\exp(-jn\pi/2)}{(n\pi/2)^2}\right] \quad (n \neq 0)$$

$$= \tfrac{1}{4}\left[\frac{1-\cos(n\pi/2)-j\sin(n\pi/2)+(jn\pi/2)[\cos(n\pi/2)-j\sin(n\pi/2)]}{(n\pi/2)^2}\right]$$

$$= \left[\frac{1 - \cos(n\pi/2) + (n\pi/2)\sin(n\pi/2)]}{(n\pi)^2}\right] = A(n) \quad (n \neq 0)$$

$$+ j\left[\frac{(n\pi/2)\cos(n\pi/2) - \sin(n\pi/2)}{(n\pi)^2}\right] = j\,B(n) \quad (n \neq 0)$$

For $n = 0$: $$F(0) = 1/T_0 \int_{-T_0/2}^{T_0/2} f_p(t)\,dt = \tfrac{1}{4}\int_{-1}^{0}(t+1)\,dt + \tfrac{1}{4}\int_{0}^{1} 1\,dt$$

$$= \left.\frac{(t+1)^2}{8}\right|_{-1}^{0} + \tfrac{1}{4} = 3/8 = A(0)$$

(b) See the Answer section for sketches of its even and odd parts. The expressions for their Fourier coefficients are as follows:

$f_{ev}(t) \Longleftrightarrow A(n)$ as defined above

$f_{od}(t) \Longleftrightarrow jB(n)$ as defined above

(c) Steps for loading the waveform using N = 256, SAMPLED, T = 4, PERIODIC:

Main menu, Create Values, Continue, Create X, Real, Neither, Center, 4 intervals, -1, 0, 1, Continue, 0, 1+t, 1, 0, Go, Neither, Go. You are now back on the main menu with the waveform correctly loaded. Plotting **X** confirms that.

To create a plot of its even part: Run ANALYSIS, Run Postprocessors, F, Copy F to F2, Set FIM to zero, Quit, Run SYNTHESIS, Plot **Y**. The

To create a plot of its odd part: Run Postprocessors, F, Copy F2 to F, Set FRE to zero, Quit, Run SYNTHESIS, Plot **Y**. The odd part appears.

(d) Steps for loading the expressions for A(n) and B(n) into the system using N = 256, SAMPLED, T = 4, PERIODIC:

Main menu, Create Values, Continue, Create F, Complex,

```
( 1 - COS(n*pi/2) + (n*pi/2) * SIN(n*pi/2) )/((n*pi)^2)
```

Note: True BASIC evaluates the operations with the following priorities: Highest ^, then /, then *, then + and -. The brackets shown above will ensure that A(n) is properly evaluated.

Alpha = 0, Division by zero, Go, B, 0.375, Go,

```
( (n*pi/2) * COS(n*pi/2) - SIN(n*pi/2) )/(n*pi)^2
```

Go, Go, Run SYNTHESIS, Plot **Y**. $f_p(t)$ will appear, with large oscillations before and after the discontinuities.

To eliminate the oscillations: Main menu, Create Values, Use the Old Problem, Accept A(n), Alpha = 5. (Use 20 if you have the patience or if you have a very fast machine.) Division by zero, Go, B, 0.375, Accept B(n), Continue aliasing with Alpha = 5, Go, Go, Run SYNTHESIS, Plot **Y**. The oscillations are now much reduced. Using larger values for Alpha will make them as small as we please.

2.31 Prove that for $f_p(t)$ real, $f_{od}(t) \Longleftrightarrow jB(n)$.

Proof: $$f_{od}(t) = \tfrac{1}{2}[f_p(t) - f_p(-t)]$$

to which we now apply the synthesis equation, continuing as

$$\ldots = \tfrac{1}{2}\sum_{n=-\infty}^{\infty} F(n)\exp(jn\omega_0 t) - \tfrac{1}{2}\sum_{n=-\infty}^{\infty} F(n)\exp(-jn\omega_0 t)$$

$$= j\sum_{n=-\infty}^{\infty} F(n)\,\frac{\exp(jn\omega_0 t) - \exp(-jn\omega_0 t)}{2j}$$

$$= j\sum_{n=-\infty}^{\infty} F(n)\sin(n\omega_0 t) = j\sum_{n=-\infty}^{\infty}[A(n) + jB(n)]\sin(n\omega_0 t)$$

$$= j\left[\sum_{n=-\infty}^{\infty} A(n)\sin(n\omega_0 t) + \sum_{n=-\infty}^{\infty} j\,B(n)\sin(n\omega_0 t)\right]$$

However $A(n)$ has been shown to be an even function of n whereas $\sin(n\omega_0 t)$ is an odd one. Thus their product is odd and so the first infinite sum in the above must be zero. We therefore continue as

$$\ldots = j \sum_{n=-\infty}^{\infty} B(n)\ j \sin(n\omega_0 t) = j \sum_{n=-\infty}^{\infty} B(n) \left[\cos(n\omega_0 t) + j \sin(n\omega_0 t) \right]$$

in which we have added nothing since the product of $B(n)$ (which is odd) with $\cos(n\omega_0 t)$ (which is even) must be odd, and so it sums to zero. We therefore continue further as

$$\ldots = \sum_{n=-\infty}^{\infty} j\, B(n) \exp(jn\omega_0 t)$$

We have thus shown that, for $f_p(t)$ real,

$$f_{od}(t) = \sum_{n=-\infty}^{\infty} j\, B(n) \exp(jn\omega_0 t) \quad \text{i.e.} \quad f_{od}(t) \Longleftrightarrow j\, B(n). \quad \text{QED}$$

2.32 Prove that for $f_p(t)$ real, $F(n)$ is purely imaginary and odd iff $f_p(t)$ is odd.

Proof: $F(n) = jB(n)$ iff $A(n) = 0$ iff $f_p(t)$ is odd.

But $B(n)$ is real and so $jB(n)$ is purely imaginary, and by Theorem 2.4 $B(n)$ is also odd. Hence $F(n)$ is purely imaginary and odd iff $f_p(t)$ is odd. QED

2.33 (a) Define

$$\left.\begin{aligned} H(n) &= \sqrt{[a(n)^2 + b(n)^2]}, \quad H(0) = a(0)/2, \\ \Phi(n) &= \arctan[b(n)/a(n)] \end{aligned}\right\} \quad (A)$$

Then $$\cos(\Phi(n)) = \frac{1}{\sec(\Phi(n))} = \frac{1}{\sqrt{[1 + \tan^2(\Phi(n))]}} = \frac{1}{\sqrt{[1 + (b(n)^2/a(n)^2]}}$$

$$= \frac{a(n)}{\sqrt{[a(n)^2 + b(n)^2]}}$$

and $$\sin(\Phi(n)) = \sqrt{[1 - \cos^2(\Phi(n))]} = \frac{b(n)}{\sqrt{[a(n)^2 + b(n)^2]}}$$

From this

$$a(n) \cos(n\omega_0 t) + b(n) \sin(n\omega_0 t)$$

$$= \sqrt{[a(n)^2 + b(n)^2]} \left[\frac{a(n)}{\sqrt{[a(n)^2 + b(n)^2]}} \cos(n\omega_0 t) + \frac{b(n)}{\sqrt{[a(n)^2 + b(n)^2]}} \sin(n\omega_0 t) \right]$$

$$= H(n) \left[\cos(\Phi(n)) \cos(n\omega_0 t) + \sin(\Phi(n)) \sin(n\omega_0 t) \right]$$

$$= H(n) \left[\cos(n\omega_0 t - \Phi(n)) \right]$$

and so the real form of the Fourier series becomes

$$f_p(t) = a(0)/2 + \sum_{n=1}^{\infty} a(n) \cos(n\omega_0 t) + \sum_{n=1}^{\infty} b(n) \sin(n\omega_0 t)$$

$$= H(0) + \sum_{n=1}^{\infty} H(n) \left[\cos(n\omega_0 t - \Phi(n)) \right] \qquad \text{QED}$$

(b) Alternatively, if we define

$$\left.\begin{aligned} H(n) &= \sqrt{[a(n)^2 + b(n)^2]}, \quad H(0) = a(0)/2, \\ \Phi(n) &= \arctan[a(n)/b(n)] \end{aligned}\right\} \quad (B)$$

then $$\cos(\Phi(n)) = \frac{b(n)}{\sqrt{[a(n)^2 + b(n)^2]}}$$

and $$\sin(\Phi(n)) = \frac{a(n)}{\sqrt{[a(n)^2 + b(n)^2]}}$$

Thus

$$a(n) \cos(n\omega_0 t) + b(n) \sin(n\omega_0 t)$$

$$= \sqrt{[a(n)^2 + b(n)^2]} \left[\frac{a(n)}{\sqrt{[a(n)^2 + b(n)^2]}} \cos(n\omega_0 t) + \frac{b(n)}{\sqrt{[a(n)^2 + b(n)^2]}} \sin(n\omega_0 t) \right]$$

$$= H(n)\ [\sin(n\omega_0 t + \Phi(n)]$$

and so the real form of the Fourier series becomes

$$f_p(t) = a(0)/2 + \sum_{n=1}^{\infty} a(n)\ \cos(n\omega_0 t) + \sum_{n=1}^{\infty} b(n)\ \sin(n\omega_0 t)$$

$$= H(0) + \sum_{n=1}^{\infty} H(n)\ [\sin(n\omega_0 t + \Phi(n)]$$

QED

Chapter 3 The Fourier Integral

3.1 (a) By the analysis equation:

$$F(\omega) = \int_{-\infty}^{\infty} f(t)\ \exp(-j\omega t)\ dt$$

Setting $\omega = 0$ gives $F(0) = \int_{-\infty}^{\infty} f(t)\ dt$ (A) QED

(b) By the synthesis equation:

$$f(t) = 1/2\pi \int_{-\infty}^{\infty} F(\omega)\ \exp(j\omega t)\ d\omega$$

Setting $t = 0$ gives $2\pi\ f(0) = \int_{-\infty}^{\infty} F(\omega)\ d\omega$ (B) QED

(c) From Exercise 2.15 (for periodic functions), the counterpart to (B) above is

$$f_p(0) = \sum_{n=-\infty}^{\infty} F(n)$$

and the counterpart to (A) is $T_0F(0) = \int_{-T_0/2}^{T_0/2} f_p(t)\ dt$

3.2 By the analysis equation:

$$F(\omega) = \int_{-\infty}^{\infty} f(t)\ \exp(-j\omega t)\ dt$$

$$= \int_{-\infty}^{\infty} [f_{ev}(t) + f_{od}(t)]\ [\cos(\omega t) - j\ \sin(\omega t)]\ dt$$

$$= \overset{A}{\int_{-\infty}^{\infty} f_{ev}(t)\ \cos(\omega t)\ dt} + \overset{B}{\int_{-\infty}^{\infty} f_{od}(t)\ \cos(\omega t)\ dt}$$

$$\overset{C}{-j \int_{-\infty}^{\infty} f_{ev}(t)\ \sin(\omega t)\ dt} \overset{D}{- j \int_{\infty}^{\infty} f_{od}(t)\ \sin(\omega t)\ dt}$$

in which B and C are equal to zero because their integrands are odd, and so we have shown that

$$F(\omega) = \int_{-\infty}^{\infty} f(t)\ \exp(-j\omega t)\ dt$$

$$= \overset{A}{\int_{-\infty}^{\infty} f_{ev}(t)\ \cos(\omega t)\ dt} \overset{D}{- j \int_{-\infty}^{\infty} f_{od}(t)\ \sin(\omega t)\ dt}$$

If $f_p(t)$ is even then $f_{od}(t)$ is zero and thus so is D. Then

$$F(\omega) = \overset{A}{\int_{-\infty}^{\infty} f_{ev}(t)\ \cos(\omega t)\ dt} = 2 \int_{0}^{\infty} f_{ev}(t)\ \cos(\omega t)\ dt \qquad \text{QED}$$

Likewise if $f_p(t)$ is odd then A is zero and so

$$F(\omega) = \overset{D}{-j \int_{-\infty}^{\infty} f_{od}(t)\ \sin(\omega t)\ dt} = -2j \int_{0}^{\infty} f_{od}(t)\ \sin(\omega t)\ dt$$

QED

3.3

(a) $f(t) = \dfrac{1}{2\pi} \int_{-\infty}^{\infty} \dfrac{1}{\beta + j\omega} \exp(j\omega t)\ d\omega \qquad (\beta > 0)$

$$\blacksquare\ F(\omega) = \frac{1}{\beta + j\omega}$$

$$F(\omega)^* = \frac{1}{\beta - j\omega} \qquad F(-\omega) = \frac{1}{\beta - j\omega}$$

Because $F(\omega)^* = F(-\omega)$ the pulse $f(t)$ is real.

- To find the Fourier transforms of $f_{ev}(t)$ and $f_{od}(t)$ we must find the real and imaginary parts of $F(\omega)$. Thus

$$F(\omega) = \frac{1}{\beta + j\omega} = \frac{1}{\beta + j\omega}\,\frac{\beta - j\omega}{\beta - j\omega} = \frac{\beta}{\beta^2 + \omega^2} - j\,\frac{\omega}{\beta^2 + \omega^2} = A(\omega) + jB(\omega)$$

and so $f_{ev}(t) \Longleftrightarrow A(\omega)$, $f_{od}(t) \Longleftrightarrow jB(\omega)$

- Because $A(\omega)$ and $B(\omega)$ are both nonzero the pulse is neither even nor odd.
- The total area under the pulse is $F(0) = 1/\beta$
- Because $F(\omega)$ is dying out like $1/\omega$ for ω large, the pulse $f(t)$ must have one or more discontinuities.

(b) $$f(t) = j\int_{-\infty}^{\infty} \frac{\omega\cos(\omega) - \sin(\omega)}{\omega^2}\exp(j\omega t)\,d\omega$$

- $$F(\omega) = j2\pi\,\frac{\omega\cos(\omega) - \sin(\omega)}{\omega^2}$$

- Testing to see if $f(t)$ is real:

$$F(\omega)^* = -j2\pi\,\frac{\omega\cos(\omega) - \sin(\omega)}{\omega^2}$$

$$F(-\omega) = j2\pi\,\frac{-\omega\cos(-\omega) - \sin(-\omega)}{(-\omega)^2} = j2\pi\,\frac{-\omega\cos(\omega) + \sin(-\omega)}{\omega^2} = F(\omega)^*$$

Thus $F(\omega)^* = F(-\omega)$ and so the pulse $f(t)$ must be real.

- $F(\omega)$ is purely imaginary, i.e. $jB(\omega) = j\,\dfrac{\omega\cos(\omega) - \sin(\omega)}{\omega^2}$

and $A(\omega) = 0$. Then $f_{ev}(t) \Longleftrightarrow A(\omega)$ and $f_{od}(t) \Longleftrightarrow jB(\omega)$

- Because $A(\omega)$ is zero the pulse $f(t)$ must be odd.
- The total area under the pulse is $F(0) = 0$ because $f(t)$ is odd. This result can also be derived by applying l'Hopital's rule to $F(\omega)$, i.e.

$$F(0) = \lim_{\omega \to 0} j2\pi\,\frac{\qquad}{\omega^2}$$

$$= \lim_{\omega \to 0} j2\pi\,\frac{\cos(\omega) - \omega\sin(\omega) - \cos(\omega)}{2\omega} = \lim_{\omega \to 0} j2\pi\,\frac{-\sin(\omega)}{2} = 0$$

- Because $F(\omega)$ is dying out like $1/\omega$ for ω large the pulse $f(t)$ must have one or more discontinuities.

(c) $$f(t) = \int_{-\infty}^{\infty} \frac{\omega\sin^2(\omega) - 2\cos(\omega)}{\omega^2}\exp(j\omega t)\,d\omega$$

- $$F(\omega) = 2\pi\,\frac{\omega\sin^2(\omega) - 2\cos(\omega)}{\omega^2}$$

- Testing to see if $f(t)$ is real: $F(\omega)^* = 2\pi\,\dfrac{\omega\sin^2(\omega) - 2\cos(\omega)}{\omega^2}$

$$F(-\omega) = 2\pi\,\frac{-\omega\sin^2(-\omega) - 2\cos(-\omega)}{(-\omega^2)} = 2\pi\,\frac{-\omega\sin^2(\omega) - 2\cos(\omega)}{(\omega^2)}$$

and so $F(\omega)^* \neq F(-\omega)$. Thus $f(t)$ is a complex function of t.

(d) $$f(t) = \frac{1}{2\pi}\int_{-\infty}^{\infty} \exp(j\omega t)\,d\omega$$

- $F(\omega) = 1$
- Testing to see if $f(t)$ is real: $F(\omega)^* = 1$ and $F(-\omega) = 1$

and so $F(\omega)^* = F(-\omega)$. Thus $f(t)$ is a real function of t.

- $f_{ev}(t) \Longleftrightarrow A(\omega) = 1$ and $f_{od}(t) \Longleftrightarrow jB(\omega) = 0$.
- $f(t)$ must be even because $B(\omega) = 0$.
- Total area $= F(0) = 1$
- $F(\omega)$ is not dying out like K/ω^p and so we cannot say whether or not it has discontinuities.

(e) $$f(t) = \frac{1}{2\pi}\int_{-\infty}^{\infty} \exp(-\beta|\omega|)\exp(j\omega t)\,d\omega \qquad (\beta > 0)$$

- $F(\omega) = \exp(-\beta|\omega|)$

- Testing to see if $f(t)$ is real:

$$F(\omega)^* = \exp(-\beta|\omega|) \quad \text{and} \quad F(-\omega) = \exp(-\beta|\omega|)$$

and so $F(\omega)^* = F(-\omega)$. Thus $f(t)$ is a real function of t.

- $f_{ev}(t) \Longleftrightarrow A(\omega) = \exp(-\beta|\omega|)$ and $f_{od}(t) \Longleftrightarrow jB(\omega) = 0$.
- $f(t)$ must be even because $B(\omega) = 0$.
- Total area $= F(0) = 1$
- $F(\omega)$ is not dying out like K/ω^p and so we cannot say whether or not it has discontinuities.

(f) $$f(t) = \frac{1}{2\pi}\int_{-\infty}^{\infty} \frac{j\omega + 2}{(j\omega)^2 + \omega + 1} \exp(j\omega t)\, d\omega$$

- $$F(\omega) = \frac{j\omega + 2}{(j\omega)^2 + \omega + 1}$$

- Testing to see if $f(t)$ is real:

$$F(\omega)^* = \frac{-j\omega + 2}{(-\omega)^2 + \omega + 1} \quad \text{and} \quad F(-\omega) = \frac{-j\omega + 2}{(-\omega)^2 - \omega + 1}$$

and so $F(\omega)^* \neq F(-\omega)$. Thus $f(t)$ is a complex function of t.

3.4 (a) $F(\omega) = Sa(\omega/4)$

- Testing to see if $f(t)$ is real: $F(\omega)^* = Sa(\omega/4)$

and $$F(-\omega) = Sa(-\omega/4) = \frac{\sin(-\omega/4)}{-\omega/4} = \frac{\sin(\omega/4)}{\omega/4} = Sa(\omega/4)$$

and so $F(\omega)^* = F(-\omega)$. Thus $f(t)$ is a real function of t.

- $$f(t) = 1/2\pi \int_{-\infty}^{\infty} Sa(\omega/4) \exp(j\omega t)\, d\omega$$

- $F(\omega)$ is real and even and so $A(\omega) = Sa(\omega/4)$

Then $$f_{ev}(t) = 1/2\pi \int_{-\infty}^{\infty} Sa(\omega/4) \exp(j\omega t)\, d\omega, \quad f_{od}(t) = 0$$

- $F(\omega) = Sa(\omega/4) = \dfrac{\sin(\omega/4)}{\omega/4}$ which is dying out like $1/\omega$ and

so $f(t)$ must be discontinuous.

- Total area under $f(t)$ is $F(0) = \lim\limits_{\omega \to 0} \dfrac{\sin(\omega/4)}{\omega/4} = 1$

(b) $F(\omega) = \exp(-j\omega/2)\, Sa(\omega/4)$

- Testing to see if $f(t)$ is real: $F(\omega)^* = \exp(j\omega/2)\, Sa(\omega/4)$

and $$F(-\omega) = \exp(j\omega/2)\, Sa(-\omega/4) = \exp(j\omega/2) \frac{\sin(-\omega/4)}{-\omega/4}$$
$$= \exp(j\omega/2) \frac{\sin(\omega/4)}{\omega/4} = \exp(j\omega/2)\, Sa(\omega/4)$$

and so $F(\omega)^* = F(-\omega)$. Thus $f(t)$ is a real function of t.

- $$f(t) = 1/2\pi \int_{-\infty}^{\infty} \exp(-j\omega/2)\, Sa(\omega/4) \exp(j\omega t)\, d\omega$$

- $$F(\omega) = \exp(-j\omega/2)\, Sa(\omega/4) = [\cos(\omega/2) - j\sin(\omega/2)]\, Sa(\omega/4)$$
$$= \cos(\omega/2)\, Sa(\omega/4) - j\sin(\omega/2)\, Sa(\omega/4) = A(\omega) + jB(\omega)$$

$$f_{ev}(t) = 1/2\pi \int_{-\infty}^{\infty} \cos(\omega/2)\, Sa(\omega/4) \exp(j\omega t)\, d\omega$$

$$f_{od}(t) = -1/2\pi \int_{-\infty}^{\infty} j\sin(\omega/2)\, Sa(\omega/4) \exp(j\omega t)\, d\omega$$

- Because neither $A(\omega)$ nor $B(\omega)$ is zero, $f(t)$ is neither even nor odd.

- $F(\omega) = \exp(-j\omega/2)\, Sa(\omega/4) = \exp(-j\omega/2) \dfrac{\sin(\omega/4)}{\omega/4}$

and so $F(\omega)$ is dying out like $1/\omega$. Thus $f(t)$ must be discontinuous.

$$F(0) = \lim_{\omega \to 0} \frac{\exp(-j\omega/2)\ \sin(\omega/4)}{\omega/4} = 1$$

(c) $$F(\omega) = 2j\left[\frac{\omega\cos(\omega) - \sin(\omega)}{\omega^2}\right]$$

- Testing to see if $f(t)$ is real:

$$F(\omega)^* = -2j\left[\frac{\omega\cos(\omega) - \sin(\omega)}{\omega^2}\right]$$

$$F(-\omega) = 2j\left[\frac{-\omega\cos(-\omega) - \sin(-\omega)}{(-\omega)^2}\right]$$

$$= 2j\left[\frac{-\omega\cos(\omega) + \sin(\omega)}{\omega^2}\right]$$

and so $F(\omega)^* = F(-\omega)$. Thus $f(t)$ is a real function of t.

- $$f(t) = \frac{j}{\pi}\int_{-\infty}^{\infty} \frac{\omega\cos(\omega) - \sin(\omega)}{\omega^2}\exp(j\omega)\ d\omega$$

- $$F(\omega) = 2j\left[\frac{\omega\cos(\omega) - \sin(\omega)}{\omega^2}\right] = jB(\omega) \quad \text{and} \quad A(\omega) = 0$$

Then, $f_{ev}(t) = 0$

and $$f_{od}(t) = 1/2\pi\int_{-\infty}^{\infty} 2j\left[\frac{\omega\cos(\omega) - \sin(\omega)}{\omega^2}\right]\exp(j\omega t)\ d\omega$$

- Because $A(\omega) = 0$ it follows that $f(t)$ is odd.
- $F(\omega)$ is dying out like $1/\omega$. Thus $f(t)$ must be discontinuous.
- Total area under $f(t)$ is zero because it is odd. This result

$$F(0) = \lim_{\omega \to 0} 2j\frac{\omega\cos(\omega) - \sin(\omega)}{\omega^2}$$

$$= \lim_{\omega \to 0} 2j\frac{\cos(\omega) - \omega\sin(\omega) - \cos(\omega)}{2\omega}$$

$$= \lim_{\omega \to 0} [-j\sin(\omega)] = 0$$

(d) $$F(\omega) = \frac{1 - \exp(-j\omega) - j\omega\exp(-j\omega)}{(j\omega)^2}$$

- Testing to see if $f(t)$ is real:

$$F(\omega)^* = \frac{1 - \exp(j\omega) + j\omega\exp(j\omega)}{(-j\omega)^2}$$

$$F(-\omega) = \frac{1 - \exp(j\omega) + j\omega\exp(j\omega)}{(-j\omega)^2}$$

and so $F(\omega)^* = F(-\omega)$. Thus $f(t)$ is a real function of t.

- $$f(t) = \frac{1}{2\pi}\int_{-\infty}^{\infty} \frac{1 - \exp(-j\omega) - j\omega\exp(-j\omega)}{(j\omega)^2}\exp(j\omega t)\ d\omega$$

- $$F(\omega) = \frac{1 - \exp(-j\omega) - j\omega\exp(-j\omega)}{(j\omega)^2}$$

$$= \frac{1 - \cos(\omega) + j\sin(\omega) - j\omega[\cos(\omega) - j\sin(\omega)]}{(j\omega)^2}$$

$$= \frac{1 - \cos(\omega) + j\sin(\omega) - j\omega\cos(\omega) - \omega\sin(\omega)}{-\omega^2}$$

$$= \frac{\cos(\omega) + \omega\sin(\omega) - 1}{\omega^2} + j\frac{\omega\cos(\omega) - \sin(\omega)}{\omega^2} = A(\omega) + jB(\omega)$$

Then, $$f_{ev}(t) = 1/2\pi\int_{-\infty}^{\infty} \frac{\cos(\omega) + \omega\sin(\omega) - 1}{\omega^2}\exp(j\omega t)\ d\omega$$

Then, $$f_{od}(t) = 1/2\pi \int_{-\infty}^{\infty} j \frac{\omega \cos(\omega) - \sin(\omega)}{\omega^2} \exp(j\omega t)\, d\omega$$

- Because neither $A(\omega) = 0$ nor $B(\omega) = 0$ it follows that $f(t)$ is neither even nor odd.
- $F(\omega)$ is dying out like $1/\omega$. Thus $f(t)$ must be discontinuous.
- Total area under $f(t)$ can be derived by applying l'Hopital's rule to $F(\omega)$ to obtain $F(0)$, i.e.

$$F(0) = \lim_{\omega \to 0} \frac{1 - \exp(-j\omega) - j\omega \exp(-j\omega)}{(j\omega)^2}$$

$$= \lim_{\omega \to 0} \frac{j \exp(-j\omega) - j \exp(-j\omega) - \omega \exp(-j\omega)}{2j(j\omega)}$$

$$= \lim_{\omega \to 0} \frac{-\omega \exp(-j\omega)}{-2\omega} = \tfrac{1}{2} = \int_{-\infty}^{\infty} f(t)\, dt$$

(e) $$F(\omega) = \frac{1 + j\omega}{1 + j\omega + (j\omega)^2}$$

- Testing to see if $f(t)$ is real:

$$F(\omega)^* = \frac{1 - j\omega}{1 - j\omega + (-j\omega)^2} \qquad F(-\omega) = \frac{1 - j\omega}{1 - j\omega + (-j\omega)^2}$$

and so $F(\omega)^* = F(-\omega)$. Thus $f(t)$ is a real function of t.

- $$f(t) = \frac{1}{2\pi} \int_{-\infty}^{\infty} \frac{1 + j\omega}{1 + j\omega + (j\omega)^2} \exp(j\omega t)\, d\omega$$

- $$F(\omega) = \frac{1 + j\omega}{1 + j\omega + (j\omega)^2}$$

$$= \frac{1 + j\omega}{(1 - \omega^2) + j\omega} \; \frac{(1 - \omega^2) - j\omega}{(1 - \omega^2) - j\omega}$$

$$= \frac{(1 - \omega^2) + \omega^2}{[illegible]} + j \frac{\omega(1 - \omega^2) - \omega}{[illegible]}$$

$$= \frac{1}{(1 - \omega^2)^2 + \omega^2} - j \frac{\omega^3}{(1 - \omega^2)^2 + \omega^2} = A(\omega) + jB(\omega)$$

Then, $$f_{ev}(t) = 1/2\pi \int_{-\infty}^{\infty} \frac{1}{(1 - \omega^2)^2 + \omega^2} \exp(j\omega t)\, d\omega$$

Then, $$f_{od}(t) = 1/2\pi \int_{-\infty}^{\infty} \frac{-j\omega^3}{(1 - \omega^2)^2 + \omega^2} \exp(j\omega t)\, d\omega$$

- Because neither $A(\omega) = 0$ nor $B(\omega) = 0$ it follows that $f(t)$ is neither even nor odd.
- $F(\omega)$ is dying out like $1/\omega$. Thus $f(t)$ must be discontinuous.
- Total area under $f(t)$ is $F(0) = 1$

(f) $F(\omega) = \cos(\omega) - j \sin(\omega/4)$

- Testing to see if $f(t)$ is real:

$$F(\omega)^* = \cos(\omega) + j \sin(\omega/4)$$

$$F(-\omega) = \cos(-\omega) - j \sin(-\omega/4) = \cos(\omega) + j \sin(\omega/4)$$

and so $F(\omega)^* = F(-\omega)$. Thus $f(t)$ is a real function of t.

- $$f(t) = \frac{1}{2\pi} \int_{-\infty}^{\infty} [\cos(\omega) - j \sin(\omega/4)] \exp(j\omega t)\, d\omega$$

- $F(\omega) = \cos(\omega) - j \sin(\omega/4) = A(\omega) + jB(\omega)$

Then, $$f_{ev}(t) = 1/2\pi \int_{-\infty}^{\infty} \cos(\omega) \exp(j\omega t)\, d\omega$$

and $$f_{od}(t) = -1/2\pi \int_{-\infty}^{\infty} j \sin(\omega/4) \exp(j\omega t)\, d\omega$$

- Because neither $A(\omega) = 0$ nor $B(\omega) = 0$ it follows that $f(t)$ is neither even nor odd.

has discontinuities.

- Total area under $f(t)$ is $F(0) = 1$

3.5 (a) $$F(\omega) = \frac{1 + 3j\omega}{1 + 5j\omega + (j\omega)^2}$$

Testing to see if $f(t)$ was real:

$$F(\omega)^* = \frac{1 - 3j\omega}{1 - 5j\omega + (-j\omega)^2} \quad \text{and} \quad F(-\omega) = \frac{1 - 3j\omega}{1 - 5j\omega + (-j\omega)^2}$$

Thus $F(\omega)^* = F(-\omega)$ and so $f(t)$ was real.

(b) $$F(\omega) = \frac{\sin(\omega) + j\cos(\omega)}{j\omega}$$

Testing to see if $f(t)$ was real: $$F(\omega)^* = \frac{\sin(\omega) - j\cos(\omega)}{-j\omega}$$

and $$F(-\omega) = \frac{\sin(-\omega) + j\cos(-\omega)}{-j\omega} = \frac{-\sin(\omega) + j\cos(\omega)}{-j\omega}$$

Thus $F(\omega)^* \neq F(-\omega)$ and so $f(t)$ was complex.

(c) $$F(\omega) = \frac{1}{1 + j\omega}$$

Testing to see if $f(t)$ was real: $$F(\omega)^* = \frac{1}{1 - j\omega} \quad \text{and} \quad F(-\omega) = \frac{1}{1 - j\omega}$$

Thus $F(\omega)^* = F(-\omega)$ and so $f(t)$ was real

(d) $F(\omega) = j\exp(-\beta|\omega|)$ $(\beta > 0)$

Testing to see if $f(t)$ was real:

$F(\omega)^* = -j\exp(-\beta|\omega|)$ and $F(-\omega) = j\exp(-\beta|-\omega|) = j\exp(-\beta|\omega|)$

Thus $F(\omega)^* \neq F(-\omega)$ and so $f(t)$ was complex.

3.6 (a)

(1) Let the two expressions be $F_1(\omega)$ and $F_2(\omega)$, and let their product be $F(\omega)$. Writing them in polar form:

$F_1(\omega) = |F_1(\omega)| \exp[j\theta_1(\omega)]$ and $F_2(\omega) = |F_2(\omega)| \exp[j\theta_2(\omega)]$

Then $F_1(\omega)\, F_2(\omega) = |F_1(\omega)| \exp[j\theta_1(\omega)]\ |F_2(\omega)| \exp[j\theta_2(\omega)]$

showing that $|F(\omega)| = |F_1(\omega)||F_2(\omega)|$ and $\theta(\omega) = \theta_1(\omega) + \theta_2(\omega)$

i.e. the magnitudes multiply and the phases add.

(2) Let the two expressions be $F_1(\omega)$ and $F_2(\omega)$, and let their quotient be $F(\omega)$. Writing them in polar form:

Then $$F_1(\omega)/F_2(\omega) = \frac{|F_1(\omega)| \exp[j\theta_1(\omega)]}{|F_2(\omega)| \exp[j\theta_2(\omega)]}$$

$$= \frac{|F_1(\omega)|}{|F_2(\omega)|} \exp[j(\theta_1(\omega) - \theta_2(\omega))] = |F(\omega)| \exp[j\theta(\omega)]$$

showing that $$|F(\omega)| = \frac{|F_1(\omega)|}{|F_2(\omega)|}$$ and $\theta(\omega) = \theta_1(\omega) - \theta_2(\omega)$

i.e. the magnitudes divide and the phases subtract.

(b)

(1) $$F(\omega) = \overbrace{\frac{\exp(j\omega) - 1 - j\omega\exp(-j\omega)}{(j\omega)^2}}^{F_1(\omega)} \overbrace{\exp(-j\omega)}^{F_2(\omega)}$$

$$= \frac{\cos(\omega) + j\sin(\omega) - 1 - j\omega\,[\cos(\omega) - j\sin(\omega)]}{(j\omega)^2} \exp(-j\omega)$$

$$= \left[\frac{1 + \omega\sin(\omega) - \cos(\omega)}{\omega^2} + j\,\frac{\omega\cos(\omega) - \sin(\omega)}{\omega^2}\right] \exp(-j\omega)$$

$$|F(\omega)| = \frac{\{[1 + \omega\sin(\omega) - \cos(\omega)]^2 + [\omega\cos(\omega) - \sin(\omega)]^2\}^{\frac{1}{2}}}{\omega^2}$$

$$\theta(\omega) = \overbrace{\tan^{-1}\left[\frac{\omega\cos(\omega) - \sin(\omega)}{1 + \omega\sin(\omega) - \cos(\omega)}\right]}^{\theta_1(\omega)} \overbrace{- \omega}^{\theta_2(\omega)}$$

(2) $$F(\omega) = \frac{j\omega - 1 + \exp(-j\omega)}{(j\omega)^2} \exp(-j3\omega)$$

$$= \frac{j\omega - 1 + \cos(\omega) - j\sin(\omega)}{(j\omega)^2}\exp(-j3\omega)$$

$$= \left[\frac{1 - \cos(\omega)}{\omega^2} + j\,\frac{\sin(\omega) - \omega}{\omega^2}\right]\exp(-j3\omega)$$

$$|F(\omega)| = \frac{\{[1 - \cos(\omega)]^2 + [\sin(\omega) - \omega]^2\}^{\frac{1}{2}}}{\omega^2}$$

$$\theta(\omega) = \tan^{-1}\left[\frac{\sin(\omega) - \omega}{1 - \cos(\omega)}\right] - 3\omega$$

(3) $$F(\omega) = \frac{(2 - j3\omega)(4 + j5\omega)(6 + j7\omega)}{(8 + j9\omega)(10 + j11\omega)(12 + j13\omega)}$$

$$|F(\omega)| = \frac{|2 - j3\omega||4 + j5\omega||6 + j7\omega|}{|8 + j9\omega||10 + j11\omega||12 + j13\omega|}$$

$$= \left[\frac{(4 + 9\omega^2)(16 + 25\omega^2)(36 + 49\omega^2)}{(64 + 81\omega^2)(100 + 121\omega^2)(144 + 169\omega^2)}\right]^{\frac{1}{2}}$$

$$\theta(\omega) = \tan^{-1}(-3\omega/2) + \tan^{-1}(5\omega/4) + \tan^{-1}(7\omega/6)$$
$$- \tan^{-1}(9\omega/8) - \tan^{-1}(11\omega/10) - \tan^{-1}(13\omega/12)$$

3.7 (a) $f(t) = \exp(-\beta t)$ $(t \geq 0)$ $(\beta > 0)$

$$I = \int_{-\infty}^{\infty} f(t)^2\,dt = \int_0^{\infty}\exp(-2\beta t)\,dt = \left.\frac{\exp(-2\beta t)}{-2\beta}\right|_0^{\infty} = 1/2\beta < \infty$$

The integral of $f(t)^2$ is finite and so the square integrability criterion guarantees the existence of $F(\omega)$.

In fact: $F(\omega) = \dfrac{1}{\beta + j\omega}$

(b) $f(t) = \exp(\beta t)$ $(t \geq 0)$ $(\beta > 0)$

$$I = \int_{-\infty}^{\infty} f(t)^2\,dt = \int_0^{\infty}\exp(2\beta t)\,dt = \left.\frac{\exp(2\beta t)}{2\beta}\right|_0^{\infty} = \infty$$

The integral of $f(t)^2$ is infinite and so the square integrability criterion does not guarantee the existence of $F(\omega)$.

In fact: $F(\omega)$ does not exist.

(c) $f(t) = t\exp(-\beta t)$ $(t \geq 0)$ $(\beta > 0)$

$$I = \int_{-\infty}^{\infty} f(t)^2\,dt = \int_0^{\infty} t^2\exp(-2\beta t)\,dt$$

$$= \left.\frac{t^2\exp(-2\beta t)}{-2\beta}\right|_0^{\infty} - \int_0^{\infty}\frac{2t\exp(-2\beta t)}{-2\beta}\,dt$$

$$= 0 - \left.\frac{2t\exp(-2\beta t)}{(-2\beta)^2}\right|_0^{\infty} + \left.\frac{2\exp(-2\beta t)}{(-2\beta)^3}\right|_0^{\infty} = 0 - 0 + 1/4\beta^3$$

The integral of $f(t)^2$ is finite and so the square integrability criterion guarantees the existence of $F(\omega)$.
In fact: $F(\omega) = 1/(\beta + j\omega)^2$

(d) $f(t) = \exp(-\beta|t|)$ $(\forall\, t)$ $(\beta > 0)$

$$I = \int_{-\infty}^{\infty}\exp(-\beta|t|)^2\,dt = 2\int_0^{\infty}\exp(-2\beta t)\,dt = \left.\frac{2\exp(-2\beta t)}{-2\beta}\right|_0^{\infty} = 1/\beta$$

The integral of $f(t)^2$ is finite and so the square integrability criterion guarantees the existence of $F(\omega)$.
In fact: $F(\omega) = 2\beta/(\beta^2 + \omega^2)$

(e) $f(t) = 1$ $(t \geq 0)$

$$I = \int_{-\infty}^{\infty} f(t)^2\,dt = \int_0^{\infty} 1^2\,dt = \infty$$

The integral of $f(t)^2$ is infinite and so the square integrability criterion does not guarantee the existence of $F(\omega)$. In fact: $F(\omega) = \pi\,\delta(\omega) + 1/j\omega$

(f) $f(t) = \cos(t) \qquad (\forall\, t)$

$$I = \int_{-\infty}^{\infty} f(t)^2\, dt = \int_{-\infty}^{\infty} \cos^2(t)\, dt = \int_{-\infty}^{\infty} \frac{1 + \cos(2t)}{2}\, dt = \infty$$

The integral of $f(t)^2$ is infinite and so the square integrability criterion does not guarantee the existence of $F(\omega)$. In fact: $F(\omega) = \pi\,\delta(\omega - 1) + \pi\,\delta(\omega + 1)$

(g) $f(t) = \sin(t) \qquad (\forall\, t)$

$$I = \int_{-\infty}^{\infty} f(t)^2\, dt = \int_{-\infty}^{\infty} \sin^2(t)\, dt = \int_{-\infty}^{\infty} \frac{1 - \cos(2t)}{2}\, dt = \infty$$

The integral of $f(t)^2$ is infinite and so the square integrability criterion does not guarantee the existence of $F(\omega)$.

In fact: $F(\omega) = \dfrac{\pi}{j}\left[\,\delta(\omega - 1) - \delta(\omega + 1)\,\right]$

(h) Finding the Fourier transform of the pulse in (a):

$$F(\omega) = \int_{0}^{\infty} f(t)\exp(-j\omega t)\, dt = \int_{0}^{\infty} \exp(-\beta t)\exp(-j\omega t)\, dt$$

$$= \int_{0}^{\infty} \exp[-(\beta + j\omega)t]\, dt = \frac{\exp(-\beta t)\exp(-j\omega t)}{-(\beta + j\omega)}\Bigg|_{0}^{\infty} = \frac{1}{\beta + j\omega}$$

3.8

(a) $$F(\omega) = \int_{-\infty}^{\infty} f(t)\exp(-j\omega t)\, dt = \int_{-\frac{1}{4}}^{\frac{1}{4}} 3\exp(-j\omega t)\, dt$$

$$= 3\,\frac{\exp(-j\omega t)}{-j\omega t}\Bigg|_{-\frac{1}{4}}^{\frac{1}{4}} = \frac{3}{2}\,\frac{\exp(-j\omega/4) - \exp(j\omega/4)}{-2j\omega/4} = (3/2)\ \mathrm{Sa}(\omega/4)$$

By (3.24): $\mathrm{Rect}(t/\tau) \Longleftrightarrow \tau\ \mathrm{Sa}(\omega\tau/2)$

In this case we are dealing with $\tau = \frac{1}{2}$ and a pulse of height 3. Thus

$$3\ \mathrm{Rect}(t/\tfrac{1}{2}) \Longleftrightarrow 3/2\ \mathrm{Sa}(\omega/4)$$

which is consistent with what we obtained above.

(b) The area under $f(t)$ equals $3/2$

and $$F(0) = \lim_{\omega \to 0} \frac{3}{2}\,\frac{\sin(\omega/4)}{\omega/4} = 3/2$$

i.e. the area under $f(t)$ does equal $F(0)$.

(c) $$\int_{-\infty}^{\infty} F(\omega)\, d\omega = 2\pi\,\frac{1}{2\pi}\int_{-\infty}^{\infty} F(\omega)\exp(j\omega t)\, d\omega\,\Bigg|_{t=0} = 2\pi\, f(0) = 6\pi$$

(d) Because $f(t)$ is discontinuous, $F(\omega)$ should decay to zero like $1/\omega$. In fact it does.

(e) By the Dirichlet-Jordan criterion:

At $t = -\frac{1}{4}$: $F(\omega)$ inverts to $\frac{1}{2}\left[\, f(-\tfrac{1}{4}^-) + f(-\tfrac{1}{4}^+)\,\right] = \frac{1}{2}[0 + 3] = 3/2$

At $t = -1/8$: $F(\omega)$ inverts to $\frac{1}{2}\left[\, f(-\tfrac{1}{8}^-) + f(-\tfrac{1}{8}^+)\,\right] = \frac{1}{2}[3 + 3] = 3$

At $t = \frac{1}{4}$: $F(\omega)$ inverts to $\frac{1}{2}\left[\, f(\tfrac{1}{4}^-) + f(\tfrac{1}{4}^+)\,\right] = \frac{1}{2}[3 + 0] = 3/2$

(f) Finding the expression for the energy spectrum $E(\omega)$:

$$E(\omega) = 1/2\pi\ |F(\omega)|^2 = \frac{1}{2\pi}\,\frac{9}{4}\,\frac{\sin^2(\omega/4)}{\omega^2/16} = \frac{9}{8\pi}\,\mathrm{Sa}^2(\omega/4)$$

[The total area under $E(\omega)$] $= \displaystyle\int_{-\infty}^{\infty} |f(t)|^2\, dt = \int_{-\frac{1}{4}}^{\frac{1}{4}} 9\, dt = 9/2$

(g) See the Answer section for the sketches.

(h) Steps for loading the pulse using N = 1024, SAMPLED, T = 2, PULSE:

Main menu, Create Values, Continue, Create X, Real, Even, 2 intervals, 0.25, Continue, 3, 0, Go, Neither, Go. You are now back

on the main menu with the pulse correctly loaded. Plotting **X** confirms that. Run ANALYSIS, Show Numbers, F, Single.

The following table confirms the correctness of $F(\omega)$. In this case the points at which the FFT values sample the ω-axis are at multiples of $\Omega = 2\pi/T = \pi$ and so we must evaluate the formula at $\omega = n\pi$. Thus at $n = 2$:

$$A(n\Omega)_{formula} = 3/2\ \frac{\sin(2\pi/4)}{2\pi/4} = 0.954929659$$

n	$A(n\ )_{FFT}$	$A(n\Omega)_{formula}$
0	1.5	1.5
1	1.3504702	1.350474475
2	0.95491767	0.954929659

3.9

(a) $$F(\omega) = \int_{-\infty}^{\infty} f(t)\exp(-j\omega t)\,dt = \int_0^{1/2} 3\exp(-j\omega t)\,dt$$

$$= 3\ \frac{\exp(-j\omega t)}{-j\omega}\Bigg|_0^{1/2} = 3\ \frac{1-\exp(-j\omega/2)}{j\omega}$$

$$= 3/2\ \exp(-j\omega/4)\ \frac{\exp(j\omega/4) - \exp(-j\omega/4)}{2j\omega/4} = \exp(-j\omega/4)\left[(3/2)\ Sa(\omega/4)\right]$$

showing that the Fourier transform of the delayed pulse is $\exp(-j\omega/4)$ times the transform of the undelayed pulse.

(b) For the magnitude spectrum:

$$|F(\omega)| = \left|\exp(-j\omega/4)\left[(3/2)\ Sa(\omega/4)\right]\right| = 3/2\ |\ Sa(\omega/4)\ |$$

which is the same as for the pulse of Exercise 3.8.

(c) Finding the expression for the energy spectrum:

$$E(\omega) = \frac{1}{2\pi}\,|\,F(\omega)\,|^2 = \frac{1}{2\pi}\,\frac{9}{4}\,Sa^2(\omega/4)$$

which is the same as for the pulse of Exercise 3.8.

(d) For sketches of the magnitude and energy spectra for this pulse: These are the same as for the pulse of Exercise 3.8.

(e) Sketching the phase spectrum for $-12\pi \le \omega \le 12\pi$:

Phase of $\exp(-j\omega/4)\left[(3/2)\ Sa(\omega/4)\right]$

= phase of $\exp(-j\omega/4)$ + phase of $(3/2)\ Sa(\omega/4)$

The phase of $\exp(-j\omega/4) = -\omega/4$

Hence: Phase of $f(t) = \theta_1(\omega) + \theta_2(\omega)$ where $\theta_1(t)$ is the same as for the pulse of Exercise 3.8

and $\theta_2(\omega) = -\omega/4$

This sketch is shown in the Answer section.

(f) Steps for loading the pulse using N = 1024, SAMPLED, T = 2, PULSE:

Main menu, Create Values, Continue, Create X, Real, Neither, Left, 2 intervals, 0.5, Continue, 3, 0, Go, Neither, Go. You are now back on the main menu with the pulse correctly loaded. Plotting **X** confirms that. Run ANALYSIS, Show Numbers, F, Single.

We obtained the following results, thereby confirming (a), (c) and (e). Note: Using $T = 2$ means that the FFT values are samples of $F(\omega)$ at $n\Omega = n2\pi/T = n\pi$.

n		FFT	formula
0	$\|F(0)\|$	1.5000000	1.500000000
1	$\|F(\pi)\|$	1.3504702	1.350474474
2	$\|F(2\pi)\|$	0.95491767	0.954929659
3	$\|F(3\pi)\|$	0.45014545	0.450158158
0	$\theta(0)$	0	0
1	$\theta(1\pi)$	-0.78539816	-0.785398164
2	$\theta(2\pi)$	-1.5707963	-1.570796327
3	$\theta(3\pi)$	-2.3561945	-2.356194491

3.10 Define $\Lambda(t/\tau)$ as a triangular pulse of width 2τ and height 1.

(a) $$F(\omega) = \int_{-\infty}^{\infty} f(t)\exp(-j\omega t)\,dt = 2\int_0^{\tau}(1 - t/\tau)\cos(\omega t)\,dt$$

$$= 2\,(1 - t/\tau)\,\frac{\sin(\omega t)}{\omega}\Bigg|_0^{\tau} - 2/\tau\ \frac{\cos(\omega t)}{\omega^2}\Bigg|_0^{\tau}$$

$$= 2/\tau\ \frac{1-\cos(\omega\tau)}{\omega^2} = 4/\tau\ \frac{\sin^2(\omega\tau/2)}{\omega^2} = \tau\ \frac{\sin^2(\omega\tau/2)}{(\omega\tau/2)^2} = \tau\ Sa^2(\omega\tau/2)$$

(b) [Area under $f(t)$] = $\frac{1}{2}(2\tau)\,1 = \tau$, and $F(0) = \tau$. Thus they are equal

(c) $F(\omega)$ goes to zero like $1/\omega^2$. Because the pulse is everywhere continuous but its first derivative is not, that is what we would expect from (3.57) and Table 3.2.

(d) The energy spectrum is

$$E(\omega) = \frac{1}{2\pi}|F(\omega)|^2 = \frac{\tau^2\,\mathrm{Sa}^4(\omega\tau/2)}{2\pi} \quad \text{which goes to zero like } 1/\omega^4$$

(e) $\int_{-\infty}^{\infty} \mathrm{Sa}^2(x)\,dx \overset{?}{=} \pi$ Consider $f(t) = \Lambda(t/\tau) \Longleftrightarrow F(\omega) = \tau\,\mathrm{Sa}^2(\omega\tau/2)$

From the synthesis equation: $f(t) = \frac{1}{2\pi}\int_{-\infty}^{\infty} F(\omega)\exp(j\omega t)\,d\omega$

Letting $\tau = 1$ and $t = 0$ gives, on the left: $f(0) = \Lambda(0) = 1$

and on the right:

$$\frac{1}{2\pi}\int_{-\infty}^{\infty}\mathrm{Sa}^2(\omega/2)\,d\omega = \frac{1}{\pi}\int_{-\infty}^{\infty}\mathrm{Sa}^2(\omega/2)\,d(\omega/2) = \frac{1}{\pi}\int_{-\infty}^{\infty}\mathrm{Sa}^2(x)\,d(x)$$

from which $\int_{-\infty}^{\infty}\mathrm{Sa}^2(x)\,dx = \pi$

(f) See Answer section for the sketches

(g) $\frac{1}{2\pi}\int_{-\infty}^{\infty}\mathrm{Sa}^2(\omega/2)\exp(j\omega t)\,d\omega = \Lambda(t/\tau)$ for $\tau = 1$

Thus the values are as follows:

t	-2	-1	$-\frac{1}{2}$	0	$\frac{1}{2}$	1	2
integral	0	0	$\frac{1}{2}$	1	$\frac{1}{2}$	0	0

(h) Steps for loading the pulse using N = 1024, SAMPLED, T = 16, PULSE: Main menu, Create Values, Continue, Create X, Real, Even, 2 intervals, 1, Continue, 1 - t, 0, Go, Neither, Go. You are now back on the main menu with the pulse correctly loaded. Plotting **X** confirms that. Run ANALYSIS, Show Numbers, F, Single.

We obtained the following values. (The sketches appearing in the Answer section were confirmed.) Note: Using T = 16 means that the FFT values are samples of $F(\omega)$ at $n\Omega = n2\pi/T = n\pi/8$.

n	$F(\omega)$	FFT	formula
0	$F(0)$	1.0000000	1.000000000
1	$F(\pi/8)$	0.98721793	0.987214831
2	$F(2\pi/8)$	0.94965312	0.949641204
3	$F(3\pi/8)$	0.88958594	0.889560882

3.11

(a) $F(\omega) = \int_{-\infty}^{\infty} f(t)\exp(-j\omega t)\,dt = \int_0^1 \exp(t)\exp(-j\omega t)\,dt$

$$= \int_0^1 \exp[(1-j\omega)t]\,dt = \left.\frac{\exp[(1-j\omega)t]}{1-j\omega}\right|_0^1 = \frac{\exp(1-j\omega)-1}{1-j\omega}$$

(b) $f(t) = \frac{1}{2\pi}\int_{-\infty}^{\infty} F(\omega)\exp(j\omega t)\,d\omega$

From a sketch of $f(t)$:

For $t = 0$: $f(0) = \frac{1}{2}$ For $t = \frac{1}{2}$: $f(\frac{1}{2}) = e^{1/2}$ For $t = 1$: $f(1) = e/2$

(c) We expect that $F(0) = \int_{-\infty}^{\infty} f(t)\,dt = \int_0^1 \exp(t)\,dt = e - 1$

In fact $F(0) = \left.\frac{\exp(1-j\omega)-1}{1-j\omega}\right|_{\omega \leftarrow 0} = e - 1$

(d) See the Answer section for the sketches of $f_{ev}(t)$ and $f_{od}(t)$. Their transforms are the real and $j \times$ imag parts of $F(\omega)$. Thus:

$$F(\omega) = \frac{\exp(1-j\omega)-1}{1-j\omega} = \frac{e\exp(-j\omega)-1}{1-j\omega}\,\frac{1+j\omega}{1+j\omega}$$

$$= \frac{[e\cos(\omega) - 1 - j\,e\sin(\omega)]\,[1 + j\omega]}{1 + \omega^2}$$

$$= \frac{[e\cos(\omega) - 1 + \omega\,e\sin(\omega)] + j[\omega\,e\cos(\omega) - \omega - e\sin(\omega)]}{1 + \omega^2} = A(\omega)+jB(\omega)$$

and so

$$f_{ev}(t) \Longleftrightarrow \frac{e\cos(\omega) - 1 + \omega\,e\sin(\omega)}{1 + \omega^2}$$

$$f_{od}(t) \Longleftrightarrow j\,\frac{\omega\,e\cos(\omega) - \omega - e\sin(\omega)}{1 + \omega^2}$$

(e) The transforms of $f_{ev}(t)$ and $f_{od}(t)$ both decay to zero like $1/\omega$ for ω large.

(f) Steps for loading the pulse using N = 1024, SAMPLED, T = 16, PULSE:

Main menu, Create Values, Continue, Create X, Real, Neither, Left, 2 intervals, 1, Continue, EXP(t), 0, Go, Neither, Go. You are now back on the main menu with the pulse correctly loaded. Plotting **X** confirms that. Run ANALYSIS, Show Numbers, F, Single.

To display a plot of $f_{ev}(t)$: Run Postprocessors, F, Copy F to F2, Set FIM to zero (i.e. zero out the imaginary part of $F(\omega)$, leaving only $A(\omega)$), Quit, Run SYNTHESIS. Plot the **Y** vector. A plot of $f_{ev}(t)$ will emerge.

To display a plot of $f_{od}(t)$: Run Postprocessors, F, Copy F2 to F, Set FRE to zero (i.e. zero out the real part of $F(\omega)$, leaving only $jB(\omega)$), Quit, Run SYNTHESIS. Plot the **Y** vector. A plot of $f_{od}(t)$ will emerge.

(g) Assuming that the pulse is still loaded as in (f) above: Run Postprocessors, F, Copy F2 to F, Quit, Show Numbers, F, Single.

We obtained the following values, which confirm the expressions derived in (a) and (d). Note: Since T = 16, the FFT spectral values lie at $\omega = n\Omega = n2\pi/T = n\pi/8$

n	item	FFT	formula
0	A(0)	1.7183168	1.718281828
1	A(π/8)	1.6633788	1.663356341
2	A(2π/8)	1.5039872	1.503999193
3	A(3π/8)	1.2557960	1.255855382
0	B(0)	0	0
1	B(π/8)	-0.38707615	-0.387042912
2	B(2π/8)	-0.74093115	-0.740877310
3	B(3π/8)	-1.0318972	-1.031845178

To verify that $A(\omega)$ decays to zero like $1/\omega$ for large ω, we examine a plot of $A(\omega)$, and see that it decays in an oscillatory fashion with peaks at n = 36, 52, 68, 84, ... where the values are as follows:

n	A(ω)	n × A(ω)
36	0.18554036	6.67945
52	0.12925377	6.72120
68	9.8750252e-2	6.71502
84	7.9558492e-2	6.68291

From the final column of the table we can infer that the peaks are dying out approximately like $1/\omega$ for large ω.

To verify that $B(\omega)$ decays to zero like $1/\omega$ for large ω, we examine a plot of $B(\omega)$, and see that it decays in an oscillatory fashion with peaks at n = 48, 64, 80. 96, ... where the values are as follows:

n	B(ω)	n × B(ω)
48	9.0241942e-2	4.33161
64	6.7379337e-2	4.31228
80	5.3536571e-2	4.28293
96	4.4221219e-2	4.24524

From the final column of the table we can infer that the peaks are dying out approximately like $1/\omega$ for large ω.

3.12

(a) $$F(\omega) = \int_{-\infty}^{\infty} f(t)\exp(-j\omega t)\,dt = 2\int_0^{\infty} \exp(-\beta t)\cos(\omega t)\,dt$$

$$= \int_0^{\infty} \exp(-\beta t)\,[\exp(j\omega t) + \exp(-j\omega t)]\,dt$$

$$= \int_0^{\infty} \exp[(-\beta+j\omega)t]\,dt + \int_0^{\infty} \exp[-\beta-j\omega t)]\,dt$$

$$= \frac{\exp[(-\beta+j\omega)t]}{-\beta + j\omega}\Bigg|_0^{\infty} + \frac{\exp[-\beta-j\omega t)]}{-\beta - j\omega}\Bigg|_0^{\infty}$$

$$= \frac{\exp(-\beta t)\,\exp(j\omega t)}{\ldots}\Bigg|^{\infty} + \frac{\exp(-\beta t)\,\exp(-j\omega t)}{\ldots}\Bigg|^{\infty}$$

$$= \frac{}{\beta - j\omega} + \frac{}{\beta + j\omega} = \frac{}{\beta^2 + \omega^2}$$

(b) $F(\omega)$ does in fact decay like $1/\omega^2$ for large ω.

(c) We know that $f_{ev}(t) \Longleftrightarrow A(\omega)$ and $f_{od}(t) \Longleftrightarrow jB(\omega)$ which is the same as

$$f_{ev}(t) = \frac{1}{2\pi}\int_{-\infty}^{\infty} A(\omega)\exp(j\omega t)\,d\omega \quad \text{and} \quad f_{od}(t) = \frac{1}{2\pi}\int_{-\infty}^{\infty} jB(\omega)\exp(j\omega t)\,d\omega$$

This means that $\displaystyle\int_{-\infty}^{\infty} A(\omega)\exp(j\omega t)\,d\omega = 2\pi f_{ev}(t)$

and $\displaystyle\int_{-\infty}^{\infty} B(\omega)\exp(j\omega t)\,d\omega = (2\pi/j)\, f_{od}(t) = 0$

since the pulse is even

d) $$E(\omega) = \frac{1}{2\pi}|F(\omega)|^2 = \frac{1}{2\pi}\left[\frac{2\beta}{\beta^2 + \omega^2}\right]^2$$ By Parseval's theorem:

$$\int_{-\infty}^{\infty} E(\omega)\,d\omega = \int_{-\infty}^{\infty}|f(t)|^2\,dt = 2\int_0^{\infty}\exp(-2\beta t)\,dt = 2\left.\frac{\exp(-2\beta t)}{-2\beta}\right|_0^{\infty} = \frac{1}{\beta}$$

(e) Steps for loading the pulse using $\beta = 2$, $N = 1024$, $T = 10$,

Main menu, Create Values, Continue, Create X, Real, Even, 1 interval, 1, Continue, EXP(-2*t), 0, Go, Neither, Go. You are now back on the main menu with the pulse correctly loaded. Plotting X confirms that. Run ANALYSIS, Show Numbers, F, Single.

We obtained the following values. Note: Since $T = 10$, the FFT spectral values lie at intervals of $\Omega = 2\pi/T = \pi/5$.

n		FFT	formula
0	F(0)	0.99998639	1.000000000
1	F(π/5)	0.91024295	0.910169838
2	F(2π/5)	0.71695604	0.716956800
3	F(3π/5)	0.52964269	0.529586854
0	E(0)	0.15915061	0.159154943
1	E(π/5)	0.13186659	0.131845409
2	E(2π/5)	8.1809773e-2	8.1809946e-2
3	E(3π/5)	4.4646364e-2	4.4636951e-2

For $f(t)$ a real function of t: $|f(t)|^2 = [f(t)]^2$ Then:

$$E\Big[f(t)\Big] = \int_{-\infty}^{\infty}|f(t)|^2\,dt = \int_{-\infty}^{\infty}[f(t)]^2\,dt = \int_{-\infty}^{\infty}[f_{ev}(t) + f_{od}(t)]^2\,dt$$

$$= \int_{-\infty}^{\infty}\Big[[f_{ev}(t)]^2 + 2 f_{ev}(t) f_{od}(t) + [f_{od}(t)]^2\Big]\,dt$$

$$= \int_{-\infty}^{\infty}[f_{ev}(t)]^2\,dt + 2\int_{-\infty}^{\infty} f_{ev}(t) f_{od}(t)\,dt + \int_{-\infty}^{\infty}[f_{od}(t)]^2\,dt$$

in which the second integrand is odd and the range of integration is symmetric, and so it integrates to zero, and so we continue as

$$\ldots = \int_{-\infty}^{\infty}[f_{ev}(t)]^2\,dt + \int_{-\infty}^{\infty}[f_{od}(t)]^2\,dt = E\Big[f_{ev}(t)\Big] + E\Big[f_{od}(t)\Big]$$

QED

(a2) Proof, using Parseval's Theorem

Let $f(t) \Longleftrightarrow F(\omega) = A(\omega) + jB(\omega)$ Then by Parseval's Theorem:

$$E\Big[f(t)\Big] = \int_{-\infty}^{\infty}|f(t)|^2\,dt = \frac{1}{2\pi}\int_{-\infty}^{\infty}|F(\omega)|^2\,d\omega$$

$$= \frac{1}{2\pi}\int_{-\infty}^{\infty}\Big[A(\omega)^2 + B(\omega)^2\Big]\,d\omega = \frac{1}{2\pi}\int_{-\infty}^{\infty}A(\omega)^2\,d\omega + \frac{1}{2\pi}\int_{-\infty}^{\infty}B(\omega)^2\,d\omega$$

but for $f(t)$ real: $f_{ev}(t) \Longleftrightarrow A(\omega)$, $f_{od}(t) \Longleftrightarrow jB(\omega)$ and so we continue as

$$\ldots = E\Big[f_{ev}(t)\Big] + E\Big[f_{od}(t)\Big]$$ QED

(b) See Answer section for the sketches.

$$E\left[f(t) \right] = \int_{-\infty}^{\infty} [f(t)]^2 \, dt = \int_{-1}^{0} (1+t)^2 \, dt + \int_{0}^{1} 1^2 \, dt$$

$$= \frac{(1+t)^3}{3} \Big|_{-1}^{0} + t \Big|_{0}^{1} = 1/3 + 1 = 4/3$$

$$E\left[f_{ev}(t) \right] = \int_{-\infty}^{\infty} [f_{ev}(t)]^2 \, dt = 2 \int_{-1}^{0} \left[\tfrac{1}{2}(2+t)^2 \right]^2 dt$$

$$= \tfrac{1}{2} \frac{(2+t)^3}{3} \Big|_{-1}^{0} = \tfrac{1}{2} \left[8/3 - 1/3 \right] = 7/6$$

$$E\left[f_{od}(t) \right] = \int_{-\infty}^{\infty} [f_{od}(t)]^2 \, dt = \int_{-1}^{1} \left[t/2 \right]^2 dt = t^3/12 = \Big|_{-1}^{1} = 1/6$$

and so, $E\left[f(t) \right] = E\left[f_{ev}(t) \right] + E\left[f_{od}(t) \right]$

3.14

(a) $$F(\omega) = \int_{-\infty}^{\infty} f(t) \exp(-j\omega t) \, dt = \int_{-1}^{0} (1+t) \exp(-j\omega t) \, dt$$

$$= \frac{(1+t)\exp(-j\omega t)}{-j\omega} \Big|_{-1}^{0} - \frac{\exp(-j\omega t)}{(-j\omega)^2} \Big|_{-1}^{0}$$

$$= \frac{1}{-j\omega} - \frac{1 - \exp(j\omega)}{(j\omega)^2} = \frac{-j\omega - 1 + \exp(j\omega)}{-\omega^2}$$

$$F_1(\omega) = \frac{1 + j\omega - \exp(j\omega)}{\omega^2}$$

(b) Area under $f(t) = \tfrac{1}{2}$ and by l'Hopital's rule:

$$F_1(0) = \lim \frac{1 + j\omega - \exp(j\omega)}{} = \lim \frac{j - j\exp(j\omega)}{}$$

$$= \lim_{\omega \to 0} \frac{-j^2 \exp(j\omega)}{2} = \tfrac{1}{2}$$

(c) See Answer section for the sketches.

(d) To find the Fourier transforms of $f_{ev}(t)$ and $f_{od}(t)$ we must split $F_1(\omega)$ into its real and imaginary parts. Thus:

$$F_1(\omega) = \frac{1 + j\omega - \exp(j\omega)}{\omega^2} = \frac{1 + j\omega - \cos(\omega) - j\sin(\omega)}{\omega^2}$$

$$= \frac{1 - \cos(\omega)}{\omega^2} + j \frac{\omega - \sin(\omega)}{\omega^2}$$

and so

$$f_{ev}(t) \Longleftrightarrow F_2(\omega) = \frac{1 - \cos(\omega)}{\omega^2} \qquad f_{od}(t) \Longleftrightarrow F_3(\omega) = j \frac{\omega - \sin(\omega)}{\omega^2}$$

(e) $F_1(\omega)$ decays like $1/\omega$, which is consistent with the fact that it has a discontinuity.

$F_2(\omega)$ decays like $1/\omega^2$, which is consistent with the fact that it is everywhere continuous

$F_3(\omega)$ decays like $1/\omega$, which is consistent with the fact that it has a discontinuity.

(f) Computing the total energy in each of the three pulses:

$$E[f(t)] = \int_{-\infty}^{\infty} [f(t)]^2 \, dt = \int_{-1}^{0} (1+t)^2 \, dt = \frac{(1+t)^3}{3} \Big|_{-1}^{0} = 1/3$$

$$E[f_{ev}(t)] = \int_{-\infty}^{\infty} [f_{ev}(t)]^2 \, dt = 2 \int_{-1}^{0} [\tfrac{1}{2}(1+t)]^2 \, dt = \tfrac{1}{2} \frac{(1+t)^3}{3} \Big|_{-1}^{0} = \frac{1}{6}$$

$$E[f_{od}(t)] = \int_{-\infty}^{\infty} [f_{od}(t)]^2 \, dt = 2 \int_{-1}^{0} [\tfrac{1}{2}(1+t)]^2 \, dt = \tfrac{1}{2} \frac{(1+t)^3}{3} \Big|_{-1}^{0} = 1/6$$

from which we see that

$$E\left[f(t) \right] = E\left[f_{ev}(t) \right] + E\left[f_{od}(t) \right]$$

(g) $$E_1 = \frac{1}{2\pi}\int_1^2 \left| \frac{1 + j\omega - \exp(j\omega)}{\omega^2} \right|^2 d\omega$$

$$E_2 = \frac{1}{2\pi}\int_1^2 \left[\frac{1 - \cos(\omega)}{\omega^2} \right]^2 d\omega \qquad E_3 = \frac{1}{2\pi}\int_1^2 \left[\frac{\omega - \sin(\omega)}{\omega^2} \right]^2 d\omega$$

(h) Total area under a Fourier transform is equal to $2\pi\, f(0)$

[Area under $F_1(\omega)$] = $2\pi\, f(0) = 2\pi \tfrac{1}{2} = \pi$

[Area under $F_2(\omega)$] = $2\pi\, f_{ev}(t) = 2\pi \tfrac{1}{2} = \pi$

[Area under $F_1(\omega)$] = $2\pi\, f_{od}(t) = 2\pi\, 0 = 0$

(i) Steps for loading the pulse using $N = 1024$, $T = 4$:

Main menu, Create Values, Continue, Create X, Real, Neither, Center, 3 intervals, -1, 0, Continue, 0, 1 + t, 0, Go, Neither, Go. You are now back on the main menu with the pulse correctly loaded. Plotting X confirms that. Run ANALYSIS, Show Numbers, F, Single.

We obtained the following results. Note: Since T = 4, the FFT spectral values lie at intervals of $\Omega = 2\pi/T = \pi/2$.

n		FFT	formula
0	A(0)	0.5	0.5
1	A(π/2)	0.40528601	0.405284734
2	A(2π/2)	0.20264491	0.202642367
3	A(3π/2)	4.5032909e-2	4.5031637e-2
0	B(0)	0	0
1	B(π/2)	0.23133177	0.231335038
2	B(2π/2)	0.31830589	0.318309886
3	B(3π/2)	0.25723351	0.257238228

(j) Using $N = 256$, SAMPLED, $T = 4$, PULSE, we synthesized the original pulse $f(t)$ by loading the expressions for $A(\omega)$ and $B(\omega)$ into the system as follows:

Main menu, Create Values, Continue, Create F, Complex,

(1 - COS(w))/w^2

Alpha = 0, Division by zero, Go, B, 0.5, Go.

(w - SIN(w))/w^2

Go, Neither, Go. You are now back on the main menu. Run SYNTHESIS, Plot Y. The plot shows the original waveform but with bad ripple before and after the discontinuity.

if you have the patience or if you have a very fast machine), Division by zero, B, 0.5, Go, Go, Accept the expression for FIM, Continue aliasing, Go, Go. Run SYNTHESIS and plot Y. The ripple is now much reduced. Using larger values for Alpha will make it as small as we please.

To synthesize $f_{ev}(t)$: Main menu, Create Values, Continue, Old-problem menu, Store to disk, Ex314 (or any other name), Go, Use the Old Problem, Accept the expression for FRE, Alpha = 0, Division by zero, B, 0.5, Go. When the expression for FIM appears clear it and enter 0, Go, Go, Run SYNTHESIS and plot Y. The plot shows $f_{ev}(t)$.

Observe that there is little or no ripple because there are no discontinuities in $f_{ev}(t)$ and so its spectrum converges like $1/\omega^2$, i.e. rapidly. In Chapter 12 we shall see why a waveform whose Fourier spectrum converges rapidly requires very little aliasing to prepare it for inversion by the FFT.

To synthesize $f_{od}(t)$: Main menu, Create Values, Continue, Old-problem menu, Fetch from disk, Ex314, Go, When the expression for FRE appears clear it and enter 0, Alpha = 0, Go, Accept the expression for FIM, Continue aliasing with Alpha = 0, Go, Go, Run SYNTHESIS and plot Y. The plot now shows $f_{od}(t)$ but with bad ripple before and after the discontinuity. The ripple can be eliminated as before by using an alias level of Alpha = 5, 10 or more.

3.15 For any real pulse $f(t) \Longleftrightarrow F(\omega)$,

(a) $f_{ev}(t) \overset{?}{\Longleftrightarrow} A(\omega)$ Proof: $f_{ev}(t) = \tfrac{1}{2}[f(t) + f(-t)]$

$$= 1/4\pi \int_{-\infty}^{\infty} F(\omega)\exp(j\omega t)\, d\omega + 1/4\pi \int_{-\infty}^{\infty} F(\omega)\exp(-j\omega t)\, d\omega$$

$$= 1/2\pi \int_{-\infty}^{\infty} F(\omega) \frac{\exp(j\omega t) + \exp(-j\omega t)}{2}\, d\omega$$

$$= 1/2\pi \int_{-\infty}^{\infty} F(\omega)\cos(\omega t)\, d\omega = 1/2\pi \int_{-\infty}^{\infty} [A(\omega) + jB(\omega)]\cos(\omega t)\, d\omega$$

in which $B(\omega)\cos(\omega t)$ is odd and so it integrates to zero

$$= 1/2\pi \int_{-\infty}^{\infty} A(\omega)\cos(\omega t)\, d\omega = 1/2\pi \int_{-\infty}^{\infty} A(\omega)\,[\cos(\omega t) + j\sin(\omega t)]\, d\omega$$

in which we have added nothing since $A(\omega)\sin(\omega t)$ is odd

$$= 1/2\pi \int_{-\infty}^{\infty} A(\omega)\ \exp(j\omega t)\ d\omega \qquad \text{i.e. } f_{ev}(t) \Longleftrightarrow A(\omega) \qquad \text{QED}$$

$f_{ev}(t) \overset{?}{\Longleftrightarrow} jB(\omega)$ Proof: $f_{od}(t) = \tfrac{1}{2}[f(t) - f(-t)]$

$$= 1/4\pi \int_{-\infty}^{\infty} F(\omega)\ \exp(j\omega t)\ d\omega - 1/4\pi \int_{-\infty}^{\infty} F(\omega)\ \exp(-j\omega t)\ d\omega$$

$$= 1/2\pi \int_{-\infty}^{\infty} F(\omega)\ j\ \frac{\exp(j\omega t) - \exp(-j\omega t)}{2j}\ d\omega$$

$$= 1/2\pi \int_{-\infty}^{\infty} F(\omega)\ \sin(\omega t)\ d\omega = 1/2\pi \int_{-\infty}^{\infty} [A(\omega) + jB(\omega)]\ j\ \sin(\omega t)\ d\omega$$

in which $A(\omega)\ \sin(\omega t)$ is odd and so it integrates to zero

$$= \frac{1}{2\pi} \int_{-\infty}^{\infty} jB(\omega)\ j\ \sin(\omega t)\ d\omega = \frac{1}{2\pi} \int_{-\infty}^{\infty} jB(\omega)\ [\cos(\omega t) + j\ \sin(\omega t)]\ d\omega$$

in which we have added nothing since $B(\omega)\ \cos(\omega t)$ is odd

$$= 1/2\pi \int_{-\infty}^{\infty} jB(\omega)\ \exp(j\omega t)\ d\omega \qquad \text{i.e. } f_{od}(t) \Longleftrightarrow jB(\omega) \qquad \text{QED}$$

(b) $F(\omega)$ is real and even iff $f(t)$ is even (?)

Proof: $F(\omega) = A(\omega)$ iff $jB(\omega) = 0$ iff $f(t)$ is even.

But $A(\omega)$ is real and even. Therefore $F(\omega)$ is real and even iff $f(t)$ is even. QED

$F(\omega)$ is purely imaginary and odd iff $f(t)$ is odd (?)

Proof: $F(\omega) = jB(\omega)$ iff $A(\omega) = 0$ iff $f(t)$ is odd.

But $B(\omega)$ is real and odd, and so $jB(\omega)$ is purely imaginary and odd. Therefore $F(\omega)$ is purely imaginary and odd iff $f(t)$ is even.

4.1 (a) Total energy in $f_1(t) = t\exp(-\beta t)\,U(t)$ $(\beta > 0)$ is

$$E[f_1(t)] = \int_{-\infty}^{\infty} [f_1(t)]^2\,dt = \int_0^{\infty} t^2\exp(-2\beta t)\,dt = 1/4\beta^3$$

(See Exercise 3.7(c).) Because the total energy is finite we are guaranteed (Square Integrability Criterion, p.84) that the Fourier transform of f1(t) exists and that it inverts back to $f_1(t)$.

(b)

$$F_1(\omega) = \int_{-\infty}^{\infty} f_1(t)\exp(-j\omega t)\,dt = \int_{-\infty}^{\infty} t\exp(-\beta t)\exp(-j\omega t)\,dt$$

$$= \int_0^{\infty} t\exp[(-\beta - j\omega)t]\,dt$$

$$= \frac{t\exp[(-\beta-j\omega)t]}{-(\beta + j\omega)}\Bigg|_0^{\infty} - \frac{\exp[(-\beta-j\omega)t]}{-[(\beta + j\omega)]^2}\Bigg|_0^{\infty}$$

$$= \frac{t\exp(-\beta t)\exp(-j\omega t)}{-(\beta + j\omega)}\Bigg|_0^{\infty} - \frac{\exp(-\beta t)\exp(-j\omega t)}{[-(\beta + j\omega)]^2}\Bigg|_0^{\infty}$$

$$= 0 - \frac{-1}{(\beta + j\omega)^2} = \frac{1}{(\beta + j\omega)^2}$$

(c) $F(\omega) = \int_{-\infty}^{\infty} f(t)\exp(-j\omega t)\,dt$ Differentiation wrt ω gives

$$\frac{d}{d\omega}F(\omega) = \frac{d}{d\omega}\int_{-\infty}^{\infty} f(t)\exp(-j\omega t)\,dt$$

in which we now assume that the order of the differentiation wrt ω and the integration wrt t can be reversed, and so we continue as

$$\ldots = \int_{-\infty}^{\infty} f(t)\frac{d}{d\omega}\Big[\exp(-j\omega t)\Big]\,dt = \int_{-\infty}^{\infty} f(t)\,(-jt)\exp(-j\omega t)\,dt$$

showing that

$$\frac{d}{d\omega}F(\omega) = -j\int_{-\infty}^{\infty} t\,f(t)\exp(-j\omega t)\,dt \qquad \text{(A)}$$

which is the same as

$$t\,f(t) \Longleftrightarrow j\frac{d}{d\omega}F(\omega) \qquad (4.108) \qquad \text{QED}$$

(d) Using Theorem 4.3 to find the Fourier transform of $f_1(t)$ we start with the known result

$$\Big[f(t) = \exp(-\beta t)\,U(t)\Big] \Longleftrightarrow \Big[F(\omega) = \frac{1}{\beta + j\omega}\Big]$$

Then, by (4.108) above:

$$\Big[t\,f(t) = t\exp(-\beta t)\,U(t)\Big] \Longleftrightarrow \Big[j\frac{d}{d\omega}F(\omega) = j\frac{d}{d\omega}\frac{1}{\beta + j\omega}\Big]$$

$$\text{But}\quad j\frac{d}{d\omega}\frac{1}{\beta + j\omega} = j\Big[j\frac{-1}{(\beta + j\omega)^2}\Big] = \frac{1}{(\beta + j\omega)^2}$$

showing that

$$t\exp(-\beta t)\,U(t) \Longleftrightarrow \frac{1}{(\beta + j\omega)^2} \qquad \text{(B)} \qquad \text{QED}$$

(e) To generalize (4.108) we differentiate (A) above once more, obtaining

$$\frac{d^2}{d\omega^2}F(\omega) = (-j)^2\int_{-\infty}^{\infty} t^2\,f(t)\exp(-j\omega t)\,dt$$

which is the same as $t^2 f(t) \Longleftrightarrow j^2\frac{d^2}{d\omega^2}F(\omega)$

From this, by the method of mathematical induction:

$$t^n f(t) \Longleftrightarrow j^n\frac{d^n}{d\omega^n}F(\omega)$$ where n is any positive integer. QED

(f) Using the corollary to find the transform of

$$f_2(t) = t^2\exp(-\beta t)\,U(t)$$

we start from (B), namely: $t \exp(-\beta t)\, U(t) \Longleftrightarrow \dfrac{1}{(\beta + j\omega)^2}$

Then $t^2 \exp(-\beta t)\, U(t) \Longleftrightarrow j \dfrac{d}{d\omega} \dfrac{1}{(\beta + j\omega)^2} = j \dfrac{-2j}{(\beta + j\omega)^3}$

showing that

$$\left[\, f_2(t) = t^2 \exp(-\beta t)\, U(t) \,\right] \Longleftrightarrow \left[\, F_2(\omega) = \frac{2}{(\beta + j\omega)^3} \,\right]$$

(g) $$F_1(\omega) = \frac{1}{(\beta + j\omega)^2} = \frac{1}{\beta^2 + 2j\beta\omega - \omega^2}$$

$$= \frac{1}{(\beta^2-\omega^2) + 2j\beta\omega} \; \frac{(\beta^2-\omega^2) - 2j\beta\omega}{(\beta^2-\omega^2) - 2j\beta\omega}$$

$$= \frac{\beta^2-\omega^2}{(\beta^2-\omega^2)^2 + (2\beta\omega)^2} - j \frac{2\beta\omega}{(\beta^2-\omega^2)^2 + (2\beta\omega)^2} = A_1(\omega) + jB_1(\omega)$$

Then $|F_1(\omega)| = [A_1(\omega)^2 + B_1(\omega)^2]^{1/2}$

and $\theta_1(\omega) = \arctan[\, B_1(\omega)/A_1(\omega) \,] = \arctan\left[\dfrac{-2\beta\omega}{\beta^2-\omega^2}\right]$

from which we see that $A_1(\omega)$ and $|F_1(\omega)|$ are even and that $B_1(\omega)$ and $\theta_1(\omega)$ are odd.

$$F_2(\omega) = \frac{2}{(\beta + j\omega)^3} = \frac{2}{\beta^3 + 3j\beta^2\omega + 3\beta(j\omega)^2 + (j\omega)^3}$$

$$= \frac{2}{(\beta^3 - 3\beta\omega^2) + j(3\beta^2\omega - \omega^3)} \; \frac{(\beta^3 - 3\beta\omega^2) - j(3\beta^2\omega - \omega^3)}{(\beta^3 - 3\beta\omega^2) - j(3\beta^2\omega - \omega^3)}$$

$$= \frac{2(\beta^3 - 3\beta\omega^2)}{(\beta^3-3\beta\omega^2)^2 + (3\beta^2\omega-\omega^3)^2} - j \frac{2(3\beta^2\omega - \omega^3)}{(\beta^3-3\beta\omega^2)^2 + (3\beta^2\omega-\omega^3)^2} = A_2(\omega) + jB_2(\omega)$$

Then $|F_2(\omega)| = [A_2(\omega)^2 + B_2(\omega)^2]^{1/2}$

and $\theta_2(\omega) = \arctan[\, B_2(\omega)/A_2(\omega) \,] = \arctan\left[\dfrac{-2(3\beta^2\omega - \omega^3)}{2(\beta^3 - 3\beta\omega^2)}\right]$

from which we see again that $A_2(\omega)$ and $|F_2(\omega)|$ are even and that $B_2(\omega)$ and $\theta_2(\omega)$ are odd.

(h) Finding the expressions for the energy spectra of $f_1(t)$ and $f_2(t)$:

We have shown earlier (see Exercise 2.9) that for any complex quantities, z and w

$|zw| = |z||w|$ and so, letting $z = w$ gives $|z^2| = |z|^2$

Letting $z = \dfrac{1}{\beta + j\omega}$ it thus follows that

$$\left| \frac{1}{(\beta + j\omega)^2} \right| = |z^2| = |z|^2 = \frac{1}{\beta^2 + \omega^2}$$

from which

$$\left| \frac{1}{(\beta + j\omega)^2} \right|^2 = |z^2|^2 = |z|^4 = \frac{1}{(\beta^2 + \omega^2)^2}$$

Then

$$E_1(\omega) = 1/2\pi\, |F_1(\omega)|^2 = 1/2\pi \left| \frac{1}{(\beta + j\omega)^2} \right|^2 = 1/2\pi \frac{1}{(\beta^2 + \omega^2)^2}$$

Starting again from $|zw| = |z||w|$ if we let $w = z^2$ then

$|z^3| = |z||z^2| = |z||z|^2 = |z|^3$

and in general, for n an integer: $|z^n| = |z|^n$

Letting $z = \dfrac{1}{\beta + j\omega}$ it thus follows that

$$\left| \frac{1}{(\beta + j\omega)^3} \right| = |z^3| = |z|^3 = \frac{1}{(\beta^2 + \omega^2)^{3/2}}$$

from which

$$\left| \frac{1}{(\beta + j\omega)^3} \right|^2 = |z^3|^2 = |z|^6 = \frac{1}{(\beta^2 + \omega^2)^3}$$

Then

$E_2(\omega) = 1/2\pi\, |F_2(\omega)|^2 = 1/2\pi \left| \dfrac{2}{\;\cdots} \right|^2 = 1/2\pi$ ...

4.2 See Answer section for the sketches. From Exercise 4.1(g) above:

$$Ev[\ f_1(t)\ ] \Longleftrightarrow = \frac{\beta^2 - \omega^2}{(\beta^2 - \omega^2)^2 + (2\beta\omega)^2}$$

$$Od[\ f_1(t)\ ] \Longleftrightarrow = -j\,\frac{2\beta\omega}{(\beta^2 - \omega^2)^2 + (2\beta\omega)^2}$$

4.3

(a) If $\int_{-\infty}^{\infty} f(t)\exp(-j\omega t)\,dt = F(\omega)$ then

$$\int_{-\infty}^{\infty} [\exp(j\omega_0 t)\ f(t)]\exp(-j\omega t)\,dt$$

$$= \int_{-\infty}^{\infty} f(t)\exp[-j(\omega - \omega_0)t]\,dt = F(\omega - \omega_0) \qquad \text{QED}$$

(b) Energy in $\exp(-\beta t)\cos(\omega_0 t)\,U(t)$ is $\int_0^{\infty}\exp(-2\beta t)\cos^2(\omega_0 t)\,dt$

in which the integrand is everywhere positive and $\exp(-2\beta t)\cos^2(\omega_0 t) < \exp(-2\beta t)$, and so we continue as

$$\ldots = \int_0^{\infty}\exp(-2\beta t)\,dt = \left.\frac{\exp(-2\beta t)}{-2\beta}\right|_0^{\infty} = 1/2\beta \qquad \text{QED}$$

Similarly, energy in $\exp(-\beta t)\sin(\omega_0 t)\,U(t)$

$$= \int_0^{\infty}\exp(-2\beta t)\sin^2(\omega_0 t)\,dt$$

in which the integrand is everywhere positive and $\exp(-2\beta t)\sin^2(\omega_0 t) < \exp(-2\beta t)$

$$\ldots = \int_0^{\infty}\exp(-2\beta t)\,dt = \left.\frac{\exp(-2\beta t)}{-2\beta}\right|_0^{\infty} = 1/2\beta \qquad \text{QED}$$

(c) $$\exp(-\beta t)\cos(\omega_0 t)\,U(t) = \exp(-\beta t)\,\frac{\exp(j\omega_0 t) + \exp(-j\omega_0 t)}{2}\,U(t)$$

$$= \tfrac{1}{2}\exp(-\beta t)\exp(j\omega_0 t)\,U(t) + \tfrac{1}{2}\exp(-\beta t)\exp(-j\omega_0 t)\,U(t)$$

in which we see $\tfrac{1}{2}\exp(-\beta t)\,U(t)$ multiplied by $\exp(\pm j\omega_0 t)$ and so we can use **frequency shift**, i.e. the result proved in Part (a), namely $\exp(\pm j\omega_0 t)\,f(t) \Longleftrightarrow F(\omega \mp \omega_0)$ and so we continue

$$\ldots \Longleftrightarrow \tfrac{1}{2}\,\frac{1}{\beta + j(\omega - \omega_0)} + \tfrac{1}{2}\,\frac{1}{\beta + j(\omega + \omega_0)}$$

$$= \frac{\beta + j\omega}{[\beta + j(\omega - \omega_0)]\,[\beta + j(\omega + \omega_0)]} = \frac{\beta + j\omega}{(\beta + j\omega)^2 + \omega_0^2} \qquad \text{(A)}$$

$$\exp(-\beta t)\sin(\omega_0 t)\,U(t) = \exp(-\beta t)\,\frac{\exp(j\omega_0 t) - \exp(-j\omega_0 t)}{2j}\,U(t)$$

$$= 1/2j\,[\exp(-\beta t)\exp(j\omega_0 t)\,U(t) - \exp(-\beta t)\exp(-j\omega_0 t)\,U(t)]$$

in which we see $1/2j\,\exp(-\beta t)\,U(t)$ multiplied by $\exp(\pm j\omega_0 t)$ and so we can use **frequency shift**, i.e. the result proved in Part (a), namely $\exp(\pm j\omega_0 t)\,f(t) \Longleftrightarrow F(\omega \mp \omega_0)$ and so we contine as

$$\Longleftrightarrow 1/2j\left[\frac{1}{\beta + j(\omega - \omega_0)} - \frac{1}{\beta + j(\omega + \omega_0)}\right]$$

$$= \frac{2j\omega_0}{[\beta + j(\omega - \omega_0)]\,[\beta + j(\omega + \omega_0)]} = \frac{2j\omega_0}{(\beta + j\omega)^2 + \omega_0^2} \qquad \text{(B)}$$

(d) Finding the inverse of $F(\omega) = \dfrac{8j\omega + 4}{(j\omega)^2 + 4j\omega + 20}$

$$\frac{8j\omega + 4}{(j\omega)^2 + 4j\omega + 20} = \frac{8j\omega + 4}{(j\omega + 2)^2 + 16} = \frac{8(j\omega + 2) - 12}{(j\omega + 2)^2 + 16}$$

$$= 8\,\frac{(j\omega + 2)}{(j\omega + 2)^2 + 16} - 3\,\frac{4}{(j\omega + 2)^2 + 16}$$

We now make use of (A) and (B) in Part (c), continuing

$$\ldots \Longleftrightarrow 8\exp(-2t)\cos(4t)\,U(t) - 3\exp(-2t)\sin(4t)\,U(t)$$

(e) $$F(\omega) = \frac{8j\omega + 4}{(j\omega)^2 + 4j\omega + 20} = \frac{8j\omega + 4}{(20-\omega^2) + 4j\omega}\,\frac{(20-\omega^2) - 4j\omega}{(20-\omega^2) - 4j\omega}$$

$$= \frac{(80 - 4\omega^2 + 32\omega^2) + j(160\omega - 16\omega - 8\omega^3)}{(20-\omega^2)^2 + 16\omega^2}$$

$$= \frac{28\omega^2 + 80}{(20-\omega^2)^2 + 16\omega^2} + j\,\frac{144\omega - 8\omega^3}{(20-\omega^2)^2 + 16\omega^2} \qquad \text{QED}$$

(f) To invert the expression in (d) using the FFT system, with N = 1024, SAMPLED, T = 8, PULSE:

Main-menu, Create Values, Continue, Create H(jw), Create, 1, 2, 4, 8, 20, 4, 1. Then LOAD using an alias level of 20. Run Postprocessors, F, Copy F2 to F, Quit. Run SYNTHESIS. Plot **Y**.

(g) Taking numerical samples from the formula for $f(t)$ in (d) and comparing them to the values for Y obtained in (f) we obtained the following:

k	t	FFT	formula
0	0.0000	3.9994600	4.000000000
8	0.0625	6.1830276	6.185498005
16	0.1250	4.3463304	4.347564937
32	0.2500	1.8992940	1.090545458
64	0.5000	-2.2285757	-2.228270413
128	1.0000	-0.40056852	-0.400422116
256	2.0000	-7.5742263e-2	-7.5681593e-2

Note: Because T = 8, and there are N = 1024 sample points in that range, it follows that each step along the time axis is equal to $T/N = T_s = 0.0078125$ seconds. Thus k = 16 corresponds to $t = 16T_s = 0.125$.

4.4 Using **time shift** (Theorem 4.2): $f(t - \tau) \Longleftrightarrow F(\omega)\exp(-j\omega\tau)$

$$f_1(t) = \text{Rect}(t + \tfrac{1}{2}) - \text{Rect}(t - \tfrac{1}{2})$$

$$\Longleftrightarrow \text{Sa}(\omega/2)\exp(j\omega\tfrac{1}{2}) - \text{Sa}(\omega/2)\exp(-j\omega\tfrac{1}{2})\cdot$$

$$= \frac{\sin(\omega/2)}{\omega/2}\,2j\,\frac{\exp(j\omega\frac{1}{2}) - \exp(-j\omega\frac{1}{2})}{2j} = 2j\,\frac{\sin^2(\omega/2)}{\omega/2}$$

and, again by **time shift**, $f_2(t) = U(t + 1) - 2\,U(t) + U(t - 1)$

$$\Longleftrightarrow [1/j\omega + \pi\delta(\omega)]\exp(j\omega 1) - 2[1/j\omega + \pi\delta(\omega)] + [1/j\omega + \pi\delta(\omega)]\exp(-j\omega 1)$$

in which Dirac-delta sampling takes place, and we continue

$$\ldots = [1/j\omega]\,[\exp(j\omega) - 2 + \exp(-j\omega)] + \pi\delta(\omega)\,[1 - 2 + 1]$$

$$= [1/j\omega]\,[\exp(j\omega/2) - \exp(-j\omega/2)]^2$$

$$= [1/j\omega](2j)^2\left[\frac{\exp(j\omega/2) - \exp(-j\omega/2)}{2j}\right]^2$$

$$= (1/j\omega)(-4)\sin^2(\omega/2) = 2j\,\frac{\sin^2(\omega/2)}{\omega/2} \qquad \text{QED}$$

4.5

(a) $$F(\omega) = \int_{-\infty}^{\infty} f(t)\exp(-j\omega t)\,dt = 2\int_0^{\infty} f(t)\cos(\omega t)\,dt \quad (\text{since } f(t) \text{ is even})$$

$$= 2\int_0^{k/2}\exp(-\beta t)\cos(\omega t)\,dt = 2\int_0^{k/2}\exp(-\beta t)\,\frac{\exp(j\omega t) + \exp(-j\omega t)}{2}\,dt$$

$$= \int_0^{k/2}\Big[\exp[(-\beta+j\omega)t] + \exp[(-\beta-j\omega)t]\Big]\,dt$$

$$= \frac{\exp[(-\beta+j\omega)t]}{-\beta + j\omega}\Bigg|_0^{k/2} + \frac{\exp[(-\beta-j\omega)t]}{-\beta - j\omega}\Bigg|_0^{k/2}$$

$$= \frac{\exp[(-\beta+j\omega)k/2] - 1}{-\beta + j\omega} + \frac{\exp[(-\beta-j\omega)k/2] - 1}{-\beta - j\omega}$$

$$= \frac{1 - \exp(-\beta k/2)\exp(j\omega k/2)}{\beta - j\omega} + \frac{1 - \exp(-\beta k/2)\exp(-j\omega k/2)}{\beta + j\omega}$$

$$= \left[\frac{1}{\beta - j\omega} + \frac{1}{\beta + j\omega}\right] - \exp(-\beta k/2)\left[\frac{\exp(j\omega k/2)}{\beta - j\omega} + \frac{\exp(-j\omega k/2)}{\beta + j\omega}\right]$$

$$= \frac{2\beta}{\beta^2 + \omega^2} - \exp(-\beta k/2)\left[\frac{\exp(j\omega k/2)}{\beta - j\omega} + \frac{\exp(-j\omega k/2)}{\beta + j\omega}\right]$$

(b) To find the transform of the double-sided decaying exponential we now let $k \longrightarrow \infty$, obtaining (since $\beta > 0$)

$$\exp(-\beta|t|) \Longleftrightarrow \frac{2\beta}{\beta^2 + \omega^2}$$

4.6

(a) $$F(\omega) = \int_{-\infty}^{\infty} f(t)\ \exp(-j\omega t)\ dt = \int_{-\infty}^{\infty} f(t)\ [\cos(\omega t) - j\ \sin(\omega t)]\ dt$$

But $f(t)$ is odd. Hence $f(t)\ \cos(\omega t)$ is also odd, and so it integrates to zero. Moreover $f(t)\ \sin(\omega t)$ is even, and so we continue

$$\ldots = -2j \int_0^{\infty} f(t)\ \sin(\omega t)\ dt = -2j \int_0^{\infty} \exp(-kt)\ \sin(\omega t)\ dt$$

$$= -2j \int_0^{\infty} \exp(-kt)\ \frac{\exp(j\omega t) - \exp(-j\omega t)}{2j}\ dt$$

$$= -\int_0^{\infty} \Big[\exp[(-k+j\omega)t] - \exp[(-k-j\omega)t] \Big]\ dt$$

$$= \frac{\exp[(-k-j\omega)t]}{-k - j\omega}\Bigg|_0^{\infty} - \frac{\exp[(-k+j\omega)t}{-k + j\omega}\Bigg|_0^{\infty}$$

$$= \frac{\exp(-kt)\ \exp(-j\omega t)}{-k - j\omega}\Bigg|_0^{\infty} - \frac{\exp(-kt)\ \exp(j\omega t)}{-k + j\omega}\Bigg|_0^{\infty}$$

$$= \frac{1}{k + j\omega} - \frac{1}{k - j\omega} = \frac{-2j\omega}{k^2 + \omega^2}$$

We now let $k \longrightarrow 0$, obtaining, as the transform of $Sgn(t)$:

$$F(\omega) = \frac{-2j\omega}{\omega^2} = 2/j\omega$$

4.7 See Answer Section for [illegible]

(c) We recall that: $\quad Rect(t/\tau) \Longleftrightarrow \tau\ Sa(\omega\tau/2)$

and (time shift): $\quad f(t - k) \Longleftrightarrow F(\omega)\ \exp(-j\omega k)$

Then, because $Rect[(t - k)/\tau]$ is $Rect(t/\tau)$ delayed by k seconds, its Fourier transform must be $\quad F_1(\omega) = \tau\ Sa(\omega\tau/2)\ \exp(-j\omega k)$

Similarly,
$$Rect\left[\frac{t}{\tau} - k\right] = Rect\left[\frac{t - k\tau}{\tau}\right]$$

is seen to be $Rect(t/\tau)$ delayed by $k\tau$ seconds, and so its Fourier transform must be $\quad F_2(\omega) = \tau\ Sa(\omega\tau/2)\ \exp(-j\omega k\tau)$

4.8 (a) $\quad Rect(t/\tau) \Longleftrightarrow \tau\ Sa(\omega\tau/2)$

To find the Fourier transform of $f(t)$ in Figure 4.37, we have

$$f(t) = Rect\left[\frac{t - \frac{1}{2}}{1}\right] - \tfrac{1}{2}\ Rect\left[\frac{t - \frac{1}{4}}{\frac{1}{2}}\right]$$

and so, making use of the time-shift property (Theorem 4.2)

$$F(\omega) = 1\ Sa(\omega 1/2)\ \exp(-j\omega\tfrac{1}{2}) - \tfrac{1}{2}\left[\tfrac{1}{2}\ Sa(\omega\tfrac{1}{2}/2)\ \exp(-j\omega\tfrac{1}{4})\right]$$

$$= Sa(\omega/2)\ \exp(-j\omega) - \tfrac{1}{4}\ Sa(\omega/4)\ \exp(-j\omega/4)$$

This can now be reduced as follows:

$$\ldots = \frac{\exp(j\omega/2) - \exp(-j\omega/2)}{2j\omega/2}\ \exp(-j\omega/2)$$

$$- \tfrac{1}{4}\ \frac{\exp(j\omega/4) - \exp(-j\omega/4)}{2j\omega/4}\ \exp(-j\omega/4)$$

$$= \frac{1 - \exp(-j\omega)}{j\omega} - \frac{\frac{1}{2} - \frac{1}{2}\exp(-j\omega/2)}{j\omega} = \frac{\frac{1}{2} + \frac{1}{2}\exp(-j\omega/2) - \exp(-j\omega)}{j\omega}$$

(b) To find $A(\omega)$, $B(\omega)$, $|F(\omega)|$ and $\theta(\omega)$ for this pulse, we must separate $F(\omega)$ into its real and imaginary parts. Thus:

$$F(\omega) = \frac{\frac{1}{2} + \frac{1}{2}\exp(-j\omega/2) - \exp(-j\omega)}{j\omega}$$

$$= \frac{\tfrac{1}{2} + \tfrac{1}{2}[\cos(\omega/2) - j\sin(\omega/2)] - [\cos(\omega) - j\sin(\omega)]}{j\omega}$$

$$= \frac{\tfrac{1}{2} + \tfrac{1}{2}\cos(\omega/2) - \cos(\omega)}{j\omega} + \frac{\sin(\omega) - \tfrac{1}{2}\sin(\omega/2)}{\omega}$$

$$= \frac{2\sin(\omega) - \sin(\omega/2)}{2\omega} + j\,\frac{2\cos(\omega) - \cos(\omega/2) - 1}{2\omega} = A(\omega) + jB(\omega)$$

Then $|F(\omega)| = [A(\omega)^2 + B(\omega)^2]^{1/2}$ and $\theta(\omega) = \arctan[B(\omega)/A(\omega)]$

We see from these that $A(\omega)$ and $|F(\omega)|$ are even, $B(\omega)$ and $\theta(\omega)$ are odd.

(c) See Answer section for the sketches. From (b) above:

$$f_{ev}(t) \Longleftrightarrow A(\omega) = \frac{2\sin(\omega) - \sin(\omega/2)}{2\omega}$$

$$f_{od}(t) \Longleftrightarrow jB(\omega) = j\,\frac{2\cos(\omega) - \cos(\omega/2) - 1}{2\omega}$$

(d) Using the FFT system to verify the sketches that were made in (c) with N = 256, SAMPLED, T = 4, PULSE:

Main menu, Create Values, Continue, Create X, Real, Neither, Left, 3 intervals, 0.5, 1, Continue, 0.5, 1, 0, Go, Neither, Go. You are now back on the main menu.

To display $f_{ev}(t)$: Run ANALYSIS. Then zero out the imaginary part of $F(\omega)$ as follows: Run Postprocessors, F, Copy F to F2, Set FIM to zero, Quit, Run SYNTHESIS, plot Y.

To display $f_{od}(t)$: Run Postprocessors, F, Copy F2 to F, Set FRE to zero, Quit, Run SYNTHESIS, plot Y.

4.9 The pulse shown in Figure 4.38, which we call $\Lambda_k(t)$, is a specially constructed **triangular** pulse that leads to the Dirac delta.

(a) Area = base × height / 2 = 2k × 1/k / 2 = 1

(b) Clearly, as $k \longrightarrow 0^+$ the base-width of the pulse becomes negligibly small and so

$$\lim_{k \to 0^+} \Lambda_k(t) = 0 \quad (t \neq 0)$$

Moreover, no matter what the value of k (as long as it is positive) the area of $\Lambda_k(t)$ is 1.

$$\text{i.e.} \quad \int_{-\infty}^{\infty} \Lambda_k(t)\, dt = 1$$

Thus $\Lambda_k(t)$ is a sequence function for $\delta(t)$.

(c) Finding the Fourier transform of $\Lambda_k(t)$ for k constant:

$$F_k(\omega) = \int_{-\infty}^{\infty} f_k(t)\exp(-j\omega t)\, dt = 2\int_0^{\infty} \Lambda_k(t)\cos(\omega t)\, dt$$

$$= 2\int_0^k \frac{1}{k^2}(k - t)\cos(\omega t)\, dt = (2/k^2)\left[(k - t)\frac{\sin(\omega t)}{\omega} - \frac{\cos(\omega t)}{\omega^2}\right]_0^k$$

$$= (2/k^2)\left[\frac{1 - \cos(\omega k)}{\omega^2}\right] = (4/k^2)\,\frac{\sin^2(\omega k/2)}{\omega^2}$$

$$= \frac{\sin^2(\omega k/2)}{(\omega k/2)^2} = \left[\frac{\sin(\omega k/2)}{\omega k/2}\right]^2$$

Then, letting k tend to its limit: $\lim_{k \to 0^+} F_k(\omega) = 1$

which is the Fourier transform of $\delta(t)$.

4.10 (a) $\exp(-2t)\ \delta(t)$ (sampling is at $t = 0$)

$= \exp(-2 \times 0)\ \delta(t) = 1\ \delta(t) = \delta(t)$

(b) $\tfrac{1}{2}\exp(-3t)\ [\delta(t + 1) + \delta(t - 1)]$

$= \tfrac{1}{2}\exp(-3t)\ \delta(t + 1)$ (sampling is at $t = -1$)

$+ \tfrac{1}{2}\exp(-3t)\ \delta(t - 1)]$ (sampling is at $t = +1$)

$= \tfrac{1}{2}\exp[-3(-1)]\ \delta(t + 1) + \tfrac{1}{2}\exp[-3(1)]\ \delta(t - 1)]$

$= \tfrac{1}{2}\exp(3)\ \delta(t + 1) + \tfrac{1}{2}\exp(-3)\ \delta(t - 1)]$

(c) $\sin(t)\ \delta(t - \pi/2)$ (sampling is at $t = \pi/2$)

$= \sin(\pi/2)\ \delta(t - \pi/2) = 1\ \delta(t - \pi/2) \equiv \delta(t - \pi/2)$

(d) $\cos(\omega/2)\ \delta(\omega - \pi)$ (sampling is at $\omega = \pi$)

$= \cos(\pi/2)\ \delta(\omega - \pi) = 0\ \delta(\omega - \pi) = 0$

(e) $\exp(j\omega)\ \delta(\omega - n\pi)$ (sampling is at $\omega = n\pi$)

$= \exp(jn\pi)\ \delta(\omega - n\pi) = (-1)^n\ \delta(\omega - n\pi)$

(f) $j\omega\ [\delta(\omega - 1) - \delta(\omega + 1)]$

$= j\omega\ \delta(\omega - 1)$ (sampling is at $\omega = 1$)

$- j\omega\ \delta(\omega + 1)$ (sampling is at $\omega = -1$)

$= j1\ \delta(\omega - 1) - j(-1)\ \delta(\omega + 1) = j\ [\delta(\omega - 1) + \delta(\omega + 1)]$

4.11

(a) $\int_{-\infty}^{\infty} \sin(\pi t)\ \delta(t - 1/2)\ dt$ (sampling is at $t = 1/2$)

$= \int_{-\infty}^{\infty} \sin(\pi/2)\ \delta(t - 1/2)\ dt = \int_{-\infty}^{\infty} 1\ \delta(t - 1/2)\ dt = 1$

(b) $\int_{-\infty}^{\infty} \cos(\pi t)\ \delta(t - 1/2)\ dt$ (sampling is at $t = 1/2$)

$= \int_{-\infty}^{\infty} \cos(\pi/2)\ \delta(t - 1/2)\ dt = \int_{-\infty}^{\infty} 0\ \delta(t - 1/2)\ dt = 0$

(c) $\int_{-\infty}^{\infty} \cos(\omega/2)\ \delta(\omega - \pi/2)\ d\omega$ (sampling is at $\omega = 1/2$)

$= \int_{-\infty}^{\infty} \cos(\pi/4)\ \delta(\omega - \pi/2)\ d\omega$ (sampling is at $\omega = \pi/2$)

$= \int_{-\infty}^{\infty} (1/\sqrt{2})\ \delta(\omega - \pi/2)\ d\omega = 1/\sqrt{2}$

(d) $\int_{-\infty}^{\infty} \omega\ \cos(\omega/2)\ \delta(\omega - \pi/2)\ d\omega$ (sampling is at $\omega = \pi/2$)

$= \int_{-\infty}^{\infty} (\pi/2)\ \cos(\pi/4)\ \delta(\omega - \pi/2)\ d\omega = \int_{-\infty}^{\infty} (\pi/2)\ (1/\sqrt{2})\ \delta(\omega - \pi/2)\ d\omega = \pi/2\sqrt{2}$

(e) $\int_{-\infty}^{\infty} \cos(2\omega)\ \delta(\omega - n\pi/2)\ \exp(j\omega t)\ d\omega$ (sampling is at $\omega = n\pi/2$)

$= \int_{-\infty}^{\infty} \cos(n\pi)\ \delta(\omega - n\pi/2)\ \exp(jn\pi t/2)\ d\omega$

$= (-1)^n\ \exp(jn\pi t/2) \int_{-\infty}^{\infty} \delta(\omega - n\pi/2)\ d\omega = (-1)^n\ \exp(jn\pi t/2)$

4.12 (a) $\delta(t - 1)$ <====> ? **We know that** $\delta(t)$ <====> 1 **and so, using time shift:**

$\delta(t - 1)$ <====> $\exp(-j\omega)$

(b) $3\delta(t + 2)$ <====> ? **We know that** $\delta(t)$ <====> 1 **and so, using time shift:**

$3\delta(t + 2)$ <====> $3\ \exp(j2\omega)$

(c) $\delta(t + 1) + 2\delta(t) + \delta(t - 1)$ <====> ? **We know that** $\delta(t)$ <====> 1 **and so, using time shift:**

$\delta(t + 1) + 2\delta(t) + \delta(t - 1)$

<====> $\exp(j\omega) + 2 + \exp(-j\omega)$

$= [\exp(j\omega/2) + \exp(-j\omega/2)]^2 = 4\cos^2(\omega/2)$

(d) $\exp(j\omega_0 t) - 2 + \exp(-j\omega_0 t)$ <====> ? **We know that** 1 <====> $2\pi\delta(\omega)$ **and so, using frequency shift:**

$\exp(j\omega_0 t) - 2 + \exp(-j\omega_0 t)$ <====> $2\pi\ \delta(\omega - \omega_0) - 4\pi\ \delta(\omega) + 2\pi\ \delta(\omega + \omega_0)$

(e) $\cos^2(\omega_0 t)$ <====> ?

$$\cos^2(\omega_0 t) = \left[\frac{\exp(j\omega_0 t) + \exp(-j\omega_0 t)}{2}\right]^2 = \frac{\exp(j2\omega_0 t) + 2 + \exp(-2j\omega_0 t)}{4}$$

We know that $1 \Longleftrightarrow 2\pi\,\delta(\omega)$, and so, using **frequency shift** this continues as

$$\Longleftrightarrow \frac{2\pi\,\delta(\omega - 2\omega_0) + 4\pi\,\delta(\omega) + 2\pi\,\delta(\omega + 2\omega_0)}{4}$$

$$= \pi\,\delta(\omega) + \tfrac{1}{2}\pi\,[\delta(\omega - 2\omega_0) + \delta(\omega + 2\omega_0)]$$

(f) $\sin^2(\omega_0 t) \Longleftrightarrow ?$

$$\sin^2(\omega_0 t) = \tfrac{1}{2} - \tfrac{1}{2}\cos(2\omega_0 t) \Longleftrightarrow \pi\,\delta(\omega) - \tfrac{1}{2}\pi\,[\delta(\omega - 2\omega_0) + \delta(\omega + 2\omega_0)]$$

(g) Adding the results of (e) and (f) we obtain

$$1 = \cos^2(\omega_0 t) + \sin^2(\omega_0 t)$$

$$\Longleftrightarrow \pi\,\delta(\omega) + \tfrac{1}{2}\pi\,[\delta(\omega - 2\omega_0) + \delta(\omega + 2\omega_0)] + \pi\,\delta(\omega) - \tfrac{1}{2}\pi\,[\delta(\omega - 2\omega_0) + \delta(\omega + 2\omega_0)] = 2\pi\,\delta(\omega)$$

4.13 (a) $\cos^2(\omega_0 t) = \dfrac{1 + \cos(2\omega_0(t)}{2}$

$$= \tfrac{1}{2} + \tfrac{1}{4}\exp(j2\omega_0 t) + \tfrac{1}{4}\exp(-j2\omega_0 t)$$

We know that $1 \Longleftrightarrow 2\pi\,\delta(\omega)$. Then, using **frequency shift**,

$1 \times \exp(j2\omega_0 t) \Longleftrightarrow 2\pi\,\delta(\omega - 2\omega_0)$ and so we continue as

$$\ldots \Longleftrightarrow \pi\delta(\omega) + \tfrac{1}{4}\,2\pi\,\delta(\omega - 2\omega_0) + \tfrac{1}{4}\,2\pi\,\delta(\omega + 2\omega_0)$$

$$= \pi\delta(\omega) + \tfrac{1}{2}\pi\,[\delta(\omega - 2\omega_0) + \delta(\omega + 2\omega_0)]$$

(b) $\sin^2(\omega_0 t) = \dfrac{1 - \cos(2\omega_0(t)}{2} = \tfrac{1}{2} - \tfrac{1}{4}\exp(j2\omega_0 t) - \tfrac{1}{4}\exp(-j2\omega_0 t)$

We know that $1 \Longleftrightarrow 2\pi\,\delta(\omega)$. Then, using **frequency shift**,
$1 \times \exp(j2\omega_0 t) \Longleftrightarrow 2\pi\,\delta(\omega - 2\omega_0)$ and so we continue as

$$\ldots \Longleftrightarrow \pi\delta(\omega) - \tfrac{1}{4}\,2\pi\,\delta(\omega - 2\omega_0) - \tfrac{1}{4}\,2\pi\,\delta(\omega + 2\omega_0)$$

$$= \pi\delta(\omega) - \tfrac{1}{2}\pi\,[\delta(\omega - 2\omega_0) + \delta(\omega + 2\omega_0)]$$

(c) $\sin(\omega_0 t)\,\cos(\omega_0 t)$

$$= \frac{\exp(j\omega_0 t) - \exp(-j\omega_0 t)}{2j}\;\frac{\exp(j\omega_0 t) + \exp(-j\omega_0 t)}{2}$$

$$= \frac{\exp(j2\omega_0 t) - \exp(-j2\omega_0 t)}{4j}$$

and so, using **frequency shift**, this transforms as

$$\Longleftrightarrow (2\pi/4j)\,[\delta(\omega - 2\omega_0) - \delta(\omega + 2\omega_0)] = \tfrac{1}{2}\pi/j\,[\delta(\omega - 2\omega_0) - \delta(\omega + 2\omega_0)]$$

(d) To show that the result in (c) is consistent with (4.91):

By (4.91): $\sin(2\omega_0 t) \Longleftrightarrow \pi/j\,[\delta(\omega - 2\omega_0) - \delta(\omega + 2\omega_0)]$ (A)

From (c) above:

$$\sin(\omega_0 t)\,\cos(\omega_0 t) \Longleftrightarrow \tfrac{1}{2}\pi/j\,[\delta(\omega - 2\omega_0) - \delta(\omega + 2\omega_0)]$$

Multiplying through by 2 gives

$$2\sin(\omega_0 t)\,\cos(\omega_0 t) = \sin(2\omega_0 t) \Longleftrightarrow \pi/j\,[\delta(\omega - 2\omega_0) - \delta(\omega + 2\omega_0)]$$

which is the same as (A)

4.14

(a) $f(t) = 1/2\pi \displaystyle\int_{-\infty}^{\infty} F(\omega)\exp(j\omega t)\,d\omega$

$$= 1/2\pi \int_{-\infty}^{\infty} \delta(\omega + 2)\exp(j\omega t)\,d\omega$$

in which Dirac delta sampling takes place, and so we continue

$$\ldots = 1/2\pi \int_{-\infty}^{\infty} \delta(\omega + 2)\exp[j(-2)t]\,d\omega = (1/2\pi)\exp[j(-2)t] \int_{-\infty}^{\infty} \delta(\omega + 2)\,dt$$

$$= (1/2\pi)\exp(-j2t)$$

Using frequency shift: $\exp(j\omega_0 t) \Longleftrightarrow 2\pi\,\delta(\omega - \omega_0)$, and so

$$(1/2\pi)\exp(-j2t) \Longleftrightarrow (1/2\pi)\,2\pi\,\delta(\omega + 2) = \delta(\omega + 2)$$

(b) $f(t) = 1/2\pi \displaystyle\int_{-\infty}^{\infty} \delta(\omega - 3)\exp(j\omega t)\,d\omega$

in which Dirac delta sampling takes place

$$\ldots = 1/2\pi \int_{-\infty}^{\infty} \delta(\omega - 3)\exp[j(3)t]\,d\omega$$

$$= (1/2\pi)\ \exp(j3t) \int_{-\infty} \delta(\omega + 2)\ \ dt = (1/2\pi)\ \exp(j3t)$$

Using frequency shift: $\exp(j\omega_0 t) \Longleftrightarrow 2\pi\ \delta(\omega - \omega_0)$, and so

$$(1/2\pi)\ \exp(j3t) \Longleftrightarrow (1/2\pi)\ 2\pi\ \delta(\omega - 3) = \delta(\omega - 3)$$

(c) $$f(t) = 1/2\pi \int_{-\infty}^{\infty} [\delta(\omega + 1) - 2\delta(\omega) + \delta(\omega - 1)]\ \exp(j\omega t)\ d\omega$$

$$= 1/2\pi \int_{-\infty}^{\infty} \delta(\omega + 1)\ \exp(j\omega t)\ d\omega - 1/2\pi \int_{-\infty}^{\infty} 2\delta(\omega)\ \exp(j\omega t)\ d\omega + 1/2\pi \int_{-\infty}^{\infty} \delta(\omega - 1)\ \exp(j\omega t)\ d\omega$$

in which Dirac delta sampling takes place, and so we continue

$$\ldots = 1/2\pi \int_{-\infty}^{\infty} \delta(\omega + 1)\ \exp[j(-1)t]\ d\omega - 1/2\pi \int_{-\infty}^{\infty} 2\delta(\omega)\ \exp[j(0)t]\ d\omega + 1/2\pi \int_{-\infty}^{\infty} \delta(\omega - 1)\ \exp[j(1)t]\ d\omega$$

$$= (1/2\pi)\ \exp(-jt) \int_{-\infty}^{\infty} \delta(\omega + 1)\ d\omega - (1/2\pi)\ 2 \int_{-\infty}^{\infty} \delta(\omega)\ d\omega + (1/2\pi)\ \exp(jt) \int_{-\infty}^{\infty} \delta(\omega - 1)\ d\omega$$

$$= (1/2\pi)\ [\exp(-jt) - 2 + \exp(jt)]$$

$$(1/2\pi)\ [\exp(-jt) - 2 + \exp(jt)]$$

$$\Longleftrightarrow (1/2\pi)[2\pi\delta(\omega+1) - 4\pi\delta(\omega) + 2\pi\delta(\omega-1)] = \delta(\omega+1) - 2\ \delta(\omega) + \delta(\omega-1)]$$

(d) $$f(t) = 1/2\pi \int_{-\infty}^{\infty} \sin(\omega)\ \exp(j\omega t)\ d\omega$$

$$= 1/2\pi \int_{-\infty}^{\infty} \frac{\exp(j\omega) - \exp(-j\omega)}{2j}\ \exp(j\omega t)\ d\omega$$

$$= 1/4\pi j \int_{-\infty}^{\infty} \Big[\exp[j(t + 1)\omega] - \exp[j(t - 1)\omega] \Big]\ d\omega$$

Then by Theorem 4.1 this continues as

$$\ldots = (1/4\pi j) \Big[2\pi\ \delta(t + 1) - 2\pi\ \delta(t - 1) \Big] = \frac{\delta(t + 1) - \delta(t - 1)}{2j}$$

Using time shift, we transform this, obtaining

$$\frac{\delta(t + 1) - \delta(t - 1)}{2j} \Longleftrightarrow \frac{1\ \exp(j\omega 1) - 1\ \exp(-j\omega 1)}{2j} = \sin(\omega)$$

(e) $$f(t) = 1/2\pi \int_{-\infty}^{\infty} \cos(2\omega)\ \exp(j\omega t)\ d\omega$$

$$= 1/2\pi \int_{-\infty}^{\infty} \frac{\exp(j2\omega) + \exp(-j\omega)}{2}\ \exp(j\omega t)\ d\omega$$

$$= 1/4\pi \int_{-\infty}^{\infty} \Big[\exp[j(t + 2)\omega] + \exp[j(t - 1)\omega] \Big]\ d\omega$$

Then by Theorem 4.1 this continues as

$$= (1/4\pi)\left[\, 2\pi\ \delta(t+2) + 2\pi\ \delta(t-2) \,\right] = \frac{\delta(t+2) + \delta(t-2)}{2}$$

Using time shift, we transform this, obtaining

$$F(\omega) = \frac{1\ \exp(j\omega 1) + 1\ \exp(-j\omega 1)}{2} = \cos(\omega)$$

(f) $\sin^2(3\omega) = \dfrac{1 - \cos(6\omega)}{2}$ $\qquad f(t) = 1/2\pi \displaystyle\int_{-\infty}^{\infty} \frac{1-\cos(6\omega)}{2} \exp(j\omega t)\ d\omega$

$$= 1/4\pi \int_{-\infty}^{\infty} \left[\exp(j\omega t) - \frac{\exp[j(t+6)\omega] + \exp[j(t-6)\omega]}{2} \right] d\omega$$

Then by Theorem 4.1 this continues as

$$= (1/4\pi)\left[\, 2\pi\ \delta(t) - \pi\ \delta(t+6) - \pi\ \delta(t-6) \,\right]$$

$$= \tfrac{1}{2}\ \delta(t) - \tfrac{1}{4}\ \delta(t+6) - \tfrac{1}{4}\ \delta(t-6)$$

Using time shift, we transform this, obtaining

$$\tfrac{1}{2}\ \delta(t) - \tfrac{1}{4}\ \delta(t+6) - \tfrac{1}{4}\ \delta(t-6)$$

$$\Longleftrightarrow \tfrac{1}{2} - \tfrac{1}{4}\ \exp(j6\omega) - \tfrac{1}{4}\ \exp(-j\omega) = \tfrac{1}{2} - \tfrac{1}{2}\cos(6\omega) = \sin^2(6\omega)$$

(g) $\cos^3(4\omega) = \left[\dfrac{\exp(j4\omega) + \exp(-j4\omega)}{2} \right]^3$

$$= \frac{\exp(j12\omega) + 3\ \exp(j4\omega) + 3\ \exp(-j4\omega) + \exp(-j12\omega)}{8} \Longleftrightarrow$$

$$\frac{1}{16\pi}\int_{-\infty}^{\infty} [\exp(j12\omega) + 3\exp(j4\omega) + 3\exp(-j4\omega) + \exp(-j12\omega)]\ \exp(j\omega t)\ d\omega$$

Then by Theorem 4.1 this continues as

$$\ldots = \frac{2\pi}{16\pi}\left[\, \delta(t+12) + 3\ \delta(t+4) + 3\ \delta(t-4) + \delta(t-12) \,\right]$$

$$= \frac{1}{8}\left[\, \delta(t+12) + 3\ \delta(t+4) + 3\ \delta(t-4) + \delta(t-12) \,\right]$$

Using time shift, we transform this, obtaining

$$F(\omega) = \frac{1}{8}\left[\exp(j12\omega) + 3\ \exp(j4\omega) + 3\ \exp(-j4\omega) + \exp(-j12\omega)\right] = \cos^3(4\omega)$$

4.15

(a) $f(t) = 1/2\pi \displaystyle\int_{-\infty}^{\infty} F(\omega)\ \exp(j\omega t)\ d\omega = 1/2\pi \int_{-\infty}^{\infty} 2\pi\delta(\omega)\ \exp(j\omega t)\ d\omega$ in which Dirac delta sampling takes place

$$\ldots = \int_{-\infty}^{\infty} \delta(\omega)\ \exp[j(0)t]\ d\omega = \int_{-\infty}^{\infty} \delta(\omega)\ 1\ d\omega = 1$$

(b) $f(t) = 1/2\pi \displaystyle\int_{-\infty}^{\infty} F(\omega)\ \exp(j\omega t)\ d\omega = 1/2\pi \int_{-\infty}^{\infty} 2\pi\delta(\omega - \omega_0)\ \exp(j\omega t)\ d\omega$ in which Dirac delta sampling takes place

$$\ldots = \int_{-\infty}^{\infty} \delta(\omega)\ \exp[j(\omega_0)t]\ d\omega = \exp(j\omega_0 t)\int_{-\infty}^{\infty} \delta(\omega)\ d\omega = \exp(j\omega_0 t)$$

(c) $f(t) = 1/2\pi \displaystyle\int_{-\infty}^{\infty} F(\omega)\ \exp(j\omega t)\ d\omega$

$$= 1/2\pi \int_{-\infty}^{\infty} \pi[\delta(\omega - \omega_0) + \delta(\omega + \omega_0)]\ \exp(j\omega t)\ d\omega$$

$$= \tfrac{1}{2}\int_{-\infty}^{\infty} \delta(\omega - \omega_0)\ \exp(j\omega t)\ d\omega + \tfrac{1}{2}\int_{-\infty}^{\infty} \delta(\omega + \omega_0)\ \exp(j\omega t)\ d\omega$$

in which Dirac delta sampling takes place

$$= \tfrac{1}{2}\int_{-\infty}^{\infty} \delta(\omega - \omega_0)\ \exp[j(\omega_0)t]\ d\omega + \tfrac{1}{2}\int_{-\infty}^{\infty} \delta(\omega + \omega_0)\ \exp[j(-\omega_0)t]\ d\omega$$

$$= \tfrac{1}{2} \exp(j\omega_0 t) \int_{-\infty}^{\infty} \delta(\omega - \omega_0) \, d\omega + \tfrac{1}{2} \exp(-j\omega_0 t) \int_{-\infty}^{\infty} \delta(\omega + \omega_0) \, d\omega$$

$$= \tfrac{1}{2} [\exp(j\omega_0 t + \exp(-j\omega_0 t)] = \cos(\omega_0 t)$$

(d) $$f(t) = 1/2\pi \int_{-\infty}^{\infty} F(\omega) \exp(j\omega t) \, d\omega$$

$$= 1/2\pi \int_{-\infty}^{\infty} \frac{\pi}{j} [\delta(\omega - \omega_0) - \delta(\omega + \omega_0)] \exp(j\omega t) \, d\omega$$

$$= \frac{1}{2j} \int_{-\infty}^{\infty} \delta(\omega - \omega_0) \exp(j\omega t) \, d\omega - \frac{1}{2j} \int_{-\infty}^{\infty} \delta(\omega + \omega_0) \exp(j\omega t) \, d\omega$$

in which Dirac delta sampling takes place

$$= \frac{1}{2j} \int_{-\infty}^{\infty} \delta(\omega - \omega_0) \exp[j(\omega_0)t] \, d\omega - \frac{1}{2j} \int_{-\infty}^{\infty} \delta(\omega + \omega_0) \exp[j(-\omega_0)t] \, d\omega$$

$$= \frac{1}{2j} \exp(j\omega_0 t) \int_{-\infty}^{\infty} \delta(\omega - \omega_0) \, d\omega - \frac{1}{2j} \exp(-j\omega_0 t) \int_{-\infty}^{\infty} \delta(\omega + \omega_0) \, d\omega$$

$$= \frac{1}{2j} [\exp(j\omega_0 t - \exp(-j\omega_0 t)] = \sin(\omega_0 t)$$

4.16 (a) We know that $1 \Longleftrightarrow 2\pi \, \delta(\omega)$ (A)

Multiplication on the left by $\exp(j\omega_0 t)$ corresponds to shifting the frequency on the right, i.e. $F(\omega)$ becomes $F(\omega - \omega_0)$. Thus, from (A),

$$\exp(j\omega_0 t) \Longleftrightarrow 2\pi \, \delta(\omega - \omega_0)$$ QED

(b) We know that $\delta(t) \Longleftrightarrow 1$ (B)

Delay on the left by τ seconds corresponds to multiplication on the right by $\exp(-j\omega\tau)$, i.e. if $f(t)$ becomes $f(t - \tau)$ then $F(\omega)$ becomes $F(\omega) \exp(-j\omega\tau)$.

Thus, from (B), $\delta(t - \tau) \Longleftrightarrow \exp(-j\omega\tau)$ QED

In Figure 4.3[illegible] we [illegible] analytical definition is

$$f_p(t) = \begin{cases} \tfrac{1}{2} & (0 < t < 1) \\ 1 & (1 < t < 2) \end{cases} \qquad f_p(t + 4) = f_p(t)$$

which is the same as the waveform in Exercise 2.23. Thus the expression for its Fourier series coefficients is

$$F_p(n) = \begin{cases} \dfrac{j}{4n\pi} \left[2(-1)^n - 1 - \exp(-jn\pi/2) \right] & (n \neq 0) \\ 3/8 & (n = 0) \end{cases}$$

Its Fourier series is

$$f_p(t) = \sum_{n=-\infty}^{\infty} F_p(n) \exp(jn\omega_0 t) \qquad (\omega_0 = 2\pi/T_0 = \pi/2)$$

Using the fact that $\exp(jn\omega_0 t) \Longleftrightarrow 2\pi \, \delta(\omega - n\omega_0)$ we obtain the required transform as

$$F(\omega) = 2\pi \sum_{n=-\infty}^{\infty} F_p(n) \, \delta(\omega - n\pi/2) \qquad \text{(A)}$$

(b) From Exercise 2.26 we have, for the coefficients of the Fourier series,

$$A(n) = \begin{cases} -(1/4\pi n) \sin(n\pi/2) & (n \neq 0) \\ 3/8 & (n = 0) \end{cases}$$

$$B(n) = \frac{1}{4n\pi} \left[2(-1)^n - 1 - \cos(n\pi/2) \right] \qquad (n \neq 0)$$

and so from (A) above, the weights of the Dirac deltas in the Fourier transform will have real and imaginary parts

$$2\pi \, A(n) = \begin{cases} -(1/2n) \sin(n\pi/2) & (n \neq 0) \\ 3\pi/4 & (n = 0) \end{cases}$$

$$2\pi B(n) = \frac{1}{2n}\left[2(-1)^n - 1 - \cos(n\pi/2)\right] \quad (n \neq 0)$$

Steps for running the problem on the FFT system with $N = 1024$, SAMPLED, $T = 4$, PULSE (remember, we intend to find a Fourier transform and so we must use PULSE):

Main Menu, Create Values, Continue, Create X, Real, Neither, Left, 3 intervals, 1, 2, Continue, 1/2, 1, 0, Go, Neither, Go. You are now back on the main menu with the waveform properly loaded. Plotting **X** confirms that. Run ANALYSIS, Show Numbers, F, Eternal. The values for the **weights** of the Dirac deltas are now on your screen.

From the expressions for $2\pi A(n)$ and $2\pi B(n)$ and the screen we obtained the values shown in the following table:

n	A_{FFT}	$A_{formula}$	B_{FFT}	$B_{formula}$
0	2.3561945	2.3561944	0	0
1	-0.49999843	-0.5000000	-1.4999953	-1.5000000
2	0	0	0.49999373	0.5000000
3	0.16666196	0.1666666	-0.49998588	-0.5000000

4.18 (a) The analytical definition is

$$f_p(t) = \begin{cases} t & (0 < t < 1) \\ 0 & (1 < t < 4) \end{cases} \qquad f_p(t + 4) = f_p(t)$$

See Answer section for the sketch.

To find the Fourier transform of the periodic function we must first find its Fourier series. The waveform is the same as the waveform in Exercise 2.29. Thus the expression for its Fourier series coefficients is

$$F_p(n) = \begin{cases} (1/n^2\pi^2)\,[\exp(-jn\pi/2)(1 + jn\pi/2) - 1] & (n \neq 0) \\ 1/8 & (n = 0) \end{cases}$$

whose real and imaginary part were shown there to be

$$A(n) = \begin{cases} (1/n^2\pi^2)\,[\cos(n\pi/2) + (n\pi/2)\sin(n\pi/2) - 1] & (n \neq 0) \\ 1/8 & (n = 0) \end{cases}$$

$B(n) = (1/n^2\pi^2)$ [illegible]

We also have $\omega_0 = 2\pi/T_0 = \pi/2$. Thus its Fourier series is

$$f_p(t) = \sum_{n=-\infty}^{\infty} F_p(n)\exp(jn\pi t/2)$$

Using the fact that $\exp(jn\omega_0 t) \Longleftrightarrow 2\pi\,\delta(\omega - n\omega_0)$ we obtain the required transform as

$$F(\omega) = 2\pi \sum_{n=-\infty}^{\infty} F_p(n)\,\delta(\omega - n\pi/2) \qquad (A)$$

(b) and (c) The power in the Fourier series is obtained from Parseval's theorem as

$$P(n) = |F(n)|^2 = A(n)^2 + B(n)^2$$

The relative energies in the Fourier transform, as a fraction of the energy at $n = 0$ are then

$$\text{ratio}(n) = P(n)/P(0).$$

Steps for running the problem on the FFT with $N = 1024$, SAMPLED, $T = 4$, PERIODIC. Note: We use PERIODIC because we wished to display values for the Fourier series coefficients of the periodic waveform. Had we wished to display values for the Fourier transform of the waveform we would have used PULSE.

Main menu, Create Values, Continue, Create X, Real Neither, Left, 2 intervals, 1, Continue, t, 0, Go, Neither, Go. You are now back on the main menu with the waveform correctly loaded. Plotting **X** confirms that. Run ANALYSIS, Show Numbers, F, Complex.

n	$P(n)_{FFT}$	$P(n)_{formula}$	$\text{ratio}_{formula}$
0	1.5625000e-2	1.5625000e-2	1
±1	1.3610696e-2	1.3610725e-2	0.871086463
±2	8.8989750e-3	8.8990695e-3	0.569540450
±3	4.2623150e-3	4.2624596e-3	0.272797417
±4	1.5829846e-3	1.5831434e-3	0.101321184
±5	7.8789692e-4	7.8805070e-4	0.050435245

4.19 Finding the total area under the function $Sa^2(x)$ using Parseval's theorem:

We know that $\text{Rect}(t/\tau) \Longleftrightarrow \tau\,Sa(\omega\tau/2)$. For $\tau = 1$ this becomes $\text{Rect}(t) \Longleftrightarrow Sa(\omega/2)$

By Parseval's theorem, total energy in a pulse is given by

$$E = \int^{\infty} |f(t)|^2\,dt = (1/2\pi)\int^{\infty} |F(\omega)|^2\,d\omega$$

For Rect(t) this becomes $\int_{-1/2}^{1/2} 1^2\, dt = (1/2\pi) \int_{-\infty}^{\infty} Sa^2(\omega/2)\, d\omega$

The first integral evaluates to 1 and the second integral can be written as follows:

$$1/\pi \int_{-\infty}^{\infty} Sa^2(\omega/2)\, d(\omega/2) = 1/\pi \int_{-\infty}^{\infty} Sa^2(x)\, dx$$

Equating these we obtain $\int_{-\infty}^{\infty} Sa^2(x)\, dx = \pi$

which is the same answer as we obtained in Exercise 3.10(e).

4.20
(a) The Dirac comb $\delta_T(t)$ discussed in Section 4.11 had its Dirac deltas located at $t = 0, \pm 1, \pm 2, \ldots$ etc. Its definition was

$$\delta_T(t) = \sum_{n=-\infty}^{\infty} \delta(t - nT_s)$$

Shifting these by $T_s/2$ gives $\delta_{TS}(t) = \sum_{n=-\infty}^{\infty} \delta(t - nT_s - T_s/2))$

which is thus the shifted Dirac comb appearing in Figure 4.40.

(b) To find its Fourier transform we use use the method of Section 4.11. First we find the Fourier series for $\delta_{TS}(t)$, and then we transform that Fourier series.

Thus, by the Fourier series analysis equation:

$$F(n) = 1/T_0 \int_0^{T_0} f_p(t) \exp(-jn\omega_0 t)\, dt \qquad (\omega_0 = 2\pi/T_0)$$

$$= 1/T_s \int_0^{T_s} \delta(t - T_s/2) \exp(-jn\omega_s t)\, dt$$

in which Dirac delta sampling takes place, and so we continue

$$\ldots = 1/T_s \int_0^{T_s} \delta(t - T_s/2) \exp[-jn\omega_s(T_s/2)]\, dt$$

But $\omega_s T_s/2 = \pi$ and so $\exp(-jn\omega_s T_s/2) = \exp(-jn\pi) = (-1)^n$ and so we continue

$$\ldots = (-1)^n\, 1/T_s \int_0^{T_s} \delta(t - T_s/2)\, dt = (-1)^n/T_s$$

The Fourier series is thus

$$f_p(t) = \sum_{n=-\infty}^{\infty} F(n) \exp(jn\omega_0 t) = \frac{1}{T_s} \sum_{n=-\infty}^{\infty} (-1)^n \exp(jn\omega_s t)$$

Using the fact that $\exp(jn\omega_s t) \Longleftrightarrow 2\pi\, \delta(\omega - n\omega_s)$ we now transform the Fourier series to obtain the Fourier transform of the shifted Dirac comb as

$$F_p(\omega) = \frac{2\pi}{T_s} \sum_{n=-\infty}^{\infty} (-1)^n \delta(\omega - n\omega_s) = \omega_s \sum_{n=-\infty}^{\infty} (-1)^n \delta(\omega - n\omega_s) \qquad \text{QED}$$

4.21 $\delta_T(t) = \frac{1}{T_0} \left[1 + 2 \sum_{n=1}^{\infty} \cos(n2\pi t/T_0) \right]$ (?)

From (4.102) the Fourier series for $\delta_T(t)$ is

$$\delta_T(t) = \frac{1}{T_0} \sum_{n=-\infty}^{\infty} \exp(jn\omega_0 t) = \frac{1}{T_0} \sum_{n=-\infty}^{\infty} [\cos(n\omega_0 t) + j \sin(n\omega_0 t)]$$

But $\sin(n\omega_0 t)$ is odd and so it sums to zero. Moreover $\cos(n\omega_0 t)$ is even and so we continue as

$$\ldots = \frac{1}{T_0} \left[1 + 2 \sum_{n=1}^{\infty} \cos(n\omega_0 t) \right] \qquad \text{QED}$$

4.22 By (4.71), to represent a time-domain Dirac delta on the system of weight μ we must load a value of $V = \mu N/T$

(a) For $2\,\delta(t)$ using N = 100, SAMPLED, T = 5, PULSE:

The weight is $\mu = 2$ and the impulse is located at $t = 0$, i.e. at $k = 0$. We must therefore load a value at $k = 0$ of

$V = 2 \times 100 / 5 = 40$

Using the FFT system: Main-menu Create Values, Continue, Create only Diracs, Time, X, 0, 2, Neither, Go. You are now back on the main menu with the waveform properly loaded. Plotting **X** confirms that. Run ANALYSIS, then Show Numbers, X and Y.

Your screen shows XRE(0) = 40 with all other values zero.

Plotting **F** shows $A(n) = 2$ for all n, i.e. the spectrum is flat across all frequencies. The phase $\theta(n)$ is zero at all frequencies. Both of these are exactly what we expect because

$2\ \delta(t) \Longleftrightarrow 2$

To see a more complete picture view PLOTS of

X vector:	You see the impulse at $t = 0$.
Real part of F:	You see all values equal to 2
Imag part of F:	All values zero. Cannot draw a plot.
Magnitude of F:	You see all values equal to 2
Phase of F:	All values zero. Cannot draw a plot.

(b) For $3\ \delta(t - 2)$ using N = 256, SAMPLED, T = 4, PULSE: The weight is $\mu = 3$ and the impulse is located at $t = 2$, i.e. at $k = 128$. We must therefore load a value at $k = 128$ of

$V = 3 \times 256 / 4 = 192$

Using the FFT system: Main-menu Create Values, Continue, Create only Diracs, Time, X, 2, 3, Neither, Go. You are now back on the main menu with the waveform properly loaded. Plotting **X** confirms that. Run ANALYSIS, Show Numbers, X and Y.

Your screen shows XRE(128) = 192 with all other values zero. Plotting **F** shows

$|F(n)| = 3$ (for all n)

However, $A(n) = 3$ (n even) and $A(n) = -3$ (n odd)

The phase appears as follows:

$\theta(n) = 0$ (n even)

$\theta(n) = \pi$ (n odd and positive)

$\theta(n) = -\pi$ (n odd and negative)

These are exactly what we expect because

$3\ \delta(t - 2) \Longleftrightarrow 3\ \exp(-j2\omega)$

"Introduction to Fourier Analysis", Instructor's Manual, Chapter 4

from which we see that $|F(n)| = |3\ \exp(-j2\omega)| = 3$

To explain $A(n)$: The FFT samples the spectrum at steps of

$n\Omega = n2\pi/T = n\pi/2$

Then $3\ \exp(-j2n\pi/2) = 3\ \exp(-jn\pi) = 3\ (-1)^n$

which accounts for why $A(n)$ has the values shown.

For the phase: The phase spectrum is zero when $A(n)$ is positive, i.e. when n is even, and $\pm\pi$ when $A(n)$ is negative, with the + or - selected so that the phase spectrum is odd.

To see a more complete picture view the PLOTS of

X vector:	You see the impulse at $t = 2$. (You also see it at $t = -2$, because the FFT is periodic with period $T = 4$)
Real part of F:	You see the alternating +1 and -1
Imag part of F:	Cannot draw a plot. All values zero
Magnitude of F:	You see the constant 3
Phase of F:	You see values of 0 and 1 $(n > 0)$ 0 and -1 $(n < 0)$

(c) For $5\ \delta(t + 1)$ using N = 240, SAMPLED, T = 12, PULSE:

The weight is $\mu = 5$ and the impulse is located at $t = -1$, i.e. at $k = -20$. However the allowable values of k run from 0 to 240, and the FFT is periodic. Thus this corresponds to a value of $k = 220$. We must therefore load a value at $k = 220$ of

$V = 5 \times 240 / 12 = 100$.

Using the FFT system: Main menu Create Values, Continue, Create only Diracs, TIME, X, -1, 5, Neither, Go. You are now back on the main menu. Run ANALYSIS, then Show Numbers, X and Y. Your screen shows XRE(220) = 100 with all other values zero. Plotting **F** shows

$|F(n)| = 5$ (for all n)

However, $A(0) = 5$ $B(0) = 0$

$A(1) = 4.3301270$ $B(1) = 2.5$

$A(2) = 2.5$ $B(2) = 4.3301270$

$A(3) = 0$ $B(3) = 5$

The phase appears as follows:

$\theta(0) = 0$

$\theta(1) = 0.52329878$

$\theta(2) = 1.0471976$

$\theta(3) = 1.5707963$

These are exactly what we expect because

$$5\ \delta(t + 1) \Longleftrightarrow 5\ \exp(j\omega)$$

from which we see that $|F(n)| = |5\ \exp(j\omega)| = 5$

To explain A(n): The FFT samples the spectrum at steps of

$$n\Omega = n2\pi/T = n\pi/6$$

Then

$$5\ \exp(jn\pi/6) = 5\ \cos(n\pi/6) + j\ \sin(n\pi/6)$$

n=0:	5	+ j
n=1:	4.3301270	+ j 2.5
n=2:	2.5	+ j 4.3301270
n=3:	0	+ j 5

which explains why A(n) has the values shown.

The phase spectrum is

$$\theta(\omega) = \arg[5\ \exp(j\omega)] = \omega$$

Sampling ω at $n\pi/6$ gives

$\theta(n) = n\pi/6 \quad \theta(0) = 0 \quad \theta(1) = \pi/6 \quad \theta(2) = 2\pi/6 \quad \theta(3) = 3\pi/6$

with the negative of these values selected so that the phase spectrum is odd.

To see a more complete picture view the PLOTS of

X vector:	You see the impulse at $t = -1$
Real part of F:	You see the $5\ \cos(n\pi/6)$
Imag part of F:	You see the $5\ \sin(n\pi/6)$
Magnitude of F:	You see the constant 5
Phase of F:	You see values of $n\pi/6$, running from $-\pi$ to π, and then repeating.

(a) For $6\pi\ \delta(\omega)$ using N = 100, SAMPLED, T = 5, PULSE

The weight is $\mu = 6\pi$ and the impulse is located at $\omega = 0$, i.e. at $n = 0$. We must therefore load a value at $n = 0$ of

$$V = 6\pi \times 5\ /\ 2\pi = 15$$

Because $3 \Longleftrightarrow 3\ 2\pi\ \delta(\omega)$ it follows that the inverse of $6\pi\ \delta(\omega)$ is 3.

Using the FFT system: Main-menu Create Values, Continue, Create only Diracs, Freq, FRE, 0, 6*PI, Neither, Go. You are now back on the main menu. Run SYNTHESIS, then Show Numbers, F

Your screen shows FRE(0) = 15 with all other values zero.

Displaying the Y vector shows YRE(k) = 3 for all values of k.

To see a more complete picture view PLOTS of

Real part of F:	You see the impulse at $\omega = 0$.
Imag part of F:	All values zero. Cannot draw a plot.
Magnitude of F:	You see the impulse at $\omega = 0$.
Phase of F:	All values zero. Cannot draw a plot.
Y vector:	All values equal to 3.

(b) $4\pi\ \delta(\omega - 12\pi)$ using N = 256, SAMPLED, T = 4, PULSE

The weight is $\mu = 4\pi$ and the impulse is located at $\omega = 12\pi$. Since the FFT samples are at the points

$$n\Omega = n2\pi/T = n\ \pi/2$$

to be at 12π we require that $n = 24$. We must therefore load a value at $n = 24$ of

$$V = 4\pi\ 4/2\pi = 8$$

Because $2\ \exp(j12\pi t) \Longleftrightarrow 2\ 2\pi\ \delta(\omega - 12\pi)$ it follows that the inverse of $4\pi\ \delta(\omega - 12\pi)$ is $2\ \exp(j12\pi t)$.

Using the FFT system: Main-menu Create Values, Continue, Create only Diracs, Freq, FRE, 12*pi, 2*PI, Neither, Go. You are now back on the main menu. Run SYNTHESIS. Show Numbers, F, Single.

Your screen shows FRE(24) = 8 with all other values zero.

Displaying the Y vector shows

YRE(0) = 2 YIM(0) = 0

YRE(1) = 1.6629392 YIM(1) = 1.1111405

YRE(2) = 0.76536686 YIM(2) = 1.8477591

YRE(3) = -0.39018064 YIM(3) = 1.9615706

These are what we expect, because

$$Y(t) = 2\exp(j12\pi t) = 2\cos(12\pi t) + j\sin(12\pi t)$$

and in this case t is in steps of

$$T_s = T/N = 0.15625$$

Thus $Y(kT_s) = 2\exp(j12\pi kT_s) = 2\cos(12\pi kT_s) + j\sin(12\pi kT_s)$

k = 0: 2 + j 0

k = 1: 1.6629392 + j 1.1111405

k = 2: 0.76536686 + j 1.8477591

k = 3: -0.39018064 + j 1.9615706

To see a more complete picture view PLOTS of

Real part of F: You see the impulse at $\omega = 12\pi$ (n = 24)

Imag part of F: All values zero. Cannot draw a plot.

Magnitude of F: You see the impulse at $\omega = 12\pi$ (n = 24)

Phase of F: All values zero. Cannot draw a plot.

YRE vector: You see the $2\cos(12\pi kT_s)$

YIM vector: You see the $2\sin(12\pi kT_s)$

(c) $(1+2j)\ \delta(\omega + 4\pi)$ using N = 250, SAMPLED, T = 12, PULSE

The weight is $\mu = 1 + 2j$ and the impulse is located at $\omega = -4\pi$. Since the FFT samples are at the points

$$n\Omega = n2\pi/T = n\pi/6$$

to be at -4π we require that n = -24. We must therefore load a value at n = -24 of

$$V = (1+2j)12/2\pi = 6/\pi + j\ 12/\pi = 1.9098593 + j\ 3.8197186$$

Because $[(1+2j)/2\pi]\exp(-j4\pi t) \Longleftrightarrow (1+j)\ \delta(\omega + 4\pi)$

it follows that the inverse of $(1+j)\ \delta(\omega + 4\pi)$ is

$$[(1+2j)/2\pi]\exp(-j4\pi t) = \frac{1}{2\pi}(1+j2)\,[\cos(4\pi t) - j\sin(4\pi t)]$$

from which $$YRE(t) = \frac{1}{2\pi}[\cos(4\pi t) + 2\sin(4\pi t)]$$

$$YIM(t) = \frac{1}{2\pi}[2\cos(4\pi t) - \sin(4\pi t)]$$

Using the FFT system: Main-menu Create Values, Continue, Create only Diracs, Freq, FRE, -4*pi, 1 FIM, -4*pi, 2, Neither, Go. You are now back on the main menu. Run SYNTHESIS. Show Numbers, F, Single. Your screen shows

FRE(-24) = 1.9098593

FIM(-24) = 3.8197186

with all other values zero, precisely as expected. Plotting **Y** shows

YRE(0) = 0.15915494 YIM(0) = 0.31830989

YRE(1) = 0.31163660 YIM(1) = 0.17185491

YRE(2) = 0.35413085 YIM(2) = -3.5253645e-2

YRE(3) = 0.27164000 YIM(3) = -0.22991996

These are what we expect, because

$$YRE(t) = \frac{1}{2\pi}[\cos(4\pi t) + 2\sin(4\pi t)]$$

$$YIM(t) = \frac{1}{2\pi}[2\cos(4\pi t) - \sin(4\pi t)]$$

and in this case t is in steps of $T_s = T/N = 0.048$

Thus $$YRE(kT_s) = \frac{1}{2\pi}[\cos(4\pi kT_s) + 2\sin(4\pi kT_s)] \qquad \text{(A)}$$

k = 0: 0.15915494

k = 1: 0.31163660

k = 2: 0.35413085

k = 3: 0.27164000

$$YIM(kT_s) = \frac{1}{2\pi} [2 \cos(4\pi kT_s) - \sin(4\pi kT_s)] \qquad (B)$$

k = 0: 0.31830989

k = 1: 0.17185491

k = 2: -3.5253645e-2

k = 3: -0.22991996

To see a more complete picture view PLOTS of

Real part of F:	You see the real part of the impulse at $\omega = -4\pi$ (n = -24)
Imag part of F:	You see the imaginary real part of the impulse at $\omega = -4\pi$ (n = -24)
Magnitude of F:	You see the modulus of the impulse at $\omega = 12\pi$ (n = -24)
Phase of F:	You see the phase (argument) of the impulse at $\omega = 12\pi$ (n = -24). Since the weight of the impulse is 1 + j2 its argument appears as arctan(2/1) = 1.0715
YRE vector:	You see a plot of (A) above
YIM vector:	You see a plot of (B) above
Modulus of Y:	All values are equal to 0.355881 because

$$\left| \frac{1 + 2j}{2\pi} \right| = 0.355881$$

4.24
(a) Using the FFT system to obtain the transform of the eternal constant 4 with N = 240, SAMPLED, T = 5, PULSE:

Main-menu, Create Values, Continue, Create X, Real, Even, 1, Continue, 4, Go, Neither, Go. You are now back on the main menu with the waveform correctly loaded. Plotting **X** confirms that. Run ANALYSIS.

The function transforms as follows:

$$4 \Longleftrightarrow 8\pi\, \delta(\omega)$$

To implement the frequency-domain Dirac delta $8\pi\, \delta(\omega)$ on the FFT we require a value of

$V = \mu\, T/2\pi = 8\pi \times 5 / 2\pi = 20$ at n = 0.

Displaying **F** (either Show Numbers or Draw Plots), we see the value A(0) = 20, with zero at all other values of n.

(b) Using the FFT system to obtain the Fourier transform of the eternal cosine $\cos(\omega_0 t)$ where $\omega_0 = 2\pi \times 4$ with N = 240, SAMPLED, T = 5, PULSE: Main-menu, Create Values, Continue, Create X, Real, Even, 1, Continue, COS(2π*4*t), Go, Neither, Go. You are now back on the main menu with the waveform correctly loaded. Plotting **X** confirms that. Run ANALYSIS.

The function transforms as follows:

$$\cos(8\pi t) \Longleftrightarrow \pi\, [\delta(\omega - 8\pi) + \delta(\omega + 8\pi)]$$

The frequency-domain sampling points of the FFT are located at

$$n\Omega = n2\pi/T = n2\pi/5$$

To be at the point $\omega = 8\pi$ we require a value for n of 20. Thus, to implement the frequency-domain Dirac delta $\pi\, \delta(\omega - 8\pi)$ on the FFT we require a value at n = 20 of

$$V = \mu\, T/2\pi = \pi 5/2\pi = 2.5$$

Similarly, to implement the frequency-domain Dirac delta $\pi\, \delta(\omega + 8\pi)$ on the FFT we require a value at n = -20 of

$$V = \mu\, T/2\pi = \pi 5/2\pi = 2.5$$

Displaying **F** (either Show Numbers or Draw Plots), we see the values A(20) = 2.5 and A(-20) = 2.5 with zero at all other values of n.

(c) Using the FFT system to obtain the Fourier transform of the eternal sine $\sin(\omega_0 t)$ where $\omega_0 = 2\pi \times 3$ with N = 240, SAMPLED, T = 5, PULSE: Main-menu, Create Values, Continue, Create X, Real, Odd, 1, Continue, SIN(2π * 3 * t), Go, Neither, Go. You are now back on the main menu with the waveform correctly loaded. Plotting **X** confirms that. Run ANALYSIS.

The function transforms as follows:

$$\sin(6\pi t) \Longleftrightarrow \pi/j\, [\delta(\omega - 6\pi) - \delta(\omega + 6\pi)] = j\pi\, [\delta(\omega + 6\pi) - \delta(\omega - 6\pi)]$$

The frequency-domain sampling points of the FFT are located at $n\Omega = n\, 2\pi/T = n\, 2\pi/5$. To be at the point $\omega = -6\pi$ we require a value for n of -15. Thus, to implement the frequency-domain Dirac delta $\pi\, \delta(\omega + 6\pi)$ on the FFT we require a value at n = -15 of

$$V = \mu\, T/2\pi = j\pi 5/2\pi = j2.5$$

This means we should have the value B(-15) = 2.5

Similarly, to implement the frequency-domain Dirac delta

$-j\pi\ \delta(\omega - 8\pi)$ on the FFT we require a value at $n = 15$ of

$$V = \mu\ T/2\pi = -j\pi \times 5\ /\ 2\pi = -j\ 2.5$$

This means we should have the value $B(15) = -2.5$

Displaying **F** (either Show Numbers or Draw Plots), we see the values $B(15) = -2.5$ and $B(-15) = 2.5$, with zero at all other values of n.

4.25 (a) Using the FFT system to find the Fourier coefficients for the periodic impulse train

$$\delta_T(t) = \sum_{n=-\infty}^{\infty} \delta(t - nT_s) \quad \text{with } N = 256\text{, SAMPLED, } T = 2\text{, PERIODIC:}$$

(We use PERIODIC because we wish to find values for the Fourier coefficients of a periodic waveform.)

The waveform has Fourier coefficients (see 4.101)

$F(n) = 1/T_s = 1/2$ for all values of n.

The central-period impulse is located at $t = 0$ and has weight 1.

For the FFT: Main-menu, Create Values, Continue, Create only Diracs, Time, X, 0, 1, Neither, Go. Run ANALYSIS.

In the time domain we expect to find a value (see 4.72)

$$V = \mu \times N\ /\ T = 1 \times 256\ /\ 2 = 128$$

in X(0) with zero in all other values of X(k). In the frequency domain we expect to find the value

$A(n) = 1/2$ for all values of n.

Displaying **X** and **F** (either Draw Plots or Show Numbers) confirms all of these results.

(b) Using the FFT system to find the Fourier coefficients for the periodic impulse train

$$\delta_{TS}(t) = \sum_{n=-\infty}^{\infty} \delta(t - nT_s - T_s/2) \quad \text{with } N = 256\text{, SAMPLED, } T = 2\text{, PERIODIC:}$$

(We use PERIODIC because we wish to find values for the Fourier coefficients of a periodic waveform.)

The waveform has Fourier coefficients (see Exercise 4.20)

$$F(n) = (-1)^n/T_s = (-1)^n/2$$

The central-period impulse is located at $t = 1$ and has weight 1.

For the FFT: Main-menu, Create Values, Continue, Create only Diracs, Time, X, 1, 1, Neither, Go. Run ANALYSIS.

In the time domain we expect to find a value (see 4.72)

$$V = \mu \times N\ /\ T = 1 \times 256\ /\ 2 = 128$$

in X(128) with zero in all other values of X(k). In the frequency domain we expect to find the value

$$A(n) = (-1)^n/2$$

for all values of n.

Displaying **X** and **F** (either Draw Plots or Show Numbers) confirms all of these results.

4.26 (a) To show that the energy contained in a Dirac delta of weight c is infinite:

A Dirac delta of weight c transforms as follows:

$$c\ \delta(t) \Longleftrightarrow c$$

Then by Parseval's theorem for pulses:

$$E = 1/2\pi \int_{-\infty}^{\infty} |F(\omega)|^2\ d\omega = 1/2\pi \int_{-\infty}^{\infty} c^2\ d\omega = \infty$$

(b) The energy contained between frequencies ω_1 and ω_2 is

$$\Delta E = 1/2\pi \int_{\omega_1}^{\omega_2} c^2\ d\omega = c^2\ \frac{\omega_2 - \omega_1}{2\pi} = c^2\ (f_2 - f_1)$$

from which the energy/unit bandwith is $\dfrac{\Delta E}{f_2 - f_1} = c^2$ joules/hertz

QED

4.27 To test for orthogonality over $(-\infty, \infty)$ of the set

$S_2 = \{ \exp(j\omega t) \mid \omega \in \mathbb{R} \}$ we must evaluate the integral

$$\int_{-\infty}^{\infty} \exp(j\omega_1 t)\ \exp(j\omega_2 t)^*\ dt$$

If the result is zero when $\omega_1 \neq \omega_2$ then the elements of S_2 are orthogonal.

In fact, since ω_2 is any real number we let it be the general frequency variable ω, obtaining:

$$\int_{-\infty}^{\infty} \exp(j\omega_1 t)\ \exp(j\omega t)^*\ dt = \int_{-\infty}^{\infty} \exp(j\omega_1 t)\ \exp(-j\omega t)\ dt$$

which is the analysis equation for the eternal pulse $\exp(j\omega_1 t)$ and so we continue

$$= 2\pi\ \delta(\omega - \omega_1)$$ which is zero for $\omega_1 \neq \omega$

Thus the elements of S_2 are indeed orthogonal over $(-\infty, \infty)$.

QED.

Chapter 5 The Method of Successive Differentiation

5.1(a)

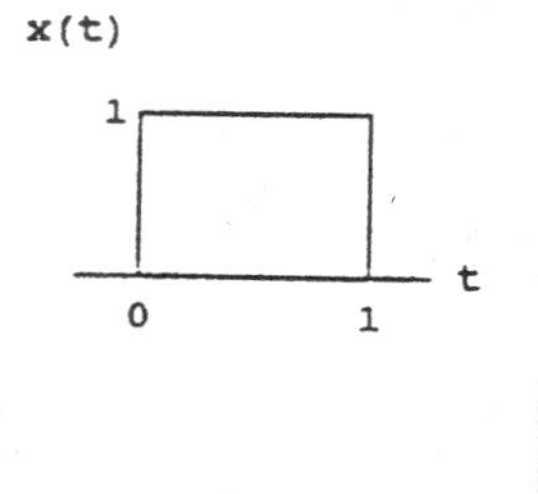

Figure 5.18.1

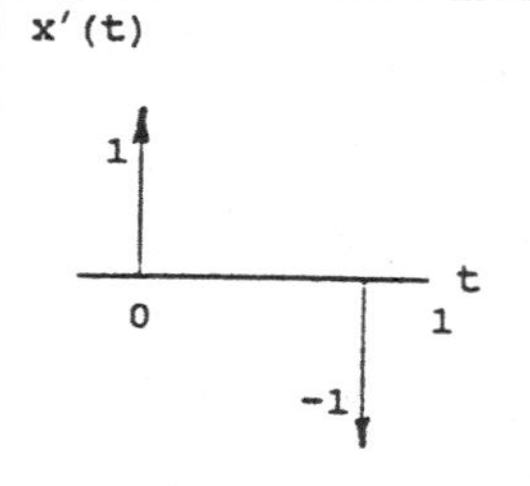

Figure 5.18.2

Because $x(t)$ has a finite span it is a Class 1 pulse.

Differentiation gives (see Figures): $x'(t) = \delta(t) - \delta(t-1)$

Transforming: $j\omega X(\omega) = 1 - \exp(-j\omega)$

from which: $$X(\omega) = \frac{1 - \exp(-j\omega)}{j\omega}$$

Because $x(t)$ is Class 1, this is its Fourier transform.

(b)

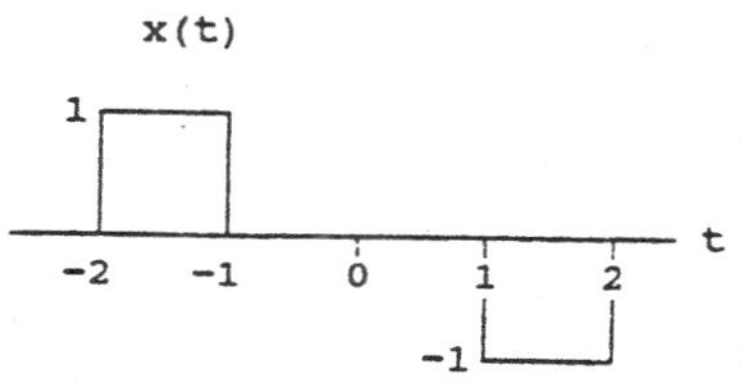

Figure 5.19.1

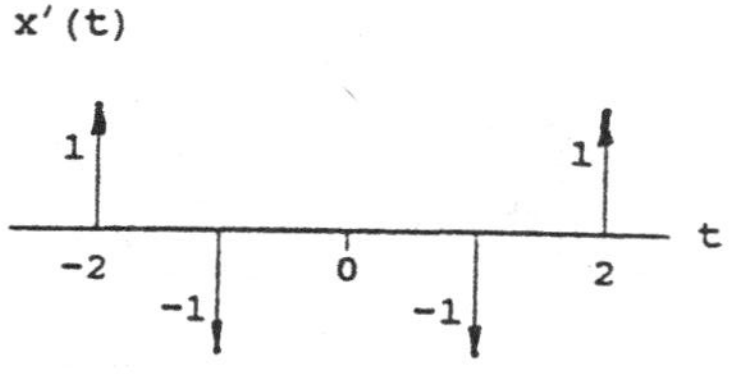

Figure 5.19.2

Because $x(t)$ has a finite span it is a Class 1 pulse. Differentiation gives (see Figures)

$$x'(t) = \delta(t+2) - \delta(t+1) - \delta(t-1) + \delta(t-2)$$

Transforming: $j\omega X(\omega) = \exp(j\omega 2) - \exp(j\omega) - \exp(-j\omega) + \exp(-j\omega 2)$

from which: $$X(\omega) = \frac{\exp(j\omega 2) - \exp(j\omega) - \exp(-j\omega) + \exp(-j\omega 2)}{j\omega}$$

Because $x(t)$ is Class 1, this is its Fourier transform.

(c)

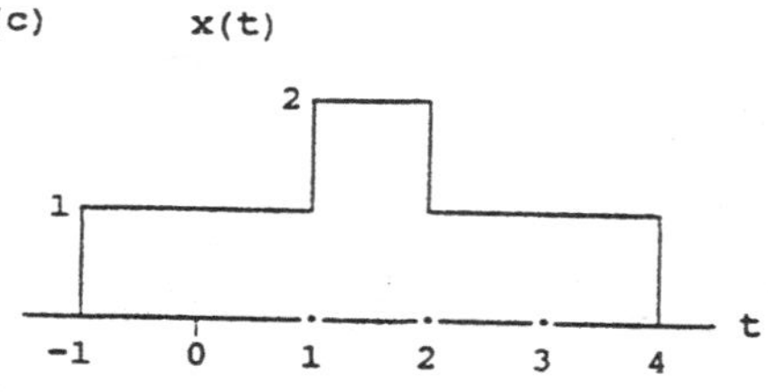

Figure 5.20.1

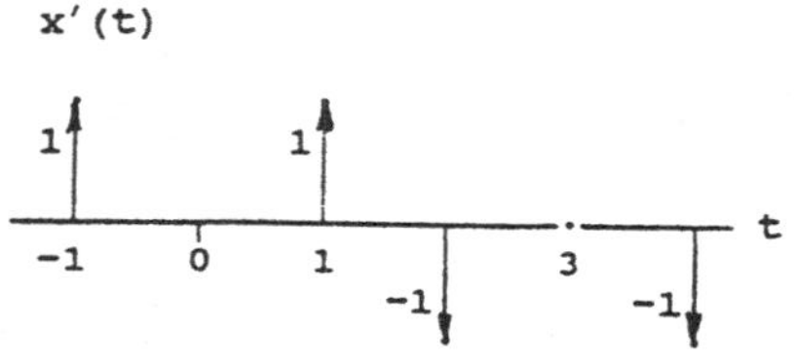

Figure 5.20.2

Because $x(t)$ has a finite span it is a Class 1 pulse. Differentiation gives (see Figures)

$$x'(t) = \delta(t+1) + \delta(t-1) - \delta(t-2) - \delta(t-4)$$

Transforming: $j\omega X(\omega) = \exp(j\omega) + \exp(-j\omega) - \exp(-j\omega 2) - \exp(-j\omega 4)$

from which: $X(\omega) = [\exp(j\omega) + \exp(-j\omega) - \exp(-j\omega 2) - \exp(-j\omega 4)]/(j\omega)$

Because $x(t)$ is Class 1, this is its Fourier transform.

(d)

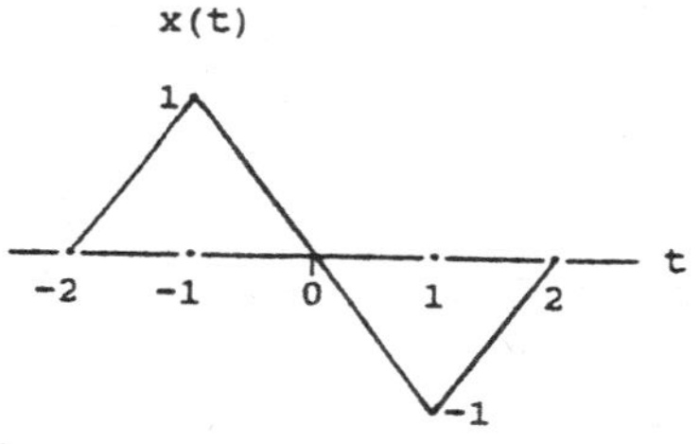

Figure 5.21.1

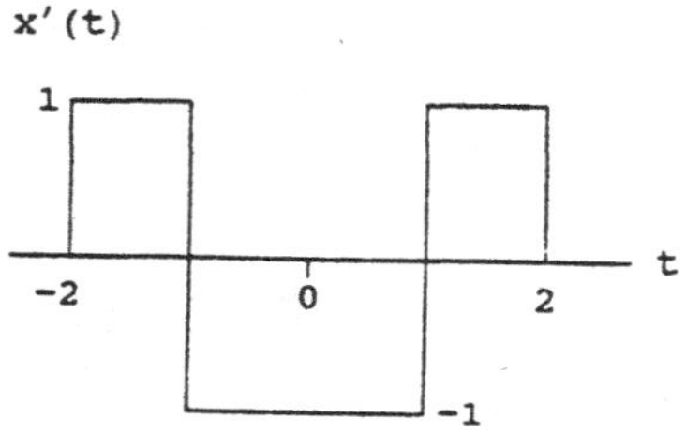

Figure 5.21.2

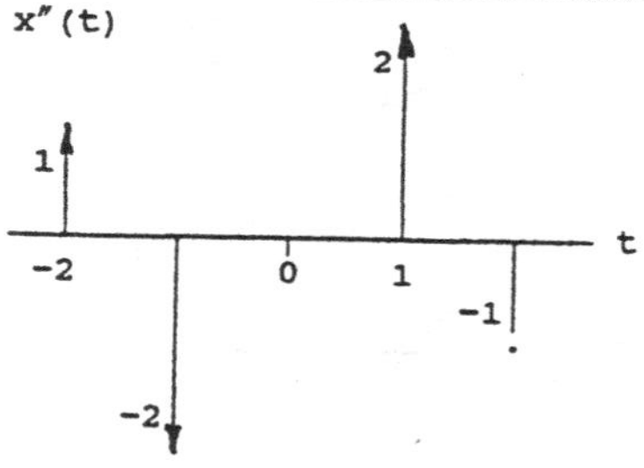

Figure 5.21.3

Because x(t) has a finite span
Differentiation gives (see Figures):

$$x''(t) = \delta(t + 2) - 2\delta(t + 1) + 2\delta(t - 1) - \delta(t - 2)$$

Transforming: $(j\omega)^2 X(\omega) = \exp(j\omega 2) - 2\exp(j\omega) + 2\exp(-j\omega) - \exp(-j\omega 2)$

from which: $X(\omega) = [\exp(j\omega 2) - 2\exp(j\omega) + 2\exp(-j\omega) - \exp(-j\omega 2)]/(j\omega)^2$

Because x(t) is Class 1, this is its Fourier transform.

(e)

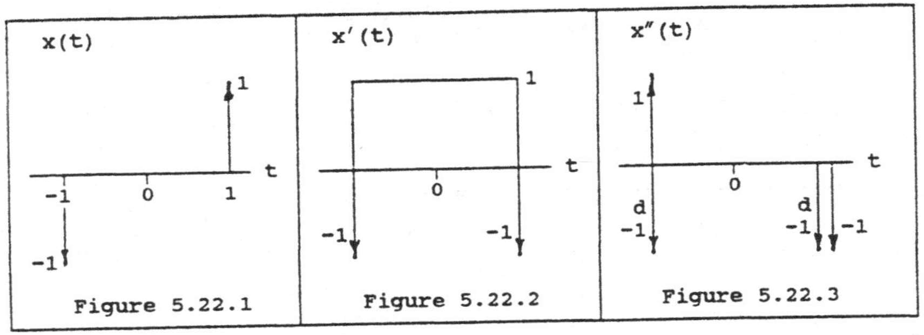

Figure 5.22.1 Figure 5.22.2 Figure 5.22.3

Because x(t) has a finite span it is a Class 1 pulse.
Differentiation gives (see Figures):

$$x''(t) = \delta(t + 1) - D\delta(t + 1) - \delta(t - 1) - D\delta(t - 1)$$

Transforming: $(j\omega)^2 X(\omega) = \exp(j\omega) - j\omega\exp(j\omega) - \exp(-j\omega) - j\omega\exp(-j\omega)$

from which: $X(\omega) = [\exp(j\omega) - j\omega\exp(j\omega) - \exp(-j\omega) - j\omega\exp(-j\omega)]/(j\omega)^2$

Because x(t) is Class 1, this is its Fourier transform.

(f) (See Figure 5.23 below.) Because x(t) has a finite span it is a Class 1 pulse. Differentiation gives (see Figures)

$$x'''(t) = 2D\delta(t + 1) - 2\delta(t + 1) + 2D\delta(t - 1) + 2\delta(t - 1)$$

Transforming:

$$(j\omega)^3 X(\omega) = 2j\omega\exp(j\omega) - 2\exp(j\omega) + 2j\omega\exp(-j\omega) + 2\exp(-j\omega)$$

from which:

$$X(\omega) = 2\,[j\omega\exp(j\omega) - \exp(j\omega) + j\omega\exp(-j\omega) + \exp(-j\omega)]/(j\omega)^3$$

Because x(t) is Class 1, this is its Fourier transform.

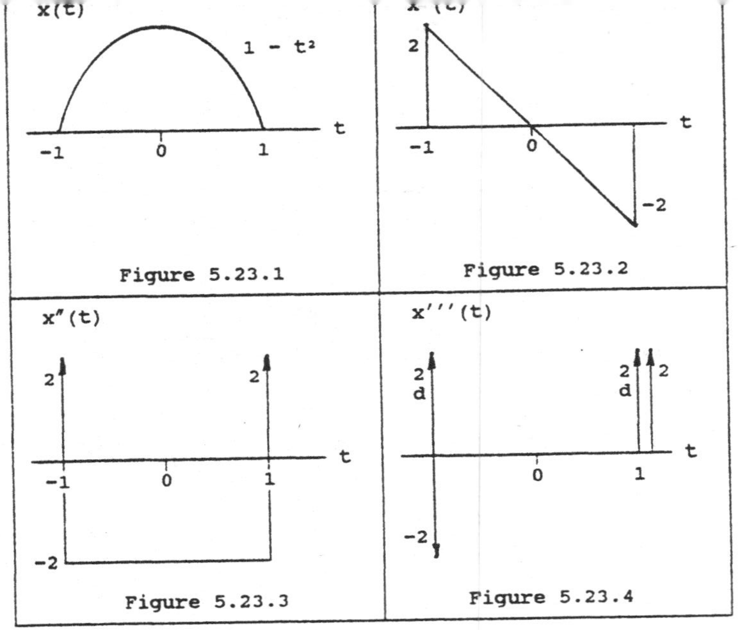

Figure 5.23.1 Figure 5.23.2

Figure 5.23.3 Figure 5.23.4

(g) (See Figure 5.24 below.) Because x(t) has a finite span it is a Class 1 pulse. Differentiation gives (see Figures)

$$x''(t) = \delta(t+1) - \delta(t) + D\delta(t-1) - D\delta(t-2) - \delta(t-3) + \delta(t+4)$$

Transforming: $(j\omega)^2 X(\omega) = \exp(j\omega) - 1 + j\omega\exp(-j\omega)$

$$- j\omega\exp(-j\omega 2) - \exp(-j\omega 3) + \exp(j\omega 4)$$

from which:

$$X(\omega) = \frac{\exp(j\omega) - 1 + j\omega\exp(-j\omega) - j\omega\exp(-j\omega 2) - \exp(-j\omega 3) + \exp(-j\omega 4)}{(j\omega)^2}$$

Because x(t) is Class 1, this is its Fourier transform.

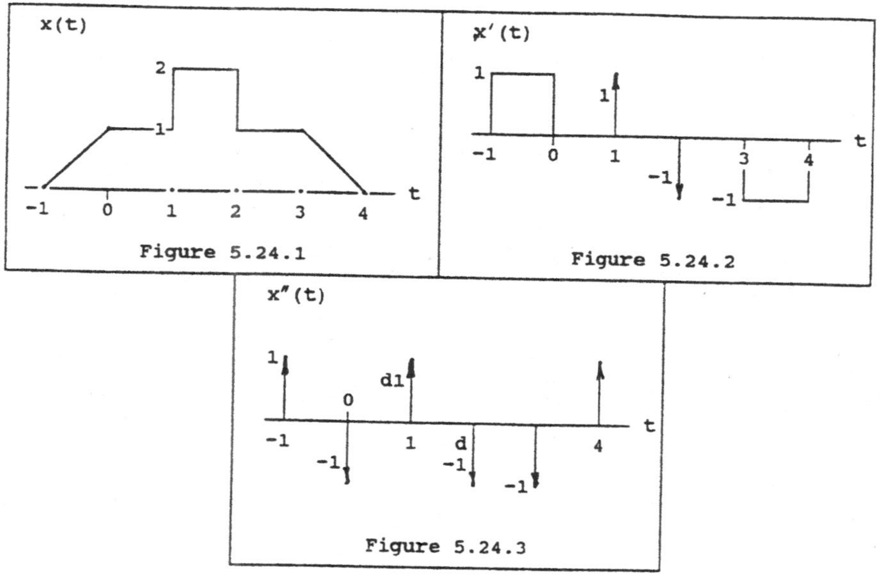

Figure 5.24.1

Figure 5.24.2

Figure 5.24.3

(h) (See Figure 5.25 below.) Because $x(t)$ has a finite span it is a Class 1 pulse. Differentiation gives

$$x''(t) = \delta(t-1) - 2D\delta(t-1) + 2D\delta(t-2) - 2D\delta(t-3) - \delta(t-3)$$

Transforming: $(j\omega)^2 X(\omega) = \exp(-j\omega) - 2j\omega \exp(-j\omega) + 2j\omega \exp(-j\omega 2)$

$$- 2j\omega \exp(-j\omega 3) - \exp(-j\omega 3)$$

from which $X(\omega)$

$$= \frac{\exp(-j\omega) - 2j\omega \exp(-j\omega) + 2j\omega \exp(-j\omega 2) - 2j\omega \exp(-j\omega 3) - \exp(-j\omega 3)}{(j\omega)^2}$$

Because $x(t)$ is Class 1, this is its Fourier transform.

(i) (See Figure 5.26 below.) Because $x(t)$ has an infinite span it is a Class 2 pulse. Its average over the entire t-axis is ½. Thus we shall have to add $\pi\,\delta(\omega)$ to the result obtained by successive differentiation.

Differentiation twice gives

$$x''(t) = \delta(t) + D\delta(t) - \delta(t-1) - \delta(t-2) + \delta(t-3)$$

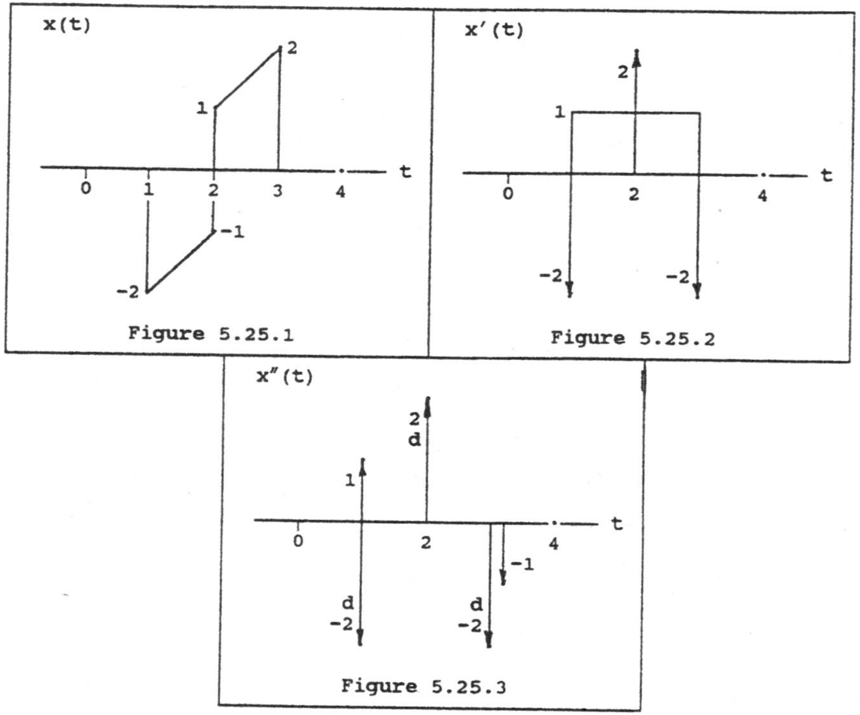

Figure 5.25.1

Figure 5.25.2

Figure 5.25.3

Transforming:

$$(j\omega)^2 X(\omega) = 1 + j\omega - \exp(-j\omega) - \exp(-j2\omega) + \exp(-j3\omega)$$

from which

$$X(\omega) = \frac{1 + j\omega - \exp(-j\omega) - \exp(-j2\omega) + \exp(-j3\omega)}{(j\omega)^2}$$

Because $x(t)$ is Class 2 with average value ½, we now add $\pi\delta(\omega)$ obtaining as the final result:

$$X(\omega) = \frac{1 + j\omega - \exp(-j\omega) - \exp(-j\omega 2) + \exp(-j\omega 3)}{(j\omega)^2} + \pi\delta(\omega)$$

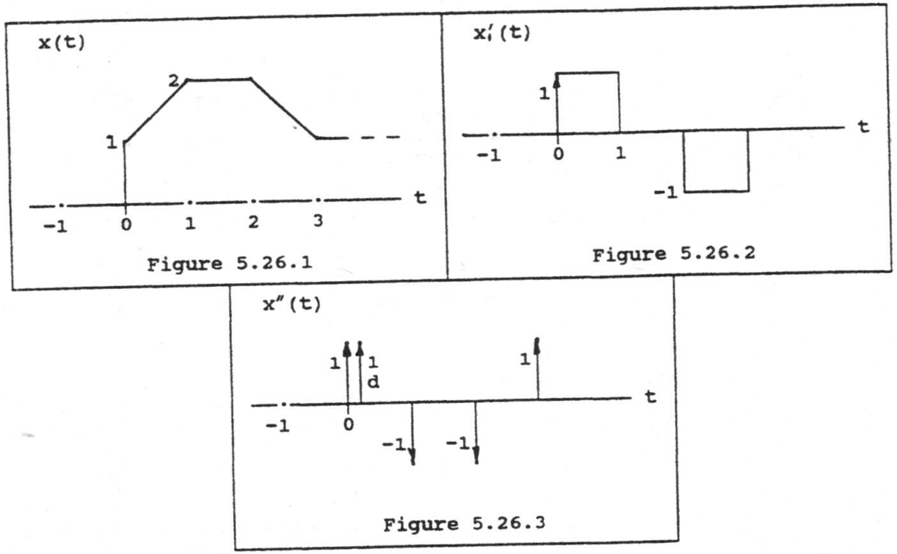

Figure 5.26.1

Figure 5.26.2

Figure 5.26.3

(j)

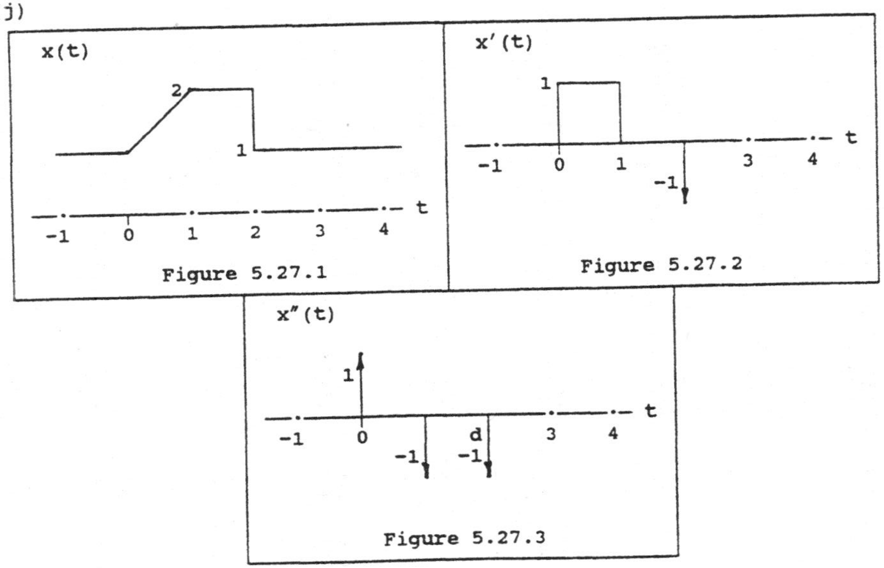

Figure 5.27.1

Figure 5.27.2

Figure 5.27.3

(j) Because $x(t)$ has an infinite span it is a Class 2 pulse. Its average over the entire t-axis is 1. Thus we shall have to add , $2\pi\delta(\omega)$ to the result obtained by successive differentiation.

Differentiation twice gives: $x''(t) = \delta(t) - \delta(t - 1) - D\delta(t - 2)$

Transforming: $(j\omega)^2 X(\omega) = 1 - \exp(-j\omega) - j\omega \exp(-j\omega 2)$

from which

$$X(\omega) = \frac{1 - \exp(-j\omega) - j\omega \exp(-j\omega 2)}{(j\omega)^2}$$

Because $x(t)$ is Class 2 with average value 1, we now add $2\pi\delta(\omega)$ obtaining as the final result:

$$X(\omega) = \frac{1 - \exp(-j\omega) - j\omega \exp(-j\omega 2)}{(j\omega)^2} + 2\pi\delta(\omega)$$

(k)

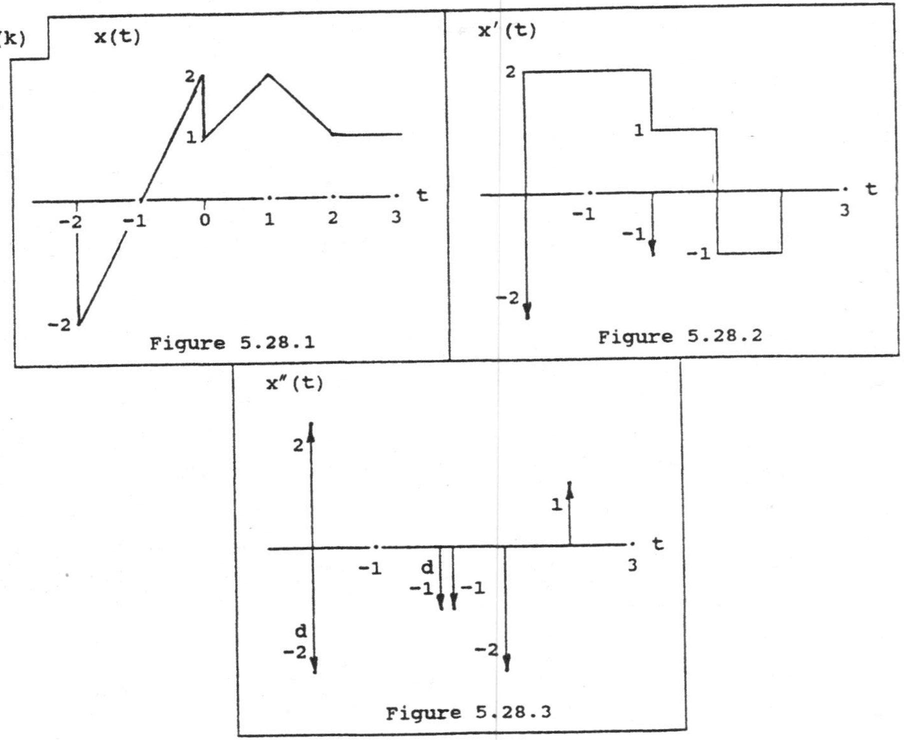

Figure 5.28.1

Figure 5.28.2

Figure 5.28.3

(k) Because $x(t)$ has an infinite span it is a Class 2 pulse. Its average over the entire t-axis is $\frac{1}{2}$. Thus we shall have to add $\pi\delta(\omega)$ to the result obtained by successive differentiation.

Differentiation twice gives:

$$x''(t) = 2\delta(t+2) - 2D\delta(t+2) - \delta(t) - D\delta(t) - 2\delta(t-2) + \delta(t-3)$$

Transforming:

$$(j\omega)^2 X(\omega) = 2\exp(j2\omega) - 2j\omega \exp(j2\omega) - 1 - j\omega - 2\exp(-j\omega) + \exp(-j3\omega)$$

from which

$$X(\omega) = \frac{2\exp(j2\omega) - 2j\omega \exp(j2\omega) - 1 - j\omega - 2\exp(-j\omega) + \exp(-j3\omega)}{(j\omega)^2}$$

Because $x(t)$ is Class 2 with average value $\frac{1}{2}$, we now add $\pi\delta(\omega)$ obtaining as the final result:

$$X(\omega) = \frac{2\exp(j2\omega) - 2j\omega \exp(j2\omega) - 1 - j\omega - 2\exp(-j\omega) - \exp(-j3\omega)}{(j\omega)^2} + \pi\delta(\omega)$$

(l) (See Figure 5.29 below.) Because $x(t)$ has an infinite span it is a Class 2 pulse. However its average over the entire t-axis is 0. Thus we shall not have to add anything to the result obtained by successive differentiation.

Differentiation twice gives:

$$x''(t) = -\delta(t+4) + \delta(t+3) + D\delta(t+3) + D\delta(t+2) + D\delta(t+1) + D\delta(t) - \delta(t) + \delta(t-1)$$

Transforming:

$$(j\omega)^2 X(\omega) = -\exp(j\omega 4) + \exp(j\omega 3) + j\omega \exp(j\omega 3) + j\omega \exp(j\omega 2) + j\omega \exp(j\omega) + j\omega - 1 + \exp(-j\omega)$$

from which

$$X(\omega) = \left[\frac{-\exp(j\omega 4) + \exp(j\omega 3) + j\omega \exp(j\omega 3) + j\omega \exp(j\omega 2) + j\omega \exp(j\omega) + j\omega - 1 + \exp(-j\omega)}{(j\omega)^2} \right]$$

Although $x(t)$ is Class 2, its average value is 0. Thus this is the final result.

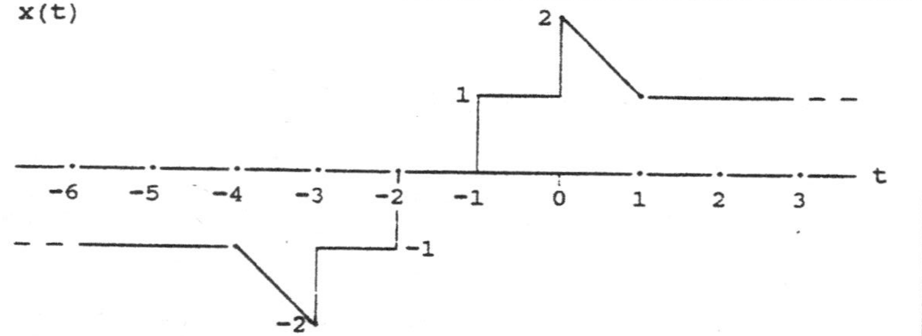

Figure 5.29.1

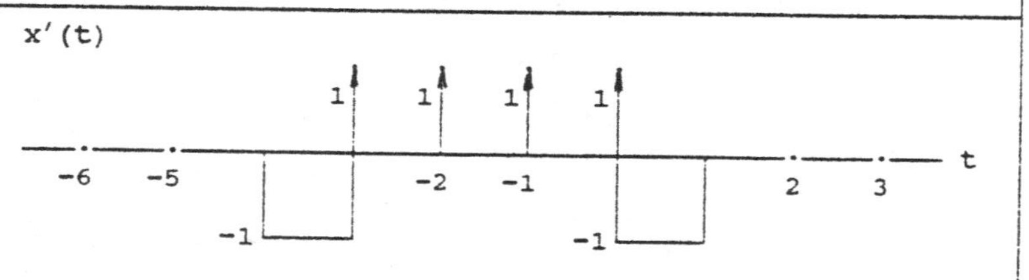

Figure 5.29.2

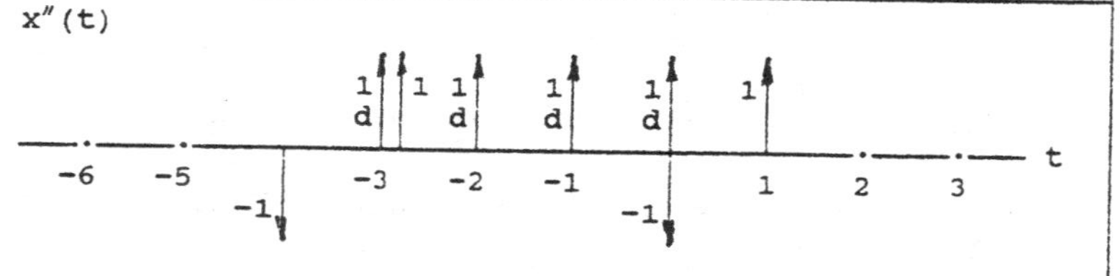

Figure 5.29.3

5.2 (a)(1) We know that $\left[\exp(-\beta t)\, U(t) \quad (\beta > 0) \right] \Longleftrightarrow \dfrac{1}{\beta + j\omega}$

Then $D\left[\exp(-\beta t)\, U(t) \right] \Longleftrightarrow j\omega \left[\dfrac{1}{\beta + j\omega} \right] = \dfrac{j\omega}{\beta + j\omega}$ (A)

(2) Differentiating $\exp(-\beta t)\, U(t)$ we use the product rule, obtaining:

$$D\left[\exp(-\beta t)\, U(t) \right] = D[\exp(-\beta t)]\, U(t) + \exp(-\beta t)\, DU(t)$$
$$= -\beta \exp(-\beta t)\, U(t) + \exp(-\beta t)\, \delta(t)$$

$$\Longleftrightarrow -\beta \frac{}{\beta + j\omega} + 1 = \frac{}{\beta + j\omega} = \frac{}{\beta + j\omega} = \frac{}{\beta + j\omega}$$

which is the same as (A).

(b) (1) We know that $\left[t \exp(-\beta t)\, U(t) \quad (\beta > 0) \right] \Longleftrightarrow \dfrac{1}{(\beta + j\omega)^2}$

Then $$D\left[t \exp(-\beta t)\, U(t) \right] \Longleftrightarrow j\omega \left[\frac{1}{(\beta + j\omega)^2} \right] = \frac{j\omega}{(\beta + j\omega)^2} \qquad \text{(B)}$$

Thus the transform shown in (B) is in fact the final answer.

(2) Differentiating $t \exp(-\beta t)\, U(t)$ we use the product rule, obtaining:

$$D\left[t \exp(-\beta t)\, U(t) \right] = D[t \exp(-\beta t)]\, U(t) + \exp(-\beta t)\, DU(t)$$

$$= [\exp(-\beta t) - \beta t \exp(-\beta t)]\, U(t) + t \exp(-\beta t)\, \delta(t)$$

$$= [\exp(-\beta t) - \beta t \exp(-\beta t)]\, U(t)$$ (final term zero due to Dirac-delta sampling)

$$\Longleftrightarrow \frac{1}{\beta + j\omega} - \frac{\beta}{(\beta + j\omega)^2} = \frac{\beta + j\omega - \beta}{(\beta + j\omega)^2} = \frac{j\omega}{(\beta + j\omega)^2}$$

which is the same as (B).

5.3 (a) Differentiating $B_k(t)$ we obtain the pair of Dirac deltas shown below:

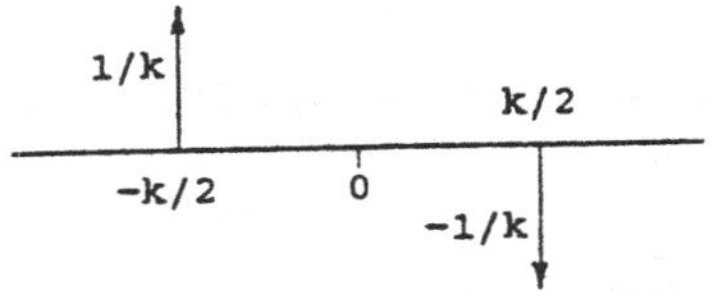

$$\text{i.e. } D\, B_k(t) = \frac{\delta(t + k/2) - \delta(t - k/2)}{k}$$

(b) Fourier transforming gives $$D\, B_k(t) \Longleftrightarrow \frac{\exp(j\omega k/2) - \exp(-j\omega k/2)}{k}$$

$$= j\omega \frac{\exp(j\omega k/2) - \exp(-j\omega k/2)}{2j\omega k/2} = j\omega\, Sa(\omega k/2)$$

i.e. $$\lim_{k \to 0} D\, B_k(t) = D\delta(t) \Longleftrightarrow j\omega$$

5.4 We know that $D\delta(t) \Longleftrightarrow j\omega$ and so

$$F(\omega) = \omega^2 = (-j\omega)^2 = -(j\omega)(j\omega)1 \Longleftrightarrow -D^2\delta(t)$$

$D^2\delta(t)$ is the second derivative of the Dirac delta, or the first derivative of the unit doublet.

5.5 (a) From Figure 5.31 we see that the analytical definition for $\Lambda_k(t)$ is

$$\Lambda_k(t) = \begin{cases} (t + k)/k^2 & (-k < t < 0) \\ (t - k)/k^2 & (0 < t < k) \end{cases}$$

Differentiation gives $$x_k(t) = \begin{cases} 1/k^2 & (-k < t < 0) \\ -1/k^2 & (0 < t < k) \end{cases}$$

which is the same as the pulse shown in Figure 5.31.

(b) Differentiating $x_k(t)$ we obtain three Dirac deltas

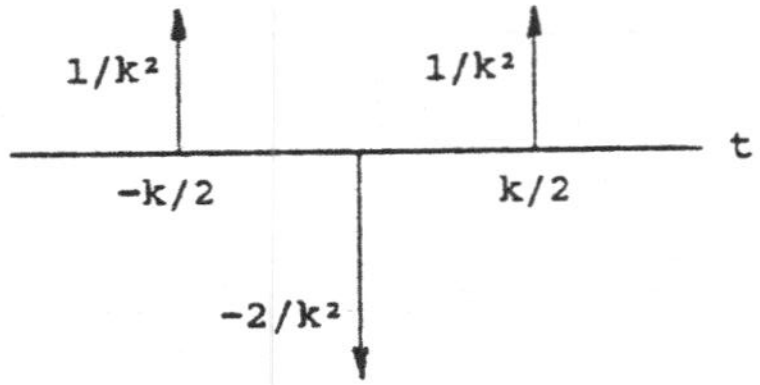

with analytical definition as follows:

$$Dx_k(t) = (1/k^2)\, \delta(t + k) - (2/k^2)\, \delta(t) + (1/k^2)\, \delta(t - 1)$$

which transform as follows:

$$Dx_k(t) \Longleftrightarrow (1/k^2) \exp(j\omega k) - (2/k^2) + (1/k^2) \exp(-j\omega k)$$

$$= \frac{1}{k^2} \left[\exp(j\omega k) - 2 + \exp(-j\omega k) \right]$$

giving $$j\omega\, X_k(\omega) = \frac{1}{k^2} \left[\exp(j\omega k) - 2 + \exp(-j\omega k) \right]$$

Rearranging this we obtain

$$X_k(\omega) = j\omega \left[\frac{\exp(j\omega k/2) - \exp(-j\omega k/2)}{2j\omega k/2} \right]^2 = j\omega\, Sa^2(\omega k/2)$$

Letting $k \longrightarrow 0$ we obtain $\lim_{k \to 0} x_k(t) = j\omega$ QED

(c) From Figure 5.31 we see that the sequence function for the doublet is an odd pulse. Then the limit as $k \longrightarrow 0$ must also be an odd pulse. Thus the doublet is real and odd.

In (b) we obtained a Fourier transform for the doublet which is real and odd. From our theorems in Chapter 3 (see the corollary to Theorem 3.3) these are consistent.

(d) Sketching $x_k(t)$ for very small k we obtain

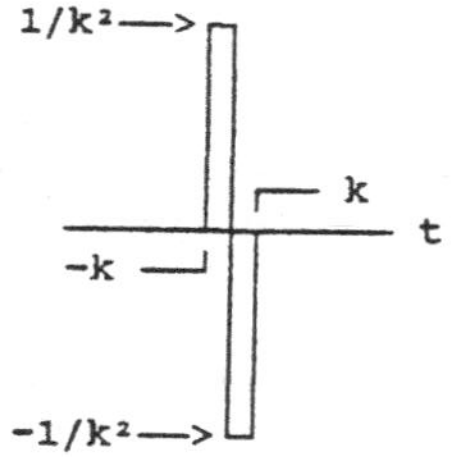

i.e. two very high, very narrow rectangular pulses, one going positive and the other negative, each of width k and height $1/k^2$. The area of each pulse is $1/k$ and so each pulse's area becomes infinite as $k \longrightarrow 0$. However the nett area under the overall pulse $x_k(t)$ is always zero for every value of k.

5.6 (a) Forming the derivative of $\mu(t)$ we obtain the pulse $\mu'(t)$ shown below.

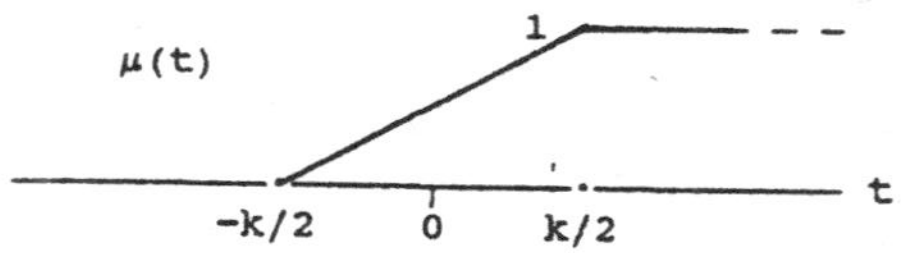

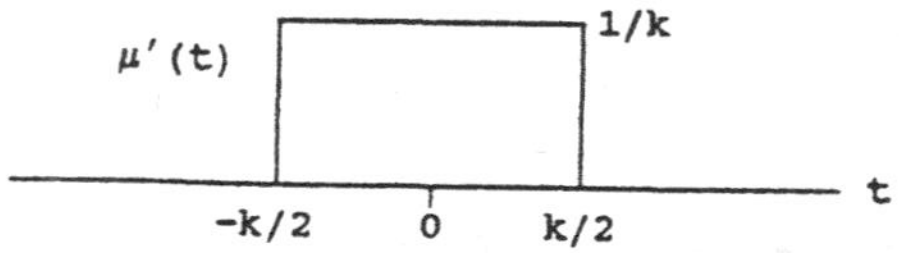

Its analytical definition is $\mu'(t) = (1/k)\ \mathrm{Rect}(t/k)$

Letting $k \longrightarrow 0$ we know that (see equation (4.37))

$$\lim_{k \to 0} \mu'(t) = \delta(t)$$

However, for $k = 0$ we have (see first figure above) $\mu(t) = U(t)$
We have thus shown that

$$D\,U(t) = \delta(t)$$ QED

5.7 (a) The pulse $f(t) = \exp(-\beta t)\ U(t)$ ($\beta > 0$) has an infinite span. To find its average over the entire t-axis we have

$$\lim_{T \to \infty} \frac{1}{2T} \int_0^T \exp(-\beta t)\, dt = \lim_{T \to \infty} \frac{\exp(-\beta t)}{-2T\beta} \Bigg|_0^T = \lim_{T \to \infty} \frac{\exp(-\beta T) - 1}{-2T\beta} = 0$$

Thus its average over the entire time line is zero.

(b) Applying the method of successive differentiation, we use the product rule as follows:

$$Df(t) = D[\exp(-\beta t)\ U(t)] = D[\exp(-\beta t)]\ U(t) + \exp(-\beta t)\ DU(t)$$

$$= -\beta \exp(-\beta t)\ U(t) + \exp(-\beta t)\ \delta(t)$$

$$= -\beta \exp(-\beta t)\ U(t) + \delta(t) \quad \text{(Dirac delta sampling)}$$

$$= -\beta f(t) + \delta(t)$$

and so $(D + \beta)\ f(t) = \delta(t)$

Transforming: $(j\omega + \beta)\ F(\omega) = 1$

i.e. $$F(\omega) = \frac{1}{j\omega + \beta} \qquad (A)$$

(c) Although the pulse is Class 2, its average over the time line is zero. Thus we do not need to add a frequency-domain Dirac delta to the result in (A). That is in fact its Fourier transform.

5.8 (a) In the figure below we show a sketch of the pulse

$f(t) = \cos(\pi t)\ \mathrm{Rect}(t)$

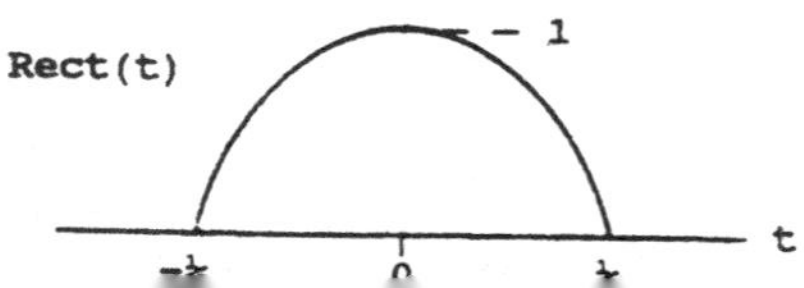

Finding its Fourier transform using the analysis equation:

$$F(\omega) = \int_{-\infty}^{\infty} f(t)\ \exp(-j\omega t)\ dt = \int_{-\infty}^{\infty} \cos(\pi t)\ \text{Rect}(t)\ \exp(-j\omega t)\ dt$$

$$= \int_{-1/2}^{1/2} \cos(\pi t)\ \exp(-j\omega t)\ dt = \int_{-1/2}^{1/2} \frac{\exp(j\pi t) + \exp(-j\pi t)}{2} \exp(-j\omega t)\ dt$$

$$= \tfrac{1}{2}\int_{-1/2}^{1/2} \left[\exp[j(\pi-\omega)t] + \exp[-j(\pi+\omega)t] \right] dt$$

$$= \tfrac{1}{2}\left[\frac{\exp[j(\pi-\omega)t]}{j(\pi-\omega)} + \frac{\exp[-j(\pi+\omega)t]}{-j(\pi+\omega)} \right]_{-1/2}^{1/2}$$

$$= \frac{\exp[j(\pi-\omega)/2] - \exp[-j(\pi-\omega)/2]}{2j(\pi-\omega)} - \frac{\exp[j(\pi+\omega)/2] - \exp[-j(\pi+\omega)/2]}{-2j(\pi+\omega)}$$

$$= \frac{\sin[(\pi-\omega)/2]}{\pi-\omega} + \frac{\sin[(\pi+\omega)/2]}{\pi+\omega} = \frac{\sin(\pi/2)\ \cos(\omega/2)}{\pi - \omega} + \frac{\sin(\pi/2)\ \cos(\omega/2)}{\pi + \omega}$$

$$= \cos(\omega/2) \left[\frac{1}{\pi - \omega} + \frac{1}{\pi + \omega} \right] = \frac{2\pi\ \cos(\omega/2)}{\pi^2 - \omega^2} \qquad \text{(A)}$$

(b) Using successive differentiation, we differentiate twice using the product rule:

$$f(t) = \cos(\pi t)\ \text{Rect}(t)$$

$$Df(t) = D[\cos(\pi t)]\ \text{Rect}(t) + \cos(\pi t)\ D[\text{Rect}(t)]$$

$$= -\pi\ \sin(\pi t)\ \text{Rect}(t) + \cos(\pi t)\ [\delta(t + \tfrac{1}{2}) - \delta(t - \tfrac{1}{2})]$$

$$= -\pi\ \sin(\pi t)\ \text{Rect}(t)$$ (final term is zero because of Dirac-delta sampling)

$$D^2 f(t) = D[-\pi\ \sin(\pi t)]\ \text{Rect}(t) - \pi\ \sin(\pi t)\ D[\text{Rect}(t)]$$

$$= -\pi^2\ \cos(\pi t)\ \text{Rect}(t) - \pi\ \sin(\pi t)\ [\delta(t + \tfrac{1}{2}) - \delta(t - \tfrac{1}{2})]$$

$$= -\pi^2\ \cos(\pi t)\ \text{Rect}(t) - \pi\ \sin(-\pi/2)\ \delta(t+\tfrac{1}{2}) + \pi\ \sin(\pi/2)\ \delta(t-\tfrac{1}{2})$$

$$= -\pi^2\ f(t) + \pi\ \delta(t + \tfrac{1}{2}) + \pi\ \delta(t - \tfrac{1}{2})$$

and so $(D^2 + \pi^2)\ f(t) = \pi\ \delta(t + \tfrac{1}{2}) + \pi\ \delta(t - \tfrac{1}{2})$

Transforming:

$$[(j\omega)^2 + \pi^2]\ F(\omega) = \pi \left[\exp(j\omega/2) + \exp(-j\omega/2) \right] = 2\pi\ \cos(\omega/2)$$

and so $F(\omega) = \dfrac{2\pi\ \cos(\omega/2)}{\pi^2 - \omega^2}$

which is the same result as in (A).

5.8 (a) In the figure below we show a sketch of the pulse

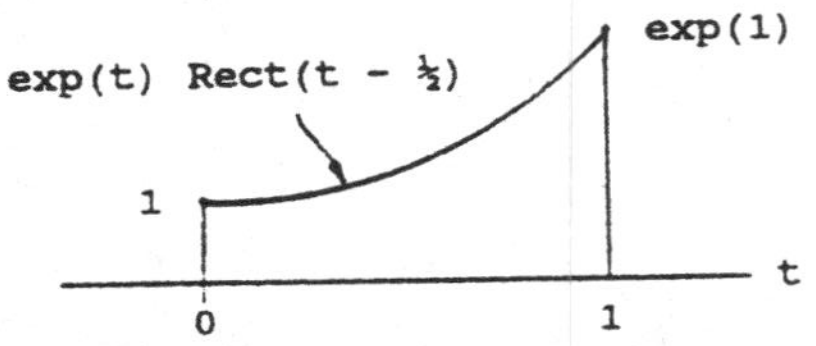

Finding its Fourier transform using successive differentiation, we differentiate using the product rule:

$$f(t) = \exp(t)\ \text{Rect}(t - \tfrac{1}{2})$$

$$Df(t) = D[\exp(t)]\ \text{Rect}(t - \tfrac{1}{2}) + \exp(t)\ D[\text{Rect}(t - \tfrac{1}{2})]$$

$$= \exp(t)\ \text{Rect}(t - \tfrac{1}{2}) + \exp(t)\ [\delta(t) - \delta(t - 1)]$$

$$= f(t) + \delta(t) - \exp(1)\ \delta(t - 1)$$ (Dirac-delta sampling)

Then $(D - 1)\ f(t) = \delta(t) - \exp(1)\ \delta(t - 1)$

Transforming:

$$(j\omega - 1)\ F(\omega) = 1 - \exp(1)\ \exp(-j\omega) = 1 - \exp(1 - j\omega)$$

$$F(\omega) = \frac{\exp(1 - j\omega) - 1}{1 - j\omega}$$

5.10 (a) A sketch of the pulse $f(t) = \exp(\beta t)\ U(-t) + \exp(-\beta t)\ U(t) + 1$ is shown below:

(b) Using successive differentiation, we split the pulse into its three components. For the first component:

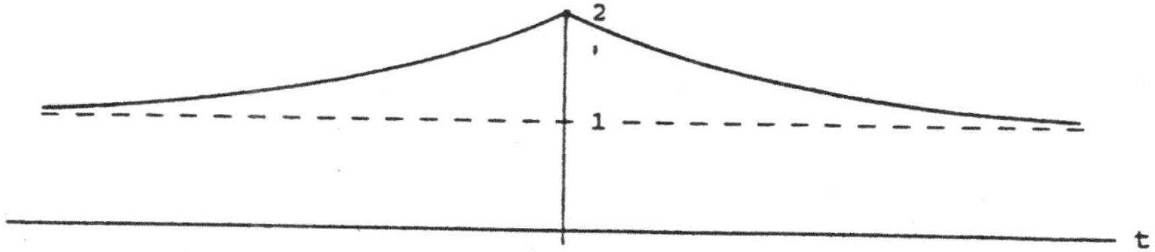

$f_1(t) = \exp(\beta t)\ U(-t) \quad (\beta > 0)$

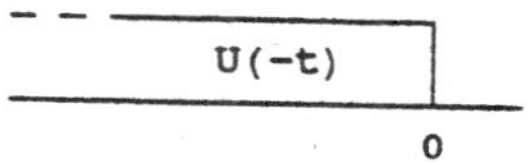

Differentiating by the product rule:

$Df_1(t) = D[\exp(\beta t)]\ U(-t) + \exp(\beta t)\ D\ U(-t)$

$= \beta \exp(\beta t)\ U(-t) + \exp(\beta t)\ [-\delta(t)] = \beta \exp(\beta t)\ U(-t) - \exp(\beta t)\ \delta(t)$

$= \beta f_1(t) - \delta(t)$ (Dirac-delta sampling)

$(D - \beta)\ f_1(t) = -\delta(t) \Longleftrightarrow (j\omega - \beta)\ F_1(\omega) = -1$

$$F_1(\omega) = \frac{1}{\beta - j\omega} \qquad \text{(A)}$$

Although $f_1(t)$ is in Class 2 its average over the entire time line is zero. (See Exercise 5.7.) Thus (A) is in fact its Fourier transform.

For the second component:

$f_2(t) = \exp(-\beta t)\ U(t) \quad (\beta > 0)$

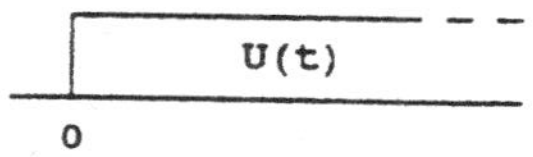

Differentiating by the product rule:

$Df_2(t) = D[\exp(-\beta t)]\ U(t) + \exp(-\beta t)\ D\ U(t)$

$= -\beta \exp(-\beta t)\ U(t) + \exp(-\beta t)\ \delta(t) = -\beta \exp(-\beta t)\ U(t) + \delta(t)$ (Dirac sampling)

$= -\beta f_2(t) + \delta(t)$

$(D + \beta)\ f_2(t) = \delta(t) \Longleftrightarrow (j\omega + \beta)\ F_2(\omega) = 1$

$$F_2(\omega) = \frac{1}{j\omega + \beta} \qquad \text{(B)}$$

Although $f_2(t)$ is also in Class 2 its average over the entire time line is zero. (See Exercise 5.7.) Thus (B) is in fact its Fourier transform.

For the third component:

$f_3(t) = 1 \qquad Df_3(t) = 0 \qquad j\omega\ F_3(\omega) = 0 \qquad F_3(\omega) = 0$

However the pulse in Class 2 and its average over the entire time line is 1. Thus to the result obtained from successive differentiation we must add the function $2\pi\delta(\omega)$, obtaining:

$F_3(\omega) = 2\pi\delta(\omega)$

Combining all three transforms:

$F(\omega) = F_1(\omega) + F_2(\omega) + F_3(\omega)$

$$= \frac{1}{\beta - j\omega} + \frac{1}{\beta + j\omega} + 2\pi\delta(\omega) = \frac{2\beta}{\beta^2 + \omega^2} + 2\pi\delta(\omega)$$

5.11 A sketch of the derivative of the pulse in Figure 5.24 is shown below.

Figure 5.24

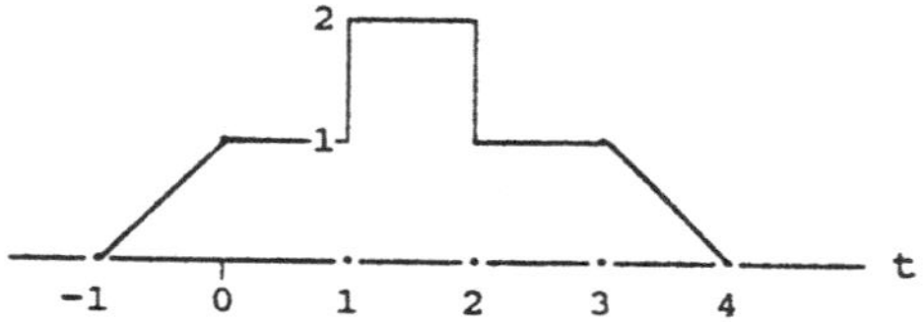

Derivative of pulse

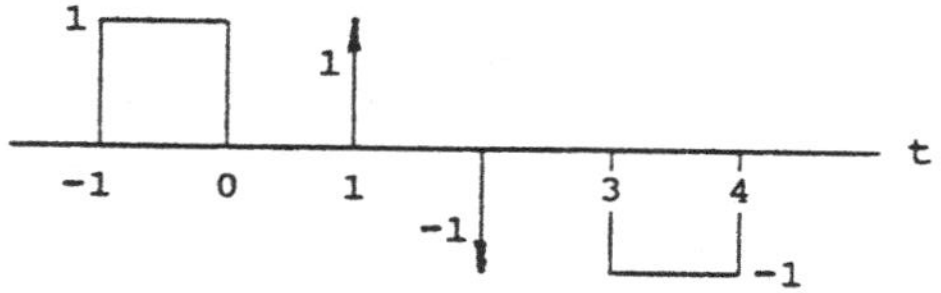

After following the directions given in the exercise we obtained a plot of Y that looks as follows:

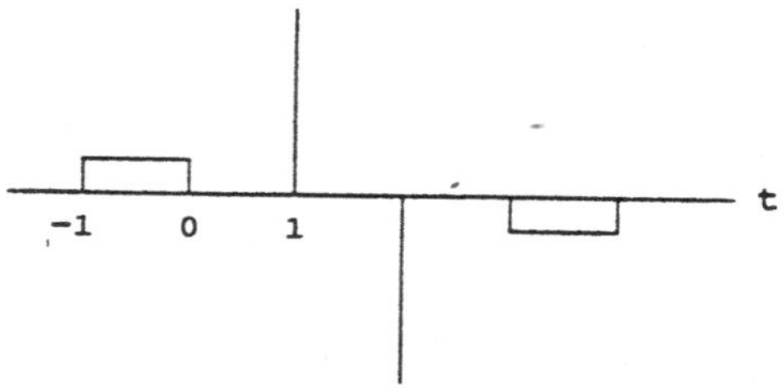

The heights of the Dirac lines were ±40.7437. We expected them to represent Dirac deltas of weights ±1.

From the formula $V = \mu N/T$ with $\mu = 1$, $N = 512$ and $T = 4\pi$, we obtain $V = 40.74366543$,

their locations are at k ... the point $t = 1$ should be at $k = 40.74366 \approx 41$. Thus the FFT has created the Dirac delta at the next integer value. Similarly for the negative-going Dirac delta.

5.12 Repeating Exercise 5.11 but using the **F** postprocessor "Forward-Diff 1st" we obtained the same numerical values, except that now the Dirac deltas are at $k = 40$ and $k = 81$. This differencing algorithm has put them on the left side of the exact values.

Repeating Exercise 5.11 but using the **F** postprocessor "Central-Diff 1st" we obtained the Dirac deltas split into two parts. For the first one, half was at $k = 40$ and half at $k = 41$. The second one was split between $k = 80$ and $k = 81$.

These are the facts of life regarding these various differencing algorithms.

5.13 To load the pulse shown in Figure 5.34, using N = 500, SAMPLED, T = 10, PULSE:

Main menu, Create Values, Continue, Create X, Real, Even, 3 intervals in the right half, 1, 2, Continue, 1, 2-t, 0, Go, Neither, Go. You are now back on the main menu. Plotting **X** shows that we have loaded the pulse correctly.

Run ANALYSIS, Run Postprocessors, F, Central-Diff 2nd, Run SYNTHESIS. Plotting **Y** shows four Dirac lines, of height ±50.

We expected Dirac deltas of weights 1. From $V = \mu N/T$ we have

$$V = 1 \times 500 / 10 = 50$$

and so the four lines have the correct lengths.

Chapter 6 Frequency-Domain Analysis

6.1 Frequency-domain proof of Theorem 6.6:

$y(t)$ is the response to $x(t)$ means that

$$Y(\omega) = H(j\omega)\ X(\omega)$$

and so, multiplying both sides by $j\omega$ means that

$$j\omega\ Y(\omega) = H(j\omega)\ j\omega\ X(\omega)$$

Inversion to the time domain then gives

$$y'(t) = 1/2\pi \int_{-\infty}^{\infty} H(j\omega) \left[j\omega\ X(\omega) \right] \exp(j\omega t)\ dt \qquad (A)$$

But $j\omega\ X(\omega) \Longleftrightarrow x'(t)$ and so (A) means that $x'(t)$ entering the network produces $y'(t)$. QED

(b) Time-domain proof: $x(t)$ entering produces the response $y(t)$. Thus the system DE is $P_1(D)\ y(t) = P_2(D)\ x(t)$.

Differentiating both sides gives

$$D\ P_1(D)\ y(t) = D\ P_2(D)\ x(t)$$

But $D\ P_1(D) = P_1(D)\ D$ and $D\ P_2(D) = P_2(D)\ D$ and so

$$P_1(D)\ Dy(t) = P_2(D)\ Dx(t)$$

which means that $x'(t)$ entering produces the response $y'(t)$.

6.2 (a) For network (a): $H(j\omega) = \dfrac{1/j\omega}{1 + 1/j\omega} = \dfrac{1}{1 + j\omega}$

(i) $[x(t) = \delta(t)] \Longleftrightarrow [X(\omega) = 1]$ $\quad Y(\omega) = H(j\omega)\ X(\omega) = \dfrac{1}{1 + j\omega}$

Inversion to the time domain gives $y(t) = \exp(-t)\ U(t)$

(ii) $[x(t) = U(t)] \Longleftrightarrow [X(\omega) = \dfrac{1}{j\omega} + \pi\delta(\omega)]$

$$Y(\omega) = H(j\omega)\ X(\omega) = \frac{1}{1 + j\omega} \left[\frac{1}{j\omega} + \pi\delta(\omega) \right]$$

$$= \left[\frac{1}{(1 + j\omega)j\omega} \right] + \frac{1}{1 + j\omega}\pi\delta(\omega) = \left[\frac{-1}{1 + j\omega} + \frac{1}{j\omega} \right] + \pi\delta(\omega) \quad \text{(Dirac-delta sampling)}$$

$$= \underbrace{\frac{-1}{1 + j\omega}}_{A} + \underbrace{\left[\frac{1}{j\omega} + \pi\delta(\omega) \right]}_{B}$$

Inverting (A) to the time domain gives $y_1(t) = -\exp(-t)\ U(t)$

and (B) gives $y_2(t) = U(t)$

and so the response is $y(t) = [1 - \exp(-t)]\ U(t)$

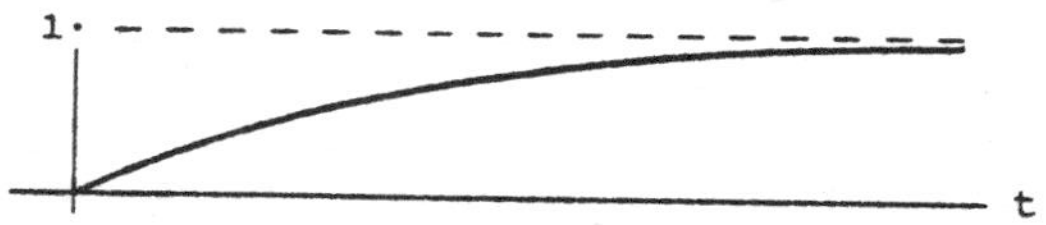

For network (b): $H(j\omega) = \dfrac{1}{1 + 1/j\omega} = \dfrac{j\omega}{1 + j\omega}$

(i) $[x(t) = \delta(t)] \Longleftrightarrow [X(\omega) = 1]$

$$Y(\omega) = H(j\omega)\ X(\omega) = \frac{j\omega}{1 + j\omega} = 1 - \frac{1}{1 + j\omega}$$

Inversion to the time domain gives $y(t) = \delta(t) - \exp(-t)\ U(t)$

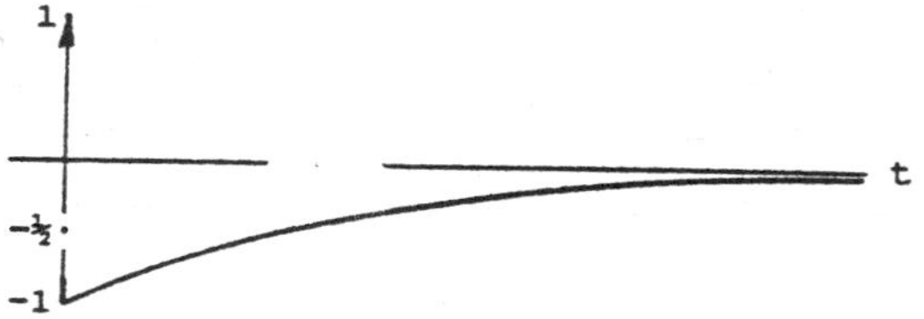

(ii) $[x(t) = U(t)] \Longleftrightarrow X(\omega) = \left[\frac{1}{j\omega} + \pi\delta(\omega) \right]$

$$Y(\omega) = H(j\omega)\ X(\omega) = \frac{j\omega}{1 + j\omega} \left[\frac{1}{j\omega} + \pi\delta(\omega) \right]$$

$$= \frac{j\omega}{j\omega(1 + j\omega)} + \frac{j\omega}{1 + j\omega}\pi\delta(\omega) = \frac{1}{1 + j\omega} + 0 \quad \text{(Dirac-delta sampling)}$$

Inverting (A) to the time domain gives $y(t) = \exp(-t)\ U(t)$

(b) The input $\delta(t)$ is the derivative of the input $U(t)$. According to Theorem 6.6 that we proved in Exercise 6.1, the response to $\delta(t)$ should be the derivative of the response to $U(t)$.

For network (a) the response to $U(t)$ was $y(t) = [1 - \exp(-t)]\ U(t)$

Then, differentiating using the product rule:

$$\frac{d}{dt} y(t) = \frac{d}{dt} \Big[[1 - \exp(-t)]\ U(t) \Big]$$

$$= \frac{d}{dt} \Big[1 - \exp(-t) \Big] U(t) + [1 - \exp(-t)] \frac{d}{dt} U(t)$$

$$= [0 + \exp(-t)]\ U(t) + [1 - \exp(-t)]\ \delta(t)$$

$$= \exp(-t)\ U(t) + [1 - 1]\ \delta(t) \quad \text{(Dirac-delta sampling)}$$

$$= \exp(-t)\ U(t)$$

which was the response to $\delta(t)$ QED

For network (b) the response to $U(t)$ was $y(t) = \exp(-t)\ U(t)$

Then, differentiating using the product rule:

$$\frac{d}{dt} y(t) = \frac{d}{dt}\Big[\exp(-t)\ U(t) \Big] = \frac{d}{dt}\Big[\exp(-t) \Big] U(t) + \exp(-t) \frac{d}{dt} U(t)$$

(Dirac-delta sampling)

QED

which was the response to $\delta(t)$

6.3 The statement $D \exp(j\omega t) = j\omega \exp(j\omega t)$ (A)

and Theorem 5.1, namely $D\ f(t) \Longleftrightarrow j\omega\ F(\omega)$ (B)

are fully equivalent.

Proof: $D \exp(j\omega t) = j\omega \exp(j\omega t)$

is true iff for any (nonzero) $F(\omega)$

$$F(\omega) \Big[D \exp(j\omega t) \Big] = F(\omega) \Big[j\omega \exp(j\omega t) \Big]$$

$$\text{iff } 1/2\pi \int_{-\infty}^{\infty} F(\omega) \left[\frac{d}{dt} \exp(j\omega t) \right] d\omega = 1/2\pi \int_{-\infty}^{\infty} F(\omega)\ j\omega \exp(j\omega t)\ d\omega$$

iff (interchanging the order of integration and differentiation on the left)

$$\frac{d}{dt} \left[1/2\pi \int_{-\infty}^{\infty} F(\omega) \exp(j\omega t)\ d\omega \right] = 1/2\pi \int_{-\infty}^{\infty} \Big[j\omega\ F(\omega) \Big] \exp(j\omega t)\ d\omega$$

iff $Df(t) \Longleftrightarrow j\omega\ F(\omega)$ QED

6.4 Solving the following CCL DE's using Fourier transforms:

(a) $y'' + 4y' + 3y = x' + 5x \quad (\forall\ t)$ where $x(t) = \exp(-2t)\ U(t)$

Transformation gives us $[(j\omega)^2 + 4j\omega + 3]\ Y(\omega) = [j\omega + 5]\ X(\omega)$

But $X(\omega) = \dfrac{1}{j\omega + 2}$ and so

$$Y(\omega) = \frac{j\omega + 5}{(j\omega)^2 + 4j\omega + 3} \frac{1}{j\omega + 2} = \frac{2}{j\omega + 1} + \frac{1}{j\omega + 3} - \frac{3}{j\omega + 2}$$

which inverts to

$$y(t) = \Big[2 \exp(-t) + \exp(-3t) - 3 \exp(-2t) \Big] U(t)$$

(b) $y'' + 4y' + 5y = 3x' + x \quad (\forall\, t)$ where $x(t) = U(t) + \exp(-t)]\, U(t)$

Transformation gives us $[(j\omega)^2 + 4j\omega + 5]\, Y(\omega) = [3j\omega + 1]\, X(\omega)$

But $$X(\omega) = \left[\frac{1}{j\omega} + \pi\delta(\omega)\right] + \frac{1}{j\omega + 1}$$

and so $$Y(\omega) = \frac{3j\omega + 1}{(j\omega)^2 + 4j\omega + 5}\left[\frac{1}{j\omega} + \pi\delta(\omega) + \frac{1}{j\omega + 1}\right]$$

$$= \frac{3j\omega + 1}{(j\omega + 2)^2 + 1}\left[\frac{1}{j\omega} + \pi\delta(\omega) + \frac{1}{j\omega + 1}\right]$$

$$= \frac{3j\omega + 1}{[(j\omega + 2)^2 + 1]j\omega} + \frac{\pi}{5}\delta(\omega) + \frac{3j\omega + 1}{[(j\omega + 2)^2 + 1](j\omega + 1)}$$

(Dirac-delta sampling) [applies to the $\frac{\pi}{5}\delta(\omega)$ term]

Expanding the first and third terms using partial fractions:

$$\frac{3j\omega + 1}{[(j\omega + 2)^2 + 1]j\omega} = \frac{1/5}{j\omega} + \frac{Aj\omega + B}{(j\omega + 2)^2 + 1}$$

$$3j\omega + 1 = \frac{1}{5}[(j\omega)^2 + 4j\omega + 5] + A(j\omega)^2 + B(j\omega)$$

$$3j\omega + 1 = \frac{(j\omega)^2}{5} + \frac{4j\omega}{5} + 1 + A(j\omega)^2 + B(j\omega)$$

$A = -1/5, \qquad B + 4/5 = 3$ and so $B = 11/5$

$$\frac{3j\omega + 1}{[(j\omega + 2)^2 + 1](j\omega + 1)} = \frac{-1}{j\omega + 1} + \frac{Pj\omega + Q}{(j\omega + 2)^2 + 1}$$

$$3j\omega + 1 = -[(j\omega)^2 + 4j\omega + 5] + Pj\omega + Q + P(j\omega)^2 + Qj\omega$$

$Q - 5 = 1, \quad Q = 6 \qquad P - 1 = 0, \quad P = 1$

and so we can now restate $Y(\omega)$ as follows:

$$Y(\omega) = \frac{1/5}{j\omega} + \frac{\pi}{5}\delta(\omega) + \frac{11/5 - (1/5)j\omega}{(j\omega + 2)^2 + 1} + \frac{j\omega + 6}{(j\omega + 2)^2 + 1} - \frac{1}{j\omega + 1}$$

$$= \frac{1}{5}\left[\frac{1}{j\omega} + \pi\,\delta(\omega)\right] - \frac{1}{j\omega + 1} + \frac{1}{5}\left[\frac{4j\omega + 41}{(j\omega + 2)^2 + 1}\right]$$

By Exercise 4.3:

$$\exp(-\beta t)\,\cos(\omega_0 t)\,U(t) \iff \frac{\beta + j\omega}{(\beta + j\omega)^2 + \omega_0^2}$$

$$\exp(-\beta t)\,\sin(\omega_0 t)\,U(t) \iff \frac{\omega_0}{(\beta + j\omega)^2 + \omega_0^2}$$

and so

$$\frac{1}{5}\left[\frac{4j\omega + 41}{(j\omega + 2)^2 + 1}\right] = \frac{4}{5}\left[\frac{j\omega + 2}{(j\omega + 2)^2 + 1}\right] + \frac{33}{5}\left[\frac{1}{(j\omega + 2)^2 + 1}\right]$$

$$\iff \frac{4}{5}\exp(-2t)\,\cos(t)\,U(t) + \frac{33}{5}\exp(-2t)\,\sin(t)\,U(t)$$

giving us, finally, $y(t) = (1/5)\,U(t) - \exp(-t)\,U(t)$

$$+ \frac{4}{5}\exp(-2t)\,\cos(t)\,U(t) + \frac{33}{5}\exp(-2t)\,\sin(t)\,U(t)$$

(c) Using the FFT system to verify the result we obtained in (a), namely:

$$y(t) = \left[\,2\exp(-t) + \exp(-3t) - 3\exp(-2t)\,\right] U(t)$$

where $x(t) = \exp(-2t)\,U(t)$, $\quad H(j\omega) = \dfrac{j\omega + 5}{(j\omega)^2 + 4j\omega + 3}$

The slowest decaying exponential in this system is $\exp(-t)\,U(t)$ which comes from the partial fraction expansion that produces $y(t)$. In order to avoid having that exponential run into the next period of the FFT we elected to use $T = 8$, which means that $\exp(-t)\,U(t)$ will have decayed to $\exp(-8) = 0.0003355$ at the end of the FFT's period. Any overflow into the next period is then negligible. The other two exponentials in this system will be even more negligible at $t = 8$.

Using $N = 1024$, SAMPLED, $T = 8$, PULSE, here are the steps for running the problem (refer to Figure 6.6):

Neither, Go. You are now back on, the main menu. Run **ANALYSIS**. The spectrum $X(\omega)$ is now in the **F** vector.

To load $H(j\omega)$ into **F2**: Main menu, Create Values, Continue, Create $H(jw)$, Create, R = 1, S = 2, 5, 1, 3, 4, 1, Load, ALpha = 20. You are now back on the main menu and $H(j\omega)$ is in **F2**.

To perform the complex multiplication of **F** and **F2**: Main menu, Run Postprocessors, F, Complex multiply F and F2, Quit. You are now on the main menu and the product $H(j\omega)\ X(\omega)$ is in **F**.

Run SYNTHESIS. The vector $y(t)$ is now in **Y**. Plot **Y**. The plot should be the same as the one shown in the following figure.

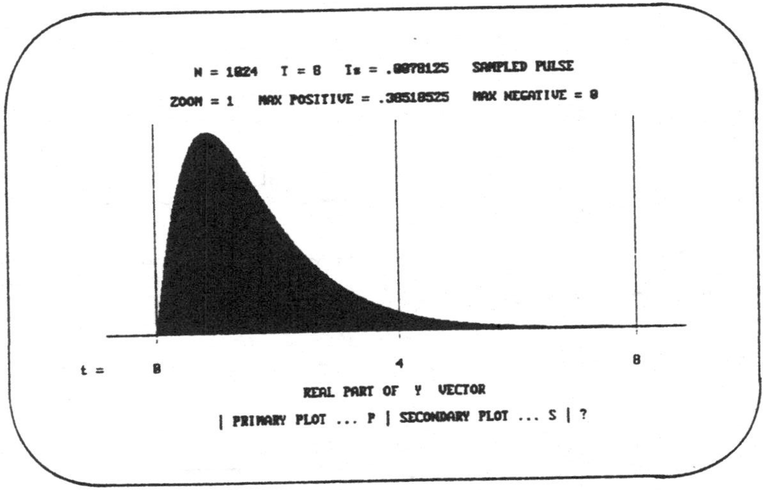

To compare the FFT values against the formula values: Main menu, Display Numbers, X.

Note: T = 8 and N = 1024, means that $T_s = T/N = 0.0078125$

Thus in the formula for $y(t)$ we must replace t by kT_s, i.e.

$$y(kT_s) = \left[\ 2\ \exp(-kT_s) + \exp(-3kT_s) - 3\ \exp(-2kT_s)\ \right] U(kT_s)$$

We can now compare FFT values against formula values derived from this expression for various values of k. The results are shown in the following table:

16	0.11647371	0.115990735	0.51
32	0.21089995	0.210376140	0.24
64	0.33296193	0.332553155	0.12
128	0.37978891	0.379540101	0.066
256	0.21829432	0.218202402	0.042
512	3.5643504e-2	3.5631034e-2	0.035

(d) Verifying the result that we obtained in (b) analytically by applying the DE to it:

$$y'' + 4y' + 3y = x' + 5x \qquad (\forall\ t) \qquad \text{where} \quad x(t) = \exp(-2t)\ U(t)$$

We obtained

$$y(t) = \left[\ 1/5 - \exp(-t) + \frac{4}{5}\exp(-2t)\cos(t) + \frac{33}{5}\exp(-2t)\sin(t)\ \right] U(t)$$

Using the product rule, we consider first the LHS of the DE:

$$Dy(t) = \left[\ 0 + \exp(-t) - \frac{8}{5}\exp(-2t)\cos(t) - \frac{4}{5}\exp(-2t)\sin(t) - \frac{66}{5}\exp(-2t)\sin(t) + \frac{33}{5}\exp(-2t)\cos(t)\ \right] U(t)$$

$$+ \left[\ 1/5 - \exp(-t) + \frac{4}{5}\exp(-2t)\cos(t) + \frac{33}{5}\exp(-2t)\sin(t)\ \right] \delta(t)$$

$$= \left[\ \exp(-t) - \frac{8}{5}\exp(-2t)\cos(t) - \frac{4}{5}\exp(-2t)\sin(t) - \frac{66}{5}\exp(-2t)\sin(t) + \frac{33}{5}\exp(-2t)\cos(t)\ \right] U(t) + \left[\ 1/5 - 1 + \frac{4}{5}\ \right] \delta(t)$$

$$= \left[\ \exp(-t) + 5\exp(-2t)\cos(t) - 14\exp(-2t)\sin(t)\ \right] U(t)$$

$$D^2y(t) = \left[\ -\exp(-t) - 10\exp(-2t)\cos(t) - 5\exp(-2t)\sin(t) + 28\exp(-2t)\sin(t) - 14\exp(-2t)\cos(t)\ \right] U(t)$$

$$+ \left[\ \exp(-t) + 5\exp(-2t)\cos(t) - 14\exp(-2t)\sin(t)\ \right] \delta(t)$$

$$= \left[-\exp(-t) - 24\exp(-2t)\cos(t) + 23\exp(-2t)\sin(t) \right] U(t) + 6\,\delta(t)$$

Then the LHS of the DE gives: $[D^2 + 4D + 5]\; y(t)$

$$= \left[\begin{array}{l} -\exp(-t) - 24\exp(-2t)\cos(t) + 23\exp(-2t)\sin(t) \\ + 4\exp(-t) + 20\exp(-2t)\cos(t) - 56\exp(-2t)\sin(t) \\ + 1 - 5\exp(-t) + 4\exp(-2t)\cos(t) + 33\exp(-2t)\sin(t) \end{array} \right] U(t) + 6\delta(t)$$

$$= [1 - 2\exp(-t)]\; U(t) + 6\,\delta(t) \qquad \text{(A)}$$

Considering now the RHS of the DE:

$$(3D + 1)\; x(t) = 3D\,[1 + \exp(-t)]\; U(t) + [1 + \exp(-t)]\; U(t)$$

$$= [-3\exp(-t)]\; U(t) + [1 + \exp(-t)]\; 3\,\delta(t) + [1 + \exp(-t)]\; U(t)$$

$$= [1 - 2\exp(-t)]\; U(t) + 6\,\delta(t)$$

which is the same as the LHS in (A). The DE balances, and so the solution that we have obtained is correct.

6.5 (a) The DE is $12y''(t) + 7y'(t) + y(t) = 2x'(t) + x(t) \qquad (\forall\, t)$

Transforming gives $\quad Y(\omega) = \dfrac{2j\omega + 1}{12(j\omega)^2 + 7j\omega + 1}$

(1) For $x(t) = U(t)$, i.e. $X(\omega) = 1/j\omega + \pi\delta(\omega)$, we have:

$$Y(\omega) = \frac{2j\omega + 1}{12(j\omega)^2 + 7j\omega + 1}\left[1/j\omega + \pi\delta(\omega) \right]$$

$$= \frac{2j\omega + 1}{(4j\omega + 1)(3j\omega + 1)j\omega} + \frac{2j\omega + 1}{12(j\omega)^2 + 7j\omega + 1}\,\pi\delta(\omega)$$

$$= \frac{-8}{4j\omega + 1} + \frac{3}{3j\omega + 1} + \frac{1}{j\omega} + \pi\delta(\omega) \qquad \text{(Dirac-delta sampling)}$$

$$= \frac{-2}{j\omega + 1/4} + \frac{1}{j\omega + 1/3} + \frac{1}{j\omega} + \pi\delta(\omega) \qquad \text{(Dirac-delta sampling)}$$

and so, after inversion,

$$y(t) = -2\exp(-t/4)\; U(t) + \exp(-t/3)\; U(t) + U(t)$$

$$= \left[-2\exp(-t/4) + \exp(-t/3) + 1 \right] U(t)$$

(2) For $x(t) = \delta(t)$, i.e. $X(\omega) = 1$, we have

$$Y(\omega) = \frac{2j\omega + 1}{12(j\omega)^2 + 7j\omega + 1}\,1 = \frac{2}{4j\omega + 1} + \frac{-1}{3j\omega + 1} = \frac{1/2}{j\omega + 1/4} - \frac{1/3}{j\omega + 1/3}$$

and so, after inversion, $y(t) = \left[1/2\exp(-t/4) - 1/3\exp(-t/3) \right] U(t)$

(b) Analytical verification.

(1) $\qquad [12D^2 + 7D + 1]\; y(t) = [2D + 1]\; x(t) \qquad (\forall\, t)$

$x(t) = U(t)$, i.e. $X(\omega) = 1/j\omega + \pi\delta(\omega)$

The solution that we obtained was

$$y(t) = \left[-2\exp(-t/4) + \exp(-t/3) + 1 \right] U(t)$$

Applying the LHS of the DE to $y(t)$, using the product rule for differentiation:

$$Dy(t) = D\left[-2\exp(-t/4) + \exp(-t/3) + 1 \right] U(t)$$

$$+ \left[-2\exp(-t/4) + \exp(-t/3) + 1 \right] DU(t)$$

$$= \left[1/2\exp(-t/4) - 1/3\exp(-t/3) + 0 \right] U(t)$$

$$+ \left[-2\exp(-t/4) + \exp(-t/3) + 1 \right] \delta(t)$$

$$= \left[1/2\exp(-t/4) - 1/3\exp(-t/3) \right] U(t)$$

$$+ \left[-2 + 1 + 1 \right] \delta(t) \qquad \text{(Dirac-delta sampling)}$$

$$= \left[1/2\exp(-t/4) - 1/3\exp(-t/3) \right] U(t)$$

$$D^2y(t) = D\left[1/2\exp(-t/4) - 1/3\exp(-t/3) \right] U(t)$$

$$+ \left[1/2\exp(-t/4) - 1/3\exp(-t/3) \right] DU(t)$$

$$= \left[-1/8\exp(-t/4) + 1/9\exp(-t/3) \right] U(t)$$

$$+ \left[1/2\exp(-t/4) - 1/3\exp(-t/3) \right] \delta(t)$$

$$= \left[-1/8 \exp(-t/4) + 1/9 \exp(-t/3) \right] U(t) + \left[1/2 - 1/3 \right] \delta(t)$$

$$= \left[-1/8 \exp(-t/4) + 1/9 \exp(-t/3) \right] U(t) + 1/6\, \delta(t)$$

The LHS of the DE then gives

$$[12D^2 + 7D + 1]\, y(t) = \left[-3/2 \exp(-t/4) + 4/3 \exp(-t/3) \right] U(t) + 2\delta(t)$$
$$+ \left[7/2 \exp(-t/4) - 7/3 \exp(-t/3) \right] U(t)$$
$$+ \left[-2 \exp(-t/4) + \exp(-t/3) + 1 \right] U(t)$$

$$= U(t) + 2\delta(t) \qquad \text{(A)}$$

The RHS of the DE gives:

$$[2D + 1]\, x(t) = [2D + 1]\, U(t) = 2\delta(t) + U(t)$$

which is the same as the LHS in (A). The DE balances, and so the solution that we have obtained is correct.

(2) $[12D^2 + 7D + 1]\, y(t) = [2D + 1]\, x(t) \quad (\forall\, t)$

$x(t) = \delta(t)$, i.e. $X(\omega) = 1$

The solution that we obtained was

$$y(t) = \left[1/2 \exp(-t/4) - 1/3 \exp(-t/3) \right] U(t)$$

Applying the LHS of the DE to $y(t)$, using the product rule for differentiation:

$$Dy(t) = D \left[1/2 \exp(-t/4) - 1/3 \exp(-t/3) \right] U(t)$$
$$+ \left[1/2 \exp(-t/4) - 1/3 \exp(-t/3) \right] DU(t)$$

$$= \left[-1/8 \exp(-t/4) + 1/9 \exp(-t/3) \right] U(t)$$
$$+ \left[1/2 \exp(-t/4) - 1/3 \exp(-t/3) \right] \delta(t)$$

$$= \left[-1/8 \exp(-t/4) + 1/9 \exp(-t/3) \right] U(t) + 1/6\, \delta(t)$$
(Dirac-delta sampling)

$$D^2 y(t) = D \left[-1/8 \exp(-t/4) + 1/9 \exp(-t/3) \right] U(t)$$
$$+ \left[-1/8 \exp(-t/4) + 1/9 \exp(-t/3) \right] DU(t) + 1/6\, D\delta(t)$$

$$= \left[1/32 \exp(-t/4) - 1/27 \exp(-t/3) \right] U(t)$$
$$+ \left[-1/8 \exp(-t/4) + 1/9 \exp(-t/3) \right] \delta(t) + 1/6\, D\delta(t)$$

$$= \left[1/32 \exp(-t/4) - 1/27 \exp(-t/3) \right] U(t)$$
$$- 1/72\, \delta(t) + 1/6\, D\delta(t)$$

The LHS of the DE then gives

$$[12D^2 + 7D + 1]\, y(t)$$
$$= \left[3/8 \exp(-t/4) - 4/9 \exp(-t/3) \right] U(t) - 1/6\, \delta(t) + 2\, D\delta(t)$$
$$+ \left[-7/8 \exp(-t/4) + 7/9 \exp(-t/3) \right] U(t) + 7/6\, \delta(t)$$
$$+ \left[1/2 \exp(-t/4) - 1/3 \exp(-t/3) \right] U(t) = \delta(t) + 2\, D\delta(t) \qquad \text{(B)}$$

The RHS gives $\quad [2D + 1]\, x(t) = [2D + 1]\, \delta(t) = 2D\delta(t) + \delta(t)$

which is the same as the LHS in (B). The DE balances, and so the solution that we have obtained is correct.

(c) Using the FFT system to verify the result we obtained in (a)(2) of this exercise:

$$y(t) = \left[1/2 \exp(-t/4) - 1/3 \exp(-t/3) \right] U(t)$$

where $x(t) = \delta(t)$, $\quad H(j\omega) = \dfrac{2j\omega + 1}{12(j\omega)^2 + 7j\omega + 1}$

The slowest decaying exponential in this system is $\exp(-t/4)\, U(t)$ which comes from the partial fraction expansion that produced

y(t). In order to avoid having that exponential run into the next period of the FFT we elected to use $T = 40$, which means that $\exp(-t/4)\ U(t)$ will have decayed to $\exp(-10) = 0.0000454$ at the end of the FFT's period. Any overflow into the next period is then negligible. The other exponential in this system will be even more negligible at $t = 10$.

Using $N = 256$, SAMPLED, $T = 40$, PULSE, here are the steps for running the problem (refer to Figure 6.6):

To create the **X** vector: Main menu, Create Values, Continue, Create Only Diracs, Time, X, 0, 1, Neither, Go. You are now back on the main menu. Run ANALYSIS. The spectrum $X(\omega)$ is now in the **F** vector.

To load $H(j\omega)$ into **F2**: Main menu, Create Values, Continue, Create H(jw), Create, R = 1, S = 2, 1, 2, 1, 7, 12, LOAD, Alpha = 20. You are now back on the main menu and $H(j\omega)$ is in **F2**.

To perform the complex multiplication of **F** and **F2**: Main menu, Run Postprocessors, F, Complex multiply F and F2, Quit. You are now on the main menu and the product $H(j\omega)\ X(\omega)$ is in **F**.

Run SYNTHESIS. The vector $y(t)$ is now in **Y**. Plot **Y**. The plot should be the same as the one shown in the following figure.

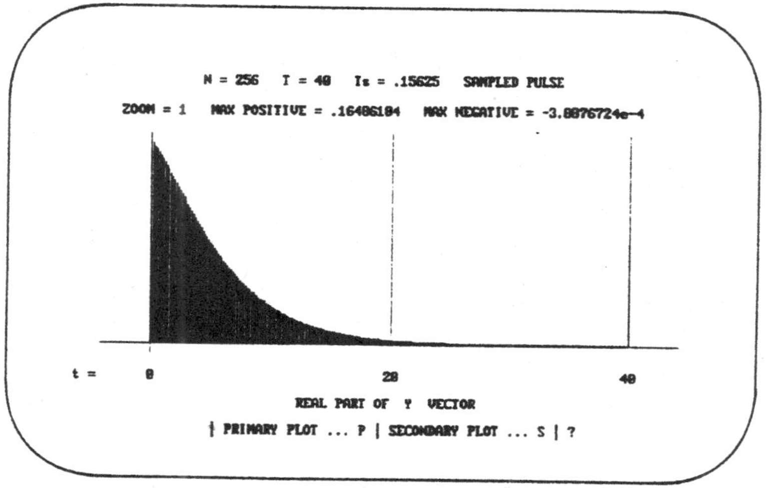

To compare the FFT values against the formula values: Main menu, Display Numbers, X.

Note: $T = 40$ and $N = 256$, means that $T_s = T/N = 0.015625$ Thus in the formula for $y(t)$ we must replace t by kT_s, i.e.

$$y(kT_s) = \left[\ 1/2\ \exp(-kT_s/4) - 1/3\ \exp(-kT_s/3)\ \right] U(kT_s)$$

We can now compare FFT values against formula values derived from this expression for various values of k. The results are shown in the following table:

k	FFT	Formula	Error (%)
4	0.15694331	0.157027215	0.053
8	0.14602587	0.146060938	0.024
16	0.12275115	0.122764645	0.011
32	8.0288064e-2	8.0293864e-2	0.0072
64	2.9147958e-2	2.9151168e-2	0.011
128	2.9449145e-3	2.9447622e-3	0.0051

6.6 (a) Deriving the CCL DE which links the input and output for the network shown in Figure 6.16:

The voltage across each of the elements is:

$$v_L(t) = L di/dt \qquad v_R(t) = Ri$$

$$v_{C1}(t) = 1/C_1 \int_{-\infty}^{t} i dt \qquad v_{C2}(t) = 1/C_2 \int_{-\infty}^{t} i dt$$

By Kirchhoff's loop-voltage law:

$$v_L(t) + v_{C1}(t) + v_{C2}(t) + v_R(t) = x(t)$$

$$v_{C2}(t) + v_R(t) = y(t)$$

Replacing these by their expressions, we obtain:

$$LDi + Ri + (1/C_1 + 1/C_2) \int_{-\infty}^{t} i\ dt = x$$

$$Ri + 1/C_2 \int_{-\infty}^{t} i\ dt = y$$

Differentiating to eliminate the integrals:

$$LD^2 i + RDi + (1/C_1 + 1/C_2)\ i = Dx \qquad \text{(A)}$$

$$RDi + (1/C_2)\ i = Dy \qquad \text{(B)}$$

$$[RD + 1/C_2]\left[LD^2 + RD + 1/C_1 + 1/C_2\right] i = [RD + 1/C_2]\, Dx$$

But these operators commute because they have constant coefficients, and so

$$\left[LD^2 + RD + 1/C_1 + 1/C_2\right][RD + 1/C_2]\, i = [RD + 1/C_2]\, Dx$$

Then, using (B):

$$\left[LD^2 + RD + 1/C_1 + 1/C_2\right] Dy = [RD + 1/C_2]\, Dx$$

Cancelling out a common D then gives us the required DE:

$$\left[LD^2 + RD + 1/C_1 + 1/C_2\right] y(t) = [RD + 1/C_2]\, x(t)$$

(b) Assuming that $x(t)$ is the complex exponential $\exp(j\omega_0 t)$, we now derive the expression for $H(j\omega_0)$ from the CCL DE, using Fourier transformation, as follows:

$$\left[x(t) = \exp(j\omega_0 t)\right] \Longleftrightarrow \left[X(\omega) = 2\pi\delta(\omega - \omega_0)\right]$$

Transforming the DE and using $X(\omega)$:

$$\left[L(j\omega)^2 + Rj\omega + 1/C_1 + 1/C_2\right] Y(\omega) = [Rj\omega + 1/C_2]\, 2\pi\delta(\omega - \omega_0)$$

from which

$$Y(\omega) = \frac{Rj\omega + 1/C_2}{L(j\omega)^2 + Rj\omega + 1/C_1 + 1/C_2}\, 2\pi\delta(\omega - \omega_0)$$

$$= \frac{Rj\omega_0 + 1/C_2}{L(j\omega_0)^2 + Rj\omega_0 + 1/C_1 + 1/C_2}\, 2\pi\delta(\omega - \omega_0) \quad \text{(Dirac-delta sampling)}$$

Then the transfer function is

$$H(j\omega_0) = \frac{Rj\omega_0 + 1/C_2}{L(j\omega_0)^2 + Rj\omega_0 + 1/C_1 + 1/C_2}$$

$$= \frac{R + 1/j\omega_0 C_2}{j\omega_0 L + R + 1/j\omega_0 C_1 + 1/j\omega_0 C_2}$$

obtaining:

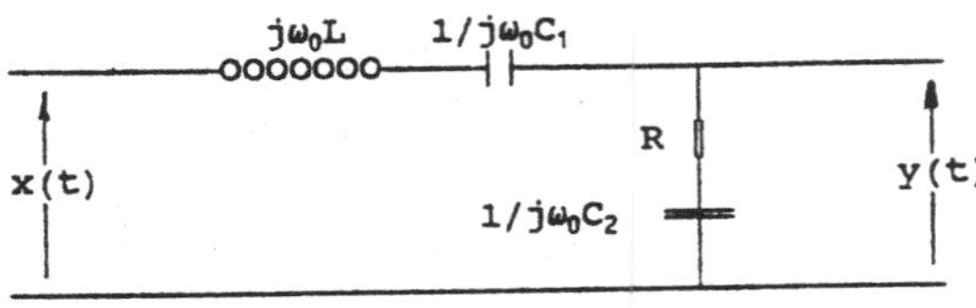

from which we obtain by inspection (voltage divider)

$$H(j\omega_0) = \frac{R + 1/j\omega_0 C_2}{j\omega_0 L + R + 1/j\omega_0 C_1 + 1/j\omega_0 C_2} \qquad \text{(C)}$$

(d) Starting from (C) we can write the CCL DE for the system as follows:

$$Y(\omega) = H(j\omega)\, X(\omega) = \frac{R + 1/j\omega C_2}{j\omega L + R + 1/j\omega C_1 + 1/j\omega C_2}\, X(\omega)$$

$$[j\omega L + R + 1/j\omega C_1 + 1/j\omega C_2]\, Y(\omega) = [R + 1/j\omega C_2]\, X(\omega)$$

$$[L(j\omega)^2 + Rj\omega + 1/C_1 + 1/C_2]\, Y(\omega) = [Rj\omega + 1/C_2]\, X(\omega)$$

Replacing $j\omega$ by D, and inverting $Y(\omega)$ and $X(\omega)$ to the time domain, gives the DE as

$$[LD^2 + RD + 1/C_1 + 1/C_2]\, y(t) = [RD + 1/C_2]\, x(t)$$

6.7 Using the method of ac circuit analysis for each of the following three LTI systems to obtain

(1) the input impedance

(2) the frequency response

(3) the CCL DE which relates the output to the input

(a)

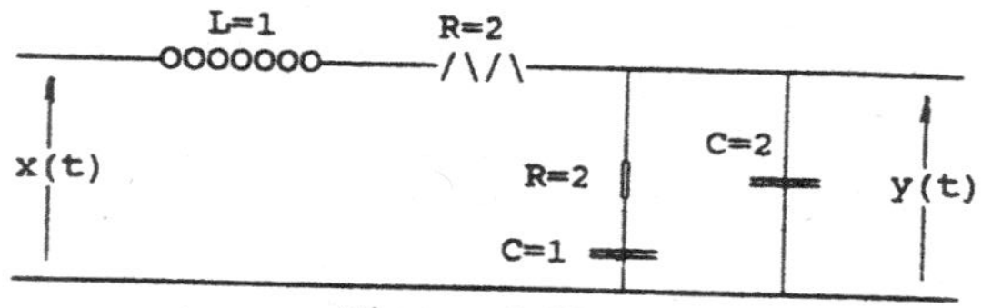

Figure 6.17

The series impedance is: $Z_{se} = j\omega 1 + 2$

The shunt impedance is:

$$Z_{sh} = \frac{1}{\dfrac{1}{2 + 1/j\omega 1} + \dfrac{1}{1/j\omega 2}} = \frac{1}{\dfrac{j\omega}{2(j\omega) + 1} + 2(j\omega)} = \frac{2(j\omega) + 1}{4(j\omega)^2 + 3(j\omega)}$$

(1) Input impedance is:

$$Z_{in} = Z_{se} + Z_{sh} = j\omega + 2 + \frac{2(j\omega) + 1}{4(j\omega)^2 + 3(j\omega)}$$

$$= \frac{4(j\omega)^3 + 11(j\omega)^2 + 8(j\omega) + 1}{4(j\omega)^2 + 3(j\omega)} \longrightarrow$$

(2) Transfer function is:

$$H(j\omega) = \frac{Z_{sh}}{Z_{in}} = \frac{\dfrac{2(j\omega) + 1}{4(j\omega)^2 + 3(j\omega)}}{\dfrac{4(j\omega)^3 + 11(j\omega)^2 + 8(j\omega) + 1}{4(j\omega)^2 + 3(j\omega)}}$$

$$= \frac{2(j\omega) + 1}{4(j\omega)^3 + 11(j\omega)^2 + 8(j\omega) + 1} \longrightarrow$$

(3) System DE:

$$Y(\omega) = H(j\omega)\ X(\omega) = \frac{2(j\omega) + 1}{4(j\omega)^3 + 11(j\omega)^2 + 8(j\omega) + 1} X(\omega)$$

$$[4(j\omega)^3 + 11(j\omega)^2 + 8(j\omega) + 1]\ Y(\omega) = [2(j\omega) + 1]\ X(\omega)$$

$$[4D^3 + 11D^2 + 8D + 1]\ y(t) = [2D + 1]\ x(t) \longrightarrow$$

(b)

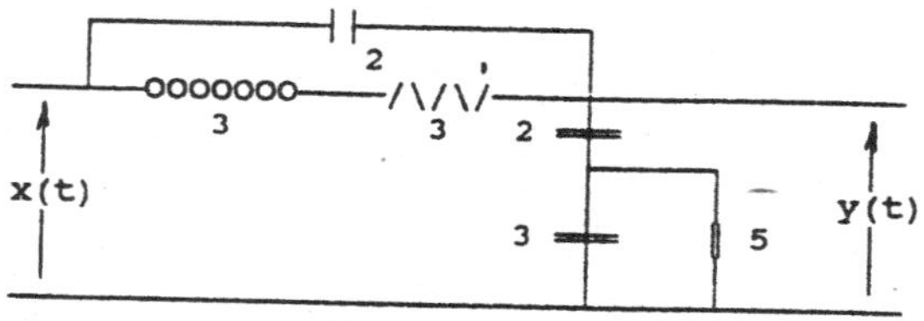

Figure 6.18

The series impedance is:

$$Z_{se} = \frac{1}{\dfrac{1}{1/j\omega 2} + \dfrac{1}{j\omega 3 + 3}} = \frac{1}{2(j\omega) + \dfrac{1}{3(j\omega) + 3}}$$

$$= \frac{3(j\omega) + 3}{6(j\omega)^2 + 6(j\omega) + 1} \longrightarrow$$

The shunt impedance is:

$$Z_{sh} = 1/j\omega 2 + \frac{1}{\dfrac{1}{1/j\omega 3} + \dfrac{1}{5}} = \frac{1}{2(j\omega)} + \frac{1}{3(j\omega) + 1/5} = \frac{25(j\omega) + 1}{30(j\omega)^2 + 2(j\omega)}$$

(1) Input impedance is:

$$Z_{in} = Z_{se} + Z_{sh} = \frac{3(j\omega) + 3}{6(j\omega)^2 + 6(j\omega) + 1} + \frac{25(j\omega) + 1}{30(j\omega)^2 + 2(j\omega)}$$

$$= \frac{240(j\omega)^3 + 252(j\omega)^2 + 37(j\omega) + 1}{180(j\omega)^4 + 192(j\omega)^3 + 42(j\omega)^2 + 2(j\omega)} \longrightarrow$$

(2) Transfer function is:

$$H(j\omega) = \frac{Z_{sh}}{Z_{in}} = \frac{\dfrac{25(j\omega) + 1}{30(j\omega)^2 + 2(j\omega)}}{\dfrac{240(j\omega)^3 + 252(j\omega)^2 + 37(j\omega) + 1}{(6(j\omega)^2 + 6(j\omega) + 1)\ (30(j\omega)^2 + 2(j\omega))}}$$

$$= \frac{150(j\omega)^3 + 156(j\omega)^2 + 31(j\omega) + 1}{240(j\omega)^3 + 252(j\omega)^2 + 37(j\omega) + 1} \longrightarrow$$

(3) System DE:

$$Y(\omega) = H(j\omega)\ X(\omega) = \frac{150(j\omega)^3 + 156(j\omega)^2 + 31(j\omega) + 1}{240(j\omega)^3 + 252(j\omega)^2 + 37(j\omega) + 1} X(\omega)$$

$$[240(j\omega)^3 + 252(j\omega)^2 + 37(j\omega) + 1]\ Y(\omega)$$
$$= [150(j\omega)^3 + 156(j\omega)^2 + 31(j\omega) + 1]\ X(\omega)$$

$$[240D^3 + 252D^2 + 37D + 1]\ y(t) = [150D^3 + 156D^2 + 31D + 1]\ x(t) \longrightarrow$$

(c)

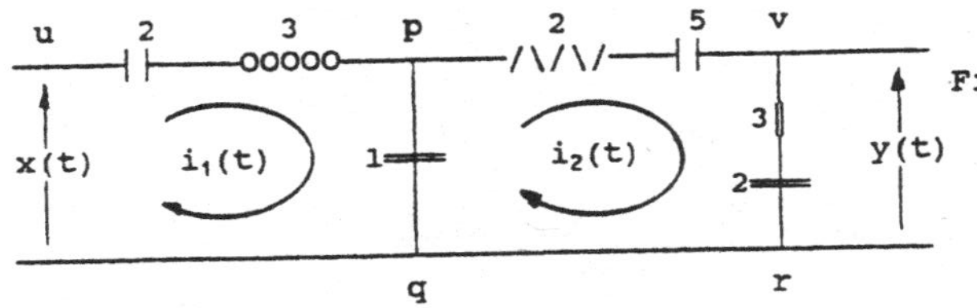

Figure 6.19

Method 1 (using matrices):

$$\left[\frac{1}{j\omega 2} + j\omega 3 + \frac{1}{j\omega 1}\right] I_1(\omega) - \frac{1}{j\omega 1} I_2(\omega) = X(\omega)$$

$$-\frac{1}{j\omega 1} I_1(\omega) + \left[\frac{1}{j\omega 1} + 2 + \frac{1}{j\omega 5} + 3 + \frac{1}{j\omega 2}\right] I_2(\omega) = 0$$

$$\begin{bmatrix} \dfrac{1 + 6(j\omega)^2 + 2}{2(j\omega)} & -\dfrac{1}{j\omega 1} \\ -\dfrac{1}{j\omega 1} & \dfrac{10 + 20(j\omega) + 2 + 30(j\omega) + 5}{10(j\omega)} \end{bmatrix} \begin{bmatrix} I_1(\omega) \\ I_2(\omega) \end{bmatrix} = \begin{bmatrix} X(\omega) \\ 0 \end{bmatrix}$$

$$\begin{bmatrix} \dfrac{6(j\omega)^2 + 3}{2(j\omega)} & -\dfrac{1}{j\omega 1} \\ -\dfrac{1}{j\omega} & \dfrac{50(j\omega) + 17}{10(j\omega)} \end{bmatrix} \begin{bmatrix} I_1(t) \\ I_2(\omega) \end{bmatrix} = \begin{bmatrix} X(\omega) \\ 0 \end{bmatrix}$$

$$\begin{bmatrix} I_1(\omega) \\ I_2(\omega) \end{bmatrix} = \Delta^{-1} \begin{bmatrix} \dfrac{50(j\omega) + 17}{10(j\omega)} & \dfrac{1}{j\omega} \\ \dfrac{1}{j\omega} & \dfrac{6(j\omega)^2 + 3}{2(j\omega)} \end{bmatrix} \begin{bmatrix} X(\omega) \\ 0 \end{bmatrix} \qquad \text{(A)}$$

where

$$\Delta = \frac{6(j\omega)^2 + 3}{2(j\omega)} \frac{50(j\omega) + 17}{10(j\omega)} - \frac{1}{(j\omega)^2} = \frac{300(j\omega)^3 + 102(j\omega)^2 + 150(j\omega) + 31}{20(j\omega)^2}$$

(1) Input impedance:

Solving for $I_1(\omega)$ from (A) gives

$$I_1(\omega) = \frac{1}{\Delta} \frac{50(j\omega) + 17}{10(j\omega)} X(\omega)$$

$$= \frac{20(j\omega)^2}{300(j\omega)^3 + 102(j\omega)^2 + 150(j\omega) + 31} \frac{50(j\omega) + 17}{10(j\omega)} X(\omega)$$

$$= \frac{100(j\omega)' + 34(j\omega)}{300(j\omega)^3 + 102(j\omega)^2 + 150(j\omega) + 31} X(\omega)$$

and so

$$Z_{in}(j\omega) = \frac{X(\omega)}{I(\omega)} = \frac{300(j\omega)^3 + 102(j\omega)^2 + 150(j\omega) + 31}{100(j\omega)^2 + 34(j\omega)} \longrightarrow$$

(2) Transfer function:

Solving for $I_2(\omega)$ from (A) gives

$$I_2(\omega) = \frac{1}{\Delta} \frac{1}{j\omega} X(\omega) = \frac{20(j\omega)^2}{300(j\omega)^3 + 102(j\omega)^2 + 150(j\omega) + 31} \frac{1}{j\omega} X(\omega)$$

$$= \frac{20(j\omega)}{300(j\omega)^3 + 102(j\omega)^2 + 150(j\omega) + 31} X(\omega)$$

Then, from the circuit in Figure 6.19:

$$Y(\omega) = [3 + 1/j\omega 2]\ I_2(\omega) = \frac{6(j\omega) + 1}{2(j\omega)} I_2(\omega)$$

$$= \frac{6(j\omega) + 1}{2(j\omega)} \frac{20(j\omega)}{300(j\omega)^3 + 102(j\omega)^2 + 150(j\omega) + 31} X(\omega)$$

$$= \frac{60(j\omega) + 10}{300(j\omega)^3 + 102(j\omega)^2 + 150(j\omega) + 31} X(\omega) \qquad \text{(B)}$$

and so

$$H(j\omega) = \frac{Y(\omega)}{X(\omega)} = \frac{60(j\omega) + 10}{300(j\omega)^3 + 102(j\omega)^2 + 150(j\omega) + 31} \longrightarrow$$

(3) The system DE: From (B),

$$Y(\omega) = \frac{60(j\omega) + 10}{300(j\omega)^3 + 102(j\omega)^2 + 150(j\omega) + 31} X(\omega) \qquad \text{and so the DE is}$$

$$[300(D)^3 + 102(D)^2 + 150(D) + 31]\ y(t) = [60(D) + 10]\ x(t) \longrightarrow$$

- - - - - - - - - - - - - - - - - - - -

Method 2 (direct): The impedance between points p and r in the figure is

$$Z_{pr} = 5 + 1/j\omega 5 + 1/j\omega 2 = \frac{50(j\omega) + 7}{10(j\omega)}$$

and so the shunt impedance between points p and q is

$$Z_{sh} = \frac{1}{1/Z_{pr} + j\omega} = \frac{1}{\dfrac{10(j\omega)}{50(j\omega) + 7} + j\omega} = \frac{50(j\omega) + 7}{j\omega(50(j\omega) + 17)}$$

The series impedance between u and p is

$$Z_{se} = 3(j\omega) + \frac{1}{2(j\omega)} = \frac{6(j\omega)^2 + 1}{2(j\omega)}$$

(1) Input impedance:

$$Z_{in} = Z_{se} + Z_{sh} = \frac{6(j\omega)^2 + 1}{2(j\omega)} + \frac{50(j\omega) + 7}{j\omega(50(j\omega) + 17)}$$

$$= \frac{300(j\omega)^3 + 102(j\omega)^2 + 150(j\omega) + 31}{100(j\omega)^2 + 34(j\omega)} \longrightarrow$$

(2) Transfer function: The voltage across the shunt impedance is

$$V_{sh}(\omega) = \frac{Z_{sh}}{Z_{in}} X(\omega) = \frac{\dfrac{50(j\omega) + 7}{j\omega(50(j\omega) + 17)}}{\dfrac{300(j\omega)^3 + 102(j\omega)^2 + 150(j\omega) + 31}{2(j\omega)(50(j\omega) + 17(j\omega))}} X(\omega)$$

$$= \frac{100(j\omega) + 14}{300(j\omega)^3 + 102(j\omega)^2 + 150(j\omega) + 31} X(\omega)$$

Then the output voltage is

$$Y(\omega) = \frac{Z_{vr}}{Z_{pr}} V_{sh} = \frac{3 + \dfrac{1}{2(j\omega)}}{\dfrac{50(j\omega) + 17}{10(j\omega)}} V_{sh} = \frac{\dfrac{6(j\omega) + 1}{2(j\omega)}}{\dfrac{50(j\omega) + 7}{10(j\omega)}} V_{sh}$$

$$= \frac{30(j\omega) + 10}{50(j\omega) + 7} \frac{100(j\omega) + 14}{300(j\omega)^3 + 102(j\omega)^2 + 150(j\omega) + 31} X(\omega)$$

$$= \frac{60(j\omega) + 10}{300(j\omega)^3 + 102(j\omega)^2 + 150(j\omega) + 31} X(\omega)$$

from which the transfer function is

$$H(j\omega) = \frac{Y(j\omega)}{X(j\omega)} = \frac{60(j\omega) + 10}{300(j\omega)^3 + 102(j\omega)^2 + 150(j\omega) + 31} \longrightarrow$$

Note: This method could be used for circuits with three and more loops. The matrix method would become extremely cumbersome beyond two loops since the inversion of the matrices in functional form would be extremely tedious.

6.8 Referring to Figure 6.4:

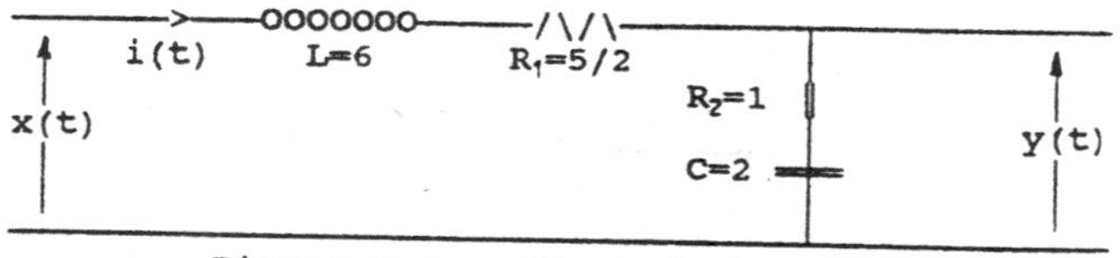

Figure 6.4: Electrical network

$$(12D^2 + 7D + 1)\, y(t) = (2D + 1)\, x(t)$$

Fourier transformation gives:

$$(12(j\omega)^2 + 7(j\omega) + 1)\, Y(\omega) = (2(j\omega) + 1)\, X(\omega)$$

For $x(t) = \cos(\omega_0 t)$, $X(\omega) = \pi[\delta(\omega - \omega_0) + \delta(\omega + \omega_0)]$.

Then the DE is

$$(12(j\omega)^2 + 7(j\omega) + 1)\, Y(\omega) = (2(j\omega) + 1)\, \pi[\delta(\omega - \omega_0) + \delta(\omega + \omega_0)]$$

Solving for $Y(\omega)$:

$$Y(\omega) = \frac{2(j\omega) + 1}{12(j\omega)^2 + 7(j\omega) + 1}\, \pi[\delta(\omega - \omega_0) + \delta(\omega + \omega_0)]$$

in which Dirac-delta sampling takes place, and so we continue

$$\ldots = \frac{2(j\omega_0) + 1}{12(j\omega_0)^2 + 7(j\omega_0) + 1}\, \pi\delta(\omega - \omega_0) + \frac{2(-j\omega_0) + 1}{12(-j\omega_0)^2 + 7(-j\omega_0) + 1}\, \pi\delta(\omega + \omega_0)$$

Direct inversion gives

$$y(t) = (1/2\pi) \int_{-\infty}^{\infty} \frac{2(j\omega_0) + 1}{12(j\omega_0)^2 + 7(j\omega_0) + 1}\, \pi\delta(\omega - \omega_0)\, \exp(j\omega t)\, d\omega + (1/2\pi) \int_{-\infty}^{\infty} \frac{-2(j\omega_0) + 1}{12(j\omega_0)^2 - 7(j\omega_0) + 1}\, \pi\delta(\omega + \omega_0)\, \exp(j\omega t)\, d\omega$$

in which Dirac-delta sampling takes place once more, and so we continue

$$\ldots = \tfrac{1}{2}\, \frac{2(j\omega_0) + 1}{12(j\omega_0)^2 + 7(j\omega_0) + 1}\, \exp(j\omega_0 t) \int_{-\infty}^{\infty} \delta(x - \omega_0)\, d\omega + \tfrac{1}{2}\, \frac{-2(j\omega_0) + 1}{12(j\omega_0)^2 - 7(j\omega_0) + 1}\, \exp(-j\omega_0 t) \int_{-\infty}^{\infty} \delta(x + \omega_0)\, d\omega$$

$$= \tfrac{1}{2}\, \frac{\ldots}{12(j\omega_0)^2 + 7(j\omega_0) + 1}\, \exp(j\omega_0 t) + \tfrac{1}{2}\, \frac{\ldots}{12(j\omega_0)^2 - 7(j\omega_0) + 1}\, \ldots$$

$$= Z + Z^* = 2\,\mathrm{Re}[Z] = \mathrm{Re}\left[\frac{2(j\omega_0) + 1}{12(j\omega_0)^2 + 7(j\omega_0) + 1}\, \exp(j\omega_0 t) \right]$$

Now,

$$\frac{2(j\omega_0) + 1}{12(j\omega_0)^2 + 7(j\omega_0) + 1} = \frac{2(j\omega_0) + 1}{1 - 12(\omega_0)^2 + 7(j\omega_0)}\, \frac{1 - 12(\omega_0)^2 - 7(j\omega_0)}{1 - 12(\omega_0)^2 - 7(j\omega_0)}$$

$$= \frac{[1 - 12(\omega_0)^2 + 14\omega_0^2] + j[2\omega_0 - 24\omega^3 - 7\omega_0]}{[1 - 12(\omega_0)^2]^2 + 49(\omega_0)^2}$$

$$= \frac{[1 + 2(\omega_0)^2] - j[5\omega_0 + 24\omega^3]}{[1 - 12(\omega_0)^2]^2 + 49(\omega_0)^2}$$

and so

$$y(t) = \mathrm{Re}\left[\frac{[1 + 2(\omega_0)^2] - j[5\omega_0 + 24\omega^3]}{[1 - 12(\omega_0)^2]^2 + 49(\omega_0)^2}\, [\cos(\omega_0 t) + j\sin(\omega_0 t)] \right]$$

$$= \frac{[1 + 2(\omega_0)^2]\cos(\omega_0 t) + [5\omega_0 + 24\omega_0^3]\sin(\omega_0 t)]}{[1 - 12(\omega_0)^2]^2 + 49(\omega_0)^2} \longrightarrow$$

(b) Applying the rules of ac circuit analysis:

$$H(j\omega_0) = \frac{1 + \dfrac{1}{2(j\omega_0)}}{6(j\omega_0) + \dfrac{7}{2} + \dfrac{1}{2(j\omega_0)}} = \frac{2(j\omega_0) + 1}{12(j\omega_0)^2 + 7(j\omega_0) + 1}$$

and so, for the input $x(t) = \cos(\omega_0 t)$:

$$y(t) = \mathrm{Re}[H(j\omega_0)\exp(j\omega_0 t)] = \mathrm{Re}\left[\frac{2(j\omega_0) + 1}{12(j\omega_0)^2 + 7(j\omega_0) + 1}\, \exp(j\omega_0 t) \right]$$

which was derived in part (a).

(c) Verifying analytically that the result obtained in (a) or (b) satisfies the system DE:

The DE is $(12D^2 + 7D + 1)\ y(t) = (2D + 1)\ x(t)$

For $x(t) = \cos(\omega_0 t)$ we must show that

$$y(t) = \frac{[1 + 2(\omega_0)^2]\cos(\omega_0 t) + [5\omega_0 + 24\omega^3]\sin(\omega_0 t)]}{[1 - 12(\omega_0)^2]^2 + 49(\omega_0)^2}$$

satisfies the DE. Applying the LHS of the DE to $y(t)$ gives

$$12\left[\frac{[1 + 2(\omega_0)^2](-\omega_0)^2\cos(\omega_0 t) + [5\omega_0 + 24\omega^3](-\omega_0)^2\sin(\omega_0 t)]}{[1 - 12(\omega_0)^2]^2 + 49(\omega_0)^2}\right]$$

$$+\ 7\left[\frac{[1 + 2(\omega_0)^2](-\omega_0)\sin(\omega_0 t) + [5\omega_0 + 24\omega^3]\ \omega_0\cos(\omega_0 t)]}{[1 - 12(\omega_0)^2]^2 + 49(\omega_0)^2}\right]$$

$$+\left[\frac{[1 + 2(\omega_0)^2]\cos(\omega_0 t) + [5\omega_0 + 24\omega^3]\sin(\omega_0 t)]}{[1 - 12(\omega_0)^2]^2 + 49(\omega_0)^2}\right]$$

$$= \frac{\begin{bmatrix} -12\omega_0^2 - 24\omega_0^4 \\ +35\omega_0^2 + 168\omega_0^4 \\ 1 + 2\omega_0^2 \end{bmatrix}\cos(\omega_0 t) + \begin{bmatrix} -60\omega_0^3 - 288\omega_0^5 \\ -7\omega_0 - 14\omega_0^3 \\ 5\omega_0 + 24\omega^3 \end{bmatrix}\sin(\omega_0 t)}{[1 - 12(\omega_0)^2]^2 + 49(\omega_0)^2}$$

$$= \frac{[1 + 25\omega_0^2 + 144\omega_0^4]\cos(\omega_0 t) - [2\omega_0 + 50\omega_0^3 + 288\omega_0^3]\sin(\omega_0 t)}{1 + 25(\omega_0)^2 + 144(\omega_0)^4}$$

$$= \cos(\omega_0 t) - 2\omega_0\sin(\omega_0 t)$$

which is what the RHS of the DE gives. Thus the DE balances and so the solution is correct.

(d) For input frequency 0.02 hertz we reduce the answer to the form $y(t) = K\cos(\omega_0 t - \theta)$ as follows:

$f_0 = 0.02 \qquad \omega_0 = 2\pi f_0 = 0.04\pi$

$$y(t) = \frac{[1 + 2(\omega_0)^2]\cos(\omega_0 t) + [5\omega_0 + 24\omega^3]\sin(\omega_0 t)]}{[1 - 12(\omega_0)^2]^2 + 49(\omega_0)^2}$$

$1.03582734\cos(\omega_0 t) + 0.67594417\sin(\omega_0 t)$

$$= 0.721037069\cos(\omega_0 t) + 0.472459250\sin(\omega_0 t) = A\cos(\omega_0 t) + B\sin(\omega_0 t)$$

$$= [A^2 + B^2]^{\frac{1}{2}}\left[\frac{A}{[A^2 + B^2]^{\frac{1}{2}}}\cos(\omega_0 t) + \frac{B}{[A^2 + B^2]^{\frac{1}{2}}}\sin(\omega_0 t)\right] = K\cos(\omega_0 t - \theta)$$

where

$K = [A^2 + B^2]^{\frac{1}{2}} = 0.862039557 \qquad \theta = \tan^{-1}[B/A] = 0.580056776$ rads

and so $\quad y(t) = 0.862039557\cos(\omega_0 t - 0.580056776) \quad \longrightarrow \quad$ (A)

(e) Running this problem using the FFT system with the following parameters (refer to Figure 6.6 in the text): N = 256, SAMPLED, T = 100, $\omega_0 = 0.04\pi$, PERIODIC:

- Loading $x(t) = \cos(\omega_0 t)$ into X: Main menu, Create Values, Continue, Create X, Real, Even, 1, Continue, COS(0.04*pi*t), Go, Neither, Go. You are now back on the main menu. Run ANALYSIS.

  Plot **X** and **FRE**. The latter shows 2 coefficients in the Fourier series, each of value ½, located at $\omega = \pm 2\pi/100$. Thus the coefficients are at $\pm 0.04\pi$, as required.

- Loading the transfer frunction: Main menu, Create Values, Continue, Create H(jw), Create, 1, 2, 1, 2, 1, 7, 12, Load, Alpha = 20. You are now back on the main menu with $H(j\omega)$ locked in **F2**. Run Postprocessors, F, Complex multiply, Quit, Run SYNTHESIS. A plot of **Y** appears below:

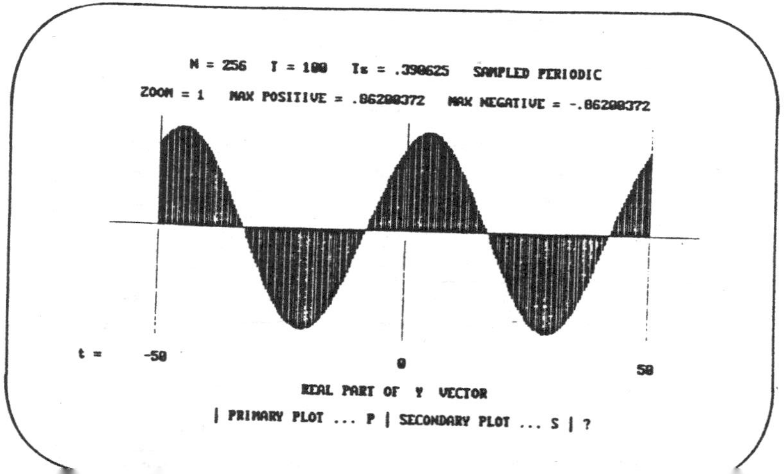

theoretical value of 0.862039557 in (A). This amounts to an error of 0.004%

From SHOW NUMBERS, the first zero crossing is at $k = 43.80945$ (using linear interpolation) and since $T_s = T/N = 100/256 = 0.390625$, the zero crossing is located at $t_c = 17.11306632$

The period of the cosine waveform is 50 seconds, and so the first zero crossing is located at 12.5 seconds after the peak, i.e. at

$$\delta t = t_c = 12.5 = 4.613066320 \text{ seconds after the peak}$$

Thus the phase delay is $\theta = (\delta t/50)2\pi = 0.579695011$ rads, compared to the theoretical value of 0.580056776. This amounts to an error of 0.06%.

6.9 For the network shown in Figure 6.20:

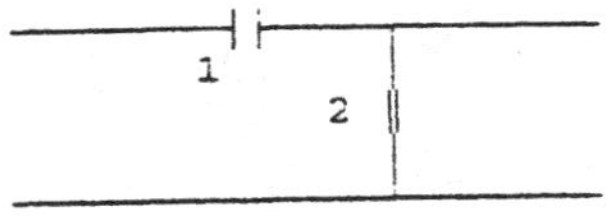

Figure 6.20

Transfer function is

$$H(j\omega) = \frac{2}{2 + \dfrac{1}{j\omega}} = \frac{2(j\omega)}{2(j\omega) + 1}$$

(a) $$y(t) = \left[\frac{2(j\omega)}{2(j\omega)+1}\right]_{\omega \leftarrow 7} 3\exp(j7t) = \left[\frac{j14}{j14+1}\right] 3\exp(j7t)$$

(b) $$y(t) = \left[\frac{2(j\omega)}{2(j\omega)+1}\right]_{\omega \leftarrow -3} 5\exp(-j3t) = \left[\frac{-j6}{-j6+1}\right] 5\exp(-j3t)$$

(c) $$y(t) = \left[\frac{2(j\omega)}{2(j\omega)+1}\right]_{\omega \leftarrow 7} 3\exp(j7t) + \left[\frac{2(j\omega)}{2(j\omega)+1}\right]_{\omega \leftarrow -3} 5\exp(-j3t)$$

$$= \left[\frac{j14}{j14+1}\right] 3\exp(j7t) + \left[\frac{-j6}{-j6+1}\right] 5\exp(-j3t)$$

(d) $$y(t) = \left[\frac{2(j\omega)}{2(j\omega)+1}\right]_{\omega \leftarrow 5\pi/2} \frac{\sin(5\pi/2)}{5\pi/2} \exp(j5\pi t/2)$$

$$= \left[\frac{j5\pi}{j5\pi+1}\right] \frac{\sin(5\pi/2)}{5\pi/2} \exp(j5\pi t/2)$$

(e) $$y(t) = \sum_{n=-\infty}^{\infty} \left[\frac{2(j\omega)}{2(j\omega)+1}\right]_{\omega \leftarrow n\pi/2} \frac{\sin(n\pi/2)}{n\pi/2} \exp(jn\pi t/2)$$

$$= \sum_{n=-\infty}^{\infty} \left[\frac{jn\pi}{jn\pi+1}\right] \frac{\sin(n\pi/2)}{n\pi/2} \exp(jn\pi t/2)$$

The $n=5$ term in the series for the output is:

$$y_5(t) = \left[\frac{j5\pi}{j5\pi+1}\right] \frac{\sin(5\pi/2)}{5\pi/2} \exp(j5\pi t/2)$$

The $n = -9$ term in the series for the output is:

$$y_{-9}(t) = \left[\frac{j(-9)\pi}{j(-9)\pi+1}\right] \frac{\sin(-9\pi/2)}{-9\pi/2} \exp(j(-9)\pi t/2)$$

(f) $$y(t) = \frac{1}{2\pi}\int_{-\infty}^{\infty} \frac{2(j\omega)}{2(j\omega)+1} \frac{2a}{a^2+\omega^2} \exp(j\omega t)\, d\omega$$

(g) $$y(t) = \sum_{n=-\infty}^{\infty} \left[\frac{2(j\omega)}{2(j\omega)+1}\right]_{\omega \leftarrow 2\pi n} \frac{2/\pi}{(1-4n^2)} \exp(j2\pi nt)$$

$$+ \frac{1}{2\pi}\int_{-\infty}^{\infty} \frac{2(j\omega)}{2(j\omega)+1} \exp(-a|\omega|) \exp(j\omega t)\, d\omega$$

$$= \sum_{n=-\infty}^{\infty} \left[\frac{2(j2\pi n)}{2(j2\pi n) + 1} \right] \frac{2/\pi}{(1 - 4n^2)} \exp(j2\pi nt)$$

$$+ \frac{1}{2\pi} \int_{-\infty}^{\infty} \frac{2(j\omega)}{2(j\omega) + 1} \exp(-a|\omega|) \exp(j\omega t)\, d\omega$$

6.10 $$x(t) = \sum_{n=-\infty}^{\infty} \frac{(-1)^n - 1}{1 + n^2} \exp(j3nt)$$

(a) $x_5(t) = -(2/26) \exp(j15t)$ and $\omega_5 = 15$

$x_{15}(t) = -(2/226) \exp(j45t)$ and $\omega_{15} = 45$

$x_{-3}(t) = -(2/10) \exp(-j9t)$ and $\omega_{-3} = -9$

$x_{-7}(t) = -(2/50) \exp(-j21t)$ and $\omega_{-7} = -21$

(b) For the network of Figure 6.16 when $L = 2$, $C_1 = 2$, $C_2 = 2$, $R = 3$:

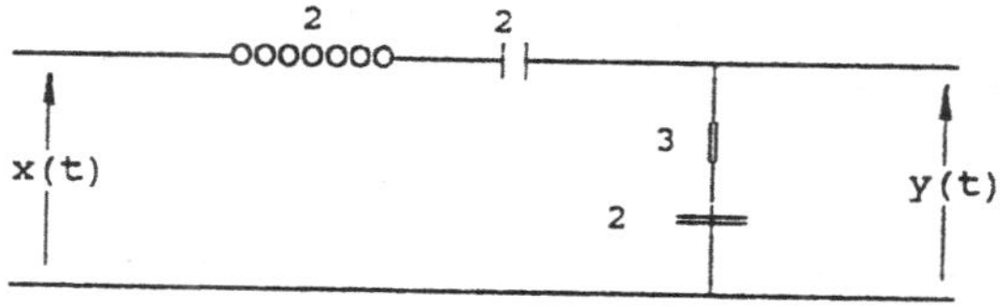

$$H(j\omega) = \frac{3 + \dfrac{1}{j\omega 2}}{j\omega 2 + \dfrac{1}{j\omega 2} + 3 + \dfrac{1}{j\omega 2}} = \frac{6(j\omega) + 1}{4(j\omega)^2 + 6(j\omega) + 2} = \tfrac{1}{2}\frac{6(j\omega) + 1}{2(j\omega)^2 + 3(j\omega) + 1}$$

(c) $$y_5(t) = \tfrac{1}{2} \left[\frac{6(j\omega) + 1}{2(j\omega)^2 + 3(j\omega) + 1} \right]_{\omega \leftarrow 15} (-1/13) \exp(j15t)$$

$$= \tfrac{1}{2} \left[\frac{6(j15) + 1}{2(j15)^2 + 3(j15) + 1} \right] (-1/13) \exp(j15t)$$

$$y_{15}(t) = \tfrac{1}{2} \left[\frac{6(j\omega) + 1}{2(j\omega)^2 + 3(j\omega) + 1} \right]_{\omega \leftarrow 45} (-1/113) \exp(j45t)$$

$$= \tfrac{1}{2} \left[\frac{6(j45) + 1}{2(j45)^2 + 3(j45) + 1} \right] (-1/113) \exp(j45t)$$

$$y_{-3}(t) = \tfrac{1}{2} \left[\frac{6(j\omega) + 1}{2(j\omega)^2 + 3(j\omega) + 1} \right]_{\omega \leftarrow -9} (-1/5) \exp(-j9t)$$

$$= \tfrac{1}{2} \left[\frac{6[j(-9)] + 1}{2[j(-9)]^2 + 3[j(-9)] + 1} \right] (-1/5) \exp(-j9t)$$

$$y_{-7}(t) = \tfrac{1}{2} \left[\frac{6(j\omega) + 1}{2(j\omega)^2 + 3(j\omega) + 1} \right]_{\omega \leftarrow -21} (-1/25) \exp(-j21t)$$

$$= \tfrac{1}{2} \left[\frac{6[j(-21)] + 1}{2[j(-21)]^2 + 3[j(-21)] + 1} \right] (-1/25) \exp(-j21t)$$

(d) $$y(t) = \tfrac{1}{2} \left[\frac{6(j15) + 1}{2(j15)^2 + 3(j15) + 1} \right] (-1/13) \exp(j15t)$$

$$+ \tfrac{1}{2} \left[\frac{6(j45) + 1}{2(j45)^2 + 3(j45) + 1} \right] (-1/113) \exp(j45t)$$

$$+ \tfrac{1}{2} \left[\frac{6[j(-9)] + 1}{2[j(-9)]^2 + 3[j(-9)] + 1} \right] (-1/5) \exp(-j9t)$$

$$+ \tfrac{1}{2} \left[\frac{6[j(-21)] + 1}{\quad} \right] (-1/25) \exp(-j21t)$$

(e) $$y(t) = \tfrac{1}{2} \sum_{n=-\infty}^{\infty} H(j\omega)\Big|_{\omega \leftarrow 3n} \left[\frac{(-1)^n - 1}{1 + n^2} \right] \exp(j3nt)$$

$$= \tfrac{1}{2} \sum_{n=-\infty}^{\infty} \left[\frac{6(j\omega) + 1}{2(j\omega)^2 + 3(j\omega) + 1} \right]_{\omega=3n} \left[\frac{(-1)^n - 1}{1 + n^2} \right] \exp(j3nt)$$

$$= \tfrac{1}{2} \sum_{n=-\infty}^{\infty} \left[\frac{6[j(3n)] + 1}{2[j(3n)]^2 + 3[j(3n)] + 1} \right] \left[\frac{(-1)^n - 1}{1 + n^2} \right] \exp(j3nt)$$

6.11 (a)

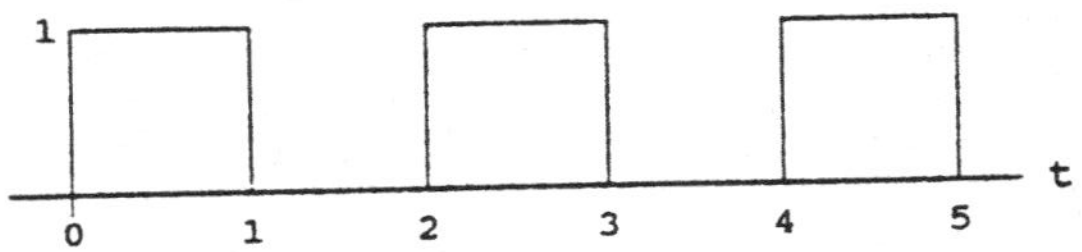

$$x_p(t) = \begin{cases} 1 & (0 < t < 1) \\ 0 & (1 < t < 2) \end{cases} \qquad x(t + 2) = x(t)$$

For the single-period defining pulse, $x(t) = \text{Rect}(t - \tfrac{1}{2})$:

$$x'(t) = \delta(t) - \delta(t - 1)$$

$$j\omega\, X(\omega) = 1 - \exp(-j\omega t)$$

$$X(\omega) = \frac{1 - \exp(-j\omega t)}{j\omega}$$

For the periodic waveform, $T_0 = 2$, $\omega_0 = \pi$, and so

$$X_p(n) = \frac{1}{T_0} X(n\omega_0) = \tfrac{1}{2} \frac{1 - \exp(-jn\pi)}{jn\pi} = j/2\pi \, \frac{(-1)^n - 1}{n} \qquad (n \neq 0)$$

Fourier series: $$x_p(t) = \tfrac{1}{2} + j/2\pi \sum_{n=-\infty}^{\infty} \frac{(-1)^n - 1}{n} \exp(jn\pi t)$$

The DE is $y'' + 4y' + 5y = 3x' + x \quad (\forall\, t)$

Fourier transformation gives

$$[(j\omega)^2 + 4j\omega + 5]\, Y(\omega) = [3(j\omega) + 1]\, X(\omega)$$

$$Y(\omega) = \frac{\text{[cut off]}}{(j\omega)^2 + 4j\omega + 5} X(\omega) \qquad H(j\omega) = \frac{\text{[cut off]}}{(j\omega)^2 + 4j\omega + 5}$$

The response is

$$y_p(t) = \sum_{n=-\infty}^{\infty} H(j\omega)\Big|_{\omega \leftarrow n\pi} X_p(n) \exp(jn\pi t)$$

$$= \tfrac{1}{2} H(0) + \sum_{\substack{n=-\infty \\ n \neq 0}}^{\infty} H(jn\pi)\, X_p(n) \exp(jn\pi t)$$

$$= 1/10 + j/2\pi \sum_{\substack{n=-\infty \\ n \neq 0}}^{\infty} \frac{3(jn\pi) + 1}{(jn\pi)^2 + 4jn\pi + 5} \, \frac{(-1)^n - 1}{n} \exp(jn\pi t)$$

(b) The Fourier series for the response is

$$y_p(t) = \sum_{n=-\infty}^{\infty} Y_p(n) \exp(jn\pi t)$$

where

$$Y_p(n) = \begin{cases} j/2\pi \, \dfrac{3(jn\pi) + 1}{(jn\pi)^2 + 4jn\pi + 5} \, \dfrac{(-1)^n - 1}{n} & (n \neq 0) \\ 1/10 & (n = 0) \end{cases}$$

$n = 0$: $\quad Y_p(0) = 1/10, \quad |Y_p(0)| = 1/10, \quad \theta(0) = 0, \quad P(0) = 1/100$

$n = 1$: $$Y_p(1) = j/2\pi \, \frac{3(j\pi) + 1}{(j\pi)^2 + 4j\pi + 5} \, \frac{(-1)^1 - 1}{n}$$

$$= -j/\pi \, \frac{3j\pi + 1}{(5 - \pi)^2 + j4\pi}$$

$$= \frac{1}{\pi} \, \frac{3\pi - j}{(5 - \pi^2) + j4\pi} \, \frac{(5 - \pi^2) - j4\pi}{(5 - \pi^2) - j4\pi}$$

$$= \frac{[3\pi(5 - \pi)^2 - 4\pi] - j[(5 - \pi^2) + 12\pi^2]}{\pi[(5 - \pi^2)^2 + 16\pi^2]}$$

$$= \frac{-58.46131089 - j113.5656484}{570.5971618}$$

$$= -0.102456365 - j0.199029466 = A(1) + jB(1)$$

$$|Y_p(1)| = [A(1)^2 + B(1)^2]^{\frac{1}{2}} = 0.223852708$$

$$\theta(1) = \tan^{-1}[B(1)/A(1)] = -2.046197808 \qquad P(1) = |Y_p(1)|^2 = 0.050110035$$

n = 2: All values zero

$$n = 3: \quad Y_p(1) = j/2\pi \frac{3(j3\pi) + 1}{(j3\pi)^2 + 4j3\pi + 5} \frac{(-1)^3 - 1}{3}$$

$$= j/2\pi \frac{j9\pi + 1}{(5 - 9\pi)^2 + j12\pi} (-2/3)$$

$$= \frac{1}{3\pi} \frac{9\pi - j}{(5 - 9\pi^2) + j12\pi} \frac{(5 - 9\pi^2) - j12\pi}{(5 - 9\pi^2) - j12\pi}$$

$$= \frac{[9\pi(5 - 9\pi^2) - 12\pi] - j[(5 - 9\pi^2) + 108\pi^2]}{3\pi[(5 - 9\pi^2)^2 + 144\pi^2]}$$

$$= \frac{-2407.835855 - 982.0908364}{79621.41973}$$

$$= -0.030241507 - 0.012334505 = A(3) + jB(3)$$

$$|Y_p(3)| = [A(3)^2 + B(3)^2]^{\frac{1}{2}} = 0.032659785$$

$$\theta(3) = \tan^{-1}[B(3)/A(3)] = -2.754317829 \qquad P(3) = |Y_p(3)|^2 = 0.001066662$$

n = -1:

$A(-1)$	$= A(1)$	because	$A(n)$	is even
$B(-1)$	$= -B(1)$	because	$B(n)$	is odd
$\|F(-1)\|$	$= \|F(1)\|$	because	$\|F(n)\|$	is even
$\theta(-1)$	$= \theta(1)$	because	$\theta(n)$	is odd
$P(-1)$	$= P(1)$	because	$P(n)$	is even

n = -2: All values zero

n = -3:

$A(-3)$	$= A(3)$	because	$A(n)$	is even
$B(-3)$	$= -B(3)$	because	$B(n)$	is odd
$\|F(-3)\|$	$= \|F(3)\|$	because	$\|F(n)\|$	is even
$\theta(-3)$	$= \theta(3)$	because	$\theta(n)$	is odd
$P(-3)$	$= P(3)$	because	$P(n)$	is even

(c) Using the FFT system to verify the numerical values that we obtained in (b), with N = 256, SAMPLED, T = 2, PERIODIC, alias-level 20:

To load the input waveform into the X vector: **Main menu, Create Values, Continue, Create X, Real, Neither, Left, 2, 1, Continue, 1, 0, Go, Neither Go. You are now back on the main menu. Run ANALYSIS.**

To load the transfer function into F2: **Main menu, Create Values, Continue, Create H(jw), Create, 1, 2, 1, 3, 5, 4, 1, Load, Alpha = 20. You are now back on the main menu.**

Run **Postprocessors, F, Complex multiply, Quit.** (You are now back on the main menu and the spectrum of the response is in **F**.) Show **Numbers, F, Complex.**

The table below shows the FFT values, the formula values as obtained above, and the relative errors of the FFT values.

		FFT	Formula	error (%)
n = 0	A(0)	0.10002714	0.10000000	0.027
	B(0)	0	0	– – –
	|F(0)|	0.10002713	0.10000000	0.027
	θ(0)	0	0	– – –
	P(0)	1.0005430e-2	0.01000000	0.054
n = 1	A(1)	-0.10243642	-0.102456365	0.019
	B(1)	-0.19903676	-0.199029466	0.0037
	|F(1)|	0.22385006	0.223852708	0.0012
	θ(1)	-2.04610370	-2.046197808	0.0046
	P(1)	5.0108849e-2	5.0110035e-2	0.0024
n = 3	A(3)	-3.0212592e-2	-3.0241507e-2	0.096
	B(3)	-1.2334692e-2	-1.2334505e-2	0.0015
	|F(3)|	3.2633500e-2	3.2659785e-2	0.080
	θ(3)	-2.7539831	-2.754317829	0.012
	P(3)	1.0649453e-3	1.0666620e-3	0.16

(d) In the figure below we show a plot of two periods of the response, as obtained from the FFT system.

6.12 To prove that when two networks are cascaded in such a way that there is no loading, the magnitudes of their frequency responses multiply and their phases add.

(a) Using polar notation, let the transfer functions be

$$H_1(j\omega) = |H_1(j\omega)| \exp[j\theta_1(\omega)] \quad \text{and} \quad H_2(j\omega) = |H_2(j\omega)| \exp[j\theta_2(\omega)]$$

Then the overall transfer function will be

$$H_{1,2}(j\omega) = H_1(j\omega)\ H_2(j\omega)$$

$$= |H_1(j\omega)| \exp[j\theta_1(\omega)]\ |H_2(j\omega)| \exp[j\theta_2(\omega)]$$

$$= |H_1(j\omega)|\ |H_2(j\omega)| \exp[j(\theta \ldots$$

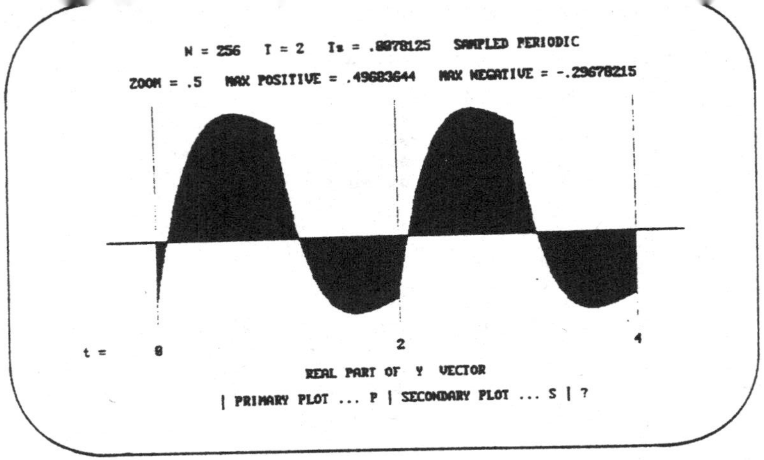

Thus the magnitude response of the overall transfer function will be

$$|H_{1,2}(j\omega)| = |H_1(j\omega)|\ |H_2(j\omega)| \longrightarrow$$

which means that the magnitudes of the frequency responses multiply.

The phase response of the overall transfer function will be

$$\theta_{1,2}(j\omega) = \theta_1(j\omega) + \theta_2(j\omega) \longrightarrow$$

which means that the phases add.

(b) In cartesian notation, let the transfer functions be

$$H_1(j\omega) = A_1(\omega) + jB_1(\omega) \quad \text{and} \quad H_2(j\omega) = A_2(\omega) + jB_2(\omega)$$

Then the overall transfer function will be

$$H_{1,2}(j\omega) = H_1(j\omega)\ H_2(j\omega)$$
$$= [A_1(\omega) + jB_1(\omega)]\ [A_2(\omega) + jB_2(\omega)]$$
$$= [A_1 A_2 - B_1 B_2] + j[A_1 B_2 + A_2 B_1] \quad (*)$$

From (*), the overall magnitude response will be

$$H_{1,2}(j\omega) = \left[[A_1 A_2 - B_1 B_2]^2 + [A_1 B_2 + A_2 B_1]^2 \right]^{\frac{1}{2}}$$

$$= \left[\begin{matrix} A_1^2 A_2^2 & - 2A_1 A_2 B_1 B_2 & + B_1^2 B_2^2 \\ A_1^2 B_2^2 & + 2A_1 A_2 B_1 B_2 & + A_2^2 B_1^2 \end{matrix} \right]$$

$$= \left[A_1^2 [A_2^2 + B_2^2] + B_1^2 [A_2^2 + B_2^2] \right]^{\frac{1}{2}}$$

$$= \left[[A_1^2 + B_1^2]\ [A_2^2 + B_2^2] \right]^{\frac{1}{2}} = |H_1(j\omega)|\ |H_2(j\omega)| \longrightarrow$$

Thus the magnitude response of the overall transfer function will be

$$|H_{1,2}(j\omega)| = |H_1(j\omega)|\ |H_2(j\omega)|$$

which means that the magnitudes of the frequency responses multiply.

From (*), the phase response of the overall transfer function will be

$$\theta_{1,2}(j\omega) = \tan^{-1} \frac{A_1 B_2 + A_2 B_1}{A_1 A_2 - B_1 B_2} = \tan^{-1} \frac{B_2/A_2 + B_1/A_1}{1 - \frac{B_1}{A_1}\frac{B_2}{A_2}}$$

$$= \tan^{-1} \frac{\tan(\theta_1) + \tan(\theta_2)}{1 - \tan(\theta_1)\tan(\theta_2)} = \tan^{-1} \left[\tan(\theta_1 + \theta_2) \right] = \theta_1 + \theta_2 \longrightarrow$$

which means that the phases add.

6.13

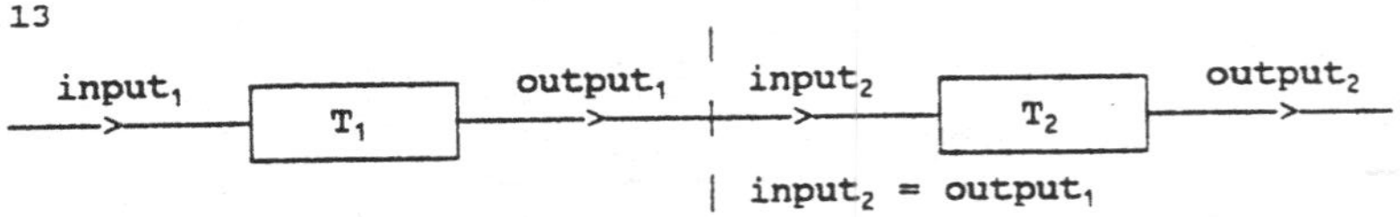

$$P_1(D) \left[output_1 \right] = P_2(D) \left[input_1 \right] \qquad (1)$$

and

$$Q_1(D) \left[output_2 \right] = Q_2(D) \left[input_2 \right] \qquad (2)$$

But

$$output_1 = input_2 \qquad (3)$$

Therefore (1) becomes

$$P_1(D) \left[input_2 \right] = P_2(D) \left[input_1 \right] \qquad (4)$$

and so, operating on (4) by Q_2,

$$Q_2(D)P_1(D)\left[\text{ input}_2 \right] = Q_2(D)P_2(D)\left[\text{ input}_1 \right] \qquad (5)$$

Now

$$Q_2(D)P_1(D) = P_1(D)Q_2(D) \qquad (6)$$

and

$$Q_2(D)P_2(D) = P_2(D)Q_2(D) \qquad (7)$$

because these are polynomials in D, and so they commute, and so, by (6) we see that (5) becomes

$$P_1(D)Q_2(D)\left[\text{ input}_2 \right] = Q_2(D)P_2(D)\left[\text{ input}_1 \right] \qquad (8)$$

But, by (2):

$$Q_2(D)\left[\text{ input}_2 \right] = Q_1(D)\left[\text{ output}_2 \right] \qquad (9)$$

and so (8) becomes

$$P_1(D)Q_1(D)\left[\text{ output}_2 \right] = Q_2(D)P_2(D)\left[\text{ input}_1 \right]$$

$$= P_2(D)Q_2(D)\left[\text{ input}_1 \right] \text{ (by (7))} \qquad (10)$$

Thus the overall transformation is

$$P_1(D)Q_1(D)\left[\text{ output}_2 \right] = P_2(D)Q_2(D)\left[\text{ input}_1 \right] \qquad \text{QED} \qquad (11)$$

(b) Reversing the order of the cascade would interchange P and Q, giving

$$Q_1(D)P_1(D)\left[\text{ output}_2 \right] = Q_2(D)P_2(D)\left[\text{ input}_1 \right] \qquad (12)$$

But these operators commute, and so (12) is the same as (11). QED

Note: This is an exceptional occurrence in the theory of linear transformations, since in general two such transformations do not commute. However polynomials in D do commute when the coefficients are constant, which is what makes these transformations commute.

(c) From the above we can infer the following: A transfer function $H_1(j\omega)$ is simply the quotient of two polynomials in $j\omega$. Clearly then, two such transfer functions will commute, i.e.

$$H_1(j\omega)\ H_2(j\omega) = H_2(j\omega)\ H_1(j\omega) \qquad (13)$$

That being the case, if $H_2(j\omega)$ follows $H_1(j\omega)$ then

$$Y_1(\omega) = H_1(j\omega)\ X_1(\omega) \qquad (14)$$

and

$$Y_2(\omega) = H_2(j\omega)\ X_2(\omega) \qquad (15)$$

But $X_2(\omega) = Y_1(\omega)$, and so, (15) becomes

$$Y_2(\omega) = H_2(j\omega)\ Y_1(\omega)$$

$$= H_2(j\omega)\ H_1(j\omega)\ X_1(\omega) \quad \text{(by (14)} \qquad (16)$$

giving as the overall transfer function

$$H_{1,2}(j\omega) = H_2(j\omega)\ H_1(j\omega) \qquad (17)$$

If we reverse the order of the cascade then $H_2(j\omega)$ and $H_1(j\omega)$ will change places. But they also commute. So nothing would be altered and (17) would still stand. i.e. the order of the cascade is immaterial. Thus the result is an LTI network with frequency response either

$$H_1(j\omega)\ H_2(j\omega) \quad \text{or} \quad H_2(j\omega)\ H_1(j\omega) \qquad (18)$$

6.14

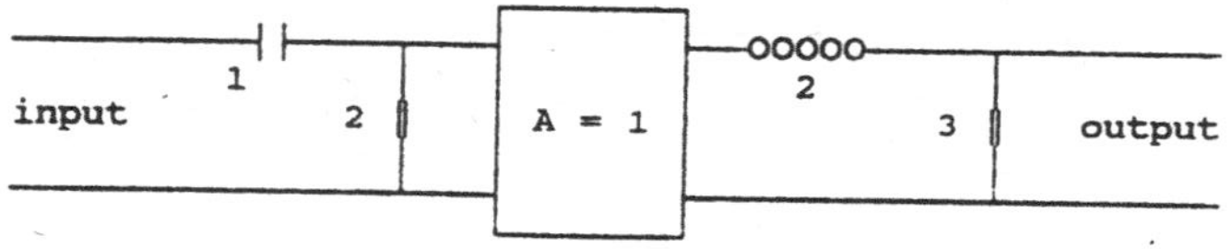

Figure 6.21

(a) The frequency response for the first network is

$$H_1(j\omega) = \frac{2}{2 + 1/j\omega} = \frac{2(j\omega)}{2(j\omega) + 1}$$

For the second network it is

$$H_2(j\omega) = \frac{3}{j\omega 2 + 3}$$

Because the networks do not load each other the overall transfer

$$H(j\omega) = H_1(j\omega)\ H_2(j\omega) = \frac{}{2(j\omega) + 1}\ \frac{}{j\omega 2 + 3} = \frac{}{4(j\omega)^2 + 8(j\omega) + 3}$$

(b)

(1) $x(t) = 3\exp(j7t)$

$$y(t) = \left[\frac{6(j\omega)}{4(j\omega)^2 + 8(j\omega) + 3}\right]_{\omega \leftarrow 7} 3\exp(j7t)$$

$$= \left[\frac{6(j7)}{4(j7)^2 + 8(j7) + 3}\right] 3\exp(j7t)$$

(2) $x(t) = 2\exp(-j5t)$

$$y(t) = \left[\frac{6(j\omega)}{4(j\omega)^2 + 8(j\omega) + 3}\right]_{\omega \leftarrow -5} 2\exp(-j5t)$$

$$= \left[\frac{6(-j5)}{4(-j5)^2 + 8(-j5) + 3}\right] 2\exp(-j5t)$$

(3) $x(t) = 3\exp(j7t) + 2\exp(-j5t)$

$$y(t) = \left[\frac{6(j\omega)}{4(j\omega)^2 + 8(j\omega) + 3}\right]_{\omega \leftarrow 7} 3\exp(j7t)$$

$$+ \left[\frac{6(j\omega)}{4(j\omega)^2 + 8(j\omega) + 3}\right]_{\omega \leftarrow -5} 2\exp(-j5t)$$

$$= \frac{6(j7)}{4(j7)^2 + 8(j7) + 3} 3\exp(j7t) + \frac{6(-j5)}{4(-j5)^2 + 8(-j5) + 3} 2\exp(-j5t)$$

(4) $x(t) = \dfrac{\sin(5\pi/2)}{5\pi/2}\exp(j5\pi t/2)$

$$y(t) = \left[\frac{6(j\omega)}{4(j\omega)^2 + 8(j\omega) + 3}\right] \left.\frac{\sin(5\pi/2)}{5\pi/2}\exp(j5\pi t/2)\right|_{\omega \leftarrow 5\pi/2}$$

$$= \left[\frac{}{4(j5\pi/2)^2 + 8(j5\pi/2) + 3}\right] \frac{}{5\pi/2}\exp(j5\pi t/2)$$

(5) $x(t) = \displaystyle\sum_{n=-\infty}^{\infty} \frac{\sin(n\pi/2)}{n\pi/2}\exp(jn\pi t/2)$

$$y(t) = \sum_{n=-\infty}^{\infty} \left[\frac{6(j\omega)}{4(j\omega)^2 + 8(j\omega) + 3}\right] \left.\frac{\sin(n\pi/2)}{n\pi/2}\exp(jn\pi t/2)\right|_{\omega \leftarrow n\pi/2}$$

$$= \sum_{n=-\infty}^{\infty} \left[\frac{6(jn\pi/2)}{4(jn\pi/2)^2 + 8(jn\pi/2) + 3}\right] \frac{\sin(n\pi/2)}{n\pi/2}\exp(jn\pi t/2)$$

$$y_5(t) = \left[\frac{6(j5\pi/2)}{4(j5\pi/2)^2 + 8(j5\pi/2) + 3}\right] \frac{\sin(5\pi/2)}{5\pi/2}\exp(j5\pi t/2)$$

$$y_{-7}(t) = \left[\frac{6(-j7\pi/2)}{4(-j7\pi/2)^2 + 8(-j7\pi/2) + 3}\right] \frac{\sin(-7\pi/2)}{-n\pi/2}\exp(-j7\pi t/2)$$

(6) $x(t) = \dfrac{1}{2\pi}\displaystyle\int_{-\infty}^{\infty} \frac{2a}{a^2 + \omega^2}\exp(j\omega t)\ d\omega$

$$y(t) = \frac{1}{2\pi}\int_{-\infty}^{\infty} \left[\frac{6(j\omega)}{4(j\omega)^2 + 8(j\omega) + 3}\right] \frac{2a}{a^2 + \omega^2}\exp(j\omega t)\ d\omega$$

(7) $x(t) = \dfrac{1}{2\pi}\displaystyle\int_{-\infty}^{\infty} \exp(-a|\omega|)\ \exp(j\omega t)\ d\omega + \sum_{n=-\infty}^{\infty} \frac{2}{\pi(1 - 4n^2)}\exp(j2\pi nt)$

$$y(t) = \frac{1}{2\pi}\int_{-\infty}^{\infty}\left[\frac{6(j\omega)}{4(j\omega)^2 + 8(j\omega) + 3}\right]\exp(-a|\omega|)\exp(j\omega t)\,d\omega$$

$$+ \sum_{n=-\infty}^{\infty}\left[\frac{6(j\omega)}{4(j\omega)^2 + 8(j\omega) + 3}\right]_{\omega \leftarrow 2\pi n}\frac{2}{\pi(1 - 4n^2)}\exp(j2\pi nt)$$

$$= \frac{1}{2\pi}\int_{-\infty}^{\infty}\left[\frac{6(j\omega)}{4(j\omega)^2 + 8(j\omega) + 3}\right]\exp(-a|\omega|)\exp(j\omega t)\,d\omega$$

$$+ \sum_{n=-\infty}^{\infty}\left[\frac{6(j2\pi n)}{4(j2\pi n)^2 + 8(j2\pi n) + 3}\right]\frac{2}{\pi(1 - 4n^2)}\exp(j2\pi nt)$$

6.15

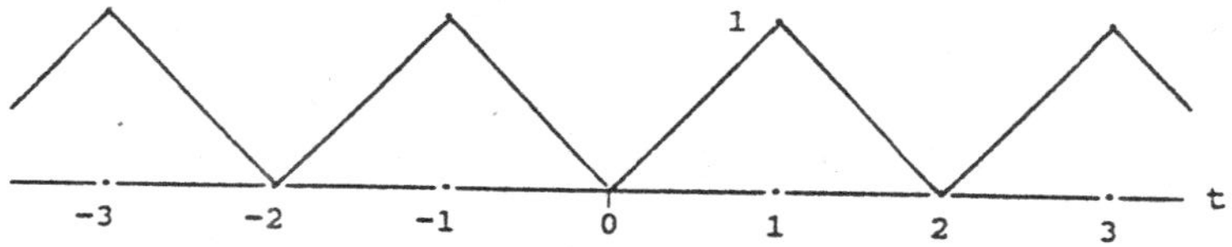

(a) The Fourier transform for the periodic function

$$x_p(t) = |t| \quad (-1 < t < 1), \qquad x_p(t + 2) = x_p(t)$$

is found as follows. First we find its Fourier series coefficients:

$$X(n) = 1/T_0 \int_{-T_0/2}^{T_0/2} x_p(t)\exp(jn\omega_0 t)\,dt$$

in which $T_0 = 2$, $\omega_0 = 2\pi/2 = \pi$, and $x_p(t)$ is even, and so we continue as

$$\ldots = 2/2\int_0^1 t\cos(n\pi t)\,dt = \left[\frac{t\sin(n\pi t)}{n\pi} + \frac{\cos(n\pi t)}{n^2\pi^2}\right]_0^1$$

$$= \frac{\cos(n\pi) - 1}{n^2\pi^2} = \frac{(-1)^n - 1}{n^2\pi^2} \qquad (n \neq 0)$$

For $n = 0$: $X(0) = 1/2$ Then the Fourier series is

$$x_p(t) = \sum_{n=-\infty}^{\infty} X(n)\exp(jn\omega_0 t) = \tfrac{1}{2} + \sum_{\substack{n=-\infty\\ n\neq 0}}^{\infty}\frac{(-1)^n - 1}{n^2\pi^2}\exp(jn\pi t) \qquad (A)$$

Fourier transforming each of the complex exponentials, the Fourier transform is

$$X_p(\omega) = \pi\delta(\omega) + \sum_{\substack{n=-\infty\\ n\neq 0}}^{\infty}\frac{(-1)^n - 1}{n^2\pi^2}2\pi\,\delta(\omega - n\pi)$$

$$= \pi\delta(\omega) + 2/\pi\sum_{\substack{n=-\infty\\ n\neq 0}}^{\infty}\frac{(-1)^n - 1}{n^2}\delta(\omega - n\pi) \longrightarrow \qquad (B)$$

(b) Sketches of $x_p(t)$, its line spectrum $X_p(n)$ and its Fourier transform $X(\omega)$ are given in the Answer section of the text.

(c) Confirming what we have done in (a) and (b) by using the FFT system with N = 1024, SAMPLED: First we use PERIODIC, T = 2, to confirm the Fourier coefficients: Main menu A, Create Values, Continue, Create X, Real, Even, 1, t, Go, Neither, Go. You are now back on the main menu. Run ANALYSIS.

From Show Numbers, F, Complex, we obtain the values shown in the following table. Clearly they confirm the correctness of the formula (A) for the **Fourier coefficients** X(n).

n	A_{FFT}	$A_{formula}$
0	0.50000000	0.500000000
1	-0.20264300	-0.202642367
2	0	0
3	-2.2516454e-2	-2.2515818e-2
4	0	0
5	-8.1063305e-3	-8.1056946e-3

We now run the problem again, but this time using PULSE rather than PERIODIC. Then: Main menu A, Create Values, Continue, Create X, Real, Even, 1, t, Go, Neither, Go. You are now back on the main menu. Run ANALYSIS.

From Show Numbers, F, Eternal, we obtain the values shown in the following table. Clearly they confirm the correctness of the formula (B) for the **weights of the Dirac deltas** in the series for $X(\omega)$.

n	A_{FFT}	$A_{formula}$
0	3.1415927	3.141592654
1	-1.2732435	-1.273239545
2	0	0
3	-1.4147506	-1.414171061
4	0	0

below, and let $y_p(t)$ be the output. The transfer function is

$$H(j\omega) = \frac{\dfrac{1}{j\omega 0.05}}{\dfrac{1}{j\omega 0.05} + 1 + j\omega 0.05} = \frac{1}{1 + 0.05(j\omega) + .0025(j\omega)^2}$$

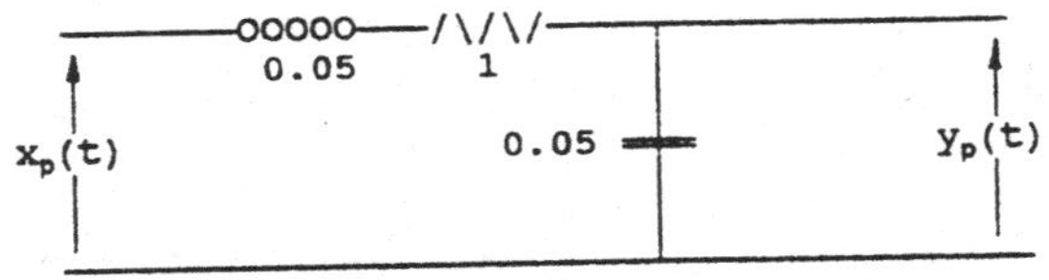

Figure 6.22: RLC integrator

from which

$$H(j\omega) = \frac{400}{(j\omega)^2 + 20(j\omega) + 400} \longrightarrow$$

(1) Starting from the Fourier series of $x_p(t)$ we have

$$y_p(t) = \sum_{n=-\infty}^{\infty} H(jn\omega_0)\, X(n)\, \exp(jn\omega_0 t) \qquad \omega_0 = 2\pi/T_0 = \pi$$

$$= H(0)/2 + \frac{1}{\pi^2} \sum_{\substack{n=-\infty \\ n\neq 0}}^{\infty} \frac{400}{(jn\pi)^2 + 20(jn\pi) + 400} \frac{(-1)^n - 1}{n^2} \exp(jn\pi t)$$

$$= \tfrac{1}{2} + \frac{1}{\pi^2} \sum_{\substack{n=-\infty \\ n\neq 0}}^{\infty} \frac{400}{(jn\pi)^2 + 20(jn\pi) + 400} \frac{(-1)^n - 1}{n^2} \exp(jn\pi t) \qquad \text{(C)}$$

(2) Starting from the Fourier transform of $x_p(t)$, we have

$$Y_p(\omega) = H(j\omega)\, X_p(\omega)$$

$$= \frac{400}{(j\omega)^2 + 20(j\omega) + 400} \left[\pi\delta(\omega) + 2/\pi \sum_{\substack{n=-\infty \\ n\neq 0}}^{\infty} \frac{(-1)^n - 1}{n^2} \delta(\omega - n\pi) \right]$$

$$= \pi\delta(\omega) + 2/\pi \sum_{\substack{n=-\infty \\ n\neq 0}}^{\infty} \frac{400}{(jn\pi)^2 + 20(jn\pi) + 400} \frac{(-1)^n - 1}{n^2} \delta(\omega - n\pi)$$

Then $y(t) = 1/2\pi \int_{-\infty}^{\infty} Y_p(\omega) \exp(j\omega t)\, d\omega$

$$= \frac{1}{2\pi} \int_{-\infty}^{\infty} \left[\pi\delta(\omega) + \frac{2}{\pi} \sum_{\substack{n=-\infty \\ n\neq 0}}^{\infty} \frac{400}{(jn\pi)^2+20(jn\pi)+400} \frac{(-1)^n-1}{n^2} \delta(\omega-n\pi) \right] \exp(j\omega t)\, d\omega$$

$$= \tfrac{1}{2} \int_{-\infty}^{\infty} \delta(\omega) \exp(j0t)\, d\omega$$

$$+ \frac{1}{\pi^2} \int_{-\infty}^{\infty} \left[\sum_{\substack{n=-\infty \\ n\neq 0}}^{\infty} \frac{400}{(jn\pi)^2+20(jn\pi)+400} \frac{(-1)^n-1}{n^2} \delta(\omega-n\pi) \right] \exp(j\omega t)\, d\omega$$

$$= \tfrac{1}{2} + \frac{1}{\pi^2} \left[\sum_{\substack{n=-\infty \\ n\neq 0}}^{\infty} \frac{400}{(jn\pi)^2+20(jn\pi)+400} \frac{(-1)^n-1}{n^2} \exp(jn\pi t) \int_{-\infty}^{\infty} \delta(\omega-n\pi)\, d\omega \right]$$

$$= \tfrac{1}{2} + \frac{1}{\pi^2} \sum_{\substack{n=-\infty \\ n\neq 0}}^{\infty} \frac{400}{(jn\pi)^2 + 20(jn\pi) + 400} \frac{(-1)^n - 1}{n^2} \exp(jn\pi t) \qquad \text{(D)}$$

which is exactly the same as (C), as it should be.

(e) From (C) or (D) the Fourier coefficients for $y_p(t)$ are as follows:

$$Y(0) = \tfrac{1}{2}$$

$$Y(n) = \frac{1}{\pi^2} \frac{400}{(jn\pi)^2 + 20(jn\pi) + 400} \frac{(-1)^n - 1}{n^2} \qquad (n \neq 0)$$

We now derive $Re[Y_p(n)]$, $Im[Y_p(n)]$, $|Y_p(n)|$, $\theta_p(n)$, $P_p(n)$ for $-3 \leq n \leq 3$:

- n = 0: $Re[Y_p(0)] = \tfrac{1}{2}$, $Im[Y_p(n)] = 0$, $|Y_p(n)| = \tfrac{1}{2}$, $\theta_p(n) = 0$
  $P_p(n) = \tfrac{1}{4}$
- n = 1:

$$Y_p(1) = \frac{1}{\pi^2} \frac{400}{(j\pi)^2 + 20(j\pi) + 400} \frac{(-1)^1 - 1}{1^2}$$

$$= \frac{-800}{\pi^2}\,\frac{1}{(400-\pi)^2 + j20\pi} = \frac{-800}{\pi^2}\,\frac{(400-\pi^2) - j20\pi}{(400-\pi^2)^2 + 400\pi^2}$$

$$= -0.202515955 + j0.032615897$$

$Re[Y_p(1)] = -0.202515955 \qquad Im[Y_p(1)] = 0.032615897$

$|Y_p(1)| = 0.205125593 \qquad \theta_p(1) = 2.981910389 \qquad P_p(1) = 4.2076508e-2$

- n = 2: All values are zero

- n = 3:

$$Y_p(3) = \frac{1}{\pi^2}\,\frac{400}{(j3\pi)^2 + 20(j3\pi) + 400}\,\frac{(-1)^3 - 1}{3^2}$$

$$= \frac{-800}{9\pi^2}\,\frac{1}{(400-9\pi^2) + j60\pi} = \frac{-800}{9\pi^2}\,\frac{(400-9\pi^2) - j60\pi}{(400-9\pi^2)^2 + 3600\pi^2}$$

$$= -2.1173619e-2 + j1.2826068e-2$$

$Re[Y_p(3)] = -2.1173619e-2 \qquad Im[Y_p(3)] = 1.2826068e-2$

$|Y_p(3)| = 2.4755407e-2 \qquad \theta_p(3) = 2.5969508 \qquad P_p(3) = 6.1283018e-4$

In the Answer section of the text we show a comparison between these formula values and values derived from the FFT system. The errors are seen to be extremely small.

For the FFT run we proceeded as follows (Refer to Figure 6.6 in the text): N = 1024, SAMPLED, T = 2, PERIODIC:

Main-menu, Create Values, Continue, Create X, Real, Even, 1, Continue, t, Go, Neither, Go. You are now back on the main menu. Run ANALYSIS. The spectrum of the periodic waveform is now in **F**.

To load the transfer function into **F2**: Main menu, Create Values, Continue, Create H(jw), Create, 0, 2, 400, 400, 20, 1, Alpha = 10. You are now back on the main menu with H(jω) in **F2** which is locked.

Run Postprocessors, F, Complex Multiply, Quit. You are now back on the main menu. Run SYNTHESIS. The response of the network is now in **Y**. (See plot of $y_p(t)$ below.)

Show Numbers, F, Complex: The values of the Fourier coefficients for $y_p(t)$ are now on your screen. They should be the same as in the table in the Answer section.

(g) From (C) we see that the coefficients for $y_p(t)$ are going to zero like $1/n^4$ for large n. That means that $y_p(t)$, $y_p'(t)$ and $y_p''(t)$ should all be continuous but that $y_p'''(t)$

(h) and (i) In the first plot we show $y_p(t)$. Observe that it is everywhere continuous. In the following three plots we show $y_p'(t)$, $y_p''(t)$ and $y_p'''(t)$. These plots were derived as follows:

The plot of $y_p'(t)$: Run Postprocessors, F, Central Diff 1st. This "differentiates" the waveform by acting on its spectrum. (Refer to Section 15.3 in the text.) QUIT. You are now back on the main menu. Run SYNTHESIS.

Plot the **Y** vector. It represents the FFT's version of $y_p'(t)$. (See figure below.) Observe that it is everywhere continuous.

Main menu, Run Postprocessors, F, Central Diff 1st. we have now differentiated twice. QUIT and Run SYNTHESIS. Then plot **Y** again. It represents the FFT's version of $y_p''(t)$. (See figure below.) Again we see that it is everywhere continuous.

Main menu, Run Postprocessors, F, Central Diff 1st. We have now differentiated thrice. QUIT and Run SYNTHESIS. Then plot **Y** again. It represents the FFT's version of $y_p'''(t)$. (See figure below.) This time we observe discontinuities.

Thus the FFT system has completely confirmed the assertion that we made in (g) above.

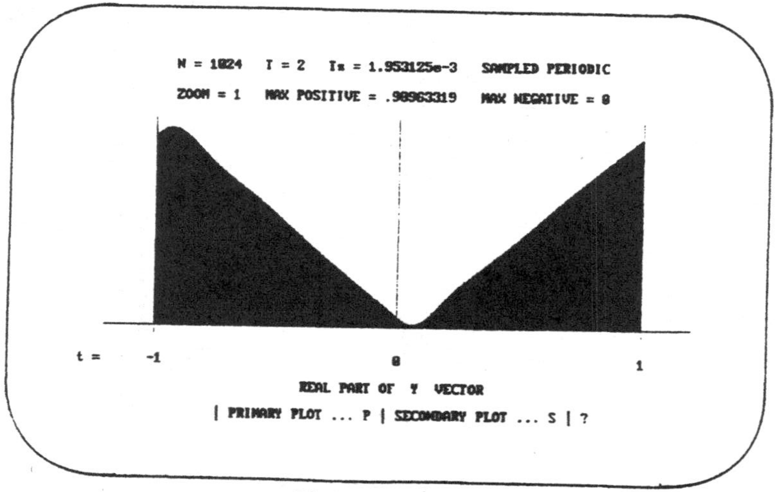

Plot of $y_p(t)$

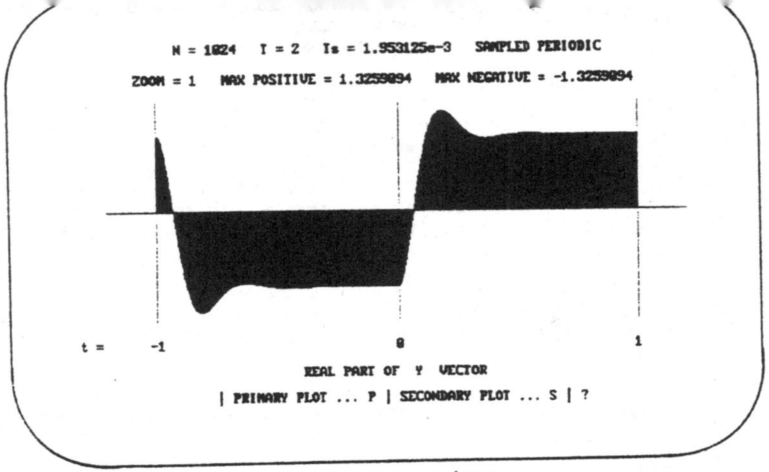

Plot of $y_p'(t)$

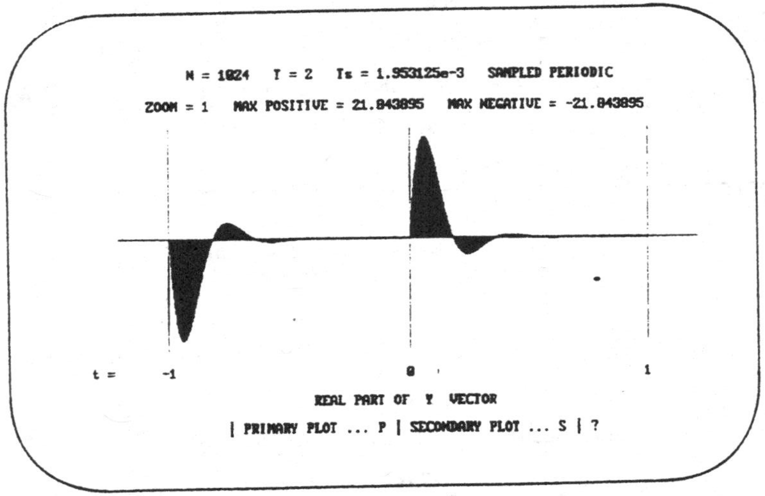

Plot of $y_p''(t)$

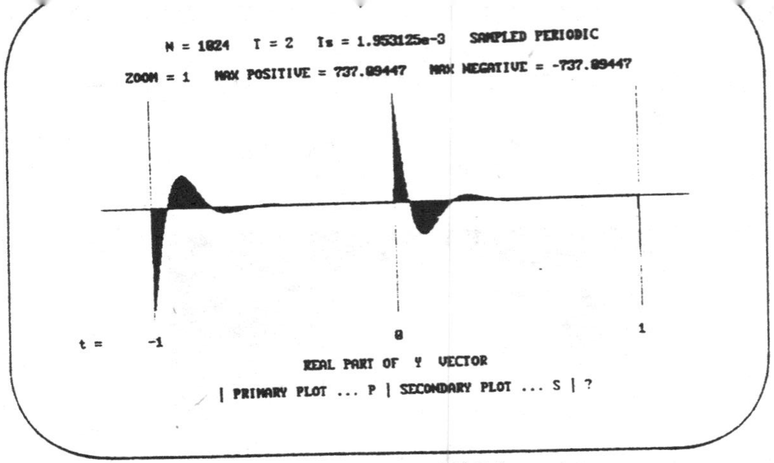

Plot of $y_p'''(t)$

6.16 The pulse that is being applied as an input to the RLC network is shown below:

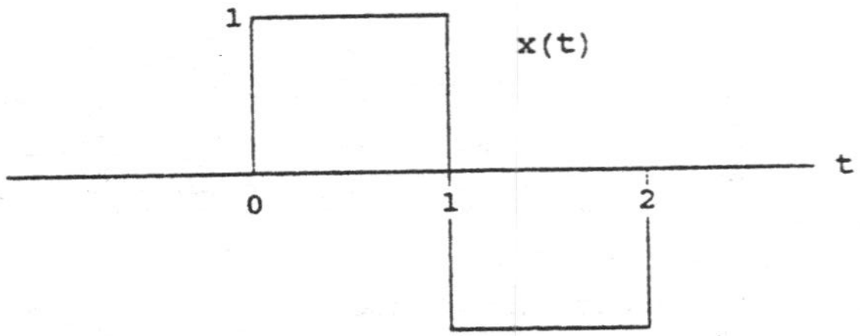

(a) Using the method of successive differentiation:

$$x'(t) = \delta(t) - 2\delta(t - 1) + \delta(t - 2)$$

Transformation gives $j\omega X(\omega) = 1 - 2\exp(-j\omega) + \exp(-j2\omega)$ from which

$$X(\omega) = \frac{1 - 2\exp(-j\omega) + \exp(-j2\omega)}{j\omega} \longrightarrow \quad (A)$$

The transfer function of the network in Figure 6.22 has been shown to be (see Exercise 6.15(d)):

$$H(j\omega) = \frac{400}{(j\omega)^2 + 20(j\omega) + 400} \longrightarrow$$

Then the transform of the response will be

$$Y(\omega) = H(j\omega)\ X(\omega)$$

$$= \frac{400}{(j\omega)^2 + 20(j\omega) + 400}\ \frac{1 - 2\exp(-j\omega) + \exp(-j2\omega)}{j\omega}$$

$$= \frac{400}{j\omega\,[(j\omega)^2 + 20(j\omega) + 400]}\ [1 - 2\exp(-j\omega) + \exp(-j2\omega)]$$

which we now invert to the time domain as follows:

$$\frac{400}{j\omega\,[(j\omega)^2 + 20(j\omega) + 400]} = \frac{1}{j\omega} - \frac{j\omega + 20}{(j\omega)^2 + 20(j\omega) + 400} \qquad \text{(B)}$$

$$= \frac{1}{j\omega} - \frac{j\omega + 10}{(j\omega + 10)^2 + 300} - \frac{10}{\sqrt{300}}\ \frac{\sqrt{300}}{(j\omega + 10)^2 + 300}$$

Then

$$\frac{j\omega + 10}{(j\omega + 10)^2 + 300} + \frac{10}{\sqrt{300}}\ \frac{\sqrt{300}}{(j\omega + 10)^2 + 300}$$

$$\Longleftrightarrow g(t) \equiv \cos(\sqrt{300}\ t)\ \exp(-10t)\ U(t) + \frac{10}{\sqrt{300}} \sin(\sqrt{300}\ t)\ \exp(-10t)\ U(t)$$

Finally

$$\frac{1}{j\omega}\,[1 - 2\exp(-j\omega) + \exp(-j2\omega)]$$

$$= \left[\frac{1}{j\omega} + \pi\delta(\omega)\right] [1 - 2\exp(-j\omega) + \exp(-j2\omega)]$$

(because $\pi\delta(\omega)\ [1 - 2\exp(-j\omega) + \exp(-j2\omega)] = 0$)

$$\Longleftrightarrow U(t) - 2U(t-1) + U(t-2) = x(t) \longrightarrow$$

and

$$-\frac{j\omega + 20}{(j\omega)^2 + 20(j\omega) + 400}\,[1 - 2\exp(-j\omega) + \exp(-j2\omega)]$$

$$\Longleftrightarrow -[g(t) - 2g(t-1) + g(t-2)] \longrightarrow$$

(1) From (A) we see that $X(\omega)$ is dying out like $1/\omega$. That means that $x(t)$ must have discontinuities. This is consistent with the figure above and the definition of $x(t)$?

(2) From (B) we see that $Y(\omega)$ is dying out like $1/\omega^3$. That means that $y(t)$, and $y'(t)$ are continuous and the $y''(t)$ has discontinuities.

(b) Finding the expressions for the real and imaginary parts of $X(\omega)$ and for its magnitude and phase spectra:

$$X(\omega) = \frac{1 - 2\exp(-j\omega) + \exp(-j2\omega)}{j\omega} = \frac{\exp(-j\omega)}{j\omega}\,[\exp(j\omega - 2 + \exp(-j\omega)]$$

$$= j\omega\exp(-j\omega)\left[\frac{\exp(j\omega/2) - \exp(-j\omega/2)}{2j\omega/2}\right]^2 = j\omega\exp(-j\omega)\ \mathrm{Sa}^2(\omega/2)$$

$$= j\omega\,[\cos(\omega) - j\sin(\omega)]\ \mathrm{Sa}^2(\omega/2) = [j\omega\cos(\omega) + \omega\sin(\omega)]\ \mathrm{Sa}^2(\omega/2)$$

$$= \omega\sin(\omega)\ \mathrm{Sa}^2(\omega/2) + j\omega\cos(\omega)\ \mathrm{Sa}^2(\omega/2)$$

from which

$$A_X(\omega) = \omega\sin(\omega)\ \mathrm{Sa}^2(\omega/2) \qquad B_X(\omega) = \omega\cos(\omega)\ \mathrm{Sa}^2(\omega/2)$$

$$|X(\omega)| = |\omega|\ \mathrm{Sa}^2(\omega/2)$$

$$\theta_X(\omega) = \arg[j\omega\exp(-j\omega)\ \mathrm{Sa}^2(\omega/2)] = \arg(j\omega) - \omega + \arg[\mathrm{Sa}^2(\omega/2)]$$

$= \arg(j\omega) - \omega$ (since $\mathrm{Sa}^2(\omega/2)$ is real and nonnegative and so its argument is everywhere zero)

For $\omega > 0$: $\arg(j\omega) = \pi/2$ and so $\theta_X(\omega) = \pi/2 - \omega$

For $\omega < 0$: $\arg(j\omega) = -\pi/2$ and so $\theta_X(\omega) = -\pi/2 - \omega$

Displaying a table of numerical samples of these functions taken at $\omega = n\omega_0$ for $-3 \le n \le 3$, where $\omega_0 = 2\pi/T$ and $T = 8$, gives us the values in the table below.

(c) Finding the expressions for the real and imaginary parts of $H(j\omega)$ and for its magnitude and phase spectra:

$$H(j\omega) = \frac{400}{(j\omega)^2 + 20(j\omega) + 400}$$

$$= \frac{400}{(400 - \omega^2) + 20(j\omega)}\ \frac{(400 - \omega^2) - 20(j\omega)}{(400 - \omega^2) - 20(j\omega)}$$

$$400\,(400 - \omega^2) \qquad 8000\omega$$

n	Variable	Formula
0	$A_X(0)$ $B_X(0)$ $\|X(0)\|$ $\theta_X(0)$	0 0 0 0
1	$A_X(1\Omega)$ $B_X(1\Omega)$ $\|X(1\Omega)\|$ $\theta_X(1\Omega)$	0.527393080 0.527393080 0.745846458 0.785398164

n	Variable	Formula
2	$A_X(2\Omega)$ $B_X(2\Omega)$ $\|X(2\Omega)\|$ $\theta_X(2\Omega)$	1.273239545 0 1.273239545 0
3	$A_X(3\Omega)$ $B_X(3\Omega)$ $\|X(3\Omega)\|$ $\theta_X(3\Omega)$	1.024624058 -1.024624058 1.448903724 -0.785398164

from which $A_H(\omega) = \dfrac{400\ (400 - \omega^2)}{(400 - \omega^2)^2 + 400\omega^2}$ $\quad B_H(\omega) = \dfrac{-8000\omega}{(400 - \omega^2) + 400\omega^2}$

$$|H(j\omega)| = \frac{400}{[(400 - \omega^2)^2 + 400\omega^2]^{1/2}}$$

$$\theta_H(\omega) = \tan^{-1}\left[B_H(\omega)/A_H(\omega) \right]$$

$$= \tan^{-1}\left[\frac{-8000\omega}{400\ (400 - \omega^2)} \right] = \tan^{-1}\left[\frac{-20\omega}{400 - \omega^2} \right]$$

Displaying a table of numerical samples of these functions taken at $\omega = n\Omega$ for $-3 \le n \le 3$, where $\Omega = 2\pi/T$ and $T = 8$:

n	Variable	Formula
0	$A_H(0)$ $B_H(0)$ $\|H(j0)\|$ $\theta_H(0)$	1.00000000 0 1.00000000 0
1	$A_H(1\Omega)$ $B_H(1\Omega)$ $\|H(j1\Omega)\|$ $\theta_H(1\Omega)$	0.999997618 -0.039330467 1.000770765 -0.039103000

n	Variable	Formula
2	$A_H(2\Omega)$ $B_H(2\Omega)$ $\|H(j2\Omega)\|$ $\theta_H(2\Omega)$	0.999961715 -0.079024271 1.003079392 -0.078863393
3	$A_H(3\Omega)$ $B_H(3\Omega)$ $\|H(j3\Omega)\|$ $\theta_H(3\Omega)$	0.999804697 -0.119444502 1.0069143070 -0.118904282

(d) Combining the results of the two tables to give a table of $|Y(\omega)|$ and $\theta_Y(\omega)$ for $\omega = n\Omega$ as in (b).

$$Y(\omega) = H(j\omega)\ X(\omega)$$

Then $\quad |Y(\omega)| = |H(j\omega)|\,|X(\omega)| \qquad \theta_Y(\omega) = \theta_H(\omega) + \theta_X(\omega)$

This gives us the following values:

$Y(\omega)$ sampled at multiples of $\Omega = \pi/4$			
Item	computed	Item	computed
$\|Y(0)\|$ $\theta(0)$	0 0	$\|Y(1\Omega)\|$ $\theta(1\Omega)$	0.746421329 0.746087864
$\|Y(2\Omega)\|$ $\theta(2\Omega)$	1.277160349 -0.078863393	$\|Y(3\Omega)\|$ $\theta(3\Omega)$	1.459056328 -0.904302446

(e) Using the FFT system on the disk, we compare the numerical values obtained from the FFT system with those that we obtained in (d).

We used $N = 512$ and $T = 8$, and for $H(j\omega)$ we used Alpha = 0. This value of Alpha gives no aliasing at all and so the values that the system calculates for $H(j\omega)$ are theoretically exactly the same as we what we would get doing it by hand starting from the CFT expression for $H(j\omega)$. Note: We are interested here in validating CFT spectral values, and not in inverting spectra to the time domain using the FFT system.

$Y(\omega)$ sampled at multiples of $\Omega = \pi/4$			
Item	FFT Value	Item	FFT Value
$\|Y(0)\|$ $\theta(0)$	0 0	$\|Y(1\Omega)\|$ $\theta(1\Omega)$	0.74641196 0.74608786
$\|Y(2\Omega)\|$ $\theta(2\Omega)$	1.27709620 -0.078863393	$\|Y(3\Omega)\|$ $\theta(3\Omega)$	1.45889150 -0.90430245

Observe that the values here are almost identical to those in the previous table. What errors exist are due solely to computational accuracy involved in the two approaches.

(f) Using the FFT system to obtain plots of $y(t)$, $y'(t)$ and $y''(t)$, we loaded the pulse using $N = 512$ and $T = 8$.

Since we now intend to invert to the time domain using the IFFT we must alias the expression for $H(j\omega)$. We elected to use Alpha = 50.

We obtained the following four plots. In the first we see the pulse $x(t)$. The second and third are of $y(t)$ and $y'(t)$, both of which are seen to be everywhere continuous. The fourth is a plot of $y''(t)$ with the time axis expanded. The discontinuity at the origin is clearly evident.

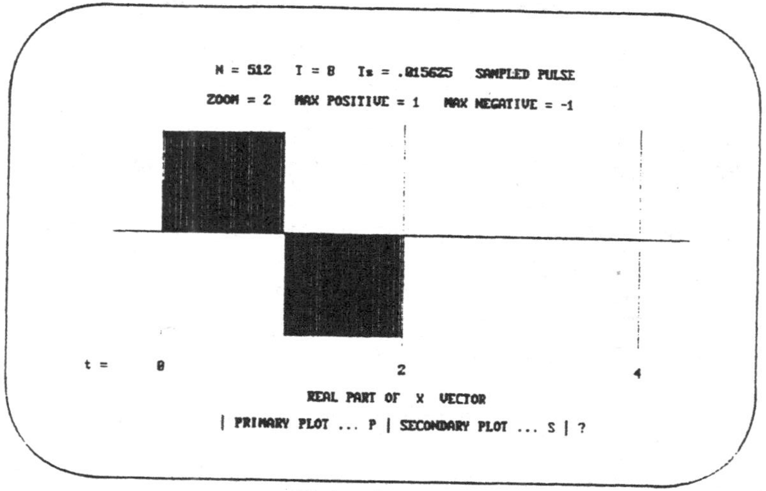

Plot of x(t)

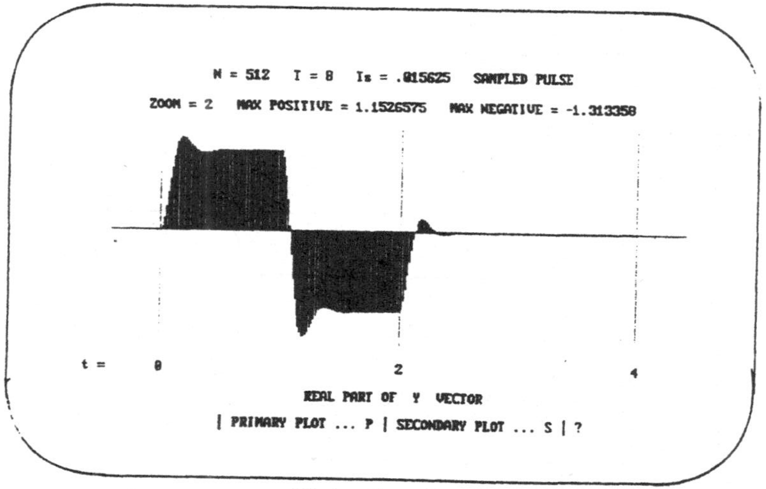

Plot of y(t)

"Introduction to Fourier Analysis", Instructor's Manual, Chapter 6

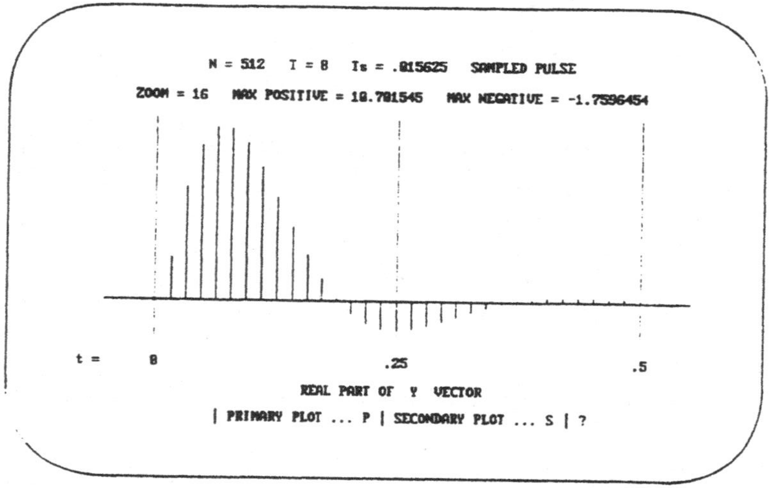

Plot of y'(t)

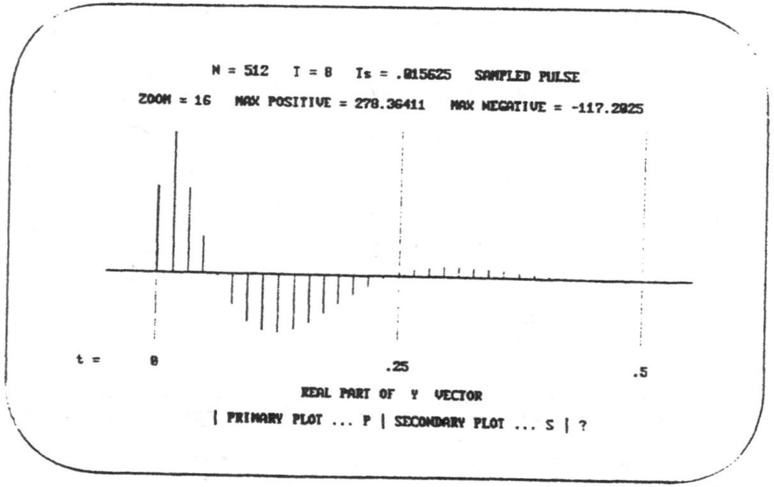

Plot of y"(t)

7.1 Deriving the impulse responses for the two networks shown in Figure 7.24.

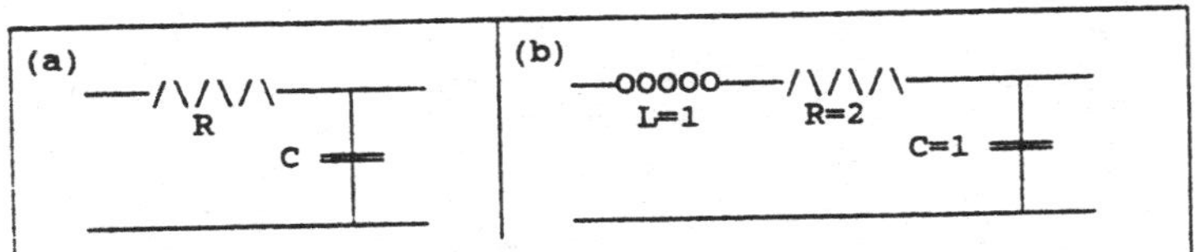

Figure 7.24

(a) $$H(j\omega) = \frac{1/j\omega C}{R + 1/j\omega C} = \frac{1}{j\omega RC + 1} = \frac{1}{RC}\,\frac{1}{j\omega + 1/RC}$$

$$\Longleftrightarrow \frac{1}{RC}\exp(-t/RC)\;U(t) = h(t)$$

(b) $$H(j\omega) = \frac{1/j\omega}{j\omega + 2 + 1/j\omega} = \frac{1}{(j\omega)^2 + 2j\omega + 1} = \frac{1}{(j\omega + 1)^2}$$

$$\Longleftrightarrow t\exp(-t)\;U(t) = h(t)$$

7.2 (a) An electrical network whose impulse response $h(t)$ is

(1) $\delta(t)$ Ans: Two perfect conductors.

Apply a Dirac delta at the input and the response will be $\delta(t)$

(2) $\frac{1}{2}\delta(t)$ Ans: A voltage divider.

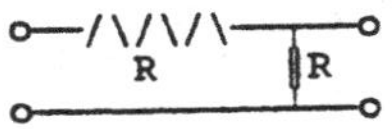

Apply a Dirac delta at the input and the response will be $\frac{1}{2}\delta(t)$

(3) $\delta(t - T)$ Ans: An ideal delay line, T seconds delay.

Apply a Dirac delta at the input and the response will be $\delta(t - T)$

(1) $\delta(t) \Longleftrightarrow 1$ $H_1(j\omega) = 1$

(2) $\frac{1}{2}\delta(t) \Longleftrightarrow \frac{1}{2}$ $H_2(j\omega) = \frac{1}{2}$

(3) $\delta(t - T) \Longleftrightarrow \exp(-j\omega T)$ $H_3(j\omega) = \exp(-j\omega T)$

(c) Using time-domain convolution to find the response $y(t)$ to a signal $x(t)$ which is applied to each of these networks:

(1) $$y(t) = \int_{-\infty}^{\infty} x(\tau)\,h(t - \tau)\,d\tau = \int_{-\infty}^{\infty} x(\tau)\,\delta(t - \tau)\,d\tau$$

$$= \int_{-\infty}^{\infty} x(t)\,\delta(t - \tau)\,d\tau = x(t)\int_{-\infty}^{\infty} \delta(t - \tau)\,d\tau = x(t) \longrightarrow$$

(2) $$y(t) = \int_{-\infty}^{\infty} x(\tau)\,h(t - \tau)\,d\tau = \int_{-\infty}^{\infty} x(\tau)\,\tfrac{1}{2}\delta(t - \tau)\,d\tau$$

$$= \int_{-\infty}^{\infty} \tfrac{1}{2}x(t)\,\delta(t - \tau)\,d\tau = \tfrac{1}{2}x(t)\int_{-\infty}^{\infty} \delta(t - \tau)\,d\tau = \tfrac{1}{2}x(t) \longrightarrow$$

(3) $$y(t) = \int_{-\infty}^{\infty} x(\tau)\,h(t - \tau)\,d\tau = \int_{-\infty}^{\infty} x(\tau)\,\delta(t - T - \tau)\,d\tau$$

$$= \int_{-\infty}^{\infty} x(t-T)\,\delta(t - T - \tau)\,d\tau = x(t-T)\int_{-\infty}^{\infty} \delta(t - T - \tau)\,d\tau = x(t-T) \longrightarrow$$

(d) Using frequency domain analysis per Chapter 6 to find the response $y(t)$ to a signal $x(t)$ which is applied to each of these networks, and verifying all three results obtained in (c) by using the inversion integral:

$$Y(\omega) = H(j\omega)\,X(\omega)$$

(1) $H(j\omega) = 1$ $Y(\omega) = X(\omega)$ $$y(t) = \int_{-\infty}^{\infty} X(\omega)\exp(j\omega t)\,d\omega = x(t)$$

(2) $H(j\omega)' = \frac{1}{2}$ $\quad Y(\omega) = \frac{1}{2}X(\omega)$ $\quad y(t) = \int_{-\infty}^{\infty} \frac{1}{2}X(\omega)\ \exp(j\omega t)\ d\omega = \frac{1}{2}x(t)$

(3) $H(j\omega) = \exp(-j\omega T)$ $\quad Y(\omega) = \exp(-j\omega T)\ X(\omega)$

$$y(t) = \int_{-\infty}^{\infty} \exp(-j\omega T)\ X(\omega)\ \exp(j\omega t)\ d\omega = \int_{-\infty}^{\infty} X(\omega)\ \exp[j\omega(t-T)]\ d\omega = X(t-T)$$

7.3 (a)(1) Solving by first using time-domain convolution followed by transformation:

$$y(t) = \int_{-\infty}^{\infty} x(\tau)\ h(t-\tau)\ d\tau = \int_{-\infty}^{\infty} \delta(\tau)\ h(t-\tau)\ d\tau$$

$$= \int_{-\infty}^{\infty} \delta(\tau)\ h(t)\ d\tau = h(t) \int_{-\infty}^{\infty} \delta(\tau)\ d\tau = h(t) \Longleftrightarrow \underline{H(j\omega)} \longrightarrow$$

(2) Transforming first and then multiplying:

$h(t) \Longleftrightarrow H(j\omega)$ $\quad x(t) = \delta(t) \Longleftrightarrow 1$

$Y(\omega) = H(j\omega)\ X(\omega) = 1\ H(j\omega) = H(j\omega) \longrightarrow$

(b) $x(t) = \cos(\omega_0 t) = \frac{1}{2}[\exp(j\omega_0 t) + \exp(-j\omega_0 t)]$

(1) Solving by first using time-domain convolution followed by transformation:

$$y(t) = \int_{-\infty}^{\infty} x(\tau)\ h(t-\tau)\ d\tau = \int_{-\infty}^{\infty} h(\tau)\ x(t-\tau)\ d\tau$$

$$= \frac{1}{2}\int_{-\infty}^{\infty} h(\tau) \Big[\exp[j\omega_0(t-\tau)] + \exp[-j\omega_0(t-\tau)] \Big]\ d\tau$$

$$= \frac{1}{2}\exp(j\omega_0 t) \int_{-\infty}^{\infty} h(\tau)\ \exp(-j\omega_0\tau)\ d\tau + \frac{1}{2}\exp(-j\omega_0 t) \int_{-\infty}^{\infty} h(\tau)\ \exp(j\omega\tau)\ d\tau$$

$$= \frac{1}{2}\exp(j\omega_0)\ H(j\omega_0) + \frac{1}{2}\exp(-j\omega_0)\ H(-j\omega_0)$$

$$\Longleftrightarrow H(j\omega_0)\ \pi\delta(\omega-\omega_0) + H(-j\omega_0)\ \pi\delta(\omega+\omega_0) \longrightarrow$$

(2) Transforming first and then multiplying:

$h(t) \Longleftrightarrow H(j\omega)$ $\quad x(t) = \cos(\omega_0 t) \Longleftrightarrow \pi[\delta(\omega-\omega_0) + \delta(\omega-\omega_0)]$

$Y(\omega) = H(j\omega)\ X(\omega) = H(j\omega)\ \pi[\delta(\omega-\omega_0) + \delta(\omega+\omega_0)]$

$$= H(j\omega_0)\ \pi\delta(\omega-\omega_0) + H(-j\omega_0)\ \pi\delta(\omega+\omega_0)] \longrightarrow$$

7.4 $\quad h(t) = \delta(t)$ and $x(t) = \delta(t)$

$$y(t) = \int_{-\infty}^{\infty} x(\tau)\ h(t-\tau)\ d\tau = \int_{-\infty}^{\infty} \delta(\tau)\ \delta(t-\tau)\ d\tau$$

$$= \int_{-\infty}^{\infty} \delta(t)\ \delta(t-\tau)\ d\tau = \delta(t) \int_{-\infty}^{\infty} \delta(t-\tau)\ d\tau = \delta(t) \longrightarrow$$

7.5

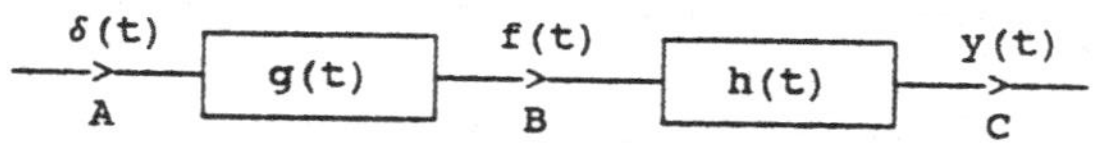

(a)

(1) Time domain: The response of Network 1 to the input $\delta(t)$ is

$$f(t) = \int_{-\infty}^{\infty} \delta(\tau)\ g(t-\tau)\ d\tau = \int_{-\infty}^{\infty} \delta(\tau)\ g(t)\ d\tau = g(t) \int_{-\infty}^{\infty} \delta(\tau)\ d\tau = \underline{g(t)} \longrightarrow$$

The response of Network 2 to the input $g(t)$ is

$$y(t) = \int_{-\infty}^{\infty} h(\tau)\ g(t-\tau)\ d\tau \longrightarrow$$

Frequency domain: $\quad F(\omega) = G(j\omega)\ X(\omega) = G(j\omega) \Longleftrightarrow g(t) \longrightarrow$

$$Y(\omega) = H(j\omega)\ F(\omega) = H(j\omega)\ G(j\omega) \Longleftrightarrow \int_{-\infty}^{\infty} h(\tau)\ g(t-\tau)\ d\tau \longrightarrow$$

(2) The impulse response

$$y(t) = \int_{-\infty}^{\infty} h(\tau)\, g(t-\tau)\, d\tau \quad \longrightarrow$$

(b) If the order of the networks is reversed the impulse response will be

$$y_2(t) = \int_{-\infty}^{\infty} g(\tau)\, h(t-\tau)\, d\tau$$

However $\displaystyle\int_{-\infty}^{\infty} g(\tau)\, h(t-\tau)\, d\tau = \int_{-\infty}^{\infty} h(\tau)\, g(t-\tau)\, d\tau$

and so the order of the cascade is immaterial.

7.6 Given that $R = L = 1$:

(a) From (7.8): $h(t) = \exp(-t)\, U(t) \Longleftrightarrow \dfrac{1}{j\omega + 1}$

which is (7.7) with $R = L = 1$. Loading exp(-t) U(t) into X and then running ANALYSIS gave the results shown in the following table, where we also show the exact values of $F(\omega)$.

n	FFT	Formula	error (%)
0	A_0 = 0.99974589 B_0 = 0	$A(0)$ = 1 $B(0)$ = 0	0.025411000 ---
1	A_1 = 0.61836033 B_1 = -0.48553128	$A(\omega_0)$ = 0.618486458 $B(\omega_0)$ = -0.485758128	0.020393041 0.046699867
2	A_2 = 0.28838505 B_2 = -0.45273859	$A(2\omega_0)$ = 0.288400439 $B(2\omega_0)$ = -0.453018350	0.005336018 0.061754761
3	A_3 = 0.15266342 B_3 = -0.35932128	$A(3\omega_0)$ = 0.152633248 $B(3\omega_0)$ = -0..359633619	0.019767384 0.086849222

Note that

$$F(\omega) = \frac{1}{1 + j\omega} = \frac{1}{1 + j\omega}\,\frac{1 - j\omega}{1 - j\omega} = \frac{1}{1 + \omega^2} - j\,\frac{\omega}{1 + \omega^2}$$

from which we have $\quad A(\omega) = \dfrac{1}{1 + \omega^2} \quad$ and $\quad B(\omega) = \dfrac{-\omega}{1 + \omega^2}$

$$A_{FFT}(n) \approx A_{FORMULA}(n\Omega) \quad \text{and} \quad B_{FFT}(n) \approx B_{FORMULA}(n\Omega)$$

Steps for loading the pulse ($N = 256$, $T = 8$): Main menu, Create Values, Continue, Create X, Real, Neither, Left, 1, Continue, EXP(-t), Go, Neither, Go. You are now back on the main menu. Run ANALYSIS

Note: Displaying the numbers in X shows that the system has loaded the following value into X_0:

$X_0 = 0.500167732$

This comes about as follows: The system always loads into X_0 the average of x(0) and x(T) from the formula that has been entered for x(t). In this case, x(t) = exp(-t) U(t) from which x(0) = 1 and x(8) = 0.000335463 whose average is 0.500167732

(b) From (7.10): $h(t) = \delta(t) - \exp(-t)\, U(t) \Longleftrightarrow \dfrac{j\omega}{j\omega + 1}$

which is (7.9) with $R = L = 1$. Loading $\delta(t)$ - exp(-t) U(t) into X and then running ANALYSIS gave the results shown in the following table, where we also show the exact values of $F(\omega)$. Note that

$$F(\omega) = \frac{j\omega}{1 + j\omega} = 1 - \frac{1}{1 + j\omega} = 1 - \frac{1}{1 + \omega^2} + j\,\frac{\omega}{1 + \omega^2}$$

from which we have $\quad A(\omega) = 1 - \dfrac{1}{1 + \omega^2} \quad$ and $\quad B(\omega) = \dfrac{\omega}{1 + \omega^2}$

Since $T = 8$, the FFT's frequency domain values are estimates of $F(\omega)$ with ω sampled at $\Omega = 2\pi/T = 0.785398164$, i.e.,

$$A_{FFT}(n) \approx A_{FORMULA}(n\Omega) \quad \text{and} \quad B_{FFT}(n) \approx B_{FORMULA}(n\Omega)$$

To create the function $\delta(t)$ - exp(-t) U(t) the sequence of steps is: Main menu, Create Values, Continue, Create X, Real, Neither, Left, 1, Continue, -EXP(-t), add a Dirac delta to X, 0, 1, Neither, Go. You are now back on the main menu. Run ANALYSIS

Note: Displaying the numbers in X shows that the system has loaded the following value into X_0:

$X_0 = 31.499832$

This comes about as follows: Absent the Dirac delta, the system always loads into X_0 the average of x(0) and x(T) from the formula

that has been entered for $x(t)$. In this case, $x(t) = -\exp(-t)\,U(t)$ from which $x(0) = -1$ and $x(8) = -0.000335463$ whose average is -0.500167732

The Dirac delta with weight 1 is loaded as a value $V = \mu N/T$ where μ is its weight (see Chapter 15) and so in this case $V = 1\times256/8 = 32$. Adding 32 to -0.500167732 gives 31.499832 which is what you see when you inspect the numbers in **X**.

n	FFT	Formula	error (%)
0	$A_0 = 2.5411104\text{e-}4$ $B_0 = 0$	$A(0) = 0$ $B(0) = 0$	--- ---
1	$A_1 = 0.38163967$ $B_1 = 0.48553128$	$A(\omega_0) = 0.381513542$ $B(\omega_0) = 0.485758128$	0.033059954 0.046699867
2	$A_2 = 0.71161495$ $B_2 = 0.45273859$	$A(2\omega_0) = 0.711599561$ $B(2\omega_0) = 0.453018350$	0.002162607 0.061754761
3	$A_3 = 0.15266342$ $B_3 = 0.84733658$	$A(3\omega_0) = 0.152633248$ $B(3\omega_0) = 0.847366752$	0.003560631 0.086849222

(c) Using the FFT system to obtain a plot of the impulse response of network (A) in Figure 7.2 starting from its frequency response:

The frequency response is $H(j\omega) = 1/(j\omega + 1)$ This is loaded as follows: Main menu, Create, Continue, Create H(jw), Create, 0, 1, 1, 1, 1, Load, Alpha = 10. You are now back on the main menu with $H(j\omega)$ in **F2**. Main menu, Run Postprocessors, F, Copy F2 to F. Quit. You are now on the main menu with $H(j\omega)$ also in **F**. Running SYNTHESIS will invert the frequency response to the time domain and give you the impulse response $h(t)$ in **Y**.

Plotting **Y** shows a single-sided decaying exponential. (See figure below.) Verifying its values gave the following table. The formula values are computed as follows: We have $N = 256$ and $T = 8$. Thus the time-domain sampling interval of the FFT is $T_s = T/N = 0.03125$. Then, to compare a value in **Y** whose k is 10 say, with the formula value, we must evaluate the formula at $t = 10T_s$, i.e. $FFT_k \approx \text{Formula}(kT_s) = \exp(-k\times0.03125)$

k	FFT	Formula	error (%)
0	0.50018480	0.500000000	0.036960000
1	0.97438092	0.969233235	0.531109058
4	0.88158784	0.882496903	0.103010288
16	0.60643653	0.606530660	0.015519364
64	0.13532149	0.135335283	0.010191873
128	1.8321785e-2	0.8315638e-2	0.033556624

The errors are all acceptably small. Using larger values for N and Alpha would make them as small as we please.

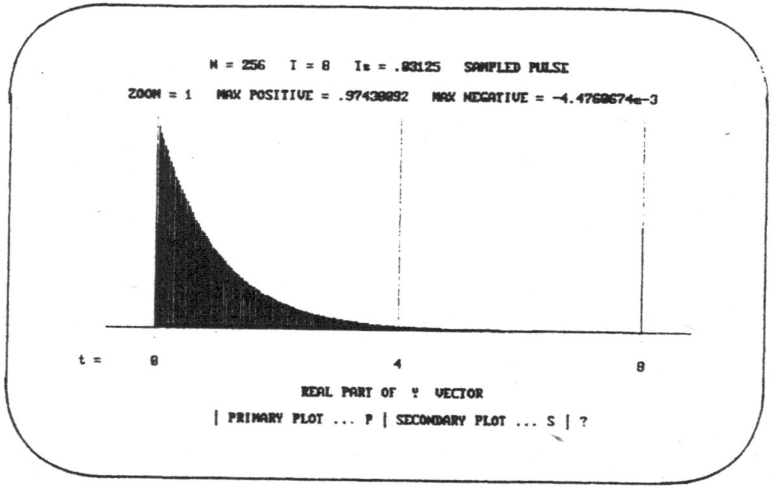

exp(-t) U(t)

7.7 (a)

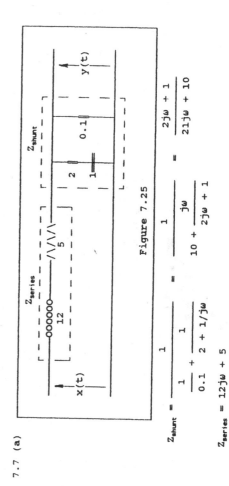

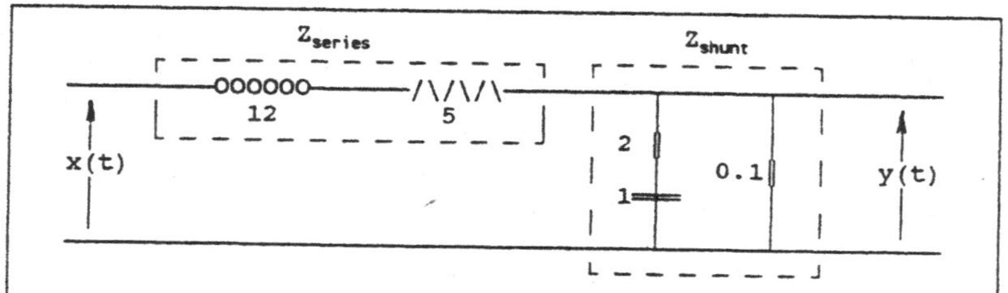

Figure 7.25

$$Z_{shunt} = \frac{1}{\dfrac{1}{0.1} + \dfrac{1}{2 + 1/j\omega}} = \frac{1}{10 + \dfrac{j\omega}{2j\omega + 1}} = \frac{2j\omega + 1}{21j\omega + 10}$$

$$Z_{series} = 12j\omega + 5$$

$$H(j\omega) = \frac{Z_{shunt}}{Z_{shunt} + Z_{series}} = \frac{\dfrac{2j\omega + 1}{21j\omega + 10}}{\dfrac{2j\omega + 1}{} + 12j\omega + 5}$$

Finding the impulse response:

$$H(j\omega) = \frac{2j\omega + 1}{(36j\omega + 17)(7j\omega + 3)} = \frac{1/11}{7j\omega + 3} - \frac{2/11}{36j\omega + 17}$$

$$= \frac{1/11}{7(j\omega + 3/7)} - \frac{2/11}{36(j\omega + 17/36)}$$

$$\Longleftrightarrow\ 1/77\ \exp(-3t/7)\ U(t) - 1/198\ \exp(-17t/36)\ U(t) \longrightarrow$$

(b) Loading $H(j\omega)$ into the FFT system with $N = 512$, $T = 16$, and inverting to the time domain we can then plot the impulse response. The sequence of steps is as follows:

Main menu, Create Values, Continue, Create H(jw), Create, 1, 2, 1, 2, 51, 227, 252, Load, Alpha = 10. You are now back on the main menu with $H(j\omega)$ in **F2**. Main menu, Run Postprocessors, F, Copy F2 to F, Quit. You are now on the main menmu with $H(j\omega)$ in **F**. Run SYNTHESIS and plot **Y**. It shows $h(t)$. (See plot below.)

Using $h(t) = 1/77\ \exp(-3t/7)\ U(t) - 1/198\ \exp(-17t/36)\ U(t)$ we verify the FFT values as shown below.

Since $N = 512$ and $T = 16$, the FFT's time-domain sampling interval is $T_s = 16/512 = 0.03125$. The values in **Y** thus correspond to $h(t)$ evaluated at $t = kT_s$.

k	FFT	Formula	error (%)
0	3.9788062e-3	3.9682539e-3	0.265916196
1	7.8868801e-3	7.8377170e-3	0.627162001
4	7.5494736e-3	7.5485744e-3	0.011911733
8	7.1845372e-3	7.1793831e-3	0.071789942
16	6.5002650e-3	6.4937005e-2	0.101089478
64	3.5514248e-3	3.5472166e-3	0.118632506
256	3.0607144e-4	3.0568837e-4	0.125313763

7.8 Convolving two copies of Rect(t)

(a) By evaluating the convolution integral

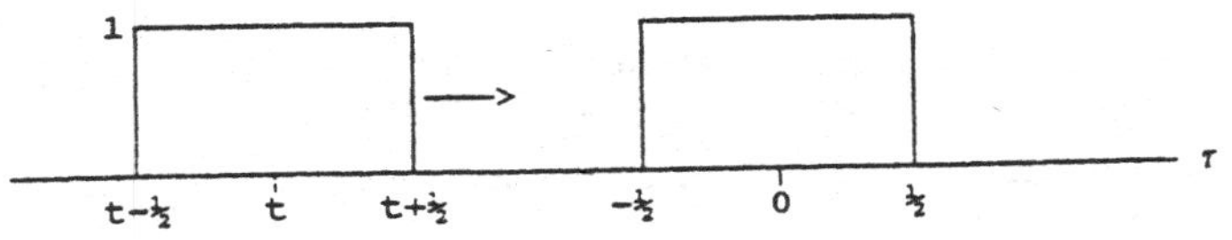

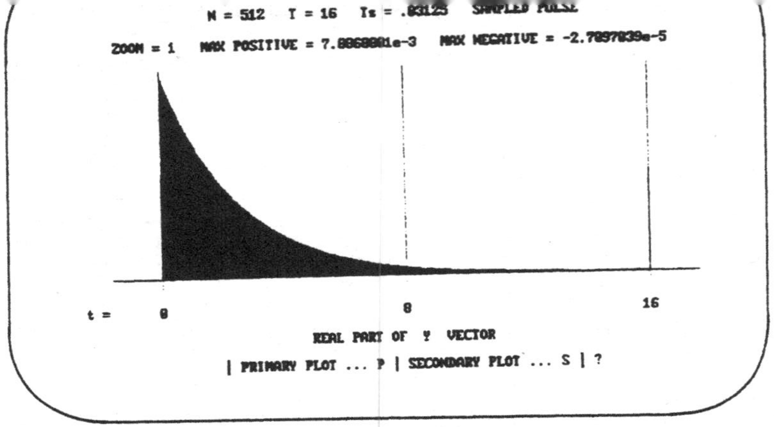

Plot of $h(t)$

$$y(t) = \int_{-\infty}^{\infty} \text{Rect}((t - \tau)\ \text{Rect}(\tau)\ d\tau$$

- For $t < -1$: $y(t) = 0$

- For $-1 < t < 0$: $y(t) = \int_{-\frac{1}{2}}^{t+\frac{1}{2}} 1\ d\tau = (t+\tfrac{1}{2}) - (-\tfrac{1}{2}) = t + 1$

- For $0 < t < 1$: $y(t) = \int_{t-\frac{1}{2}}^{\frac{1}{2}} 1\ d\tau = (\tfrac{1}{2}) - (t-\tfrac{1}{2}) = 1 - t$

- For $t > 1$: $y(t) = 0$

Result: $$y(t) = \begin{cases} 0 & (t < -1) \\ 1 + t & (-1 < t < 0) \\ 1 - t & (0 < t < 1) \\ 0 & (t > 1) \end{cases}$$

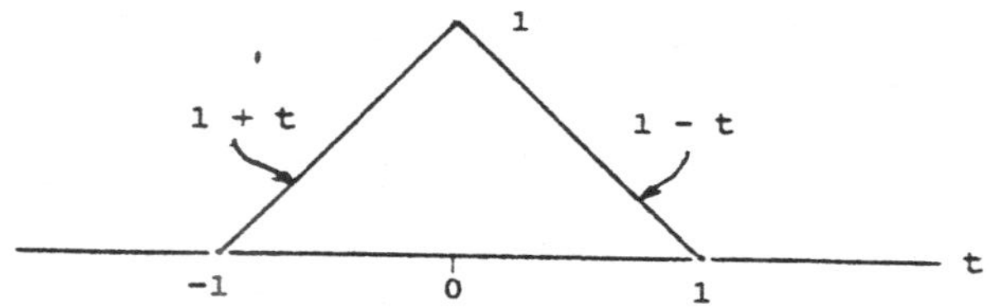

i.e., the result is a triangular pulse of width 2 and height 1, centred on the origin.

(b) Using the graphical method gives the same result

(c) (1) Using the FFT system (N = 1024, T = 4) we load the pulses as follows:

Main menu, Create Values, Continue, Create X, Real, Even, 2, .5, Continue, 1, 0, Go, Neither, Go. You are now back on the main menu and Rect(t) is in **X**. Main Menu, Run Postprocessors, X, Copy X to X2, Quit. You are now on the main menu and Rect(t) is in both **X** and **X2**.

Run CONVOLUTION. The convolution product is now in **Y**. Plotting it shows a triangular pulse like the one above.

(2) Using T = 1.5 we ran the problem again. **Aliasing is now present**, resulting from the fact that the triangular pulse, which is 2 seconds wide, cannot fit into a 1.5-second window, and so the triangular pulses overlap into the adjacent windows (remember that the FFT is periodic in both time and frequency domains).

7.9 (a) Convolving $x(t) = \text{Rect}(t)$ with the delayed Dirac delta $h(t) = \delta(t - 1)$:

$$x(t) * h(t) = \int_{-\infty}^{\infty} g(\tau)\, h(t - \tau)\, d\tau = \int_{-\infty}^{\infty} \text{Rect}(\tau)\, \delta(t - 1 - \tau)\, d\tau$$

$$= \int_{-\infty}^{\infty} \text{Rect}(t - 1)\, \delta(t - 1 - \tau)\, d\tau = \text{Rect}(t - 1) \int_{-\infty}^{\infty} \delta(t - 1 - \tau)\, d\tau$$

$$= \text{Rect}(t - 1)$$

(b) Convolving $x(t)$ with the delayed Dirac delta $h(t) = \delta(t - 1)$:

$$x(t) * h(t) = \int_{-\infty}^{\infty} x(\tau)\, h(t - \tau)\, d\tau = \int_{-\infty}^{\infty} x(\tau)\, \delta(t - 1 - \tau)\, d\tau$$

$$= \int_{-\infty}^{\infty} x(t - 1)\, \delta(t - 1 - \tau)\, d\tau = x(t - 1) \int_{-\infty}^{\infty} \delta(t - 1 - \tau)\, d\tau = x(t - 1)$$

(c) Convolving Rect(t) and $\delta(t - 1)$ using the FFT system with N = 256, T = 4: The sequence of steps is as follows:

Main menu, Create Values, Continue, Create X, Real, Even, 2, 1/2, Continue, 1, 0, Go, Add a Dirac delta to X2, 1, 1, Neither, Go. You are now back on the main menu. Plotting **X** shows Rect(t). A plot of **X2** shows a single line at t = 1 of height 64. According to Chapter 15, a time-domain Dirac delta of weight μ appears as a line of height $V = \mu N/T$. In this case $\mu = 1$, N = 256 and T = 4, and so V = 64. Running CONVOLUTION and plotting **Y**, the screen shows Rect(t - 1).

7.10 (a)

$$h(t) = \begin{cases} 1 & (0 < t < 1) \\ 0 & \text{otherwise} \end{cases}$$

$$x(t) = e^{-t}\, U(t)$$

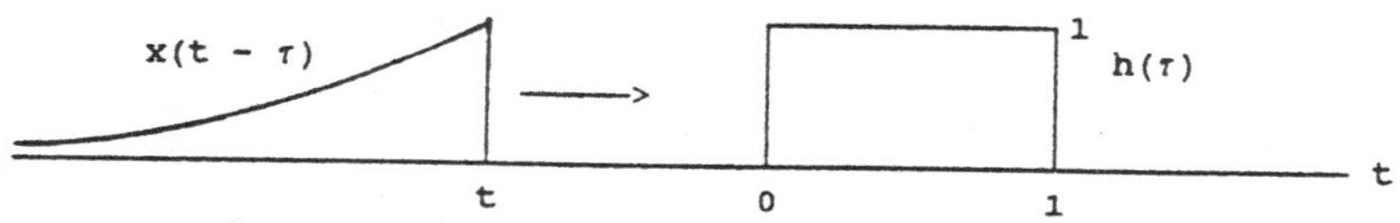

$$y(t) = \int_{-\infty}^{\infty} x(t - \tau)\, h(\tau)\, d\tau$$

- For $t < 0$: $y(t) = 0$

- For $0 < t < 1$: $y(t) = \int_0^t x(t - \tau)\, h(\tau)\, d\tau = \int_0^t e^{-(t - \tau)}\, d\tau$

$$= e^{-t} \int_0^t e^{\tau}\, d\tau = e^{-t}\, e^{\tau} \Big|_0^t = e^{-t}[e^t - 1] = 1 - e^{-t}$$

• For t > 1: $$y(t) = \int_0 x(t - \tau)\, h(\tau)\, d\tau = \int_0 e^{-(t-\tau)}\, d\tau$$

$$= e^{-t} \int_0^1 e^{\tau}\, d\tau = e^{-t}\, e^{\tau} \Big|_0^1 = e^{-t}\, [e^1 - 1]$$

$$y(t) = \begin{cases} 0 & (t < 0) \\ 1 - e^{-t} & (0 < t < 1) \\ e^{-t}[e - 1] & (1 < t) \end{cases}$$

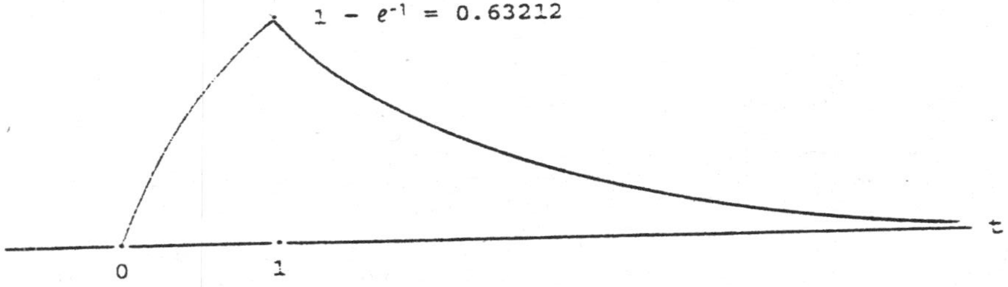

(b) Validating the result using the FFT system with N = 1024, T = 8, the sequence of steps is as follows: Main menu, Create Values, Continue, Create X and X2, Yes, Neither, Left, 1, Continue, EXP(-t), Go, Neither, Left, 2, 1, Continue, 1, 0, Go, Neither, Go. You are now on the main menu. Plotting **X** and **X2** shows that the two expressions are loaded correctly. Run CONVOLUTION and plot **Y**. (See following figure.) It looks the same as the sketch above. The peak value is 0.630171 which compares well with the theoretical peak of $1 - e^{-1} = 0.63212$.

(c) Fourier transforming the result obtained in (a):

$$y(t) = \begin{cases} 0 & (t < 0) \\ 1 - e^{-t} & (0 < t < 1) \\ e^{-t}[e - 1] & (1 < t) \end{cases}$$

$$Y(\omega) = \int_{-\infty}^{\infty} y(t)\, \exp(-j\omega t)\, dt$$

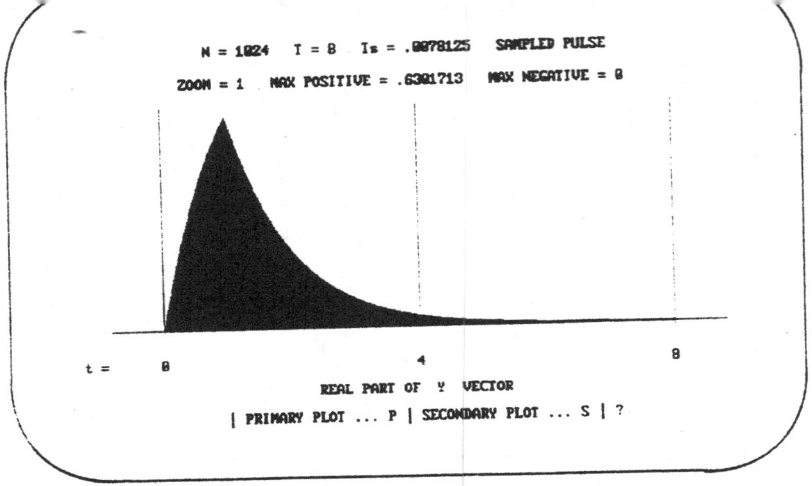

$$= \int_0^1 \exp(-j\omega t)\, dt - \int_0^1 \exp[-(1 + j\omega)t]\, dt + (e - 1) \int_1^{\infty} \exp[-(1 + j\omega t)]\, dt$$

$$= \frac{\exp(-j\omega t)}{-j\omega} \Big|_0^1 - \frac{\exp[-(1 + j\omega)t]}{-(1 + j\omega)} \Big|_0^1 + (e - 1) \frac{\exp[-(1 + j\omega)t]}{-(1 + j\omega)} \Big|_1^{\infty}$$

$$= \frac{\exp(-j\omega) - 1}{-j\omega} + \frac{\exp[-(1 + j\omega)] - 1}{1 + j\omega} + (e - 1) \frac{\exp[-(1 + j\omega)]}{1 + j\omega}$$

$$= \frac{\exp(-j\omega) - 1}{-j\omega} - \frac{1}{1 + j\omega} + \frac{\exp(-j\omega)}{1 + j\omega}$$

$$= [1 - \exp(-j\omega)] \left[\frac{1}{j\omega} - \frac{1}{1 + j\omega} \right] = [1 - \exp(-j\omega)] \frac{1}{j\omega\, (1 + j\omega)}$$

$$= \frac{1 - \exp(-j\omega)}{j\omega} \frac{1}{1 + j\omega} \longrightarrow \qquad \text{(A)}$$

Transforming h(t) and x(t):

$$h(t) = \text{Rect}(t-1) \Longleftrightarrow \frac{1-\exp(-j\omega)}{j\omega} = H(j\omega)$$

$$x(t) = \exp(-t)\,U(t) \Longleftrightarrow \frac{1}{1+j\omega} = X(\omega)$$

and so $H(j\omega)\,X(\omega) = \dfrac{1-\exp(-j\omega)}{j\omega}\,\dfrac{1}{1+j\omega} \longrightarrow$ (Same as (A))

Thus $Y(\omega) = H(j\omega)\,X(\omega)$.

(d) Loading the expression for $Y(\omega)$ obtained in (c) into **F**:

$$Y(\omega) = \frac{1-\exp(-j\omega)}{j\omega}\,\frac{1}{1+j\omega} = \frac{1}{\omega}\,\frac{1-\cos(\omega)+j\sin(\omega)}{-\omega+j}$$

$$= \frac{1}{\omega}\,\frac{1-\cos(\omega)+j\sin(\omega)}{-\omega+j}\,\frac{-\omega-j}{-\omega-j}$$

$$= \frac{1}{\omega}\,\frac{\sin(\omega)-\omega[1-\cos(\omega)]}{\omega^2+1} + \frac{j}{\omega}\,\frac{\cos(\omega)-1-\omega\sin(\omega)}{\omega^2+1}$$

$$= A(\omega) + jB(\omega)$$

The steps for loading these expressions (N = 1024 and T = 8): Main menu, Create Values, Continue, Create F, Complex,

(SIN(w) - w + w*COS(w))/(w^3 + w), Alpha = 5,

Division by zero, Let the system find the value (displays 0.9999999), Go,

(COS(w) - 1 - w*SIN(w))/(w^3 + w),

Continue aliasing, Go, Neither, Go. You are now back on the main menu. Run SYNTHESIS and plot Y. It looks very similar to the figure displayed above.

Displaying the numbers in Y gave the following table where we also show the formula values obtained from the convolution. Note: The FFT's time domain sampling interval is T_s = T/N = 0.0078125 and so we must evaluate the formula for y(t) at multiples of T_s, i.e. at t = k×0.0078125

$$y(t) = \begin{cases} 0 & (t<0) \\ 1-e^{t} & (0<t<1) \\ e^{-t}[e-1) & (1<t) \end{cases}$$

k	FFT	Formula	error (%)
1	8.3539434e-3	7.7820617e-3	7.348717100
8	6.1128492e-2	6.0586937e-2	0.893847461
32	0.22164816	0.221199217	0.202958720
64	0.39381895	0.393469340	0.088853098
128	0.63226060	0.632120559	0.022154192
256	0.23262207	0.232544158	0.033504260
512	3.1481866e-2	3.1471429e-2	0.033161919

7.11 (a) $x(t) = \exp(-t)\,U(t)$ $\quad x_2(t) = \begin{cases} 1-t & (0<t<1) \\ 0 & \text{otherwise} \end{cases}$

$$y(t) = \int_{-\infty}^{\infty} x(t-\tau)\,x_2(\tau)\,d\tau$$

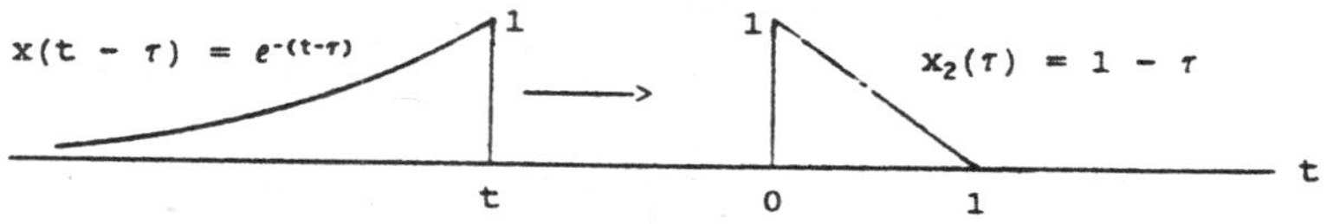

For $t < 0$: $\quad y(t) = 0$

For $0 < t < 1$: $\quad y(t) = \displaystyle\int_0^t e^{-(t-\tau)}(1-\tau)\,d\tau$

$$= e^{-t}\int_0^t e^{\tau}(1-\tau)\,d\tau = e^{-t}\Big[e^{\tau}(1-\tau)+e^{\tau}\Big]_0^t$$

$$= e^{-t}\,e^{\tau}(2-\tau)\Big|_0^t = e^{-t}\,[e^{t}(2-t)-2] = 2-t-2e^{-t}$$

For $1 < t$: $\quad y(t) = \displaystyle\int_0^1 e^{-(t-\tau)}(1-\tau)\,d\tau$

$$= e^{-t}\int_0^1 e^{\tau}(1-\tau)\,d\tau = e^{-t}\Big[e^{\tau}(1-\tau)+e^{\tau}\Big]^1$$

$$= e^{-t} e^{\tau}(2 - \tau) \Big|_0 = e^{-t}\,[e(2 - 1) - 2] = e^{1-t} - 2e^{-t}$$

$$y(t) = \begin{cases} 0 & (t < 0) \\ 2 - t - 2e^{-t} & (0 < t < 1) \\ e^{1-t} - 2e^{-t} & (1 < t) \end{cases}$$

(b) Fourier transforming $y(t)$: $\quad Y(\omega) = \int_{-\infty}^{\infty} y(t)\,\exp(-j\omega t)\,dt$

$$= \int_0^1 [2 - t - 2e^{-t}]\,\exp(-j\omega t)\,dt + \int_1^{\infty} (e - 2)\,e^{-t}\exp(-j\omega t)\,dt$$

$$= \frac{(2 - t)\,\exp(-j\omega t)}{-j\omega} + \frac{\exp(-j\omega t)}{(-j\omega)^2} - \frac{2\,\exp[-(1+j\omega)t]}{-(1 + j\omega)}\Bigg|_0^1$$

$$+ (e - 2)\,\frac{\exp[-(1+j\omega)t]}{-(1 + j\omega)}\Bigg|_1^{\infty}$$

$$= \frac{\exp(-j\omega) - 2}{-j\omega} + \frac{\exp(-j\omega) - 1}{(j\omega)^2} - \frac{2\,\exp[-(1+j\omega)] - 2}{-(1 + j\omega)}$$

$$- (e - 2)\,\frac{\exp[-(1+j\omega)]}{-(1 + j\omega)}$$

$$= \frac{2 - \exp(-j\omega)}{j\omega} + \frac{\exp(-j\omega) - 1}{(j\omega)^2} + \frac{2e^{-1}\exp(-j\omega)}{1 + j\omega} - \frac{2}{1 + j\omega}$$

$$+ \frac{\exp(-j\omega)}{1 + j\omega} - \frac{2e^{-1}\exp(-j\omega)}{1 + j\omega}$$

$$= \frac{2 - \exp(-j\omega)}{j\omega} + \frac{\exp(-j\omega) - 1}{(j\omega)^2} - \frac{2 - \exp(-j\omega)}{1 + j\omega}$$

$$= [2 - \exp(-j\omega)]\left[\frac{1}{j\omega} - \frac{1}{1 + j\omega}\right] + \frac{\exp(-j\omega) - 1}{(j\omega)^2}$$

$$= \frac{}{j\omega\,(1 + j\omega)} + \frac{}{(j\omega)^2}$$

$$= \frac{2j\omega - j\omega\,\exp(-j\omega) + [\exp(-j\omega) - 1](1 + j\omega)}{(j\omega)^2\,(1 + j\omega)}$$

$$= \frac{2j\omega - j\omega\,\exp(-j\omega) + \exp(-j\omega) + j\omega\,\exp(-j\omega) - 1 - j\omega}{(j\omega)^2\,(1 + j\omega)}$$

$$= \frac{j\omega - 1 + \exp(-j\omega)}{(j\omega)^2}\;\frac{1}{(1 + j\omega)} \quad \longrightarrow \quad (A)$$

Transforming $x(t)$ and $x_2(t)$:

$$x(t) = \exp(-t)\,U(t) \iff \frac{1}{1 + j\omega}$$

$x_2(t) = 1 - t \quad x_2'(t) = \delta(t) - \mathrm{Rect}(t - \tfrac{1}{2}) \quad x_2''(t) = j\omega - 1 + \exp(-j\omega)$

$$X_2(\omega) = \frac{j\omega - 1 + \exp(-j\omega)}{(j\omega)^2}$$

and so $\quad X_2(\omega)\,X(\omega) = \dfrac{j\omega - 1 + \exp(-j\omega)}{(j\omega)^2}\;\dfrac{1}{1 + j\omega} \quad \longrightarrow$

which is the same as (A) above.

(c) Loading the two expressions in (a) into **X** and **X2**, with N = 1024 and T = 8, and runnning CONVOLUTION: Main menu, Create Values, Continue, Create X and X2, Real, Neither, Left, 1, Continue, EXP(-t), Go, Neither, Left, 2, 1, Continue, 1 - t, 0, Go, Neither, Go. You are now back on the main menu. Plotting **X** and **X2** shows that we have loaded the pulses correctly. Run CONVOLUTION, plot **Y**. We see the FFT's representation of the convolution product.

Displaying the numbers for **Y** gives the following table where we also show the values computed from the time domain convolution formula obtained in (a). Note: Since the FFT's time-domain sampling interval is $T_s = T/N = 8/1024 = .0078125$, we must compute values from the formula at multiples of that number.

$$y(t) = \begin{cases} 0 & (t < 0) \\ 2 - t - 2e^{-t} & (0 < t < 1) \\ e^{1-t} - 2e^{-t} & (1 < t) \end{cases}$$

k	FFT	Formula	error (%)
16	0.11018130	0.110006195	0.159177399
32	0.19252027	0.192398434	0.063324840
64	0.28698603	0.286938681	0.016501435
128	0.26423603	0.264241118	0.001925363
256	9.7207004e-2	9.7208874e-2	0.001924516
512	1.3155537e-2	1.3155790e-2	0.001927592

7.12 (a) Input is $x(t)\ U(t)$ and impulse response is $h(t)\ U(t)$

Then the response will be: $y(t)$ = input * impulse response

$$= \int_{-\infty}^{\infty} \left[x(\tau)\ U(\tau) \right] \left[h(t-\tau)\ U(t-\tau) \right] d\tau$$

$$= \int_{-\infty}^{\infty} \left[x(\tau)\ h(t-\tau) \right] \left[U(\tau)\ U(t-\tau) \right] d\tau$$

We now sketch the expression $U(\tau)\ U(t-\tau)$ on the τ axis, for the two cases $t > 0$ and $t < 0$:

(1) $t > 0$:

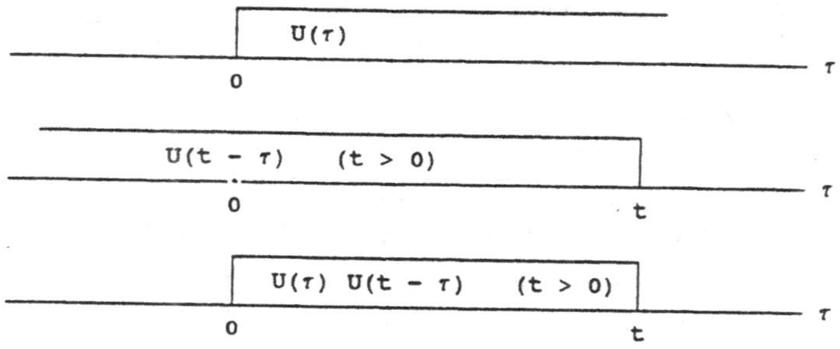

(2) $t < 0$: $U(\tau)$ is the same

We see then that $U(\tau)\ U(t-\tau) = 1$ from 0 to t if $t > 0$

Thus our integral $y(t) = \int_{-\infty}^{\infty} \left[x(\tau)\ h(t-\tau) \right] \left[U(\tau)\ U(t-\tau) \right] d\tau$

is equal to $\int_0^t x(\tau)\ h(t-\tau)\ d\tau$ if $t > 0$,

and is equal to zero if $t < 0$. We can therefore write all of this as

$$y(t) = \left[\int_0^t x(\tau)\ h(t-\tau)\ d\tau \right] U(t) \longrightarrow$$

(b) For $h(t) = \exp(-2t)\ U(t)$ and input $x(t) = \exp(-3t)\ U(t)$ we have

$$y(t) = \left[\int_0^t x(\tau)\ h(t-\tau)\ d\tau \right] U(t)$$

$$= \left[\int_0^t e^{-3\tau}\ e^{-2(t-\tau)}\ d\tau \right] U(t) = \left[e^{-2t} \int_0^t e^{-\tau}\ d\tau \right] U(t)$$

$$= \left[e^{-2t}\ \ (-e^{-\tau}) \Big|_0^t \right] U(t) = \left[e^{-2t}\ \ (1 - e^{-t}) \right] U(t)$$

$$= [e^{-2t} - e^{-3t}]\ U(t) \longrightarrow$$

(c) Working in the frequency domain:

$$x(t) = \exp(-3t)\ U(t) \iff \frac{1}{j\omega + 3} = X(\omega)$$

$$Y(\omega) = H(j\omega)\ X(\omega) = \frac{1}{j\omega + 2}\,\frac{1}{j\omega + 3} = \frac{1}{j\omega + 2} - \frac{1}{j\omega + 3}$$

$$\iff e^{-2t}\ U(t) - e^{-3t}\ U(t) \longrightarrow$$

7.13 For the circuit in Figure 7.26:

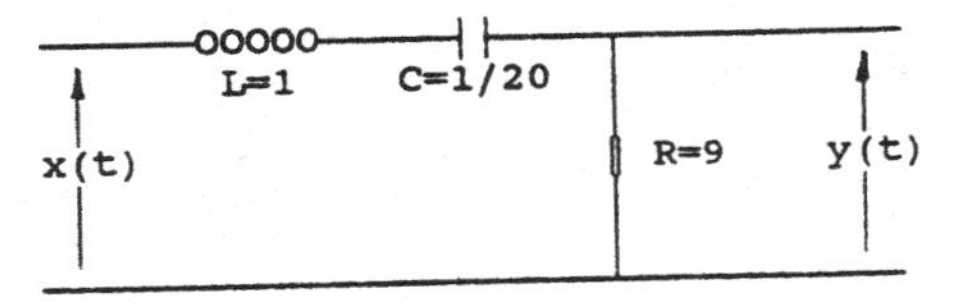

Figure 7.26

(a) Finding the impulse response h(t):

$$H(j\omega) = \frac{R}{j\omega L + R + 1/j\omega C} = \frac{9}{j\omega + 9 + 20/j\omega} = \frac{9j\omega}{(j\omega)^2 + 9j\omega + 20}$$

$$= \frac{9j\omega}{(j\omega + 5)(j\omega + 4)} = \frac{45}{j\omega + 5} - \frac{36}{j\omega + 4}$$

$$\iff [45e^{-5t} - 36e^{-4t}]\ U(t) = h(t)$$

(b) Finding $y(t)$ by time-domain convolution for the case where the input is $x(t) = \exp(-t)\ U(t)$

$$y(t) = x(t) * h(t) = \int_{-\infty}^{\infty} x(\tau)\ h(t - \tau)\ d\tau$$

$$= \left[\int_0^t x(\tau)\ h(t - \tau)\ d\tau\right] U(t)$$

(by the result of Exercise 7.12(a)

$$= \left[\int_0^t e^{-\tau}\ [45e^{-5(t-\tau)} - 36e^{-4(t-\tau)}]\ d\tau\right] U(t)$$

$$= \left[45e^{-5t}\int_0^t e^{-\tau}e^{5\tau}\ d\tau\right] U(t) - \left[36e^{-4t}\int_{-\infty}^{\infty} e^{-\tau}e^{4\tau}\ d\tau\right] U(t)$$

$$= \left[45e^{-5t}\int_0^t e^{4\tau}\ d\tau\right] U(t) - \left[36e^{-4t}\int_{-\infty}^{\infty} e^{3\tau}\ d\tau\right] U(t)$$

$$= \left[45e^{-5t}\ \frac{e^{4\tau}}{4}\Big|_0^t - 36e^{-4t}\ \frac{e^{3\tau}}{3}\Big|_0^t\right] U(t)$$

$$= [(45/4)e^{-5t}(e^{4t} - 1) - 12e^{-4t}(e^{3t} - 1)]\ U(t)$$

$$= [(45/4)e^{-t} - (45/4)e^{-5t} - 12e^{-t} + 12e^{-4t}]\ U(t)$$

$$= [12e^{-4t} - (45/4)e^{-5t} - (3/4)e^{-t}]\ U(t) \longrightarrow$$

(c) Finding $y(t)$ by working in the frequency domain:

$$h(t) = [45e^{-5t} - 36e^{-4t}]\ U(t) \iff \frac{45}{j\omega + 5} - \frac{36}{j\omega + 4} = H(j\omega)$$

$$x(t) = x(t) = \exp(-t)\ U(t) \iff \frac{1}{j\omega + 1} = X(\omega)$$

$$Y(\omega) = H(j\omega)\ X(\omega) = \left[\frac{45}{j\omega + 5} - \frac{36}{j\omega + 4}\right]\frac{1}{j\omega + 1}$$

$$= \frac{9j\omega}{(j\omega + 5)(j\omega + 4)(j\omega + 1)} = \frac{12}{j\omega + 4} - \frac{45/4}{j\omega + 5} - \frac{3/4}{j\omega + 1}$$

$$\iff [12e^{-4t} - (45/4)e^{-5t} - (3/4)e^{-t}]\ U(t) \longrightarrow$$

(d) Transforming the result obtained in (b):

$$[12e^{-4t} - (45/4)e^{-5t} - (3/4)e^{-t}]\ U(t)$$

$$\Longleftrightarrow \frac{12}{j\omega + 4} - \frac{45/4}{j\omega + 5} - \frac{3/4}{j\omega + 1} = \frac{9j\omega}{(j\omega + 5)(j\omega + 4)}\ \frac{1}{j\omega + 1}$$

$$= H(j\omega)\ X(\omega) \longrightarrow$$

7.14 Using Fourier transformation to prove that

$$\int_{-\infty}^{\infty} x(\tau)\ g(t - \tau)\ d\tau = \int_{-\infty}^{\infty} g(\tau)\ x(t - \tau)\ d\tau$$

$$\int_{-\infty}^{\infty} x(\tau)\ g(t - \tau)\ d\tau \Longleftrightarrow X(\omega)G(\omega) = G(\omega)X(\omega) \Longleftrightarrow \int_{-\infty}^{\infty} g(\tau)\ x(t - \tau)\ d\tau$$

QED

7.15 By frequency shift: $x(t)\ \cos(\omega_0 t)$

$$= x(t)\ [\tfrac{1}{2}\exp(j\omega_0 t) + \tfrac{1}{2}\exp(-j\omega_0 t)] \Longleftrightarrow \tfrac{1}{2}X(\omega - \omega_0) + \tfrac{1}{2}X(\omega + \omega_0)$$

7.16 Steps for creating the functions with $N = 1024$, $T = 8$:

Main menu, Create Values, Continue, Create X and X2, Real, Even, 2, 1/2, Continue, 1, 0, Go, Even, 1, Continue, COS(16*pi*t), Go, Neither, Go. You are now back on the main menu. Run Postprocessors, X, Real multiply X and X2, Quit. You are now on the main menu. Plotting **X** shows a Rect pulse that has been multiplied by a cosine. Run ANALYSIS. Plot **FRE**. We see the double Sa spectrum, with the Sa's not symmetric because they are running into one another.

Displaying the numbers gives the values shown below where we also show values taken from the expression

$$Y(\omega) = \tfrac{1}{2}\,\mathrm{Sa}\left[\frac{\omega - \omega_0}{2}\right] + \tfrac{1}{2}\,\mathrm{Sa}\left[\frac{\omega + \omega_0}{2}\right]$$

with sampling of ω at multiples of $\Omega = 2\pi/T = 2\pi/8 = \pi/4$.

n	FFT	Formula	error (%)
1	-2.4105317e-4	-2.3797195e-4	1.294782852
17	-4.4039580e-3	-4.3515684e-3	1.203924544
33	-1.0796245e-2	-1.0694495e-2	0.951422469
62	0.44336665	0.443012791	0.079875572
63	0.48360570	0.483411083	0.040259027
64	0.5	0.5	0
65	0.49082397	0.491024793	0.040898729
66	0.45670685	0.457083668	0.082439655

7.17 (a) Loading the pulse shown in Figure 7.27 into **X** and obtaining a plot of its magnitude spectrum with N = 1024 and T = 16. (We will also be needing the function $\cos(2\pi 8t)$ in **X2** and so we will create both of them in the first place.)

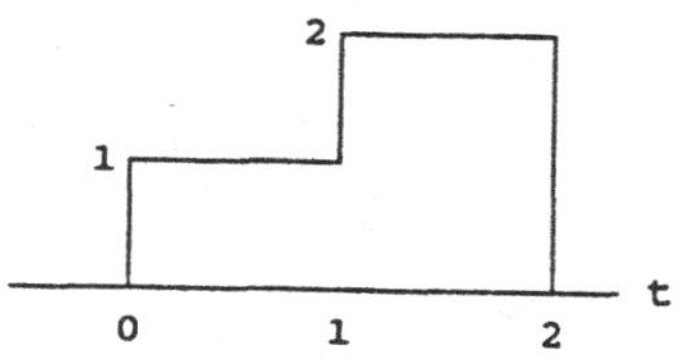

Figure 7.27: x(t)

Main menu, Create Values, Continue, Create X and X2, Real, Neither, Left, 3, 1, 2, Continue, 1, 2, 0, Go, Even, 1, Continue, COS(2*pi*8*t), Go, Neither, Go. You are now back on the main menu.

The eternal-cosine function $\cos(2\pi 8t)$ is already in **X2**. Using the REAL-MULTIPLY package in the **X** postprocessor we multiply **XRE** and **X2RE**.

Run ANALYSIS. We obtain a magnitude spectrum which has two copies of the original magnitude spectrum, halved in height, one at $\omega = 2\pi 8$ and the other at $\omega = -2\pi 8$. This is also shown in the figure.

(b) Because the transform of $\sin(\omega_0 t)$ consists of Dirac deltas with **opposite** signs (as against $\cos(\omega_0 t)$ where the signs are the same) if we multiply a Rect pulse by $\sin(\omega_0 t)$ we expect to obtain a spectrum made up of two Sa's, but now with their signs opposite.

Repeating the plot of Figure 7.23 (N = 1024, T = 8) but this time using an eternal sine instead of a cosine, the sequence of steps is as follows:

Main menu, Create Values, Continue, Create X and X2, Real, Even, 2, .5, Continue, 1, 0, Go, Odd, 1,

correctly,

Run Postprocessors, X, Real multiply X and X2, Quit. You are now back on the main menu. Plotting X shows a plot of Rect(t) SIN(16*pi*t), as required.

Run ANALYSIS and plot FIM. We see that the spectrum is now made up of two Sa's, one posistive and the other negative. (See printout.)

7.18 (a) Using convolution to find the Fourier transform of the single radar pulse $x(t)$ shown in Figure 7.28.

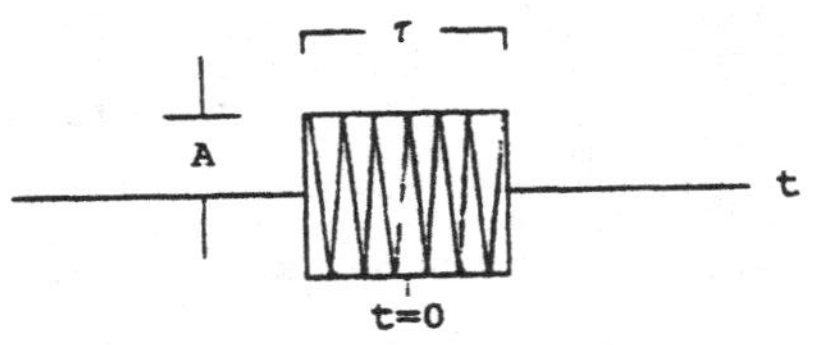

Figure 7.28: Radar pulse $x(t)$

$$x(t) = A\,\mathrm{Rect}(t/\tau)\cos(\omega_0 t)$$

To find $X(\omega)$ we observe that $x(t)$ is a multiplication of two functions $f(t) = A\,\mathrm{Rect}(t/\tau)$ and $g(t) = \cos(\omega_0 t)$ in the time domain, which corresponds to convolution in the frequency domain. Thus:

$$X(\omega) = \frac{1}{2\pi} F(\omega) * G(\omega) = \frac{1}{2\pi}\int_{-\infty}^{\infty} F(\theta)\, G(\omega - \theta)\, d\theta$$

$$= \frac{1}{2\pi}\int_{-\infty}^{\infty} \Big[A\tau\, \mathrm{Sa}(\theta\tau/2) \Big] \Big[\pi[\delta(\omega - \omega_0 - \theta) + \delta(\omega + \omega_0 - \theta] \Big]\, d\theta$$

$$= \frac{A\tau}{2}\int_{-\infty}^{\infty} \mathrm{Sa}[(\omega - \omega_0)\tau/2]\, \delta(\omega - \omega_0 - \theta)\, d\theta$$

$$+ \frac{A\tau}{2}\int_{-\infty}^{\infty} \mathrm{Sa}[(\omega + \omega_0)\tau/2]\, \delta(\omega - \omega_0 + \theta)\, d\theta$$

$$= \frac{A\tau}{2} \mathrm{Sa}[(\omega - \omega_0)\tau/2] \int_{-\infty}^{\infty} \delta(\omega - \omega_0 - \theta)'\, d\theta$$

$$+ \frac{A\tau}{2} \mathrm{Sa}[(\omega + \omega_0)\tau/2] \int_{-\infty}^{\infty} \delta(\omega - \omega_0 + \theta)\, d\theta$$

$$= \frac{A\tau}{2} \Big[\mathrm{Sa}[(\omega - \omega_0)\tau/2] + \mathrm{Sa}[(\omega + \omega_0)\tau/2] \Big] \longrightarrow$$

(b) Using the inversion integral to find the Fourier inverse of the above result:

$$x(t) = \frac{1}{2\pi}\int_{-\infty}^{\infty} X(\omega)\exp(j\omega t)\, d\omega$$

$$= \frac{1}{2\pi}\int_{-\infty}^{\infty} \frac{A\tau}{2} \Big[\mathrm{Sa}[(\omega - \omega_0)\tau/2] + \mathrm{Sa}[(\omega + \omega_0)\tau/2] \Big] \exp(j\omega t)\, d\omega$$

$$= \frac{A\tau}{4\pi}\int_{-\infty}^{\infty} \mathrm{Sa}[(\omega - \omega_0)\tau/2] \exp(j\omega t)\, d\omega \quad \text{———} \quad I_1$$

$$+ \frac{A\tau}{4\pi}\int_{-\infty}^{\infty} \mathrm{Sa}[(\omega + \omega_0)\tau/2] \exp(j\omega t)\, d\omega \quad \text{———} \quad I_2$$

In I_1: Let $\omega - \omega_0 = z$. Then $\omega = z + \omega_0$ and $d\omega = dz$, and so:

$$I_1 = \frac{A}{4\pi}\int_{-\infty}^{\infty} \tau\, \mathrm{Sa}(z\tau/2) \exp(jzt) \exp(j\omega_0 t)\, dz$$

$$= \frac{A}{2}\exp(j\omega_0 t)\, \frac{1}{2\pi}\int_{-\infty}^{\infty} \tau\, \mathrm{Sa}(z\tau/2)\exp(jzt)\, dz = \frac{A}{2}\exp(j\omega_0 t)\, \mathrm{Rect}(t/\tau)$$

Similarly, $I_2 = (A/2)\exp(-j\omega_0 t)\,\mathrm{Rect}(t/\tau)$, and so

$$x(t) = I_1 + I_2 = \frac{\exp(j\omega_0 t) + \exp(-j\omega_0 t)}{2} A\ \text{Rect}(t/\tau)$$

$$= A\ \text{Rect}(t/\tau)\ \cos(\omega_0 t) \longrightarrow$$

7.19 (a) Finding the Fourier transform of the infinite train of radar pulses $x_p(t)$ shown in Figure 7.29.

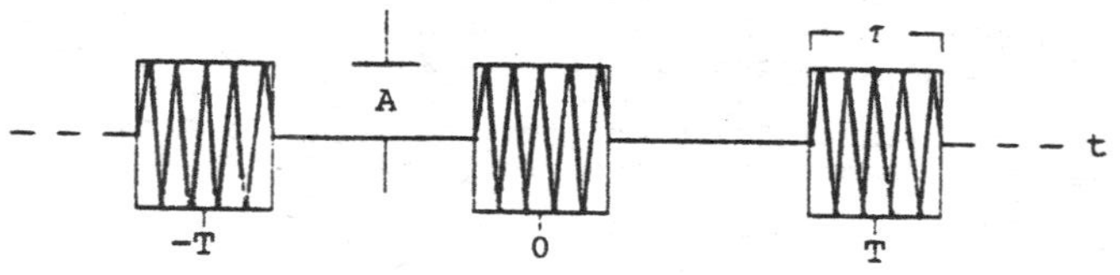

Figure 7.29: Radar pulse train $x_p(t)$

There are two ways to carry this out:

Method 1: We assume that an RF oscillator is putting out $\cos(\omega_0 t)$ which is being gated (multiplied) by an **independently operated** periodic train of Rect pulses.

The envelope of the pulse train is the periodic function

$$f_p(t) = \begin{cases} 0 & (-T/2 < t < -\tau/2) \\ A & (-\tau/2 < t < \tau/2) \\ 0 & (\tau/2 < t < T/2) \end{cases}$$

We need to find its Fourier transform. First we shall find its Fourier series and then we shall find the transform of that series.

Step (1): Finding the Fourier series:

$$F(n) = 1/T \int_{-\tau/2}^{\tau/2} A \exp(-jn\omega_1 t)\ dt \qquad \omega_1 = 2\pi/T$$

$$= \frac{A}{T} \frac{\exp(-jn\omega_1 t)}{-jn\omega_1} \Bigg|_{-\tau/2}^{\tau/2} = \frac{\tau A}{T} \frac{\exp(jn\omega_1 \tau/2) - \exp(-jn\omega_1\tau/2)}{2j\ n\omega_1\tau/2}$$

$$= \frac{\tau A}{T} \text{Sa}(n\omega_1\tau/2)$$

Then the required Fourier series of the envelope is

$$f_p(t) = \sum_{n=-\infty}^{\infty} F(n) \exp(jn\omega_1 t) = \frac{\tau A}{T} \sum_{n=-\infty}^{\infty} \text{Sa}(n\omega_1\tau/2) \exp(jn\omega_1 t) \longrightarrow$$

Step (2): Fourier-transforming the Fourier series:

$$\exp(jn\omega_1 t) \Longleftrightarrow 2\pi\delta(\omega - n\omega_1)$$

and so $$F(\omega) = \frac{\tau A}{T} \sum_{n=-\infty}^{\infty} \text{Sa}(n\omega_1\tau/2)\ 2\pi\delta(\omega - n\omega_1)$$

$$= \frac{2\pi\tau A}{T} \sum_{n=-\infty}^{\infty} \text{Sa}(n\omega_1\tau/2)\ \delta(\omega - n\omega_1) \longrightarrow$$

The RF carrier is assumed to be $g(t) = \cos(\omega_0 t)$ and so the train of radar pulses is $x(t) = f_p(t)\ g(t)$. This is seen to be a product of two functions in the time domain, and so its Fourier transform can be found by using convolution in the frequency domain. Thus

$$X(\omega) = \frac{1}{2\pi} F(\omega) * G(\omega) = \frac{1}{2\pi} \int_{-\infty}^{\infty} F(\theta)\ G(\omega - \theta)\ d\theta$$

$$= \frac{1}{2\pi} \int_{-\infty}^{\infty} \left[\frac{2\pi\tau A}{T} \sum_{n=-\infty}^{\infty} \text{Sa}(n\omega_1\tau/2)\ \delta(\theta - n\omega_1) \right] \left[\pi[\delta(\omega-\omega_0-\theta) + \delta(\omega+\omega_0-\theta] \right] d\theta$$

$$= \frac{\tau A\pi}{T} \sum_{n=-\infty}^{\infty} \text{Sa}(n\omega_1\tau/2) \left[\delta(\omega - \omega_0 - n\omega_1) \int_{-\infty}^{\infty} \delta(\omega - \omega_0 - \theta)\ d\theta \right.$$

$$\left. + \delta(\omega + \omega_0 - n\omega_1) \int_{-\infty}^{\infty} \delta(\omega + \omega_0 - \theta)\ d\theta \right]$$

$$= \frac{\tau A\pi}{T} \sum_{n=-\infty}^{\infty} \text{Sa}(n\omega_1\tau/2) \left[\delta(\omega - \omega_0 - n\omega_1) + \delta(\omega + \omega_0 - n\omega_1) \right]$$

(b) Inverting this expression to the time domain:

$$x(t) = \frac{1}{2\pi} \int_{-\infty}^{\infty} X(\omega)\ \exp(j\omega t)\ d\omega$$

$$= \frac{1}{2\pi}\int_{-\infty}^{\infty} \frac{\tau A\pi}{T}\sum_{n=-\infty}^{\infty} Sa(n\omega_1\tau/2) \left[\delta(\omega-\omega_0-n\omega_1) + \delta(\omega+\omega_0-n\omega_1) \right] \exp(j\omega t)\ d\omega$$

$$= \frac{\tau A}{2T}\sum_{n=-\infty}^{\infty} Sa(n\omega_1\tau/2) \int_{-\infty}^{\infty} \left[\delta(\omega-\omega_0-n\omega_1) + \delta(\omega+\omega_0-n\omega_1) \right] \exp(j\omega t)\ d\omega$$

$$= \frac{\tau A}{2T}\sum_{n=-\infty}^{\infty} Sa(n\omega_1\tau/2) \left[\int_{-\infty}^{\infty} \exp[j(\omega_0 + n\omega_1)t]\ \delta(\omega - \omega_0 - n\omega_1)\ d\omega + \int_{-\infty}^{\infty} \exp[j(-\omega_0 + n\omega_1)t]\ \delta(\omega + \omega_0 - n\omega_1)\ d\omega \right]$$

$$= \frac{\tau A}{2T}\sum_{n=-\infty}^{\infty} Sa(n\omega_1\tau/2)\ \exp(jn\omega_1 t) \left[\exp(j(\omega_0 t)) \int_{-\infty}^{\infty} \delta(\omega - \omega_0 - n\omega_1)\ d\omega + \exp(j(-\omega_0 t)) \int_{-\infty}^{\infty} \delta(\omega + \omega_0 - n\omega_1)\ d\omega \right]$$

$$= \frac{\tau A}{T}\sum_{n=-\infty}^{\infty} Sa(n\omega_1\tau/2)\ \exp(jn\omega_1 t)\ \tfrac{1}{2}\left[\exp(j(\omega_0 t)) + \exp(j(-\omega_0 t)) \right]$$

$$= \left[\frac{\tau A}{T}\sum_{n=-\infty}^{\infty} Sa(n\omega_1\tau/2)\ \exp(jn\omega_1 t) \right] \cos(\omega_0 t) = f_p(t)\ \cos(\omega_0 t)$$

——————————>

Method 2: It is assumed that the reader is already familiar with the use of Theorem 9.3, which we shall be using in what follows.

We are assuming that the RF oscillator starts up whenever a Rect pulse commences, thereafter putting out $\cos(\omega_0 t)$. The net result is that each burst of RF is an **exact replica** of the previous one.

The signal is now a repetition of the pulse

$$f(t) = A\cos(\omega_0 t) \quad (-\tau/2 < t < \tau/2) \quad \text{period } T \qquad (1)$$

Using Theorem 9.3 we are able to find the Fourier coefficients (and hence the Fourier series and hence the Fourier transform) of such a periodically repeated pulse.

The complete analytical definition of the pulse is

$$f(t) = A\ Rect(t - \tau/2)\ \cos(\omega_0 t)$$

i.e. it is comprised of the product of

$$\left[g(t) = A\ Rect(t - \tau/2) \right] \Longleftrightarrow \left[G(\omega) = A\tau\ Sa(\omega\tau/2)\ \exp(-j\omega\tau/2) \right]$$

and the RF carrier

$$\left[h(t) = \cos(\omega_0 t) \right] \Longleftrightarrow \left[H(\omega) = \pi[\delta(\omega - \omega_0) + \delta(\omega + \omega_0)] \right]$$

which have been multiplied in the time domain to give $f(t)$. Thus $F(\omega)$ can be obtained by the use of convolution in the frequency domain, i.e.,

$$F(\omega) = (1/2\pi)\ G(\omega) * H(\omega) = 1/2\pi \int_{-\infty}^{\infty} G(\theta)\ H(\omega - \theta)\ d\theta$$

$$= \frac{A\tau}{2}\int_{-\infty}^{\infty} Sa(\theta\tau/2)\ \exp(-j\theta\tau/2)\ [\delta(\omega - \omega_0 - \theta) + \delta(\omega + \omega_0 - \theta)]\ d\theta$$

$$= \frac{A\tau}{2}\int_{-\infty}^{\infty} Sa[(\omega - \omega_0)\tau/2]\ \exp[-j(\omega - \omega_0)\tau/2]\ \delta(\omega - \omega_0 - \theta)\ d\theta + \frac{A\tau}{2}\int_{-\infty}^{\infty} Sa[(\omega + \omega_0)\tau/2]\ \exp[-j(\omega + \omega_0)\tau/2]\ \delta(\omega + \omega_0 - \theta)\ d\theta$$

$$= \frac{A\tau}{2} Sa[(\omega - \omega_0)\tau/2]\ \exp[-j(\omega - \omega_0)\tau/2] \int_{-\infty}^{\infty} \delta(\omega - \omega_0 - \theta)\ d\theta + \frac{A\tau}{2} Sa[(\omega + \omega_0)\tau/2]\ \exp[-j(\omega + \omega_0)\tau/2] \int_{-\infty}^{\infty} \delta(\omega + \omega_0 - \theta)\ d\theta$$

$$= \frac{A\tau}{2}\left[Sa[(\omega - \omega_0)\tau/2]\ \exp[-j(\omega - \omega_0)\tau/2] + Sa[(\omega + \omega_0)\tau/2]\ \exp[-j(\omega + \omega_0)\tau/2] \right]$$

Step 2: Finding the Fourier coefficients of an eternal train of such pulses, assuming that the period is T: By Theorem 9.3 the expression for the Fourier coefficients will be

$$F_p(n) = (1/T)\ F(n\omega_1) \qquad \text{where } \omega_1 = 2\pi/T$$

$$= \frac{A\tau}{2T}\Big[Sa[(\omega - \omega_0)\tau/2]\ \exp[-j(\omega - \omega_0)\tau/2] + Sa[(\omega + \omega_0)\tau/2]\ \exp[-j(\omega + \omega_0)\tau/2] \Big]_{\omega = n\omega_1}$$

$$= \frac{A\tau}{2T}\Big[Sa[(n\omega_1 - \omega_0)\tau/2]\ \exp[-j(n\omega_1 - \omega_0)\tau/2] + Sa[(n\omega_1 + \omega_0)\tau/2]\ \exp[-j(n\omega_1 + \omega_0)\tau/2] \Big]$$

Step 3: Assembling the Fourier series

$$f_p(t) = \sum_{n=-\infty}^{\infty} F_p(n)\ \exp(jn\omega_1 t)$$

$$= \frac{A\tau}{2T}\sum_{n=-\infty}^{\infty}\left[\begin{array}{l} Sa[(n\omega_1 - \omega_0)\tau/2]\ \exp[-j(n\omega_1 - \omega_0)\tau/2] \\ + Sa[(n\omega_1 + \omega_0)\tau/2]\ \exp[-j(n\omega_1 + \omega_0)\tau/2] \end{array}\right] \exp(jn\omega_1 t)$$

Step 4: Fourier transforming the Fourier series

Since $\exp(jn\omega_1 t) \Longleftrightarrow 2\pi\delta(\omega - n\omega_1)$ we have

$$F(\omega) = \sum_{n=-\infty}^{\infty} F_p(n)\ 2\pi\ \delta(\omega - n\omega_1)$$

$$= \frac{A\tau}{2T}\sum_{n=-\infty}^{\infty}\left[\begin{array}{l} Sa[(n\omega_1 - \omega_0)\tau/2]\ \exp[-j(n\omega_1 - \omega_0)\tau/2] \\ + Sa[(n\omega_1 + \omega_0)\tau/2]\ \exp[-j(n\omega_1 + \omega_0)\tau/2] \end{array}\right] 2\pi\ \delta(\omega - n\omega_1)$$

$$= \frac{\pi A\tau}{T}\sum_{n=-\infty}^{\infty}\left[\begin{array}{l} Sa[(n\omega_1 - \omega_0)\tau/2]\ \exp[-j(n\omega_1 - \omega_0)\tau/2] \\ + Sa[(n\omega_1 + \omega_0)\tau/2]\ \exp[-j(n\omega_1 + \omega_0)\tau/2] \end{array}\right] \delta(\omega - n\omega_1)$$

————>

7.20 (a) Finding the Fourier transform of $f(t) = x(t)\sin(\omega_0 t)$ by convolution:

$$x(t) \Longleftrightarrow X(\omega)$$

and

Using frequency domain convolution:

$$x(t)\ y(t) \Longleftrightarrow (1/2\pi)\ X(\omega) * Y(\omega) = (1/2\pi)\int_{-\infty}^{\infty} X(\theta)\ Y(\omega - \theta)\ d\theta$$

$$= (1/2\pi)\int_{-\infty}^{\infty} X(\theta)\ \frac{\pi}{j}\ [\delta(\omega - \omega_0 - \theta) - \delta(\omega + \omega_0 - \theta)]\ d\theta$$

$$= (1/2j)\int_{-\infty}^{\infty} X(\omega - \omega_0)\ \delta(\omega - \omega_0 - \theta)\ d\theta - (1/2j)\int_{-\infty}^{\infty} X(\omega + \omega_0)\ \delta(\omega + \omega_0 - \theta)\ d\theta$$

$$= (1/2j)\ X(\omega - \omega_0)\int_{-\infty}^{\infty} \delta(\omega - \omega_0 - \theta)\ d\theta - (1/2j)\ X(\omega + \omega_0)\int_{-\infty}^{\infty} \delta(\omega + \omega_0 - \theta)\ d\theta$$

$$= (1/2j)\ [X(\omega - \omega_0) - X(\omega + \omega_0)] \quad \longrightarrow$$

(b) Using the frequency-shift property:

$$x(t)\sin(\omega_0 t) = x(t)\ \frac{1}{2j}\ [\exp(j\omega_0 t) - \exp(-j\omega_0 t)]$$

$$\Longleftrightarrow (1/2j)\ [X(\omega - \omega_0) - X(\omega + \omega_0)] \quad \longrightarrow$$

7.21 (a)(1) Finding the Fourier transform of $f(t) = \cos^2(\omega_0 t)$ using the analysis equation:

$$F(\omega) = \int_{-\infty}^{\infty} \cos^2(\omega_0 t)\ \exp(-j\omega t)\ dt = \int_{-\infty}^{\infty} \tfrac{1}{2}[1 + \cos(2\omega_0 t)]\ \exp(-j\omega t)\ dt$$

$$= \int_{-\infty}^{\infty} [\tfrac{1}{2} + \tfrac{1}{4}\exp(j2\omega_0 t) + \tfrac{1}{4}\exp(-j2\omega_0 t)]\ \exp(-j\omega t)\ dt$$

$$= \frac{1}{2}\int_{-\infty}^{\infty} \exp(-j\omega t)\,dt + \frac{1}{4}\int_{-\infty}^{\infty} \exp[-j(\omega - 2\omega_0)t]\,dt + \frac{1}{4}\int_{-\infty}^{\infty} \exp[-j(\omega + 2\omega_0)t]\,dt$$

$$= \frac{1}{2}\, 2\pi\delta(\omega) + \frac{1}{4}\, 2\pi\delta(\omega - 2\omega_0) + \frac{1}{4}\, 2\pi\delta(\omega + 2\omega_0)$$

$$= \frac{\pi}{2}\left[\delta(\omega - 2\omega_0) + 2\delta(\omega) + \delta(\omega + 2\omega_0)\right]$$

(a)(2) Using frequency-domain convolution:

$$f(t) = \cos(\omega_0 t) \Longleftrightarrow F(\omega) = \pi[\delta(\omega - \omega_0) + \delta(\omega + \omega_0)]$$

$\cos(\omega_0 t)\ \cos(\omega_0 t)$

$$\Longleftrightarrow (1/2\pi)\ F(\omega) * F(\omega) = (1/2\pi)\int_{-\infty}^{\infty} F(\theta)\ F(\omega - \theta)\ d\theta$$

$$= \frac{1}{2\pi}\int_{-\infty}^{\infty} \pi[\delta(\theta - \omega_0) + \delta(\theta + \omega_0)]\ \pi[\delta(\omega - \omega_0 - \theta) + \delta(\omega + \omega_0 - \theta)]\ d\theta$$

$$= \frac{\pi}{2}\int_{-\infty}^{\infty} \delta(\theta - \omega_0)\ \delta(\omega - \omega_0 - \theta)\ d\theta + \frac{\pi}{2}\int_{-\infty}^{\infty} \delta(\theta + \omega_0)\ \delta(\omega - \omega_0 - \theta)\ d\theta$$

$$+ \frac{\pi}{2}\int_{-\infty}^{\infty} \delta(\theta - \omega_0)\ \delta(\omega + \omega_0 - \theta)\ d\theta + \frac{\pi}{2}\int_{-\infty}^{\infty} \delta(\theta + \omega_0)\ \delta(\omega + \omega_0 - \theta)\ d\theta$$

$$= \frac{\pi}{2}\int_{-\infty}^{\infty} \delta(\omega - \omega_0 - \omega_0)\ \delta(\omega - \omega_0 - \theta)\ d\theta + \frac{\pi}{2}\int_{-\infty}^{\infty} \delta(\omega - \omega_0 + \omega_0)\ \delta(\omega - \omega_0 - \theta)\ d\theta$$

$$+ \frac{\pi}{2}\int_{-\infty}^{\infty} \delta(\omega + \omega_0 - \omega_0)\ \delta(\omega + \omega_0 - \theta)\ d\theta + \frac{\pi}{2}\int_{-\infty}^{\infty} \delta(\omega + \omega_0 + \omega_0)\ \delta(\omega + \omega_0 - \theta)\ d\theta$$

$$= \frac{\pi}{2}\,\delta(\omega - 2\omega_0)\int_{-\infty}^{\infty} \delta(\omega - \omega_0 - \theta)\ d\theta + \frac{\pi}{2}\,\delta(\omega)\int_{-\infty}^{\infty} \delta(\omega - \omega_0 - \theta)\ d\theta$$

$$+ \frac{\pi}{2}\,\delta(\omega)\int_{-\infty}^{\infty} \delta(\omega + \omega_0 - \theta)\ d\theta + \frac{\pi}{2}\,\delta(\omega + 2\omega_0)\int_{-\infty}^{\infty} \delta(\omega + \omega_0 - \theta)\ d\theta$$

$$= \frac{\pi}{2}\,\delta(\omega - 2\omega_0) + \frac{\pi}{2}\,\delta(\omega) + \frac{\pi}{2}\,\delta(\omega) + \frac{\pi}{2}\,\delta(\omega + 2\omega_0)$$

$$= \frac{\pi}{2}\left[\delta(\omega - 2\omega_0) + 2\delta(\omega) + \delta(\omega + 2\omega_0)\right] \longrightarrow$$

(b) Proving that $2\sin(\omega_0 t)\cos(\omega_0 t) = \sin(2\omega_0 t)$ by using frequency-domain convolution to find the Fourier transform of the LHS, and then showing that it is equal to the Fourier transform of the RHS.

$$f(t) = \sin(\omega_0 t) \Longleftrightarrow F(\omega) = \frac{\pi}{j}[\delta(\omega - \omega_0) - \delta(\omega + \omega_0)]$$

$$g(t) = \cos(\omega_0 t) \Longleftrightarrow G(\omega) = \pi[\delta(\omega - \omega_0) + \delta(\omega + \omega_0)]$$

$2\sin(\omega_0 t)\ \cos(\omega_0 t)$

$$\Longleftrightarrow 2(1/2\pi)\ F(\omega) * G(\omega) = \frac{1}{\pi}\int_{-\infty}^{\infty} F(\theta)\ G(\omega - \theta)\ d\theta$$

$$= \frac{1}{\pi}\int_{-\infty}^{\infty} \frac{\pi}{j}[\delta(\theta - \omega_0) - \delta(\theta + \omega_0)]\ \pi[\delta(\omega - \omega_0 - \theta) + \delta(\omega + \omega_0 - \theta)]\ d\theta$$

$$= \frac{\pi}{j}\int_{-\infty}^{\infty} \delta(\theta - \omega_0)\ \delta(\omega - \omega_0 - \theta)\ d\theta + \frac{\pi}{j}\int_{-\infty}^{\infty} \delta(\theta + \omega_0)\ \delta(\omega - \omega_0 - \theta)\ d\theta$$

$$- \frac{\pi}{j}\int_{-\infty}^{\infty} \delta(\theta - \omega_0)\ \delta(\omega + \omega_0 - \theta)\ d\theta - \frac{\pi}{j}\int_{-\infty}^{\infty} \delta(\theta + \omega_0)\ \delta(\omega + \omega_0 - \theta)\ d\theta$$

$$= \frac{\pi}{j}\int_{-\infty}^{\infty} \delta(\omega - \omega_0 - \omega_0)\ \delta(\omega - \omega_0 - \theta)\ d\theta + \frac{\pi}{j}\int_{-\infty}^{\infty} \delta(\omega - \omega_0 + \omega_0)\ \delta(\omega - \omega_0 - \theta)\ d\theta$$

$$-\frac{\pi}{j}\int_{-\infty}^{\infty} \delta(\omega+\omega_0-\omega_0)\ \delta(\omega+\omega_0-\theta)\ d\theta - \frac{\pi}{j}\int_{-\infty}^{\infty} \delta(\omega+\omega_0+\omega_0)\ \delta(\omega+\omega_0-\theta)\ d\theta$$

$$= \frac{\pi}{j}\delta(\omega - 2\omega_0)\int_{-\infty}^{\infty} \delta(\omega-\omega_0-\theta)\ d\theta + \frac{\pi}{j}\delta(\omega)\int_{-\infty}^{\infty}\delta(\omega-\omega_0-\theta)\ d\theta$$

$$-\frac{\pi}{j}\delta(\omega)\int_{-\infty}^{\infty}\delta(\omega+\omega_0-\theta)\ d\theta - \frac{\pi}{j}\delta(\omega+2\omega_0)\int_{-\infty}^{\infty}\delta(\omega+\omega_0-\theta)\ d\theta$$

$$= \frac{\pi}{j}\delta(\omega-2\omega_0) + \frac{\pi}{j}\delta(\omega) - \frac{\pi}{j}\delta(\omega) - \frac{\pi}{j}\delta(\omega+2\omega_0)$$

$$= \frac{\pi}{j}\left[\delta(\omega-2\omega_0) - \delta(\omega+2\omega_0)\right] \longrightarrow$$

$$\sin(2\omega_0 t) \iff = \frac{\pi}{j}\left[\delta(\omega-2\omega_0) - \delta(\omega+2\omega_0)\right] \longrightarrow$$

7.22 (a) Transforming $f(t) = x(t)\cos^2(\omega_0 t)$ using frequency shift:

$$x(t)\cos^2(\omega_0 t) = x(t)[\tfrac{1}{2} + \tfrac{1}{2}\cos(2\omega_0 t)]$$

$$= x(t)[\tfrac{1}{2} + \tfrac{1}{4}\exp(j2\omega_0 t) + \tfrac{1}{4}\exp(-j2\omega_0 t)]$$

$$= \tfrac{1}{2}x(t) + \tfrac{1}{4}x(t)\exp(j2\omega_0 t) + \tfrac{1}{4}x(t)\exp(-j2\omega_0 t)]$$

$$\iff \tfrac{1}{2}X(\omega) + \tfrac{1}{4}X(\omega-2\omega_0) + \tfrac{1}{4}X(\omega+2\omega_0)$$

We see three copies of the original spectrum, one shifted by $2\omega_0$ to the right, one shifted by $2\omega_0$ to the left and the third **unshifted**, i.e. in the baseband. To extract $x(t)$ all we need to do is to pass $f(t)$ through a lowpass filter which passes the baseband portion and excludes the two shifted spectra.

(b) To show that if we demodulate $x(t)\cos(\omega_0 t)$ by multiplication with $\sin(\omega_0 t)$ we get zero in the base-band:

$$f(t) = [x(t)\cos(\omega_0 t)]\sin(\omega_0 t) = \tfrac{1}{2}x(t)\sin(2\omega_0 t)$$

$$= (1/4j)[x(t)\exp(j2\omega_0 t) - x(t)\exp(-j2\omega_0 t)]$$

$$\iff (1/4j)[X(\omega-2\omega_0) - X(\omega+2\omega_0)]$$

Observe that now there is nothing in the baseband, and so nothing would come out of the lowpass filter.

If the phase of the demodulating cosine in (a) drifts then it tends towards becoming a sine, and so the magnitude of the output falls. For this reason it is best that the demodulating cosine be phase locked to the incoming cosine carrier, if that is possible.

7.23 (a) Proving that

$$\cos(\omega_0 t)\ U(t) \iff \frac{j\omega}{(j\omega)^2 + (\omega_0)^2} + \tfrac{1}{2}\pi\left[\delta(\omega-\omega_0) + \delta(\omega+\omega_0)\right]$$

The LHS is the time-domain product of $\cos(\omega_0 t)$ and $U(t)$, and so we can transform it by applying frequency-domain convolution, as follows:

$$f(t) = \cos(\omega_0 t) \iff F(\omega) = \pi[\delta(\omega-\omega_0) + \delta(\omega+\omega_0)]$$

$$g(t) = U(t) \iff \pi\delta(\omega) + 1/j\omega$$

$$f(t)\ g(t) \iff (1/2\pi)\ F(\omega) * G(\omega) = (1/2\pi)\int_{-\infty}^{\infty} F(\theta)\ G(\omega-\theta)\ d\omega$$

$$= (1/2\pi)\int_{-\infty}^{\infty}\pi[\delta(\theta-\omega_0) + \delta(\theta+\omega_0)]\ [\pi\delta(\omega-\theta) + 1/j(\omega-\theta)]\ d\theta$$

$$= \tfrac{1}{2}\int_{-\infty}^{\infty}\delta(\theta-\omega_0)\ [\pi\delta(\omega-\theta) + 1/(j\omega-\theta)]\ d\theta$$

$$+ \tfrac{1}{2}\int_{-\infty}^{\infty}\delta(\theta+\omega_0)\ [\pi\delta(\omega-\theta) + 1/j(\omega-\theta)]\ d\theta$$

$$= \tfrac{1}{2}\int_{-\infty}^{\infty}\delta(\theta-\omega_0)\ [\pi\delta(\omega-\omega_0) + 1/j(\omega-\omega_0)]\ d\theta$$

$$+ \tfrac{1}{2}\int_{-\infty}^{\infty}\delta(\theta+\omega_0)\ [\pi\delta(\omega+\omega_0) + 1/j(\omega+\omega_0)]\ d\theta$$

$$= \tfrac{1}{2}\,[\pi\delta(\omega-\omega_0) + 1/j(\omega-\omega_0)] \int_{-\infty}^{\infty} \delta(\theta-\omega_0)\, d\theta$$

$$+ \tfrac{1}{2}\,[\pi\delta(\omega+\omega_0) + 1/j(\omega+\omega_0)] \int_{-\infty}^{\infty} \delta(\theta+\omega_0)\, d\theta$$

$$= \tfrac{1}{2}\,[\pi\delta(\omega-\omega_0) + 1/j(\omega-\omega_0)] + \tfrac{1}{2}\,[\pi\delta(\omega+\omega_0) + 1/j(\omega+\omega_0)]$$

$$= \tfrac{1}{2}\,\frac{1}{j(\omega-\omega_0)} + \tfrac{1}{2}\,\frac{1}{j(\omega+\omega_0)} + \tfrac{1}{2}\pi[\delta(\omega-\omega_0) + \delta(\omega+\omega_0)]$$

$$= \frac{j\omega}{(j\omega)^2 + \omega_0^2} + \tfrac{1}{2}\pi[\delta(\omega-\omega_0) + \delta(\omega+\omega_0)] \longrightarrow \quad (A)$$

(b) From (A) we see that $\cos(\omega_0 t)\ U(t)$ transforms to the sum of two terms, its imaginary part being the **odd function**

$$P(\omega) = \frac{j\omega}{(j\omega)^2 + (\omega_0)^2} \qquad (B)$$

and its real part being the **even function** which is one half the transform of $\cos(\omega_0 t)$ i.e., $\tfrac{1}{2}\pi[(\omega-\omega_0) + \delta(\omega+\omega_0)]$.

Sketches of $\cos(\omega_0 t)\ U(t)$, its even part $\tfrac{1}{2}\cos(\omega_0 t)$, and its odd part $\tfrac{1}{2}\cos(\omega_0 t)\ \mathrm{Sgn}(t)$, are shown in the Answer section.

Since (B) is the imaginary part of the transform, it must invert to the odd part of $\cos(\omega_0 t)\ U(t)$, which we see from the figure is

$$p(t) = \begin{cases} \tfrac{1}{2}\cos(\omega_0 t) & (t > 0) \\ 0 & (t = 0) \\ -\tfrac{1}{2}\cos(\omega_0 t) & (t < 0) \end{cases}$$

(c) From the figure, the odd part of $\cos(\omega_0 t)\ U(t)$ is $p(t)$ above, whose transform is given in (B). Thus, doubling $p(t)$ we obtain

$$\cos(\omega_0 t)\ \mathrm{sgn}(t) \Longleftrightarrow \frac{2j\omega}{(j\omega)^2 + (\omega_0)^2}$$

$$\sin(\omega_0 t)\ U(t) \Longleftrightarrow \frac{\omega_0}{(j\omega)^2 + (\omega_0)^2} + \frac{\pi}{2j}\left[\delta(\omega-\omega_0) - \delta(\omega+\omega_0)\right]$$

The LHS is the time-domain product of $\sin(\omega_0 t)$ and $U(t)$, and so we can transform it by applying frequency-domain convolution, as follows:

$$f(t) = \sin(\omega_0 t) \Longleftrightarrow F(\omega) = (\pi/j)[\delta(\omega-\omega_0) - \delta(\omega+\omega_0)]$$

$$g(t) = U(t) \Longleftrightarrow \pi\delta(\omega) + 1/j\omega$$

$$f(t)\ g(t) \Longleftrightarrow (1/2\pi)\ F(\omega) * G(\omega) = (1/2\pi) \int_{-\infty}^{\infty} F(\theta)\ G(\omega-\theta)\, d\omega$$

$$= (1/2\pi) \int_{-\infty}^{\infty} (\pi/j)[\delta(\theta-\omega_0) - \delta(\theta+\omega_0)]\ [\pi\delta(\omega-\theta) + 1/j(\omega-\theta)]\, d\theta$$

$$= 1/2j \int_{-\infty}^{\infty} \delta(\theta-\omega_0)\ [\pi\delta(\omega-\theta) + 1/j(\omega-\theta)]\, d\theta$$

$$- 1/2j \int_{-\infty}^{\infty} \delta(\theta+\omega_0)\ [\pi\delta(\omega-\theta) + 1/j(\omega-\theta)]\, d\theta$$

$$= 1/2j \int_{-\infty}^{\infty} \delta(\theta-\omega_0)\ [\pi\delta(\omega-\omega_0) + 1/j(\omega-\omega_0)]\, d\theta$$

$$- 1/2j \int_{-\infty}^{\infty} \delta(\theta+\omega_0)\ [\pi\delta(\omega+\omega_0) + 1/j(\omega+\omega_0)]\, d\theta$$

$$= (1/2j)\ [\pi\delta(\omega-\omega_0) + 1/j(\omega-\omega_0)] \int_{-\infty}^{\infty} \delta(\theta-\omega_0)\, d\theta$$

$$- (1/2j)\ [\pi\delta(\omega+\omega_0) + 1/j(\omega+\omega_0)] \int_{-\infty}^{\infty} \delta(\theta+\omega_0)\, d\theta$$

$$= (1/2j)\ [\pi\delta(\omega - \omega_0) + 1/j(\omega - \omega_0)] - (1/2j)\ [\pi\delta(\omega + \omega_0) + 1/j(\omega + \omega_0)]$$

$$= \frac{1}{2j}\,\frac{1}{j(\omega - \omega_0)} - \frac{1}{2j}\,\frac{1}{j(\omega + \omega_0)} + (\pi/2j)[\delta(\omega - \omega_0) - \delta(\omega + \omega_0)]$$

$$= \frac{\omega_0}{(j\omega)^2 - \omega_0^2} + (\pi/2j)\ [\delta(\omega - \omega_0) - \delta(\omega + \omega_0)] \longrightarrow \quad (A)$$

(b) From (A) we see that $\sin(\omega_0 t)\ U(t)$ transforms to the sum of two terms, its real part being the **even function**

$$P(\omega) = \frac{\omega_0}{(j\omega)^2 + (\omega_0)^2} \qquad (B)$$

and its imaginary part being the **odd function** which is one half the transform of $\sin(\omega_0 t)$ i.e., $(\pi/2j)\ [(\omega - \omega_0) - \delta(\omega + \omega_0)]$

Sketches of $\sin(\omega_0 t)\ U(t)$, its odd part $\frac{1}{2}\sin(\omega_0 t)$, and its even part $\frac{1}{2}\sin(\omega t)\ Sgn(t)$, are shown in the Answer section.

Since (B) is the real part of the transform, it must invert to the even part of $\sin(\omega_0 t)\ U(t)$, which we see from the figure is

$$p(t) = \begin{cases} \frac{1}{2}\sin(\omega_0 t) & (t > 0) \\ 0 & (t = 0) \\ -\frac{1}{2}\sin(\omega_0 t) & (t < 0) \end{cases}$$

(c) From the figure, the even part of $\sin(\omega_0 t)\ U(t)$ is $p(t)$ above, whose transform is given in (A). Thus, doubling $p(t)$ we obtain

$$\sin(\omega_0 t)\ sgn(t) \Longleftrightarrow \frac{2\omega_0}{(j\omega)^2 + (\omega_0)^2}$$

7.25 (a) Using frequency-domain convolution to find the Fourier transform of $\cos(\omega_0 t)\ Sgn(t)$:

$$f(t) = \cos(\omega_0 t) \Longleftrightarrow F(\omega) = \pi[\delta(\omega - \omega_0) + \delta(\omega + \omega_0)]$$

$$g(t) = Sgn(t) \Longleftrightarrow G(\omega) = 2/j\omega$$

$$f(t)\ g(t) \Longleftrightarrow (1/2\pi)\ F(\omega) * G(\omega) = (1/2\pi)\int_{-\infty}^{\infty} F(\theta)\ G(\omega - \theta)\ d\theta$$

$$= (1/2\pi)\int_{-\infty}^{\infty} \pi[\delta(\theta - \omega_0) + \delta(\theta + \omega_0)]\ [2/j(\omega - \theta)]\ d\theta$$

"Introduction to Fourier Analysis", Instructor's Manual, Chapter 7

$$= \int_{-\infty}^{\infty} \delta(\theta - \omega_0)\ 1/j(\omega - \theta)\ d\theta + \int_{-\infty}^{\infty} \delta(\theta + \omega_0)\ 1/j(\omega - \theta)\ d\theta$$

$$= \int_{-\infty}^{\infty} \delta(\theta - \omega_0)\ 1/j(\omega - \omega_0)\ d\theta + \int_{-\infty}^{\infty} \delta(\theta + \omega_0)\ 1/j(\omega + \omega_0)\ d\theta$$

$$= 1/j(\omega - \omega_0)\int_{-\infty}^{\infty} \delta(\theta - \omega_0)\ d\theta + 1/j(\omega + \omega_0)\int_{-\infty}^{\infty} \delta(\theta + \omega_0)\ d\theta$$

$$= \frac{1}{j(\omega - \omega_0)} + \frac{1}{j(\omega + \omega_0)} = \frac{2j\omega}{(j\omega)^2 + \omega_0^2} \longrightarrow$$

(b) Using frequency-domain convolution to find the Fourier transform of $\sin(\omega_0 t)\ Sgn(t)$:

$$f(t) = \sin(\omega_0 t) \Longleftrightarrow F(\omega) = (\pi/j)[\delta(\omega - \omega_0) - \delta(\omega + \omega_0)]$$

$$g(t) = Sgn(t) \Longleftrightarrow G(\omega) = 2/j\omega$$

$$f(t)\ g(t) \Longleftrightarrow (1/2\pi)\ F(\omega) * G(\omega) = (1/2\pi)\int_{-\infty}^{\infty} F(\theta)\ G(\omega - \theta)\ d\theta$$

$$= (1/2\pi)\int_{-\infty}^{\infty} (\pi/j)[\delta(\theta - \omega_0) - \delta(\theta + \omega_0)]\ [2/j(\omega - \theta)]\ d\theta$$

$$= 1/j\int_{-\infty}^{\infty} \delta(\theta - \omega_0)\ 1/j(\omega - \theta)\ d\theta - 1/j\int_{-\infty}^{\infty} \delta(\theta + \omega_0)\ 1/j(\omega - \theta)\ d\theta$$

$$= 1/j\int_{-\infty}^{\infty} \delta(\theta - \omega_0)\ 1/j(\omega - \omega_0)\ d\theta - 1/j\int_{-\infty}^{\infty} \delta(\theta + \omega_0)\ 1/j(\omega + \omega_0)\ d\theta$$

$$= 1/j\left[\frac{1}{\ldots}\int^{\infty} \delta(\theta - \omega_0)\ d\theta - \frac{1}{\ldots}\int^{\infty} \delta(\theta + \omega_0)\ d\theta\right]$$

$$= 1/j\left[\frac{1}{j(\omega-\omega_0)} - \frac{1}{j(\omega+\omega_0)}\right] = \frac{2\omega_0}{(j\omega)^2+\omega_0^2} \longrightarrow$$

7.26

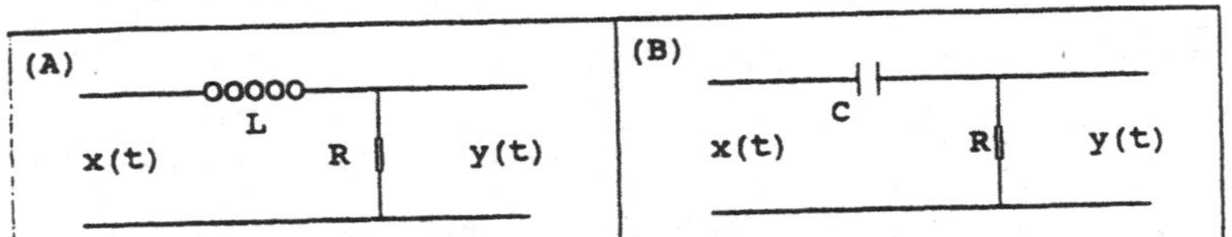

(A) $x(t) = \cos(\omega_0 t)\ U(t)$ $\quad L = R = 1 \quad \omega_0 = 2$

(B) $x(t) = \sin(\omega_0 t)\ U(t)$ $\quad R = C = 1 \quad \omega_0 = 3$

Figure 7.30

(A)(a) The frequency transfer function is:

$$H(j\omega) = \frac{R}{j\omega L + R} = \frac{1}{j\omega + 1}$$

From this we see that the CCL DE is

$$(D + 1)\ y(t) = x(t)$$

(b) and (c): From Exercise (7.23), using the fact that $\omega = 2$:

$$X(\omega) = \frac{j\omega}{(j\omega)^2 + 4} + \tfrac{1}{2}\pi\left[\delta(\omega - 2) + \delta(\omega + 2)\right]$$

and so $\quad Y(\omega) = H(j\omega)\ X(\omega)$

$$= \frac{1}{j\omega + 1}\left[\frac{j\omega}{(j\omega)^2 + 4} + \tfrac{1}{2}\pi\left[\delta(\omega - 2) + \delta(\omega + 2)\right]\right]$$

$$= \frac{1}{j\omega + 1}\frac{j\omega}{(j\omega)^2 + 4} + \frac{\pi/2}{j\omega + 1}\delta(\omega - 2) + \frac{\pi/2}{j\omega + 1}\delta(\omega + 2)$$

$$= \frac{1}{5}\frac{j\omega}{(j\omega)^2 + 4} + \frac{2}{5}\frac{2}{(j\omega)^2 + 4} - \frac{1}{5}\frac{1}{j\omega + 1} + \frac{\pi}{2}\frac{\delta(\omega - 2)}{j2 + 1} + \frac{\pi}{2}\frac{\delta(\omega + 2)}{-j2 + 1}$$

$$= \frac{1}{5}\frac{j\omega}{(j\omega)^2 + 4} + \frac{2}{5}\frac{2}{(j\omega)^2 + 4} - \frac{1}{5}\frac{1}{j\omega + 1}$$

$$+ \frac{\pi}{2}\frac{1 - 2j}{5}\delta(\omega - 2) + \frac{\pi}{2}\frac{1 + 2j}{5}\delta(\omega + 2)$$

$$= \frac{1}{5}\frac{j\omega}{(j\omega)^2 + 4} + \frac{\ }{10}[\delta(\omega \ldots$$

$$+ \frac{2}{5}\frac{2}{(j\omega)^2 + 4} + \frac{\pi}{5j}\left[\delta(\omega - 2) + \delta(\omega - 2)\right] - \frac{1}{5}\frac{1}{j\omega + 1}$$

$$\Longleftrightarrow \frac{1}{5}\cos(2t)\ U(t) + \frac{2}{5}\sin(2t)\ U(t) - \frac{1}{5}\exp(-t)\ U(t) = y(t) \longrightarrow$$

(d) $(D + 1)\ y(t) = Dy(t) + y(t)$

$$y(t) = \frac{1}{5}\cos(2t)\ U(t) + \frac{2}{5}\sin(2t)\ U(t) - \frac{1}{5}\exp(-t)\ U(t)$$

$$Dy(t) = \frac{-2}{5}\sin(2t)\ U(t) + \frac{4}{5}\cos(2t)\ U(t) + \frac{1}{5}\exp(-t)\ U(t)$$

$$+ (1/5)\ \delta(t) \qquad\qquad - (1/5)\ \delta(t)$$

from which we obtain $\quad (D + 1)y(t) = \cos(2t)\ U(t)$

whose RHS is exactly equal to $x(t)$.

(B)(a) The frequency transfer function is:

$$H(j\omega) = \frac{R}{R + 1/j\omega C} = \frac{1}{1 + 1/j\omega} = \frac{j\omega}{j\omega + 1}$$

From this we see that the CCL DE is $\quad (D + 1)\ y(t) = Dx(t)$

(b) and (c): From Exercise (7.24), using the fact that $\omega = 3$:

$$X(\omega) = \frac{3}{(j\omega)^2 + 9} + \pi/2j\left[\delta(\omega - 3) - \delta(\omega + 3)\right]$$

and so $\quad Y(\omega) = H(j\omega)\ X(\omega)$

$$= \frac{j\omega}{j\omega + 1}\left[\frac{3}{(j\omega)^2 + 9} + \pi/2j\left[\delta(\omega - 3) - \delta(\omega + 3)\right]\right]$$

$$= \frac{j\omega}{j\omega + 1}\frac{3}{(j\omega)^2 + 9} + \frac{\omega\pi/2}{j\omega + 1}\delta(\omega - 3) - \frac{\omega\pi/2}{j\omega + 1}\delta(\omega + 3)$$

$$= \frac{3}{10}\frac{j\omega}{(j\omega)^2 + 9} + \frac{9}{10}\frac{3}{(j\omega)^2 + 9} - \frac{3}{10}\frac{1}{j\omega + 1} + \frac{3\pi/2}{j3 + 1}\,\delta(\omega - 3) + \frac{3\pi/2}{-j3 + 1}\,\delta(\omega + 3)$$

$$= \frac{3}{10}\frac{j\omega}{(j\omega)^2 + 9} + \frac{9}{10}\frac{3}{(j\omega)^2 + 9} - \frac{3}{10}\frac{1}{j\omega + 1} + \frac{3\pi}{20}(1 - j3)\,\delta(\omega - 3) + \frac{3\pi}{20}(1 + j3)\,\delta(\omega - 3)$$

$$= \frac{3}{10}\frac{j\omega}{(j\omega)^2 + 9} + \frac{3\pi}{20}\left[\delta(\omega - 3) + \delta(\omega - 3)\right] + \frac{9}{10}\frac{3}{(j\omega)^2 + 9} + \frac{9\pi}{20j}\left[\delta(\omega - 3) - \delta(\omega - 3)\right] - \frac{3}{10}\frac{1}{j\omega + 1}$$

$$\Longleftrightarrow \frac{3}{10}\cos(3t)\,U(t) + \frac{9}{10}\sin(3t)\,U(t) - \frac{3}{10}\exp(-t)\,U(t) = y(t)$$

(d) $(D + 1)\,y(t) = Dy(t) + y(t)$

$$y(t) = \frac{3}{10}\cos(3t)\,U(t) + \frac{9}{10}\sin(3t)\,U(t) - \frac{3}{10}\exp(-t)\,U(t)$$

$$Dy(t) = \frac{-9}{10}\sin(3t)\,U(t) + \frac{27}{10}\cos(3t)\,U(t) + \frac{3}{10}\exp(-t)\,U(t) + (3/10)\,\delta(t) - (3/10)\,\delta(t)$$

from which we obtain

$$(D + 1)y(t) = 3\cos(3t)\,U(t)$$

The RHS of the CCL DE is

$$Dx(t) = D[\sin(3t)\,U(t)] = 3\cos(3t)\,U(t) + \sin(3t)\,\delta(t) = 3\cos(3t)\,U(t)$$

thus validating the solution.

7.27(a)

(1) First we note the following: Recall from the proof of Theorem 4.1 that (4.63), namely

$$\int_{-\infty}^{\infty} \exp(j\omega t)\,d\omega = 2\pi\,\delta(t) \qquad (4.63)$$

was based on the assumption that **the Fourier analysis and synthesis transformations were each other's inverses**, and so on that account we cannot use it to prove that very same proposition.

However (4.63) can also be derived by a **totally independent method** as well [see e.g. Papoulis (2)] and so in fact we **are** entitled to use it in order to prove the proposition, as we shall now proceed to do.

From (4.63):

$$\delta(t - \tau) = 1/2\pi \int_{-\infty}^{\infty} \exp[j\omega(t - \tau)]\,d\omega$$

and so (7.73) becomes

$$x(t) = \int_{-\infty}^{\infty} x(\tau)\,\delta(t - \tau)\,d\tau = \int_{-\infty}^{\infty} x(\tau)\left[1/2\pi \int_{-\infty}^{\infty} \exp[j\omega(t - \tau)]\,d\omega\right]d\tau$$

$$= \int_{-\infty}^{\infty} x(\tau)\left[1/2\pi \int_{-\infty}^{\infty} \exp(j\omega t)\,\exp(-j\omega\tau)\,d\omega\right]d\tau$$

(2) Interchanging the order of integration, we now obtain

A, B, C

$$x(t) = \frac{1}{2\pi}\int_{-\infty}^{\infty}\left[\int_{-\infty}^{\infty} x(\tau)\,\exp(-j\omega\tau)\,d\tau\right]\exp(j\omega t)\,d\omega \qquad (7.75)$$

(3) Then (7.75) is seen to [illegible] (C) yielding the original function (A). Thus the two transformations (B) and (C) are each other's inverses. ■

(b) From (4.63), the complex exponential transforms as follows:

$$\exp(jzt) \Longleftrightarrow 2\pi\,\delta(\omega - z) \qquad (7.76)$$

Then, starting from

$$G(\omega) = \int_{-\infty}^{\infty} \left[\frac{1}{2\pi} \int_{-\infty}^{\infty} F(z)\,\exp(jzt)\,dz \right] \exp(-j\omega t)\,dt$$

we interchange the order of integration, continuing

$$\ldots = \frac{1}{2\pi} \int_{-\infty}^{\infty} \left[\int_{-\infty}^{\infty} \exp(jzt)\,\exp(-j\omega t)\,dt \right] F(z)\,dz$$

But by (7.76) the inside bracket is $2\pi\,\delta(\omega - z)$ and so we continue

$$\ldots = \frac{1}{2\pi} \int_{-\infty}^{\infty} 2\pi\,\delta(\omega - z)\,F(z)\,dz = F(\omega) \int_{-\infty}^{\infty} \delta(\omega - z)\,dz = F(\omega)$$

We have thus shown that $G(\omega)$ is in fact $F(\omega)$, i.e.,

$$\overset{A}{F(\omega)} = \overbrace{\int_{-\infty}^{\infty} \left[\overbrace{\frac{1}{2\pi} \int_{-\infty}^{\infty} F(z)\,\exp(jzt)\,dz}^{C} \right] \exp(-j\omega t)\,dt}^{B}$$

In (C) we see $F(\omega)$ being inverted to the time domain and in (B) the result is then transformed back to the frequency domain giving us back $F(\omega)$ at (A).

Once again we have shown that synthesis and analysis are each other's inverses.

$$f(n) = \sum_{k=-\infty}^{\infty} g(k)\,h(n - k)$$

Let $n - k = p$
Then $k = n - p$
$k = \infty$ means $p = -\infty$
$k = -\infty$ means $p = \infty$

$$= \sum_{p=\infty}^{-\infty} g(n - p)\,h(p) = \sum_{p=-\infty}^{\infty} g(n - p)\,h(p) = \sum_{k=-\infty}^{\infty} h(k)\,g(n - k)$$

(b) Given that $g(n) = 0$ and $h(n) = 0$ for $n < 0$:

$$f(n) = \sum_{k=-\infty}^{\infty} g(k)\,h(n - k) = \sum_{k=0}^{\infty} g(k)\,h(n - k) \qquad \text{because } g(k) = 0 \text{ for } k < 0:$$

$$= \sum_{k=0}^{n} g(k)\,h(n - k) \qquad \text{because } h(n - k) = 0 \text{ for } k > n\ 0:$$

(c) $g(n) = 1 \quad (1 \le n \le 6)$ and $h(n) = 1 \quad (1 \le n \le 6)$

From (b) above:

$$f(n) = \sum_{k=0}^{n} g(k)\,h(n - k)$$

$n = 0$: $f(0) = \sum_{k=0}^{0} g(k)\,h(n - k) = 0$

```
       0 1 1 1 1 1 1 0
0 1 1 1 1 1 1 0
```

$n = 1$: $f(1) = \sum_{k=0}^{1} g(k)\,h(n - k) = 0$

```
       0 1 1 1 1 1 1 0
 0 1 1 1 1 1 1 0
```

$n = 2$: $f(2) = \sum_{k=0}^{2} g(k)\,h(n - k) = 1$

```
       0 1 1 1 1 1 1 0
  0 1 1 1 1 1 1 0
```

$n = 3$: $f(3) = \sum_{k=0}^{3} g(k)\,h(n - k) = 2$

```
       0 1 1 1 1 1 1 0
   0 1 1 1 1 1 1 0
```

$n = 4$: $f(4) = \sum_{k=0}^{4} g(k)\,h(n - k) = 3$

```
       0 1 1 1 1 1 1 0
    0 1 1 1 1 1 1 0
```

$n = 5$: $f(5) = \sum_{k=0}^{5} g(k)\,h(n - k) = 4$

```
       0 1 1 1 1 1 1 0
     0 1 1 1 1 1 1 0
```

$n = 6:\ ,\ f(6) = \sum_{k=0}^{6} g(k)\ h(n-k) = 5$

```
  0 1 1 1 1 1 1 0
0 1 1 1 1 1 1 0
```

$n = 7:\quad f(7) = \sum_{k=0}^{7} g(k)\ h(n-k) = 6$

```
0 1 1 1 1 1 1 0
0 1 1 1 1 1 1 0
```

$n = 8:\quad f(8) = \sum_{k=0}^{8} g(k)\ h(n-k) = 5$

```
0 1 1 1 1 1 1 0
  0 1 1 1 1 1 1 0
```

$n = 9:\quad f(9) = \sum_{k=0}^{9} g(k)\ h(n-k) = 4$

```
0 1 1 1 1 1 1 0
    0 1 1 1 1 1 1 0
```

$n = 10:\ f(10) = \sum_{k=0}^{10} g(k)\ h(n-k) = 3$

```
0 1 1 1 1 1 1 0
      0 1 1 1 1 1 1 0
```

$n = 11:\ f(11) = \sum_{k=0}^{11} g(k)\ h(n-k) = 2$

```
0 1 1 1 1 1 1 0
        0 1 1 1 1 1 1 0
```

$n = 12:\ f(12) = \sum_{k=0}^{12} g(k)\ h(n-k) = 1$

```
0 1 1 1 1 1 1 0
          0 1 1 1 1 1 1 0
```

$n = 13:\ f(13) = \sum_{k=0}^{13} g(k)\ h(n-k) = 0$

```
0 1 1 1 1 1 1 0
            0 1 1 1 1 1 1 0
```

Plotting $f(n)$ we obtain the following:

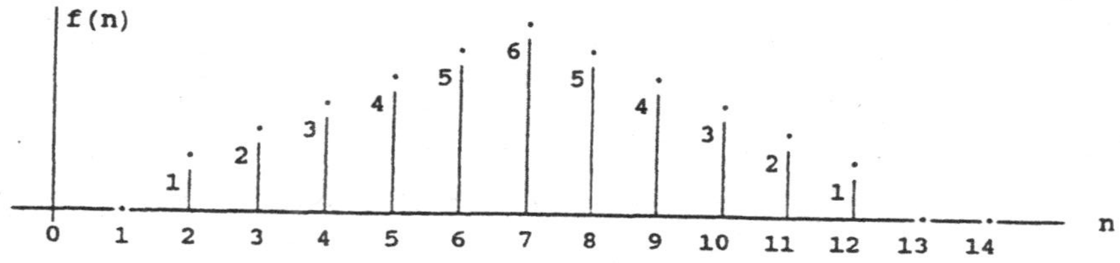

(d) Modifying the values shown in Figure 7.31 to 1/6 we obtain, by the above procedure:

$$f(n) = \sum_{k=0}^{n} g(k)\ h(n-k)$$

n	0	1	2	3	4	5	6	7	8	9	10	11	12	13	14
f(n)	0/—	0/—	1/—	2/—	3/—	4/—	5/—	6/—	5/—	4/—	3/—	2/—	1/—	0/—	0/—

The top row of this table shows the values that a pair of dice can assume. The bottom row shows the probability of that value occurring. A plot of $f(n)$ would be the same as that shown above except that the values are now multiplied by 1/36.

(e) Obtaining the same plots that were obtained in (c) and (d) by running this problem on the FFT system with $N = 14$, DISCRETE:

Main menu, Create Values, Continue, Create X and X2, Real, 3, 0, 6, Continue, 0, 1/6, 0, Go, 3, 0, 6, Continue, 0, 1/6, 0, Go, Go. You are now back on the main menu. Plotting **X** and **X2** confirms that we have entered the values correctly. Run CONVOLUTION and plot **Y**. We see in the display the same values that we obtained in the above table.

7.29 (1) Using $N = 1024$ and $T = 8$, the steps to load Rect(t) into **X** and $\cos(16\pi t)$ into **X2** are as follows:

Main menu, Create values, Continue, Create both X and X2, Real, Even, 2, .5, Continue, 1, 0, Go, Even, 1, Continue, COS(16*pi*t), Go, Neither Go. You are now back on the main menu. Plot **X** and **X2** to verify that the functions have been loaded correctly.

(2) Run ANALYSIS. This places the FFT spectrum of **X** in **F**. (Both **X** and **F** are shown in the following figures.)

(3) Now use the system to **band-limit** this spectrum. (Go into the **F** postprocessor via main-menu G where you will see a package called LOW-PASS FILTER. Use $n_{max} = 56$. This will cut off all frequency components in **F** for $|\omega| > 14\pi$.)

(4) After running LOW-PASS FILTER, inspect the FFT spectrum to observe how the spectrum of the pulse is now zero outside of the interval $-14\pi \le \omega \le 14\pi$. (This band-limited spectrum is shown in the following figures.)

(5) Invert this **band-limited** spectrum using SYNTHESIS and inspect a plot. You will see some distortion of the original Rect pulse caused by the band-limiting, which will depend on how severely you have band-limited.

(6) Now move **Y** into **X** using the package "COPY Y to X" in the **X** postprocessor. Inspect a plot of the signal that is now in **X**.

(7) Now use the REAL-MULTIPLY package in the **X** postrocessor to multiply this band-limited Rect(t) pulse by the cosine in **X2**.

(8) Run ANALYSIS and inspect **FRE**. You will observe that the two copies of the spectrum of the band-limited Rect(t) no longer run into each other. They are perfectly symmetrical wrt their center frequencies.

8.1 (a) This is the synthesis equation, multiplied by 2π and evaluated at $t = -1$, i.e.

$$\int_{-\infty}^{\infty} F(\omega)\ \exp(-j\omega)\ d\omega = 2\pi\ 1/2\pi \int_{-\infty}^{\infty} F(\omega)\ \exp(j\omega t)\ d\omega \Big|_{t=-1}$$

and so the answer is $2\pi\ f(t)$ at $t = -1$. From the figure we see that it is 2π. ———>

(b) $$\int_{-\infty}^{\infty} F(\omega)\ \exp(j\omega)\ d\omega = 2\pi\ 1/2\pi \int_{-\infty}^{\infty} F(\omega)\ \exp(j\omega t)\ d\omega \Big|_{t=1}$$

$$= 2\pi\ f(t) \Big|_{t=1} = 2\pi\ \tfrac{1}{2} = \pi$$ ———>

(c) $\int_{-\infty}^{\infty} |F(\omega)|^2 d\omega$ is equal to 2π times the total energy in the pulse (Parseval's theorem).

We can obtain the energy from the time-domain statement as follows:

$$E = \int_{-\infty}^{\infty} |f(t)|^2\ dt = 4 \int_{0}^{1} (2 - t)^2\ dt = 4\ \frac{-(2 - t)^3}{3} \Bigg|_0^1 = \frac{4}{3}(8 - 1) = \frac{28}{3}$$

Thus the final answer is $2\pi\ 28/3 = 56\pi/3$ ———>

(d) $\int_{-\infty}^{\infty} |A(\omega)|^2 d\omega$ is equal to 2π times the total energy in the even part of the pulse (Parseval's theorem).

We can obtain the energy in the even part of the pulse from the time-domain statement as follows:

$$E = \int_{-\infty}^{\infty} |f_{ev}(t)|^2\ dt = 4 \int_{1}^{2} (t/2)^2\ dt + 2 \int_{0}^{1} (2 - t)^2\ dt$$

$$= 2\ \frac{t^3}{12} \Bigg|_1^2 - 2\ \frac{(2 - t)^3}{3} \Bigg|_0^1 = \frac{1}{6}(8 - 1) - \frac{1}{3}(1 - 8) = \frac{7}{6} + \frac{7}{3} = \frac{21}{6}$$

Thus the final answer is $2\pi\ 21/6 = 42\pi/3$ ———>

(d) $\int_{-\infty}^{\infty} |B(\omega)|^2 d\omega$ is equal to 2π times the total energy in the odd part of the pulse (Parseval's theorem).

We can obtain the energy in the odd part of the pulse from the time-domain statement as follows:

$$E = \int_{-\infty}^{\infty} |f_{od}(t)|^2\ dt = 4 \int_{1}^{2} (-t/2)^2\ dt = 4\ \frac{t^3}{12} \Bigg|_1^2 = \frac{1}{3}(8 - 1) = 7/3$$

Thus the final answer is $2\pi\ 7/3 = 14\pi/3$ ———>

8.2 (a) $f(t) \Longleftrightarrow F(\omega)$ $\quad \therefore f(\alpha t) \Longleftrightarrow \frac{1}{|\alpha|} F(\omega/\alpha)$ (reciprocal scaling)

$\therefore f[\alpha(t - \tau)] \Longleftrightarrow \frac{1}{|\alpha|} F(\omega/\alpha)\ \exp(-j\omega\tau)$ (time shift)

(b) $f(t) \Longleftrightarrow F(\omega)$ $\quad \therefore f(t - \tau) \Longleftrightarrow F(\omega)\ \exp(-j\omega\tau)$ (time shift)

$\therefore f[\alpha t - \tau)] \Longleftrightarrow \frac{1}{|\alpha|} F(\omega/\alpha)\ \exp(-j\omega\tau/\alpha)$ (reciprocal scaling)

(c)(1) $f[\alpha(t - \tau)] \Longleftrightarrow \int_{-\infty}^{\infty} f[\alpha(t - \tau)]\ \exp(-j\omega t)\ dt$

Let $\alpha > 0$
Let $\alpha(t - \tau) = z$
Then $t = z/\alpha + \tau$ and $dt = dz/\alpha$

$$= \int_{-\infty}^{\infty} f(z)\ \exp[-j\omega(z/\alpha + \tau)]\ dz/\alpha = \frac{1}{\alpha} \int_{-\infty}^{\infty} f(z)\ \exp[-j(\omega/\alpha)z]\ \exp(-j\omega\tau)\ dz$$

$$= \frac{1}{\alpha} \exp(-j\omega\tau)\ F(\omega/\alpha)$$

Now let $\alpha < 0$. Then $\alpha = -|\alpha|$. Let $\alpha(t - \tau) = z$.
Then $t = z/\alpha + \tau = -z/|\alpha| + \tau$ and so $dt = -dz/|\alpha|$.

$$f[\alpha(t - \tau)] \Longleftrightarrow \int_{-\infty}^{\infty} f[\alpha(t - \tau)]\ \exp(-j\omega t)\ dt$$

$$= \int_{\infty}^{-\infty} f(z)\ \exp[-j\omega(z/\alpha + \tau)]\ (-dz/|\alpha|)$$

$$= \frac{1}{|\alpha|}\int_{-\infty}^{\infty} f(z)\ \exp[(-j(\omega/\alpha)z]\ \exp(-j\omega\tau)\ dz = \frac{1}{|\alpha|}\exp(-j\omega\tau)\ F(\omega/\alpha)$$

And so, for $\alpha > 0$ or $\alpha < 0$: $f[\alpha(t - \tau)] \Longleftrightarrow = \frac{1}{|\alpha|}\exp(-j\omega\tau)\ F(\omega/\alpha)$

QED

(c)(2) $f(\alpha t - \tau) \Longleftrightarrow \int_{-\infty}^{\infty} f(\alpha t - \tau)\ \exp(-j\omega t)\ dt$ Let $\alpha > 0$
Let $\alpha t - \tau = z$
Then $t = (z + \tau)/\alpha$ and $dt = dz/\alpha$

$$= \int_{-\infty}^{\infty} f(z)\ \exp[-j\omega(z+\tau)/\alpha]\ dz/\alpha = \frac{1}{\alpha}\int_{-\infty}^{\infty} f(z)\ \exp[-j(\omega/\alpha)z]\ \exp(-j\omega\tau/\alpha)\ dz$$

$$= \frac{1}{\alpha}\exp(-j\omega\tau/\alpha)\ F(\omega/\alpha)$$

Now let $\alpha < 0$. Then $\alpha = -|\alpha|$. Let $\alpha t - \tau = z$.
Then $t = (z + \tau)/\alpha = -(z + \tau)/|\alpha|$ and so $dt = -dz/|\alpha|$.

$$f(\alpha t - \tau) \Longleftrightarrow \int_{-\infty}^{\infty} f(\alpha t - \tau)\ \exp(-j\omega t)\ dt$$

$$= \int_{\infty}^{-\infty} f(z)\ \exp[-j\omega(z + \tau)/\alpha]\ (-dz/|\alpha|)$$

$$= \frac{1}{|\alpha|}\int_{-\infty}^{\infty} f(z)\ \exp[(-j(\omega/\alpha)z]\ \exp(-j\omega\tau/\alpha)\ dz = \frac{1}{|\alpha|}\exp(-j\omega\tau/\alpha)\ F(\omega/\alpha)$$

And so, for $\alpha > 0$ or $\alpha < 0$: $f(\alpha t-\tau) \Longleftrightarrow = \frac{1}{|\alpha|}\exp(-j\omega\tau/\alpha)\ F(\omega/\alpha)$

QED

8.3 $f(t) \Longleftrightarrow F(\omega)$ $\therefore f(-t) \Longleftrightarrow F(-\omega)$ (reciprocal scaling)

$\therefore F(-t) \Longleftrightarrow 2\pi\ f(\omega)$ (duality)

$\therefore F(-t)\ \exp(j\tau t) \Longleftrightarrow 2\pi\ f(\omega - \tau)$ (frequency shift)

$\therefore 2\pi\ f(t - \tau) \Longleftrightarrow 2\pi\ F(\omega)\ \exp(-j\omega\tau)$ (duality)

8.4 $f(t) \Longleftrightarrow F(\omega)$ $\therefore f(-t) \Longleftrightarrow F(-\omega)$ (reciprocal scaling)

$\therefore F(-t) \Longleftrightarrow 2\pi\ f(\omega)$ (duality)

$\therefore F[-(t + \omega_0)] \Longleftrightarrow 2\pi\ f(\omega)\ \exp(j\omega_0\omega)$ (time shift)

$\therefore 2\pi\ f(t)\ \exp(j\omega_0 t) \Longleftrightarrow 2\pi\ F(\omega - \omega_0)$ (duality)

$\therefore f(t)\ \exp(j\omega_0 t) \Longleftrightarrow F(\omega - \omega_0)$ (duality)

8.5 (a) $\sin(t\tau/2) \Longleftrightarrow \pi/j\ [\delta(\omega - \tau/2) - \delta(\omega + \tau/2)]$

$\therefore \pi/j\ [\delta(t - \tau/2) - \delta(t + \tau/2)] \Longleftrightarrow 2\pi\ \sin(-\omega\tau/2)$ (duality)

$\therefore \delta(t + \tau/2) - \delta(t - \tau/2) \Longleftrightarrow 2j\ \sin(\omega\tau/2)$ QED

(b) $\delta(t + \tau/2) - \delta(t - \tau/2)$

$$\Longleftrightarrow \int_{-\infty}^{\infty} \delta(t + \tau/2)\ \exp(-j\omega t)\ dt - \int_{-\infty}^{\infty} \delta(t - \tau/2)\ \exp(-j\omega t)\ dt$$

$$= \exp(j\omega\tau/2) - \exp(-j\omega\tau/2) = 2j\ \sin(\omega\tau/2) \longrightarrow$$

8.6 (a) $2\ \cos(t\tau) \Longleftrightarrow 2\pi\ [\delta(\omega - \tau) + \delta(\omega + \tau)]$

$\therefore 2\pi\ [\delta(t - \tau) + \delta(t + \tau)] \Longleftrightarrow 4\pi\ \cos(-\omega\tau)$ (duality)

$\therefore \delta(t + \tau) + \delta(t - \tau) \Longleftrightarrow 2\ \cos(\omega\tau)$ QED

(b) $\delta(t + \tau) + \delta(t - \tau)$

$$\Longleftrightarrow \int_{-\infty}^{\infty} \delta(t + \tau)\ \exp(-j\omega t)\ dt + \int_{-\infty}^{\infty} \delta(t - \tau)\ \exp(-j\omega t)\ dt$$

$$= \exp(j\omega\tau) + \exp(-j\omega\tau) = 2\ \cos(\omega\tau/2) \longrightarrow$$

8.7 $\exp(-\beta t)\ U(t) \Longleftrightarrow 1/(j\omega + \beta)$ $(\beta > 0)$

(a) $\exp(-\beta t)\cos(\omega_0 t)U(t) = \frac{1}{2}[\exp(-\beta t)\exp(j\omega_0 t) + \exp(-\beta t)\exp(-j\omega_0 t)]U(t)$

$$\Longleftrightarrow \frac{\frac{1}{2}}{j(\omega - \omega_0) + \beta} + \frac{\frac{1}{2}}{j(\omega + \omega_0) + \beta} = \frac{j\omega + \beta}{(j\omega + \beta)^2 + \omega_0^2}$$ QED

(b)

$\exp(-\beta t)\sin(\omega_0 t)U(t) = (1/2j)[\exp(-\beta t)\exp(j\omega_0 t) - \exp(-\beta t)\exp(-j\omega_0 t)]U(t)$

$$\Longleftrightarrow \frac{1/2j}{j(\omega - \omega_0) + \beta} - \frac{1/2j}{j(\omega + \omega_0) + \beta} = \frac{\omega_0}{(j\omega + \beta)^2 + \omega_0^2} \qquad \text{QED}$$

(c) $$\frac{8j\omega + 4}{(j\omega)^2 + 4j\omega + 20} = \frac{8(j\omega + 2) - 12}{(j\omega + 2)^2 + 16} = \frac{8(j\omega + 2)}{(j\omega + 2)^2 + 16} - \frac{3 \times 4}{(j\omega + 2)^2 + 16}$$

$$\Longleftrightarrow 8\cos(4t)\exp(-2t)U(t) - 3\sin(4t)\exp(-2t)U(t) \longrightarrow$$

8.8 $\exp(-\beta t)\ U(t) \Longleftrightarrow 1/(j\omega + \beta) \qquad (\beta > 0)$

$\cos(\omega_0 t) \Longleftrightarrow \pi[\delta(\omega - \omega_0) + \delta(\omega + \omega_0)]$

$\therefore \exp(-\beta t)\cos(\omega_0 t)\ U(t)$

$$\Longleftrightarrow 1/2\pi \int_{-\infty}^{\infty} \frac{1}{j\theta + \beta}\, \pi[\delta(\omega - \omega_0 - \theta) + \delta(\omega + \omega_0 - \theta)]\, d\theta$$

$$= \tfrac{1}{2}\int_{-\infty}^{\infty} \frac{1}{j(\omega-\omega_0) + \beta}\,\delta(\omega-\omega_0-\theta)\, d\theta + \tfrac{1}{2}\int_{-\infty}^{\infty} \frac{1}{j(\omega+\omega_0) + \beta}\,\delta(\omega+\omega_0-\theta)\, d\theta$$

$$= \tfrac{1}{2}\frac{1}{j(\omega-\omega_0)+\beta}\int_{-\infty}^{\infty}\delta(\omega-\omega_0-\theta)\, d\theta + \tfrac{1}{2}\frac{1}{j(\omega+\omega_0)+\beta}\int_{-\infty}^{\infty}\delta(\omega+\omega_0-\theta)\, d\theta$$

$$= \tfrac{1}{2}\frac{1}{j(\omega - \omega_0) + \beta} + \tfrac{1}{2}\frac{1}{j(\omega + \omega_0) + \beta} = \frac{j\omega + \beta}{(j\omega + \beta)^2 + \omega_0^2} \qquad \text{QED}$$

8.9 (a) $\displaystyle\int_{-\infty}^{\infty} \frac{1}{t^2}\, dt = 2\int_{0}^{\infty} \frac{1}{t^2}\, dt = -2/t \Big|_0^{\infty} = \infty$ Thus $1/t$ is not square integrable over $-\infty < t < \infty$

(b) By direct transformation: $\displaystyle\int_{-\infty}^{\infty} \frac{1}{t}\exp(-j\omega t)\, dt$ This is a very hard integral to evaluate.

(c) By duality: $\mathrm{Sgn}(t) \Longleftrightarrow 2/j\omega \qquad \therefore 2/jt \Longleftrightarrow 2\pi\ \mathrm{Sgn}(-\omega)$

$\therefore 1/t \Longleftrightarrow j\pi\ \mathrm{Sgn}(-\omega) \qquad \therefore 1/t \Longleftrightarrow \pi/j\ \mathrm{Sgn}(\omega)$

with height 1 (Figure 8.9).

(a)

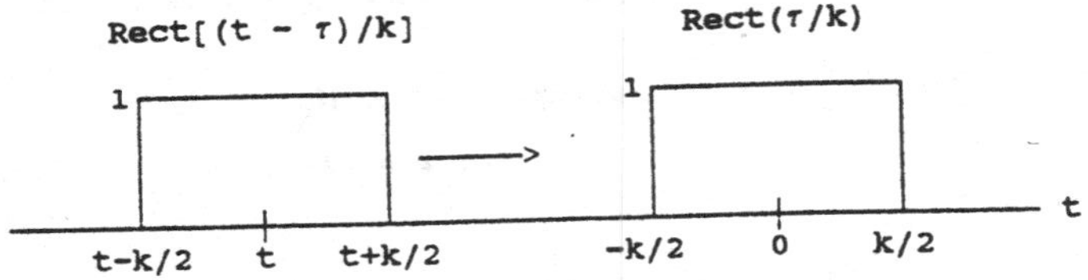

$$f(t) = \int_{-\infty}^{\infty} \mathrm{Rect}[(t - \tau)/k]\ \mathrm{Rect}(\tau/k)\, d\tau$$

- For $t < -k$: $\quad f(t) = 0$
- For $-k < t < 0$: $\quad f(t) = \displaystyle\int_{-k/2}^{t+k/2} 1\, d\tau = (t + k/2) + k/2 = k + t$
- For $0 < t < k$: $\quad f(t) = \displaystyle\int_{t-k/2}^{k/2} 1\, d\tau = k/2 - (t - k/2) = k - t$
- For $k < t$: $\quad f(t) = 0$

Thus: $$f(t) = \begin{cases} 0 & (t < -k) \\ k + t & (-k < t < 0) \\ k - t & (0 < t < k) \\ 0 & (k < t) \end{cases}$$ This is a triangular pulse of width 2k and height k.

and so $\mathrm{Rect}(t/k) * \mathrm{Rect}(t/k) = k\,\Lambda(t/k)$

(b) $\mathrm{Rect}(t/k) \Longleftrightarrow k\,\mathrm{Sa}(\omega k/2)$

$\therefore k\,\Lambda(t/k) \Longleftrightarrow k\,\mathrm{Sa}(\omega k/2)\ k\,\mathrm{Sa}(\omega k/2) = k^2\ \mathrm{Sa}^2(\omega k/2)$

$\therefore \Lambda(t/k) \Longleftrightarrow k\,\mathrm{Sa}^2(\omega k/2)$ QED

8.11 (a) $\Lambda(t/k) \Longleftrightarrow k\,\mathrm{Sa}^2(\omega k/2)$

$\therefore k\,\mathrm{Sa}^2(tk/2) \Longleftrightarrow 2\pi\,\Lambda(-\omega/k) = 2\pi\,\Lambda(\omega/k) \qquad (\Lambda \text{ is even})$

$\therefore k/2\pi\ \mathrm{Sa}^2(tk/2) \Longleftrightarrow \Lambda(\omega/k) \longrightarrow$

(b) $\Lambda(t/k) \Longleftrightarrow k\, Sa^2(\omega k/2)$

$\therefore k\, Sa^2(tk/2) \Longleftrightarrow 2\pi\, \Lambda(-\omega/k) = 2\pi\, \Lambda(\omega/k)$

$\therefore Sa^2(tk/2) \Longleftrightarrow 2\pi/k\, \Lambda(\omega/k)$

Setting $k = 2$ gives: $Sa^2(t) \Longleftrightarrow \pi\, \Lambda(\omega/2)$ ———>

8.12 $\exp(-\beta t)\, U(t) \Longleftrightarrow 1/(j\omega + \beta)$

Let $\alpha = \sigma/\beta$. Then $f(\alpha t) \Longleftrightarrow 1/\alpha\, F(\omega/\alpha)$ means $f(\sigma t/\beta) \Longleftrightarrow \beta/\sigma\, F(\omega\beta/\sigma)$

Thus $\exp(-\beta t)\, U(t) \Longleftrightarrow 1/(j\omega + \beta)$ means

$$\exp(-\sigma t)\, U(t) \Longleftrightarrow \frac{\beta}{\sigma}\frac{1}{j\omega\beta/\sigma + \beta} = \frac{\beta}{\sigma}\frac{\sigma}{\beta(j\omega + \sigma)} = \frac{1}{j\omega + \sigma}$$ ———>

8.13 $Rect(t) \Longleftrightarrow Sa(\omega/2)$ $\therefore Sa(t/2) \Longleftrightarrow 2\pi\, Rect(-\omega) = 2\pi\, Rect(\omega)$

$\therefore Sa(\alpha t/2) \Longleftrightarrow 2\pi/|\alpha|\, Rect(\omega/\alpha)$

Letting $\alpha = 6$ gives: $Sa(3t) \Longleftrightarrow \pi/3\, Rect(\omega/6)$ ———>

8.14 Multiplication in the frequency domain corresponds to convolution in the time domain.

$\exp(-j\omega t_0) \Longleftrightarrow \delta(t - t_0)$ and $F(\omega) \Longleftrightarrow f(t)$

Thus $$\exp(-j\omega t_0)\, F(\omega) \Longleftrightarrow \int_{-\infty}^{\infty} f(\tau)\, \delta(t - t_0 - \tau)\, d\tau$$

$$= \int_{-\infty}^{\infty} f(t - t_0)\, \delta(t-t_0-\tau)\, d\tau = f(t - t_0)\int_{-\infty}^{\infty} \delta(t-t_0-\tau)\, d\tau = f(t - t_0)$$ QED

8.15 (a) $\sin(\omega_0 t)\, Rect(t/\tau)$

$= 1/2j\, [\exp(j\omega_0 t)\, Rect(t/\tau) - \exp(-j\omega_0 t)\, Rect(t/\tau)]$

$\Longleftrightarrow 1/2j\, [\tau\, Sa[(\omega - \omega_0)\tau/2] - \tau\, Sa[(\omega + \omega_0)\tau/2)]$

$= \tau/2j\, [Sa[(\omega - \omega_0)\tau/2) - Sa[(\omega + \omega_0)\tau/2)]$ ———>

(b) $\cos(\omega_0 t)\, f(t) = \tfrac{1}{2}\, [\exp(j\omega_0 t)\, f(t) + \exp(-j\omega_0 t)\, f(t)]$

$\Longleftrightarrow \tfrac{1}{2}\, [F(\omega - \omega_0) + F(\omega + \omega_0)]$ ———>

8.16 (a) $$f(t)\, Rect(t/\tau) \Longleftrightarrow 1/2\pi \int_{-\infty}^{\infty} F(\theta)\, \tau\, Sa[(\omega - \theta)\tau/2]\, d\theta$$

(b) Let $B(t) = 1/k\, Rect(t/k)$

Then $B(t - t_0) = 1/k\, Rect[(t - t_0)/k]$

$\Longleftrightarrow 1/k\, k\, Sa(\omega k/2)\, \exp(-j\omega t_0) = Sa(\omega k/2)\, \exp(-j\omega t_0)$

$$\therefore f(t)\, B(t - t_0) \Longleftrightarrow 1/2\pi \int_{-\infty}^{\infty} F(\theta)\, Sa[(\omega - \theta)k/2]\, \exp[-j(\omega - \theta)t_0]\, d\theta \quad (A)$$

As $k \longrightarrow 0$: $B(t - t_0) \longrightarrow \delta(t - t_0)$ and $Sa(\omega k/2) \longrightarrow 1$.

Thus the LHS of (A) gives:

$$\lim_{k \to 0} f(t)\, B(t - t_0) = f(t)\, \delta(t - t_0) \quad (B)$$

And the RHS of (A) gives:

$$\lim_{k \to 0} 1/2\pi \int_{-\infty}^{\infty} F(\theta)\, Sa[(\omega - \theta)k/2]\, \exp[-j(\omega - \theta)t_0]\, d\theta$$

$$= 1/2\pi \int_{-\infty}^{\infty} F(\theta)\, \exp[-j(\omega - \theta)t_0]\, d\theta = \exp(-j\omega t_0)\; 1/2\pi \int_{-\infty}^{\infty} F(\theta)\, \exp(j\theta t_0)\, d\theta$$

$$= \exp(-j\omega t_0)\, f(t_0) \Longleftrightarrow f(t_0)\, \delta(t - t_0) \quad (C)$$

Combining (B) and (C) we obtain

$f(t)\, \delta(t - t_0) = f(t_0)\, \delta(t - t_0)$ QED

8.17 (a)

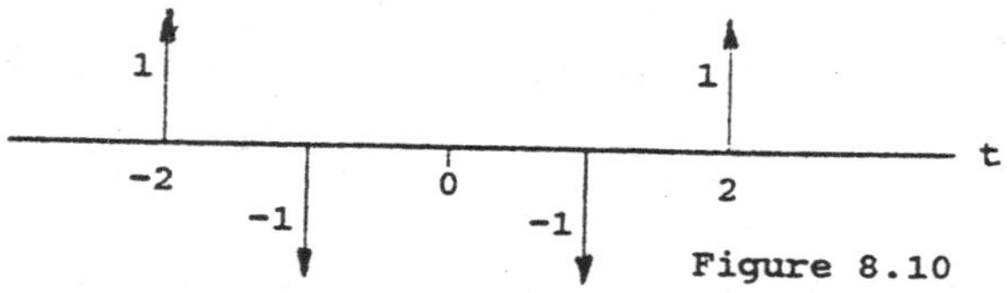

Figure 8.10

The group of impulses is

$$X(\omega) = \exp(j\omega 2) - \exp(j\omega) - \exp(-j\omega) + \exp(-j\omega 2)$$

$$= 2\cos(2\omega) - 2\cos(\omega) \longrightarrow$$

(b)(1) $\exp(-j\omega\tau) \Longleftrightarrow \delta(t-\tau)$

(2) $\cos(\omega\tau) = \frac{1}{2}\exp(j\omega\tau) + \frac{1}{2}\exp(-j\omega\tau) \Longleftrightarrow \frac{1}{2}\delta(t+\tau) + \frac{1}{2}(t-\tau)$

(3) $j\sin(\omega\tau) = \frac{1}{2}\exp(j\omega\tau) - \frac{1}{2}\exp(-j\omega\tau) \Longleftrightarrow \frac{1}{2}\delta(t+\tau) - \frac{1}{2}\delta(t-\tau)$

(4) $\cos^2(\omega\tau) = \frac{1}{2}[1+\cos(2\omega\tau)] = \frac{1}{2} + \frac{1}{4}\exp(j2\omega\tau) + \frac{1}{4}\exp(-j2\omega\tau)$

$$\Longleftrightarrow \frac{1}{2}\delta(t) + \frac{1}{4}\delta(t+2\tau) + \frac{1}{4}\delta(t-2\tau)$$

(5) $4j\cos(p\omega)\sin(q\omega) = 2j\sin[(p+q)\omega] - 2j\sin[(p-q)\omega]$

$= \exp[j(p+q)\omega] - \exp[-j(p+q)\omega] - \exp[j(p-q)\omega] + \exp[-j(p-q)\omega]$

$\Longleftrightarrow \delta(t+p+q) - \delta(t-p-q) - \delta(t+p-q) + \delta(t-p+q)$

(6) $\cos(\omega\tau)\,F(\omega) = \frac{1}{2}\exp(j\omega\tau)\,F(\omega) + \frac{1}{2}\exp(-j\omega\tau)\,F(\omega)$

$$\Longleftrightarrow \frac{1}{2}f(t+\tau) + \frac{1}{2}f(t-\tau)$$

(7) $\left[\exp(j\omega\tau) + \exp(j2\omega\tau)\right]^2 = \exp(j\omega 2\tau) + 2\exp(j\omega 3\tau) + \exp(j\omega 4\tau)$

$$\Longleftrightarrow \delta(t+2\tau) + 2\delta(t+3\tau) + \delta(t+4\tau)$$

8.18 (a) $\mathrm{Rect}(t) \Longleftrightarrow \mathrm{Sa}(\omega/2)$ $\quad\therefore \mathrm{Rect}(t/\tau) \Longleftrightarrow \tau\,\mathrm{Sa}(\omega\tau/2)$

$\therefore \mathrm{Rect}[(t+\tau/2)/\tau] \Longleftrightarrow \tau\,\mathrm{Sa}(\omega\tau/2)\exp(j\omega\tau/2)$

and $\mathrm{Rect}[(t-\tau/2)/\tau] \Longleftrightarrow \tau\,\mathrm{Sa}(\omega\tau/2)\exp(-j\omega\tau/2)$

$\therefore f(t) \Longleftrightarrow \tau\,\mathrm{Sa}(\omega/2)[\exp(j\omega\tau/2) - \exp(-j\omega\tau/2)]$

$= 2j\tau\,\mathrm{Sa}(\omega\tau/2)\sin(\omega\tau/2) \longrightarrow$ (A)

(b) $f'(t) = \delta(t+\tau) - 2\delta(t) + \delta(t-\tau) \Longleftrightarrow \exp(j\omega\tau) - 2 + \exp(-j\omega\tau)$

$$= [\exp(\omega\tau/2) - \exp(-j\omega\tau/2)]^2 = -4\sin^2(\omega\tau/2)$$

(c) From (A): $f'(t) \Longleftrightarrow j\omega\,[2j\tau\,\mathrm{Sa}(\omega\tau/2)\sin(\omega\tau/2)]$

$$= j\omega\,2j\tau\,\frac{\sin^2(\omega\tau/2)}{\omega\tau/2} = -4\sin^2(\omega\tau/2)$$

8.19

(a) The frequency response of the filter is $H(\omega) = \mathrm{Rect}(\omega/\omega_0)$
Thus the response to $x(t) \Longleftrightarrow X(\omega)$ will be

(1) In the frequency domain: $Y(\omega) = \mathrm{Rect}(\omega/\omega_0)\,X(\omega)$ Then, inverting to the time domain: $y(t) = 1/2\pi \int_{-\infty}^{\infty} \mathrm{Rect}(\omega/\omega_0)\,X(\omega)\exp(j\omega t)\,d\omega$

$$= 1/2\pi \int_{-\omega_0/2}^{\omega_0/2} X(\omega)\exp(j\omega t)\,d\omega \longrightarrow$$

(2) By convolution in the time domain: Inverting $H(\omega) = \mathrm{Rect}(\omega/\omega_0)$:

$\mathrm{Rect}(t/\omega_0) \Longleftrightarrow \omega_0\,\mathrm{Sa}(\omega\omega_0/2)$ $\quad\therefore \omega_0\,\mathrm{Sa}(t\omega_0/2) \Longleftrightarrow 2\pi\,\mathrm{Rect}(\omega/\omega_0)$
$\therefore \mathrm{Rect}(\omega/\omega_0) \Longleftrightarrow \omega_0/2\pi\,\mathrm{Sa}(\omega_0 t/2)$ Then, by time domain convolution:

$$y(t) = \int_{-\infty}^{\infty} x(\tau)\,h(t-\tau)\,d\tau = \int_{-\infty}^{\infty} x(\tau)\,\frac{\omega_0}{2\pi}\,\mathrm{Sa}[\omega_0(t-\tau)/2]\,d\tau$$

(b) Now assume that $x(t) = \delta(t)$. Then $X(\omega) = 1$, and (a)(1) gives:

$$y(t) = 1/2\pi \int_{-\omega_0/2}^{\omega_0/2} X(\omega)\exp(j\omega t)\,d\omega = 1/2\pi \int_{-\omega_0/2}^{\omega_0/2} \exp(j\omega t)\,d\omega$$

$$= \frac{1}{2\pi}\,\frac{\exp(j\omega t)}{jt}\Bigg|_{-\omega_0/2}^{\omega_0/2} = \frac{\omega_0}{2\pi}\,\frac{\exp(j\omega_0 t/2) - \exp(j\omega_0 t/2)}{2j\omega_0 t/2} = \frac{\omega_0}{2\pi}\,\mathrm{Sa}(\omega_0 t/2)$$

(a)(2) gives: $y(t) = \int_{-\infty}^{\infty} \delta(\tau)\,\omega_0/2\pi\,\mathrm{Sa}[\omega_0(t-\tau)/2]\,d\tau$

$$= \int_{-\infty}^{\infty} \delta(\tau)\,\frac{\omega_0}{2\pi}\,\mathrm{Sa}(\omega_0 t/2)\,d\tau = \frac{\omega_0}{2\pi}\,\mathrm{Sa}(\omega_0 t/2)\int_{-\infty}^{\infty}\delta(\tau)\,d\tau = \frac{\omega_0}{2\pi}\,\mathrm{Sa}(\omega_0 t/2)$$

8.20 $H(\omega) = 1 + 2k\cos(\omega T) = 1 + k\exp(j\omega T) + k\exp(-j\omega T)$

(a) $h(t) = \delta(t) + k\,\delta(t+T) + k\,\delta(t-T)$

(b) $$y(t) = x(t) * h(t) = \int_{-\infty}^{\infty} x(\tau)\,[\delta(t-\tau) + k\,\delta(t+T-\tau) + k\,\delta(t-T-\tau)]\,d\tau$$

$$= \int_{-\infty}^{\infty} x(t)\,\delta(t-\tau)\,d\tau + \int_{-\infty}^{\infty} k\,x(t+T)\,\delta(t+T-\tau)\,d\tau + \int_{-\infty}^{\infty} k\,x(t-T)\,\delta(t-T-\tau)]\,d\tau$$

$$= x(t)\int_{-\infty}^{\infty}\delta(t-\tau)\,d\tau + k\,x(t+T)\int_{-\infty}^{\infty}\delta(t+T-\tau)\,d\tau + k\,x(t-T)\int_{-\infty}^{\infty}\delta(t-T-\tau)]\,d\tau$$

$$= x(t) + k\,x(t+T) + k\,x(t-T) \longrightarrow$$

(c) $$y(t) = 1/2\pi\int_{-\infty}^{\infty} X(\omega)\,[1 + k\exp(j\omega T) + k\exp(-j\omega T)]\exp(j\omega t)\,d\omega$$

$$= 1/2\pi\int_{-\infty}^{\infty}\{X(\omega)\exp(j\omega t) + k\,X(\omega)\exp[j\omega(t+T)] + k\,X(\omega)\exp[j\omega(t-T)]\}\,d\omega$$

$$= x(t) + k\,x(t+T) + k\,x(t-T) \longrightarrow$$

See Answers for the sketch.

8.21 From the figure: $y(t) = 2\tau\,\Lambda(t/2\tau) - \tau\,\Lambda(t/\tau)$

$\Longleftrightarrow (2\tau)^2\,Sa^2(\omega 2\tau/2) - \tau^2\,Sa^2(\omega\tau/2) = \tau^2[4\,Sa^2(\omega\tau) - Sa^2(\omega\tau/2)]$

(b) By the method of successive differentiation: Differentiating $f(t)$ twice gives four Dirac deltas:

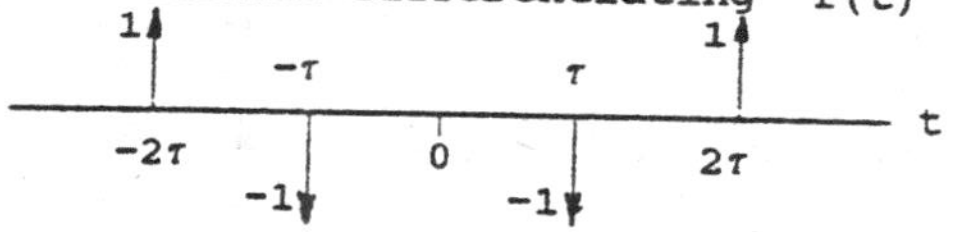

$$f''(t) = \delta(t+2\tau) - \delta(t+\tau) - \delta(t-\tau) + \delta(t-2\tau)$$

$$(j\omega)^2\,F(\omega) = \exp(j\omega 2\tau) - \exp(j\omega\tau) - \exp(-j\omega\tau) + \exp(-j\omega 2\tau)$$

$$F(\omega) = [\exp(j\omega 2\tau) - \exp(j\omega\tau) - \exp(-j\omega\tau) + \exp(j\omega 2\tau)]/(j\omega)^2$$

$$= \{[\exp(j\omega 2\tau) + \exp(j\omega 2\tau)] - [\exp(j\omega\tau) + \exp(-j\omega\tau)]\}/(j\omega)^2$$

$$= 2\,[\cos(\omega 2\tau) - \cos(\omega\tau)]/(j\omega)^2$$

$$= 2\,[1 - 2\sin^2(\omega\tau) - 1 + 2\sin^2(\omega\tau/2)]/(j\omega)^2$$

$$= \tau^2\frac{4\sin^2(\omega\tau)}{(\omega\tau)^2} - \tau^2\frac{\sin^2(\omega\tau/2)}{(\omega\tau/2)^2} = \tau^2[4\,Sa^2(\omega\tau) - Sa^2(\omega\tau/2)]$$

8.22 $x(t) = \exp(-2t)\,U(t)$ and $h(t) = \exp(-3t)\,U(t)$

(a) $$y(t) = x(t) * h(t) = \int_{-\infty}^{\infty} x(\tau)\,h(t-\tau)\,d\tau$$

$$= \int_{-\infty}^{\infty}\exp(-2\tau)\,U(\tau)\exp[-3(t-\tau)]\,U(t-\tau)\,d\tau$$

$$= \int_{-\infty}^{\infty}\exp(-2\tau)\exp[-3(t-\tau)]\,U(\tau)\,U(t-\tau)\,d\tau \qquad (A)$$

For $t > 0$:

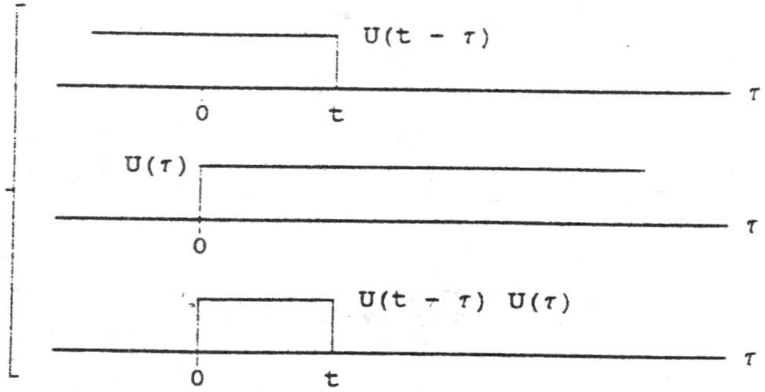

For $t < 0$: $U(t-\tau)\,U(\tau) = 0$. Thus

$$U(\tau)\,U(t-\tau) = \begin{cases} 1 \quad (0 < \tau < t) & \text{if } t > 0 \\ 0 & \text{if } t < 0 \end{cases}$$

$$\ldots = \left[\int_0^t \exp(-2\tau)\exp[-3(t - \tau)]\, d\tau \right] U(t) = \left[\exp(-3t) \int_0^t \exp(\tau)\, d\tau \right] U(t)$$

$$= \left[\exp(-3t)\ \exp(\tau) \right]_0^t U(t) = \left[\exp(-3t)\ [\exp(t) - 1] \right] U(t)$$

$$= [\exp(-2t) - \exp(-3t)]\ U(t) \longrightarrow$$

(b) $[\exp(-2t) - \exp(-3t)]\ U(t) \Longleftrightarrow \dfrac{1}{j\omega + 2} - \dfrac{1}{j\omega + 3} = \dfrac{1}{(j\omega + 2)(j\omega + 3)}$

(c) $x(t) = \exp(-2t)\ U(t) \Longleftrightarrow X(\omega) = 1/(j\omega + 2)$

$h(t) = \exp(-3t)\ U(t) \Longleftrightarrow H(\omega) = 1/(j\omega + 3)$

$x(t) * h(t) \Longleftrightarrow X(\omega)\ H(\omega) = \dfrac{1}{(j\omega + 2)(j\omega + 3)}$

(d) $y(t) = x(t) * H(t) = [\exp(-2t) - \exp(-3t)]\ U(t) \longrightarrow$ (B)

See the Answer section for the sketch.

(e) Confirming the above result using CONVOLUTION on the disk with N = 512, T = 4: Main menu, Create Values, Continue, Create X and X2, Real, Neither, Left, 1, Continue, EXP(-2*t), Go, Neither, Left, 1, Continue, EXP(-3*t), Go, Neither, Go. You are now back on the main menu. Plotting **X** and **X2** shows that the pulses have been correctly entered.

Run CONVOLUTION.

However as soon as CONVOLUTION is started the system alerts you that the **span restriction is being violated**. (Refer to Chapter 14, and to Section 17.4 in README 17.) Either:

(1) Select CONTINUE REGARDLESS, and accept possible aliasing errors

(2) Start again with a larger value for T, say T = 8

Because of time-domain aliasing, the convolution product in (1) will be less accurate than in (2). We elected to start again with T = 8. After entering the two pulses and displaying plots we note that they have both died down to negligibly small values at the end of the FFT window, i.e. at T = 4. We then displayed the numbers (Main menu, Show Numbers) where we see that for at least 50% of the window the two pulses are less than 1% of their peak value, and so now they will pass the span restriction test as applied to exponentially decaying pulses.

In the following table we show the [illegible] from the formula derived in (B). Note that $T_s = 8/512 = 0.015625$, and so the formula must be evaluated at $t = kT_s$ to be compared to an FFT value on line k of the table.

k	FFT	Formula	error (%)
1	1.5026982e-2	1.5026568e-2	0.002751793
2	2.8903392e-2	2.8902701e-2	0.002389396
4	5.3468965e-2	5.3467784e-2	0.002208059
8	9.1513443e-2	9.1511504e-2	0.002118531
16	0.13416689	0.134164107	0.002074325
32	0.14475225	0.144749281	0.002051064
64	8.5549961e-2	8.5548214e-2	0.002041153
128	1.5837209e-2	1.5836886e-2	0.002035059
256	3.2932511e-4	3.2931841e-4	0.002032835

The agreement between the FFT values and the formula values is remarkably good all across the range. The plot below shows y(t) as obtained from the FFT.

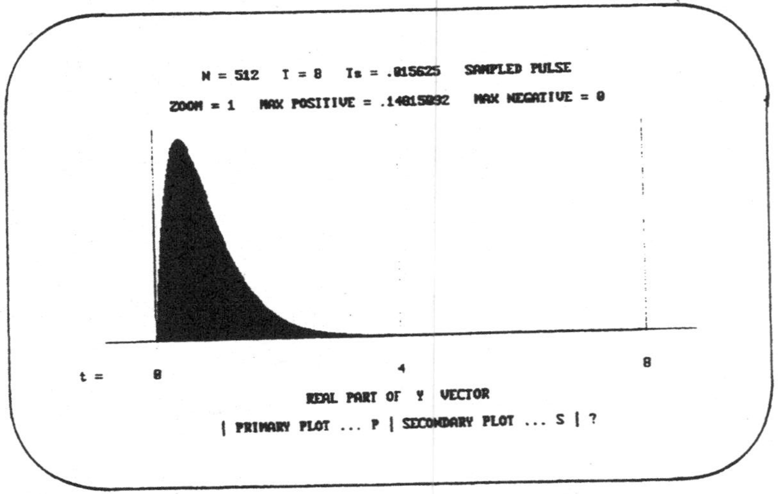

8.23 (a) $\dfrac{1}{2\pi}\displaystyle\int_{-\infty}^{\infty} A(\omega)\exp(j\omega t)\, d\omega$ is equal to the even part of the pulse.

and $\dfrac{1}{2\pi}\displaystyle\int_{-\infty}^{\infty} jB(\omega)\exp(j\omega t)\, d\omega$ is equal to the odd part of the pulse.

We obtain these from Figure 8.13. The results are in the Answer section.

(b) $$F(0) = \int_{-\infty}^{\infty} f(t)\ \exp(-j\omega t)\ dt \Big|_{\omega=0} = \int_{-\infty}^{\infty} f(t)\ dt = \text{area of pulse.}$$

The area of the pulse is seen by inspection to be 5.

Similarly, A(0) and jB(0) are the areas of the even and odd parts. Since the odd part has zero area it follows that the area of the even part must always equal the area of the original pulse, i.e. 5.

(c) $$\int_{-\infty}^{\infty} F(\omega)\ d\omega = 2\pi \frac{1}{2\pi}\int_{-\infty}^{\infty} F(\omega)\ \exp(j\omega t)\ d\omega \Big|_{t=0} = 2\pi\ f(t)\Big|_{t=0} = 2\pi$$

$$\int_{-\infty}^{\infty} A(\omega)\ d\omega = 2\pi\ f_{ev}(t) \Big|_{t=0} = 2\pi \quad \text{and} \quad \int_{-\infty}^{\infty} jB(\omega)\ d\omega = 2\pi\ f_{od}(t) \Big|_{t=0} = 0$$

(d) $$\int_{-\infty}^{\infty} |F(\omega)|^2 d\omega = 2\pi(\text{energy in pulse}) = 2\pi \int_{-\infty}^{\infty} |f(t)|^2\ dt$$

$$= 2\pi \left[2\int_{-1}^{0} (1+t)^2\ dt + 1 + 4 + 1 \right] = 2\pi \left[2\frac{(1+t)^3}{3}\Big|_{-1}^{0} + 6 \right] = \frac{40\pi}{3}$$

$$\int_{-\infty}^{\infty} |A(\omega)|^2 d\omega = 2\pi(\text{energy in even part of pulse}) = 2\pi \int_{-\infty}^{\infty} |f_{ev}(t)|^2\ dt$$

$$= 2\times 2\pi \left[\int_{0}^{2} (1 - t/2)^2\ dt + 5/4 \right] = 4\pi \left[\frac{-2(1-t/2)^3}{3}\Big|_{0}^{2} + 5/4 \right] = \frac{23\pi}{3}$$

$$\int_{-\infty}^{\infty} |B(\omega)|^2 d\omega = 2\pi(\text{energy in odd part of pulse}) = 2\pi \int_{-\infty}^{\infty} |f_{od}(t)|^2\ dt$$

$$= 2\times 2\pi \left[2\int_{0}^{1} (t/2)^2\ dt + 5/4 \right] = 4\pi \left[\frac{4\ (t/2)^3}{3}\Big|_{0}^{1} + 5/4 \right] = \frac{17\pi}{3}$$

Observe that E[pulse] = E[even part] + E[odd part]

(e) Sketching $$\frac{1}{2\pi}\int_{-\infty}^{\infty} F(\omega)' \frac{2\sin(\omega)}{\omega} \exp(j\omega t)\ d\omega$$

First we note that $\dfrac{2\sin(\omega)}{\omega}$ <===> Rect(t/2) and so what this integral represents is the time domain convolution of f(t) with Rect(t/2). We can carry that out graphically. All we need to find is the value of the area of the product of the two pulses at each of the critical points, and then we can sketch in the curve between those values.

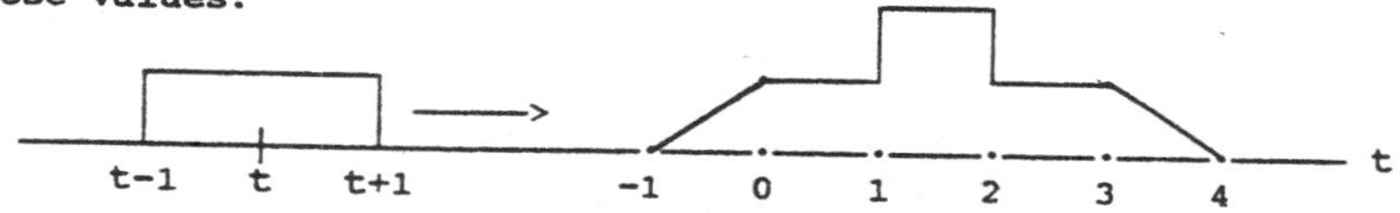

- t = -2: value = 0,
- t = -1: value = ½,
- t = 0: value = 3/2,
- t = 1: value = 3,
- t = 2: value = 3,
- t = 3: value = 3/2,
- t = 4: value = ½,
- t = 5: value = 0.

The resulting curve is shown in the Answer section.

(f) Evaluating $$\int_{-\infty}^{\infty} F(\omega) \frac{2\sin(\omega)}{\omega} \exp(j2\omega)\ d\omega$$

This is 2π times the value of the curve in (e) at the point t = 1, namely, $2\pi \times 3 = 6\pi$.

(g) Using the FFT system to obtain confirmation of the results from (a), (d), (e) and (f) with N = 512, T = 16. To load the pulses f(t) from Figure 8.13, and Rect(t/2), the steps are as follows:

Main menu, Create Values, Continue, Create X and X2, Real, Neither, Center, 7 intervals, -1, 0, 1, 2, 3, 4, Continue, 0, 1+t, 1, 2, 1, 4-t, Go, Even, 2, 1, Continue, 1, 0, Go, Neither, Go. You are now back on the main menu. Plotting **X** and **X2** shows that the two pulses have been entered correctly.

To obtain the energy in the pulse: Run **ANALYSIS**, Run Postprocessors, F, Energy, 256. The system shows 6.65137. The exact value is 20/3 or 6.66666. The error is 0.23%

To display the even part of f(t): Run **ANALYSIS**, Run Postprocessors, F, Set FIM to zero, Quit, Run **SYNTHESIS**. Plotting **Y** displays the even part. To obtain the energy in the even part: Run Postprocessors, F, Energy, 256. The system shows 3.82568. The exact value is 23/6 or 3.83333. The error is 0.20%

To display the odd part of f(t): Run ANALYSIS, Run Postprocessors, F, Set FRE to zero, Quit, Run SYNTHESIS. Plotting Y displays the odd part. To obtain the energy in the odd part: Run Postprocessors, F, Energy, 256. The system shows 2.82568. The exact value is 17/6 or 2.83333. The error is 0.27%

To display the curve for (e): Run CONVOLUTION. Plot Y.

8.24 The ideal low pass filter has response 1 for $-\frac{1}{2} < \omega < \frac{1}{2}$ and zero otherwise. Thus $H(\omega) = \text{Rect}(\omega)$.

Input is $x(t) = \delta(t - 1)$. Thus $X(\omega) = \exp(-j\omega)$.

(a) In the frequency domain the response is $Y(\omega) = H(\omega)\,X(\omega)$, i.e.

$$Y(\omega) = \text{Rect}(\omega)\exp(-j\omega) \longrightarrow$$

(b) Finding the time domain expression for $y(t)$:

$$\text{Rect}(t) \Longleftrightarrow \text{Sa}(\omega/2) \quad \therefore \text{Sa}(t/2) \Longleftrightarrow 2\pi\,\text{Rect}(\omega)$$

$$\therefore \text{Rect}(\omega) \Longleftrightarrow 1/2\pi\ \text{Sa}(t/2)$$

$$\therefore \text{Rect}(\omega)\exp(-j\omega) \Longleftrightarrow 1/2\pi\ \text{Sa}[(t-1)/2] = y(t) \longrightarrow$$

(c) The pulse $y(t)$ is now sampled (multiplied) by a second unit impulse at $t = 1$ which we call $w(t)$, and the result is called $z(t)$. Finding the time domain expression for $z(t)$:

$$z(t) = y(t)\,w(t) = 1/2\pi\ \text{Sa}[(t-1)/2]\ \delta(t-1)$$

$$= 1/2\pi\ \text{Sa}(0)\ \delta(t-1) = 1/2\pi\ \delta(t-1) \longrightarrow$$

(d) Writing the frequency domain expression for the operation that takes place in (c) in terms of $Y(\omega)$ and $W(\omega)$:

Multiplication in the time domain corresponds to convolution in the frequency domain. Thus: $w(t) = \delta(t-1) \Longleftrightarrow \exp(-j\omega)$ and so

$$Z(\omega) = 1/2\pi\ Y(\omega) * W(\omega) = 1/2\pi \int_{-\infty}^{\infty} \text{Rect}(\theta)\exp(-j\theta)\exp[-j(\omega-\theta)]\,d\theta$$

$$= 1/2\pi \int_{-\infty}^{\infty} \text{Rect}(\theta)\exp(-j\omega)\,d\theta = 1/2\pi\exp(-j\omega)\int_{-\infty}^{\infty}\text{Rect}(\theta)\,d\theta = 1/2\pi\exp(-j\omega)$$

$$\Longleftrightarrow 1/2\pi\ \delta(t-1) \longrightarrow$$

which agrees with (c).

8.25 $y(t) = t\exp(-\beta t)\,U(t)$ then

$$y(t) * U(t) = \int_{-\infty}^{\infty} y(\tau)\,U(t-\tau)\,d\tau = \int_{-\infty}^{\infty} [\tau\exp(-\beta\tau)\,U(\tau)]\,U(t-\tau)\,d\tau$$

However (see Exercise 8.22): $U(\tau)\,U(t-\tau) = \begin{cases} 1 \quad (0 < \tau < t) & \text{if } t > 0 \\ 0 & \text{if } t < 0 \end{cases}$

and so the integral continues

$$\ldots = \left[\int_0^t \tau\exp(-\beta\tau)\,d\tau\right] U(t) = \left[\frac{\tau\exp(-\beta\tau)}{-\beta} - \frac{\exp(-\beta\tau)}{(-\beta)^2}\right]_0^t U(t)$$

$$= \left[\frac{t\exp(-\beta t)}{-\beta} - \frac{\exp(-\beta t) - 1}{(-\beta)^2}\right] U(t) = \left[\frac{-\beta t\exp(-\beta t) - \exp(-\beta t) + 1}{\beta^2}\right] U(t)$$

$$= 1/\beta^2\left[1 - (1 + \beta t)\exp(-\beta t)\right] U(t)$$

QED

8.26 For the two pulses $x(t)$ and $h(t)$ shown in Figure 8.14:

(a) Finding $x(t) * h(t)$ analytically by evaluating the convolution integral:

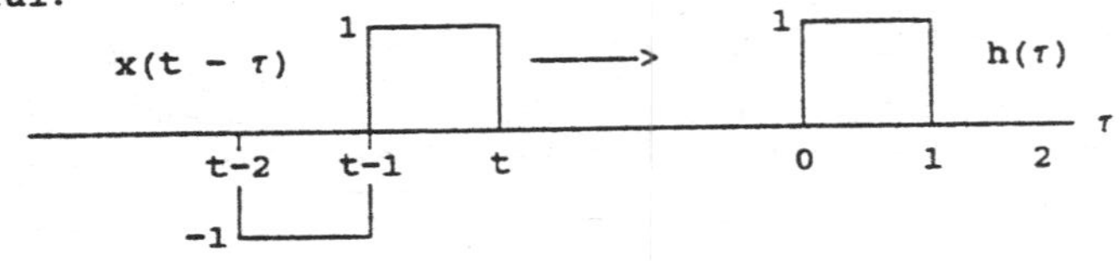

$$y(t) = \int_{-\infty}^{\infty} x(t-\tau)\,h(\tau)\,d\tau$$

- $t < 0$: $y(t) = 0 \longrightarrow$ $\qquad 0 < t < 1$: $y(t) = \int_0^t 1\,d\tau = t \longrightarrow$

- $1 < t < 2$: $y(t) = \int_0^{t-1} (-1)\,d\tau + \int_{t-1}^{1} 1\,d\tau = [-t + 1] + [1 - (t - 1)]$

$$= 3 - 2t \longrightarrow$$

- $2 < t < 3$: $y(t) = \int_{t-2} (-1)\, d\tau = (t - 2) - 1 = t - 3 \longrightarrow$
- $3 < t$: $y(t) = 0 \longrightarrow$

A plot of $y(t)$ is shown in the Answer section.

(b) Finding $x(t) * h(t)$ graphically we obtain the following values at the critical points:

- $t = 0$: $y(t) = 0$ - $t = 1$: $y(t) = 1$ - $t = 2$: $y(t) = -1$
- $t = 3$: $y(t) = 0$

Connecting these points by straight lines we obtain the same result as in (a).

(c) $x(t)$ is the input to an LTI system whose impulse response is $h(t)$ and output is $y(t)$. Finding the expression for $Y(\omega)$:

$$h(t) = \text{Rect}(t - \tfrac{1}{2}) \Longleftrightarrow H(\omega) = \text{Sa}(\omega/2)\exp(-j\omega/2)$$

$$x(t) = \text{Rect}(t - \tfrac{1}{2}) - \text{Rect}(t - 3/2)$$

$$\Longleftrightarrow \text{Sa}(\omega/2)\exp(-j\omega/2) - \text{Sa}(\omega/2)\exp(-j3\omega/2)$$

$$= \text{Sa}(\omega/2)\,[\exp(-j\omega/2) - \exp(-j3\omega/2)]$$

$$Y(\omega) = H(\omega)X(\omega) = \text{Sa}^2(\omega/2)\,[\exp(-j\omega) - \exp(-j2\omega)] \longrightarrow$$

(d) Loading the pulses with N = 512, T = 4: Create Values, Continue, Create both X and X2, Real, Neither, Left, 3, 1, 2, Continue, 1, -1, 0. Go, Neither, Left, 2, 1, Continue, 1, 0, Go, Neither, Go. You are now on the main menu. Plotting **X** and **X2** shows that the pulses have been entered correctly. Run CONVOLUTION, and plot **Y**. We see the same pulse as we obtained in (a) or (b).

(e) $$Y(\omega) = \text{Sa}^2(\omega/2)\,[\cos(\omega) - j\sin(\omega) - \cos(2\omega) + j\sin(2\omega)]$$

$$= \text{Sa}^2(\omega/2)\,[\cos(\omega) - \cos(2\omega)] + j\,\text{Sa}^2(\omega/2)\,[\sin(2\omega) - \sin(\omega)]$$

$$= A(\omega) + jB(\omega) \qquad \text{(A)}$$

Using the FFT system to invert the expression for $Y(\omega)$:

Main menu, Create Values, Continue, Create F, Complex,
  (SIN(w/2)^2*(COS(w) - COS(2*w)))/(w/2)^2,
Alpha = 0, Division by zero, B, 3.7011e-6, Go, Go,
  (SIN(w/2)^2*(SIN(2*w) - SIN(w)))/(w/2)^2,
Continue aliasing with the Alpha = 0, Go, Neither, Go. You are now on the main menu. Run SYNTHESIS and plot **Y**. We see a plot that is almost identical to the one obtained in (a) or (b).

Note: Because $y(t)$ is everywhere continuous we know that its spectrum $Y(\omega)$ is dying out like $1/\omega^2$. This is confirmed in (A)

(See also a plot of the magnitude spectrum of $Y(\omega)$ after CONVOLUTION has been run.) Thus we knew that we did not need much aliasing, and in fact Alpha = 0 was sufficient.

8.27 (a) $x(t) = \text{Rect}(t + \tfrac{1}{2}) + 2\,\text{Rect}[(t - 2)/2] + 3/2\,\text{Rect}[(t - 3.5)/3]$

$$X(\omega) = \text{Sa}(\omega/2)\exp(j\omega/2) + 4\text{Sa}(\omega)\exp(-j2\omega) + 9/2\,\text{Sa}(3\omega/2)\exp(-j\omega 3.5)$$

(b) $$x'(t) = \delta(t + 1) + \delta(t) - \tfrac{1}{2}(t - 2) - 3/2\,\delta(t - 5)$$

$$j\omega X(\omega) = \exp(j\omega) + 1 - \tfrac{1}{2}\exp(-j\omega 2) - 3/2\exp(-j\omega 5)$$

$$X(\omega) = [\exp(j\omega) + 1 - \tfrac{1}{2}\exp(-j\omega 2) - 3/2\exp(-j\omega 5)]/j\omega$$

$$= \frac{\exp(j\omega) - 1}{j\omega} + 2\,\frac{1 - \exp(-j\omega 2)}{j\omega} + 3/2\,\frac{\exp(-j\omega 2) - \exp(-j\omega 5)}{j\omega}$$

$$= \frac{\exp(j\omega/2) - \exp(-j\omega/2)}{2j\omega/2}\exp(j\omega/2) + 4\,\frac{\exp(j\omega) - \exp(-j\omega)}{2j\omega}\exp(-j\omega)$$

$$+ 9/2\,\frac{\exp(j3\omega/2) - \exp(-j3\omega 2)}{2j3\omega/2}\exp(-j7\omega/2)$$

$$= \text{Sa}(\omega/2)\exp(j\omega/2) + 4\,\text{Sa}(\omega)\exp(-j\omega) + 9/2\,\text{Sa}(3\omega/2)\exp(-j\omega 3.5)$$

(c) Sketches of the even and the odd parts of $x(t)$ are in the Answers.

$$X(\omega) = [\exp(j\omega) + 1 - \tfrac{1}{2}\exp(-j\omega 2) - 3/2\exp(-j\omega 5)]/j\omega$$

$$= [\cos(\omega) + j\sin(\omega) + 1 - \tfrac{1}{2}\cos(-2\omega) + \tfrac{1}{2}j\sin(2\omega)$$

$$- 3/2\cos(5\omega) + 3/2\,j\sin(5\omega)]/j\omega$$

$$= [\sin(\omega) + \tfrac{1}{2}\sin(2\omega) + 3/2\sin(5\omega)]/\omega \longrightarrow A(\omega)$$

$$- j\,[\cos(\omega) + 1 - \tfrac{1}{2}\cos(-2\omega) - 3/2\cos(5\omega)]/\omega \longrightarrow jB(\omega)$$

$$x_{ev}(t) \Longleftrightarrow A(\omega) \qquad x_{od}(t) \Longleftrightarrow jB(\omega) \longrightarrow$$

(d) Using the FFT system with N = 480, T = 12, to create plots of the even and odd parts of the pulse: Main menu, Create Values, Continue, Create X, Real, Neither, Center, 5 intervals, -1, 0, 2, 5, Continue, 0, 1, 2, 3/2, 0, Go, Neither, Go. You are now back on the main menu. Plotting **X** shows that the pulse has been entered correctly.

To display the even part of the pulse: Run ANALYSIS, Run Postprocessors, F, Set FIM to zero, Quit, Run SYNTHESIS. Plotting **Y** shows the even part.

To display the odd part of the pulse: Run ANALYSIS, Run Postprocessors, F, Set FRE to zero, Quit, Run SYNTHESIS. Plotting Y

$x'(t) = \delta(t) - 4\,\delta(t-1) + 6\,\delta(t-2) - 4\,\delta(t-3) + \delta(t-4)$

$$j\omega\, X(\omega) = 1 - 4\exp(-j\omega) + 6\exp(-j2\omega) - 4\exp(-j3\omega) + \exp(-j4\omega)$$

$$X(\omega) = [1 - 4\exp(-j\omega) + 6\exp(-j2\omega) - 4\exp(-j3\omega) + \exp(-j4\omega)]/j\omega$$

(b) $x(t) = \mathrm{Rect}(t-\tfrac{1}{2}) - 3\mathrm{Rect}(t-3/2) + 3\mathrm{Rect}(t-5/2) - \mathrm{Rect}(t-7/2)$

$$X(\omega) = \mathrm{Sa}(\omega/2)[\exp(-j\omega/2) - 3\exp(-j3\omega/2) + 3\exp(-j5\omega/2) - \exp(-j7\omega/2)] \longrightarrow$$

$$= \sin(\omega/2)\,[\exp(-j\omega/2) - 3\exp(-j3\omega/2) + 3\exp(-j5\omega/2) - \exp(-j7\omega/2)]/\omega/2$$

$$= [\exp(j\omega/2) - \exp(-j\omega/2)]\,[\exp(-j\omega/2) - 3\exp(-j3\omega/2) + 3\exp(-j5\omega/2) - \exp(-j7\omega/2)]/2j\omega/2$$

$$= [1 - 3\exp(-j\omega) + 3\exp(-j2\omega) - \exp(-j3\omega)]/j\omega$$
$$+ [-\exp(-j\omega) + 3\exp(-j2\omega) - 3\exp(-j3\omega) + \exp(-j4\omega)]/j\omega$$

$$= [1 - 4\exp(-j\omega) + 6\exp(-j2\omega) - 4\exp(-j3\omega) + \exp(-j4\omega)]/j\omega$$

(c) Finding $y(t) = x(t) * h(t)$ graphically, where $h(t)$ is the same function as in Exercise 8.26 above.

We require the values at each of the critical points:

- $t = 0$: $y(t) = 0$
- $t = 1$: $y(t) = 1$
- $t = 2$: $y(t) = -3$
- $t = 3$: $y(t) = 3$
- $t = 4$: $y(t) = -1$
- $t = 5$: $y(t) = 0$

Between these values we know that $y(t)$ is linear. Thus connecting them by a series of straight lines gives us the plot of $y(t)$ shown in the Answers.

(d) Using CONVOLUTION in the FFT system with N = 300, T = 6:

Main menu, Create Values, Continue, Create X and X2, Real, Neither Left, 5 intervals, 1, 2, 3, 4, Continue, 1, −3, 3, −1, 0, Go, Neither, Left, 2, 1, Continue, 1, 0, Go, Neither, Go. You are now back on the main menu. Plotting **X** and **X2** confirms that the pulses have been entered correctly.

Run CONVOLUTION. Plotting **Y** shows almost the same pulse as derived in (c) above.

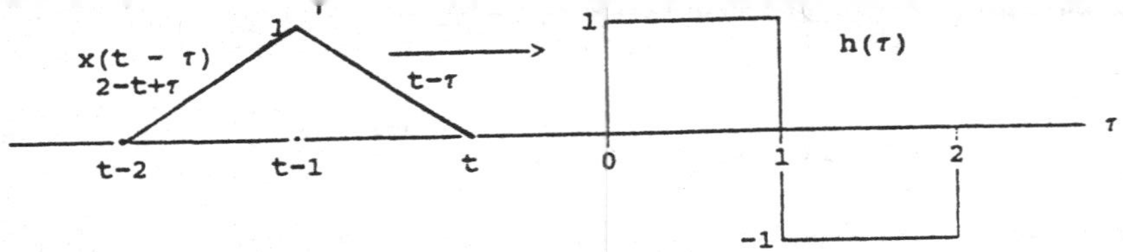

- For $t < 0$: $y(t) = 0$
- For $0 < t < 1$:

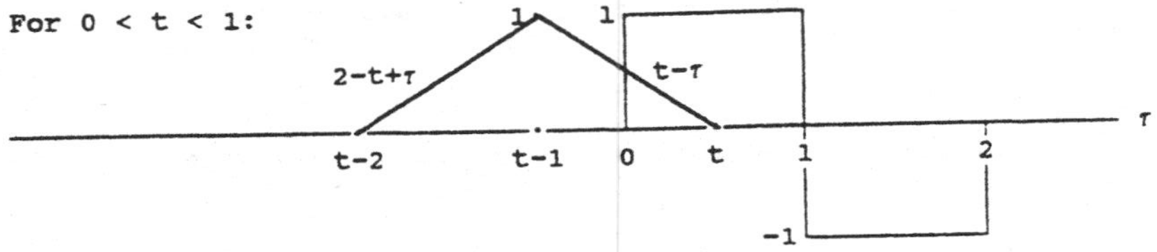

$$y(t) = \int_0^t (t-\tau)\, d\tau = t^2/2 \longrightarrow$$

- For $1 < t < 2$:

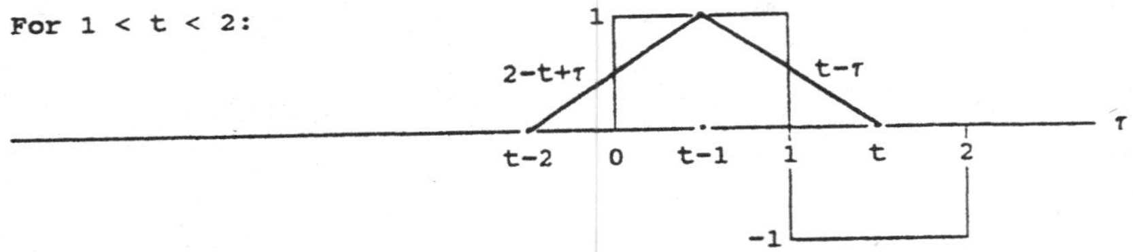

$$y(t) = \int_0^{t-1} (2-t+\tau)\, d\tau + \int_{t-1}^{1} (t-\tau)\, d\tau - \int_1^t (t-\tau)\, dt = -3t^2/2 + 4t - 2 \longrightarrow$$

- For $2 < t < 3$:

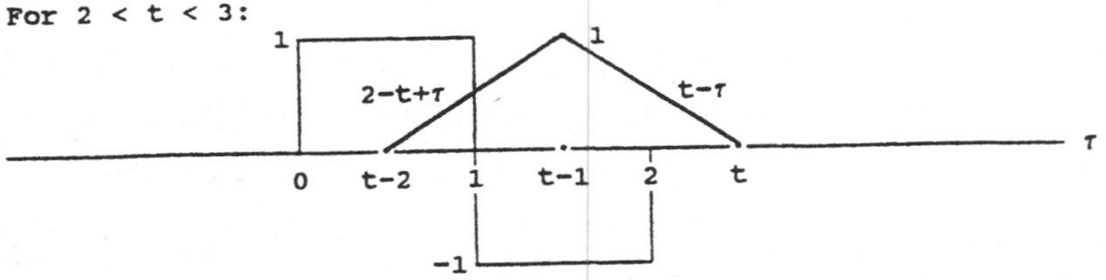

$$y(t) = \int_{t-2}^{1} (2-t+\tau)\, d\tau - \int_{1}^{t-1} (2-t+\tau)\, d\tau - \int_{t-1}^{2} (t-\tau)\, dt = 3t^2/2 - 8t + 10 \longrightarrow$$

- For $3 < t < 4$:

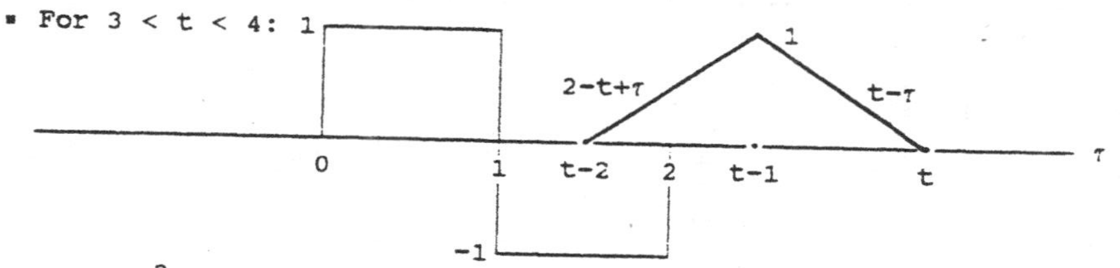

$$y(t) = -\int_{t-2}^{2} (2-t+\tau)\, d\tau = -t^2/2 + 4t - 8 \longrightarrow$$

$$y(t) = \begin{cases} 0 & (t < 0) \\ t^2/2 & (0 < t < 1) \\ -3t^2/2 + 4t - 2 & (1 < t < 2) \\ 3t^2/2 - 8t + 10 & (2 < t < 3) \\ -t^2/2 + 4t - 8 & (3 < t < 4) \\ 0 & (4 < t) \end{cases}$$

A sketch of $y(t)$ appears in the Answers.

(b) Examining the continuity of $y(t)$ and its derivatives:

From the sketch we see that $y(t)$ is everywhere continuous.

$$y'(t) = \begin{cases} 0 & (t < 0) \\ t & (0 < t < 1) \\ -3t + 4 & (1 < t < 2) \\ 3t - 8 & (2 < t < 3) \\ -t + 4 & (3 < t < 4) \\ 0 & (4 < t) \end{cases}$$

At $t = 0$: $y'_{left} = 0 = y'_{right}$
At $t = 1$: $y'_{left} = 1 = y'_{right}$
At $t = 2$: $y'_{left} = -2 = y'_{right}$
At $t = 3$: $y'_{left} = 1 = y'_{right}$
At $t = 4$: $y'_{left} = 0 = y'_{right}$

Thus $y'(t)$ is everywhere continuous.

$$y''(t) = \begin{cases} 0 & (t < 0) \\ 1 & (0 < t < 1) \\ -3 & (1 < t < 2) \\ t & (2 < t < 3) \\ -1 & (3 < t < 4) \\ 0 & (4 < t) \end{cases}$$

Thus $y''(t)$ has discontinuities.

Thus $y(t)$ and $y'(t)$ are everywhere continuous but $y''(t)$ has discontinuities. Thus we expect that the transform of $y(t)$ will die out like $1/\omega^3$.

(c) To find $Y(\omega)$ we have: $Y(\omega) = X(\omega)\, H(\omega)$

$x(t) = \Lambda(t-1) \Longleftrightarrow Sa^2(\omega/2)\, \exp(-j\omega)$

$h(t) = Rect(t-\tfrac{1}{2}) - Rect(t-3/2) \Longleftrightarrow Sa(\omega/2)[\exp(-j\omega/2) - \exp(-j3\omega/2)]$

$Y(\omega) = j\omega\, Sa^4(\omega/2)\, \exp(-j2\omega)$

Thus $Y(\omega)$ dies out like $1/\omega^3$, thereby confirming what we learned in (b).

(d) Using CONVOLUTION in the FFT system with $N = 512$, $T = 8$:

Main menu, Create values, Continue, Create X and X2, Real, Neither, Left, 3 intervals, 1, 2, Continue, t, 2-t, 0, Go, Neither, Left, 3 intervals, 1, 2, Continue, 1, -1, 0, Go, Neither, Go. You are now back on the main menu. Plotting **X** and **X2** confirms that the pulses have been entered correctly.

Run CONVOLUTION. Plot **Y**. It is almost identical to the sketch of $y(t)$.

8.30 (a) Sketches of $y'(t)$ and $y''(t)$ are shown in the Answers.

(b) Validation using The FFT system with $N = 512$, $T = 8$: The pulses were entered as described in Exercise 8.29. Run CONVOLUTION.

(1) To form $y'(t)$: Run Postprocessors, F, Central-diff 1st, Quit, Run SYNTHESIS, Plot **Y**. The plot is almost identical to the sketch of $y'(t)$.

(2) To form $y''(t)$: Run Postprocessors, F, Central-diff 1st, Quit, Run SYNTHESIS, Plot **Y**. The plot is identical to the sketch of $y''(t)$.

Now we repeat the procedure, but this time we differentiate the input $x(t)$ first.

(3) Run ANALYSIS. $X(\omega)$ is now in **F**. Run Postprocessors, F, Central-diff 1st, Run SYNTHESIS. $x'(t)$ is now in **Y**. Plotting **Y** confirms that. Run Postprocessors, X, Copy Y to X. $x'(t)$ is now in **X** and $h(t)$ is in **X2**. Plotting **X** and **X2** confirms that.

Run CONVOLUTION. Plot **Y**. the plot is identical to the earlier plot of $y'(t)$ from (1). Thus if $y(t)$ is the reponse to $x(t)$ then $y'(t)$ is the response to $x'(t)$.

(4) Start again. Load the pulses by "Using the Old Problem": Main menu, Create Values, Continue, Use the Old Problem, accept the expressions and return to the main menu. Run ANALYSIS, Run Postprocessors, F, Central-diff 2nd, Quit, Run SYNTHESIS, Plot **Y**. It displays three Dirac deltas which form $x''(t)$.

Run Postprocessors, X, Copy Y to X, Quit, Run CONVOLUTION, plot Y. It is identical to $y''(t)$ from (2). Thus if $y(t)$ is the reponse to $x(t)$ then $y''(t)$ is the response to $x''(t)$.

8.31 From (4.15) (see Figure 4.6): $\exp(-\beta|t|) \Longleftrightarrow 2A(\omega) = \dfrac{2\beta}{\beta^2 + \omega^2}$

Using duality: $\dfrac{2\beta}{\beta^2 + t^2} \Longleftrightarrow 2\pi \exp(-\beta|\omega|)$

$\therefore \dfrac{1}{\beta^2 + t^2} \Longleftrightarrow \pi/\beta \exp(-\beta|\omega|)$ QED

8.32 From (4.15) (see Figure 4.7):

$$\exp(-\beta|t|)\ \mathrm{Sgn}(t) \Longleftrightarrow 2jB(\omega) = \frac{-j2\omega}{\beta^2 + \omega^2}$$

Using duality: $\dfrac{-j2t}{\beta^2 + t^2} \Longleftrightarrow 2\pi \exp(-\beta|\omega|)\ \mathrm{Sgn}(-\omega)$

$\therefore \dfrac{t}{\beta^2 + t^2} \Longleftrightarrow \dfrac{2\pi}{-j2} \exp(-\beta|\omega|)\ \mathrm{Sgn}(-\omega) = -j\pi \exp(-\beta|\omega|)\ \mathrm{Sgn}(\omega)$ QED

8.33 Using frequency-domain differentiation, i.e. If $f(t) \Longleftrightarrow F(\omega)$ then $t\,f(t) \Longleftrightarrow j\,dF(\omega)/d\omega$. From Exercise 8.31:

$$\frac{1}{\beta^2 + t^2} \Longleftrightarrow \pi/\beta \exp(-\beta|\omega|) \quad \therefore \frac{t}{\beta^2 + t^2} \Longleftrightarrow j\frac{d}{d\omega}\,\pi/\beta \exp(-\beta|\omega|) \qquad \text{(A)}$$

For $\omega > 0$: $|\omega| = \omega$, and so we continue from (A):

$$\ldots = j\frac{d}{d\omega}\,\pi/\beta \exp(-\beta\omega) = -j\pi \exp(-\beta\omega) = -j\pi \exp(-\beta|\omega|)$$

For $\omega < 0$: $|\omega| = -\omega$, and so we continue from (A):

$$\ldots = j\frac{d}{d\omega}\,\pi/\beta \exp(\beta\omega) = j\pi \exp(\beta\omega) = j\pi \exp(-\beta|\omega|)$$

Thus for ω positive or negative: $\dfrac{t}{\beta^2 + t^2} = -j\pi \exp(-\beta|\omega|)\ \mathrm{Sgn}(\omega)$ QED

8.34

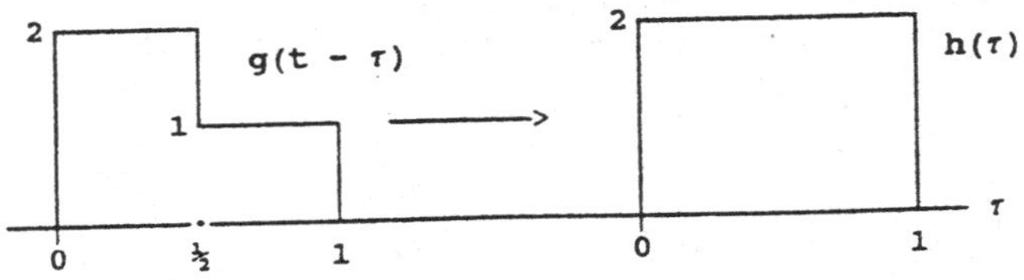

(a) In order to sketch the convolution ... Figure 8.18, we require the values at the following critical points:

$t = 0$: $y(t) = 0$, $t = \frac{1}{2}$: $y(t) = 1$, $t = 1$: $y(t) = 3$,

$t = 1.5$: $y(t) = 2$, $t = 2$: $y(t) = 0$

Between these points $y(t)$ is linear. Thus connecting them by straight lines will give the required sketch for $y(t)$. The result is shown in the Answers.

(b) Finding $F(\omega)$ from the sketch: First we form $f'(t)$.

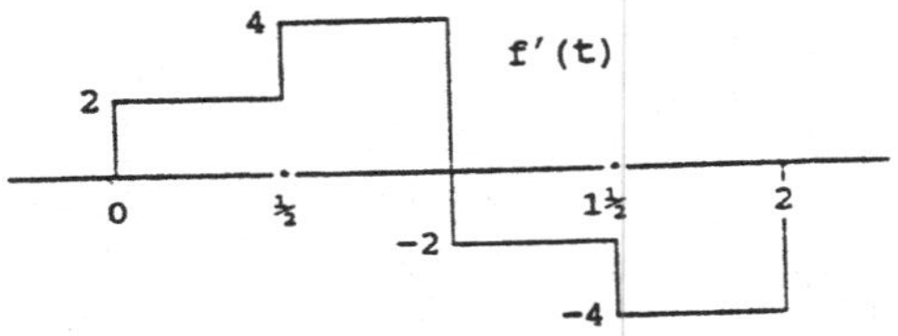

Then:

$f''(t) = 2\,\delta(t) + 2\,\delta(t - \frac{1}{2}) - 6\,\delta(t - 1) - 2\,\delta(t - 3/2) + 4\,\delta(t - 2)$

$(j\omega)^2 F(\omega) = 2 + 2\exp(-j\omega/2) - 6\exp(-j\omega) - 2\exp(-j3\omega/2) + 4\exp(-j2\omega)$

$F(\omega) = 2[1 + \exp(-j\omega/2) - 3\exp(-j\omega) - \exp(-j3\omega/2) + 2\exp(-j2\omega)]/(j\omega)^2$ (A)

(c) $F(\omega) = G(\omega)\ H(\omega)$

(d) Finding $G(\omega)$ and $H(\omega)$:

$g'(t) = \delta(t) + \delta(t - \frac{1}{2}) - 2\,\delta(t - 1)$

$j\omega\,G(\omega) = 1 + \exp(-j\omega/2) - 2\exp(-j\omega)$

$G(\omega) = [1 + \exp(-j\omega/2) - 2\exp(-j\omega)]/j\omega$ ⟶

$h'(t) = 2\,\delta(t) - 2\,\delta(t - 1)$

$j\omega\,H(\omega) = 2 - 2\exp(-j\omega)$

$H(\omega) = 2[1 - \exp(-j\omega)]/j\omega$ ⟶

(e) Multiplying $G(\omega)$ and $H(\omega)$:

$G(\omega)\ H(\omega) = \Big[\,[1 + \exp(-j\omega/2) - 2\exp(-j\omega)]\ 2[1 - \exp(-j\omega)]\,\Big]/(j\omega)^2$

$= 2[1 + \exp(-j\omega/2) - 2\exp(-j\omega) - \exp(-j\omega) - \exp(-j3\omega/2) + 2\exp(-j2\omega)]/(j\omega)^2$

$= 2[1 + \exp(-j\omega/2) - 3\exp(-j\omega) - \exp(-j3\omega/2) + 2\exp(-j2\omega)]/(j\omega)^2$ which is the same as (A).

8.35 **Proof of the integration property.** As shown in the text:

$$\int_{-\infty}^{t} f(\tau)\,d\tau = \int_{-\infty}^{\infty} f(\tau)\,U(t-\tau)\,d\tau \qquad \text{(A)}$$

The RHS of (A) is seen to be a time-domain convolution product. Transforming, we obtain:

$$\Longleftrightarrow F(\omega)\,[1/j\omega + \pi\delta(\omega)] = F(\omega)/j\omega + \pi\,F(0)\,\delta(\omega)$$

and so we have proved that

$$\int_{-\infty}^{t} f(\tau)\,d\tau \Longleftrightarrow \frac{1}{j\omega}\,F(\omega) + \pi\,F(0)\,\delta(\omega) \qquad \text{QED}$$

8.36 $\delta(t) \Longleftrightarrow 1$, and so, by the integration property

$$\int_{-\infty}^{t} \delta(\tau)\,d\tau \Longleftrightarrow F(\omega)/j\omega + \pi\,F(0)\,\delta(\omega) = 1/j\omega + \pi\delta(\omega)$$

This is the same as the transform of $U(t)$, namely $1/j\omega + \pi\delta(\omega)$

8.37 (a) Applying successive differentiation to $g(t) = \Lambda(t)$:

$$g''(t) = \delta(t+1) - 2\,\delta(t) + \delta(t-1)$$

$$(j\omega)^2\,G(\omega) = \exp(j\omega) - 2 + \exp(-j\omega) = [\exp(j\omega/2) - \exp(-j\omega/2)]^2$$

$$G(\omega) = [\exp(j\omega/2) - \exp(-j\omega/2)]^2/(j\omega)^2 \qquad \text{(A)}$$

(b) To use the integration property we require the transform of $f(t)$:

By successive differentiation:

$$f'(t) = \delta(t+1) - 2\,\delta(t) + \delta(t-1)$$

$$j\omega\,F(\omega) = \exp(j\omega) - 2 + \exp(-j\omega) = [\exp(j\omega/2) - \exp(-j\omega/2)]^2$$

$$F(\omega) = [\exp(j\omega/2) - \exp(-j\omega/2)]^2/j\omega$$

Then $$\int_{-\infty}^{t} f(\tau)\,d\tau \Longleftrightarrow F(\omega)/j\omega + \pi\,F(0)\,\delta(\omega)$$

$$= [\exp(j\omega/2) - \exp(-j\omega/2)]^2/(j\omega)^2 + \pi\,F(0)\,\delta(\omega)$$

In order to find $F(0)$ we use l'Hôpital's rule:

$$F(0) = \lim_{\omega \to 0} \frac{[\exp(j\omega/2) - \exp(-j\omega/2)]^2}{j\omega}$$

$$= \lim_{\omega \to 0} \frac{2[\exp(j\omega/2) - \exp(-j\omega/2)][j/2\,\exp(j\omega/2) + j/2\,\exp(-j\omega/2)]}{j} = 0$$

Thus $$g(t) = \int_{-\infty}^{t} f(\tau)\,d\tau \Longleftrightarrow [\exp(j\omega/2) - \exp(-j\omega/2)]^2/(j\omega)^2$$

which is the same as (A).

8.38 We know that $\Lambda(t/\tau) \Longleftrightarrow \tau\,Sa^2(\omega\tau/2)$

Then by duality: $\tau\,Sa^2(t\tau/2) \Longleftrightarrow 2\pi\,\Lambda(-\omega/\tau) = 2\pi\,\Lambda(\omega/\tau)$

Let $\tau = 1$. Then $Sa^2(t/2) = 2\pi\,\Lambda(\omega)$

We now apply reciprocal scaling.

Then $Sa^2(\alpha t/2) = 2\pi/|\alpha|\,\Lambda(\omega/\alpha)$

Letting $\alpha = 2$: $Sa^2(t) \Longleftrightarrow \pi\,\Lambda(\omega/2)$ ———>

8.39 (a) Redrawing Figure 8.21:

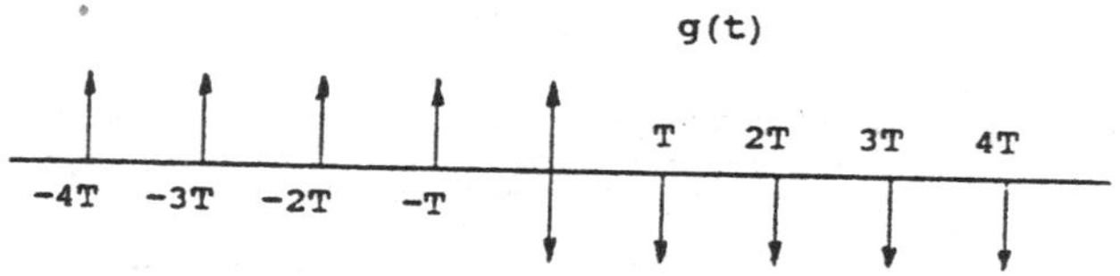

There are five Dirac deltas in each group but the central ones cancel. From this figure we see that

$$g(t) = \frac{1}{8}\left[\sum_{n=0}^{4} \delta(t+nT) - \sum_{n=0}^{4} \delta(t-nT)\right]$$

(b) $$x(t) * h(t) = \int_{-\infty}^{\infty} x(t-\tau)\,h(\tau)\,d\tau$$

$$= \int_{-\infty}^{\infty} x(t-\tau)\,\frac{1}{8}\left[\sum_{n=0}^{4} \delta(\tau+nT) - \sum_{n=0}^{4} \delta(\tau-nT)\right] d\tau$$

$$= \frac{1}{8}\sum_{n=0}^{4} x(t + nT) \int_{-\infty}^{\infty} \delta(\tau + nT)\, d\tau - \frac{1}{8}\sum_{n=0}^{4} x(t - nT) \int_{-\infty}^{\infty} \delta(\tau - nT)\, d\tau$$

$$= \frac{1}{8}\sum_{n=0}^{4} x(t + nT) - \frac{1}{8}\sum_{n=0}^{4} x(t - nT) = y(t) \text{ as given.} \qquad \text{QED}$$

(c) By long division: $\dfrac{1 - a^5}{1 - a} = 1 + a + a^2 + a^3 + a^4 + a^5$ QED

$$g(t) = \frac{1}{8}\left[\sum_{n=0}^{4} \delta(t + nT) - \sum_{n=0}^{4} \delta(t - nT)\right]$$

$$\Longleftrightarrow G(\omega) = \frac{1}{8}\left[\sum_{n=0}^{4} \exp(j\omega nT) - \sum_{n=0}^{4} \exp(-j\omega nT)\right]$$

$$= \frac{1}{8}\left[\frac{1 - \exp(j\omega T)^5}{1 - \exp(j\omega T)} - \frac{1 - \exp(-j\omega T)^5}{1 - \exp(-j\omega T)}\right] = \frac{1}{8}\left[Z + Z^*\right] = \tfrac{1}{4}j\ \text{Im}(Z)$$

where $Z = \dfrac{1 - \exp(j\omega T)^5}{1 - \exp(j\omega T)} = \dfrac{1 - \exp(j5\omega T)}{1 - \exp(j\omega T)}$

$$= \frac{1 - \cos(5\omega T) - j\sin(5\omega T)}{1 - \cos(\omega T) - j\sin(\omega T)}\ \frac{1 - \cos(\omega T) + j\sin(\omega T)}{1 - \cos(\omega T) + j\sin(\omega T)}$$

and so $\text{Im}(Z) = \dfrac{\sin(\omega T)[1 - \cos(5\omega T)] - \sin(5\omega T)[1 - \cos(\omega T)]}{[1 - \cos(\omega T)]^2 + \sin^2(\omega T)}$

$$= \frac{\sin(\omega T) - \sin(5\omega T) + \sin(5\omega T)\cos(\omega T) - \cos(5\omega T)\sin(\omega T)}{1 - 2\cos(\omega T) + \cos^2(\omega T) + \sin^2(\omega T)}$$

$$= \frac{\sin(\omega T) - \sin(5\omega T) + \sin(4\omega T)}{2 - 2\cos(\omega T)} = \frac{\sin(\omega T) - \sin(5\omega T) + \sin(4\omega T)}{4\sin^2(\omega T/2)}$$

Thus $G(\omega) = j\,\dfrac{\sin(\omega T) + \sin(4\omega T) - \sin(5\omega T)}{16\sin^2(\omega T/2)}$ ⟶

(d) Instead of just 4 impulses in each sum $g(t)$ with weights 1/8 we now assume that there are m impulses each with weight 1/2m. Then

$$g(t) = \frac{1}{8}\left[\sum_{n=0}^{m} \delta(t + nT) - \sum_{n=0}^{m} \delta(t - nT)\right]$$

$$\Longleftrightarrow G(\omega) = \frac{1}{8}\left[\sum_{n=0}^{m} \exp(j\omega nT) - \sum_{n=0}^{m} \exp(-j\omega nT)\right]$$

$$= \frac{1}{8}\left[\frac{1 - \exp(j\omega T)^{m+1}}{1 - \exp(j\omega T)} - \frac{1 - \exp(-j\omega T)^{m+1}}{1 - \exp(-j\omega T)}\right] = \frac{1}{8}\left[Z + Z^*\right] = \tfrac{1}{4}j\ \text{Im}(Z)$$

where $Z = \dfrac{1 - \exp(j\omega T)^{m+1}}{1 - \exp(j\omega T)} = \dfrac{1 - \exp[j(m+1)\omega T]}{1 - \exp(j\omega T)}$

$$= \frac{1 - \cos[(m+1)\omega T] - j\sin[(m+1)\omega T]}{1 - \cos(\omega T) - j\sin(\omega T)}\ \frac{1 - \cos(\omega T) + j\sin(\omega T)}{1 - \cos(\omega T) + j\sin(\omega T)}$$

and so

$$\text{Im}(Z) = \frac{\sin(\omega T)\{1 - \cos[(m+1)\omega T]\} - \sin[(m+1)\omega T][1 - \cos(\omega T)]}{[1 - \cos(\omega T)]^2 + \sin^2(\omega T)}$$

$$= \frac{\sin(\omega T) - \sin[(m+1)\omega T] + \sin[(m+1)\omega T]\cos(\omega T) - \cos[(m+1)\omega T]\sin(\omega T)}{1 - 2\cos(\omega T) + \cos^2(\omega T) + \sin^2(\omega T)}$$

$$= \frac{\sin(\omega T) - \sin[(m+1)\omega T] + \sin(m\omega T}{2 - 2\cos(\omega T)} = \frac{\sin(\omega T) - \sin[(m+1)\omega T] + \sin(m\omega T)}{4\sin^2(\omega T/2)}$$

and so $G(\omega) = j\left[\dfrac{\sin(\omega T) + \sin(m\omega T) - \sin[(m+1)\omega T]}{4m\sin^2(\omega T/2)}\right]$ ⟶

(e) Using the FFT system to verify the expression for $G(\omega)$ that we obtained in (c), with N = 256 and T_{FFT} = 16. In the formula we use T = 1, then:

$$G(\omega) = j\,\frac{\sin(\omega) + \sin(4\omega) - \sin(5\omega)}{16\sin^2(\omega T/2)}$$

We loaded this expression into the **F** vector as follows:

Main menu, Create values, Continue, Create F, Imaginary, (SIN(w)+SIN(4*w)-SIN(5*w))/(16*SIN(w/2)^2), Alpha = 0.

However, on both our DOS and our MAC machines, the system then created an **FIM** vector with **very large values at every 16th step**. (To see them do a Show Numbers, F, Single and take a look at FIM at n = 16, 32, and so on. Indeed, if CLEAN is on, then those are the **only** values that are displayed. Go back to the main menu, go into SETUP and toggle CLEAN off. Then inspect **FIM** once more. You will now see **all** of the values including the bad ones.)

These spurious values come about because of the division taking place in the above expression. Ideally the parser would report "division by zero" at each of the points where, mathematically, the denominator is zero, i.e. every 16th value. (The numerator is also mathematically zero there.) We could then deal with those cases manually, by stating that the quotient was zero, as it is. However neither the numerator nor the denominator is **precisely** zero at those points and so a division does in fact take place, producing these incredibly large errors. Of course the spectrum as it then stands is worthless, and certainly will not invert to give eight Dirac deltas.

To fix the problem:

Method (A): After loading the expression, return to the main menu take option C, Edit from Keyboard, and set those incorrect values all to zero manually, one at a time. Then run SYNTHESIS and plot **Y**. You will see the **eight Dirac deltas correctly displayed**.

Method (B): Use the alternative expression

$$G(\omega) = j\,\frac{\sin(\omega) + \sin(4\omega) - \sin(5\omega)}{8[1 - \cos(\omega T)]}$$

using the following steps: Main menu, Create values, Continue, Create F, Imaginary, (SIN(w)+SIN(4*w)-SIN(5*w))/(8*(1-COS(w))), Alpha = 0.

The system now **does** detect zeros in the denominator at every 16th step and so "divide by zero" events do occur. In each case let the system find the value for you or else enter zero. You are now back on the main menu.

<u>Either way</u>: After entering the spectrum, run SYNTHESIS and plot **Y**. The eight Dirac functions appear correctly, each with height 2, which can be explained as follows:

From the expression for $g(t)$ in (c) we see that the weights of the Dirac deltas are all 1/8. A time-domain Dirac delta of weight μ appears as a line of height $\mu N/T$ on the FFT, and so in this case the height should be

$$\frac{1}{8} \times 256 / 16 = 2 \quad \longrightarrow$$

9.1

$$\delta_T(t) = \frac{1}{T_s}\sum_{n=-\infty}^{\infty} \exp(jn\omega_s t)$$

$$x_s(t) = x(t)\ \delta_T(t) = x(t)\ \frac{1}{T_s}\sum_{n=-\infty}^{\infty} \exp(jn\omega_s t) = \frac{1}{T_s}\sum_{n=-\infty}^{\infty} x(t)\ \exp(jn\omega_s t) \qquad (A)$$

Then, by the frequency-shift property,

$$x(t)\ \exp(jn\omega_s t) \iff X(\omega - n\omega_s)$$

and so (A) transforms as follows:

$$x_s(t) \iff X_s(\omega) = \frac{1}{T_s}\sum_{n=-\infty}^{\infty} X(\omega - n\omega_s) \qquad \text{which is the same as (9.10)}$$

9.2
(a) Steps to load the pulse Rect(t) with N = 1024, T = 4:

Main menu, Create Values, Continue, Create X, Real, Even, 2, .5, Continue, 1, 0, Go, Neither, Go. You are now back on the main menu. Plotting **X** confirms that Rect(t) is in **X**. Run ANALYSIS and plot **FRE**, using Zoom = 4. You see a single copy of Sa(ω/2) with a max value of 1.

(b) To carry out the impulse sampling: Main menu, Run Postprocessors, X, Sample X, 4. You have now retained every fourth value in **X**.

To make the sampling look like impulse sampling: On the FFT, a Dirac delta of weight μ appears as a line of height V = μN/T. Thus for μ = 1 we have V = 1024/4 = 256. While still in the **X** postprocessor take the option "Multiply X by a Constant". Use 256. Quit. You are now back on the main menu with Rect(t) impulse-sampled at every fourth FFT step.

Run ANALYSIS and plot **FRE**. There are now **four** copies of Sa in the display. According to (9.10) they should each have a **maximum** value of $1/T_s$. In our case the FFT sampling interval is T/N = .00390625. Then the sampling interval which corresponds to four of these is T_s = 4 T/N = 0.015625. This is the interval at which the Rect in **X** has been sampled. By (9.10) it follows that the four Sa's should now each have max value $1/T_s$ = 64. The plot of **FRE** confirms that.

(c) To repeat: Main menu, Create Values, Continue, Use the old Problem, Go, Go, Go, Go. You are now back on the main menu with Rect(t) in **X**. Run Postprocessors, X, Sample, 32, Multiply X by a Constant, 256, Quit. You are now back on the main menu with Rect(t) in **X**, impulse sampled at every 32nd FFT step. Plotting **X** confirms that.

Run ANALYSIS and plot **FRE**. In this case T_s is 32 FFT steps which are each T/N = 0.00390625, and so T_s = 32T/N = 0.125. There should be 32 Sa's, each with a max value of $1/T_s$ = 8. Even though they are now heavily aliased, all of this is confirmed from the plot.

9.3
(a) A sketch of the band-limited function X(ω) = cos(πω) Rect(ω) is shown in the Answers.

(b) Inverse transforming X(ω):

$$\cos(\pi\omega)\ \text{Rect}(\omega) \iff \frac{1}{2\pi}\int_{-\infty}^{\infty} \cos(\pi\omega)\ \text{Rect}(\omega)\ \exp(j\omega t)\ d\omega$$

$$= \frac{1}{2\pi}\int_{-\frac{1}{2}}^{\frac{1}{2}} \tfrac{1}{2}[\exp(j\pi\omega) + \exp(-j\pi\omega)]\ \exp(j\omega t)\ d\omega$$

$$= \frac{1}{4\pi}\int_{-\frac{1}{2}}^{\frac{1}{2}} \{\exp[j\omega(t+\pi)] + \exp[j\omega(t-\pi)]\}\ d\omega$$

$$= \frac{1}{4\pi}\left[\frac{\exp[j\omega(t+\pi)]}{j(t+\pi)} + \frac{\exp[j\omega(t-\pi)]}{j(t-\pi)}\right]_{-\frac{1}{2}}^{\frac{1}{2}}$$

$$= \frac{1}{4\pi}\left[\frac{\sin[(t+\pi)/2]}{(t+\pi)/2} + \frac{\sin[(t-\pi)/2]}{(t-\pi)/2}\right]$$

$$= \frac{1}{4\pi}\left[\frac{\cos(t/2)}{(t+\pi)/2} - \frac{\cos(t/2)}{(t-\pi)/2}\right] = \frac{\cos(t/2)}{2\pi}\left[\frac{1}{t+\pi} - \frac{1}{t-\pi}\right] = \frac{\cos(t/2)}{\pi^2 - t^2}$$

QED

(c) A sketch of this function for the range $-7\pi \le t \le 7\pi$ is shown in the Answers. The indeterminacy at $t = \pm\pi$ is handled by l'Hôpital's rule as follows:

$$x(\pi) = \lim_{t \to \pm\pi} \frac{\cos(t/2)}{\pi^2 - t^2} = \lim_{t \to \pm\pi} \frac{-\frac{1}{2}\sin(t/2)}{-2t} = 1/4\pi \longrightarrow$$

(d) From the sketch in (a) we see that the highest frequency in x(t) is $\omega_H = \frac{1}{2}$. Then, by the Nyquist sampling criterion, $\omega_{min} = 2\omega_H = 1$. Then $T_{max} = 2\pi/\omega_{min} = 2\pi$. This is the largest sampling interval that we can use and still recover the original pulse from the resulting spectrum.

(e) A sketch of the pulse after impulse sampling at steps of $T = 2\pi$ is shown in the Answers.

(f) According to (9.10): $x_s(t) \Longleftrightarrow X_s(\omega) = \frac{1}{T_s} \sum_{n=-\infty}^{\infty} X(\omega - n\omega_s)$ and so

in this case $\omega_s = 1$, and $1/T_s = 1/2\pi$. The sketch in the Answers shows the spectra repeated with spacing $\omega_s = 1$ and max value $1/2\pi$.

(g) Sampling x(t) at intervals $\frac{1}{2}T_{max}$ means that $T_s = \pi$. Then $\omega_s = 2$. and $1/T_s = 1/\pi$. The sketch in the Answers shows the spectra repeated with spacing $\omega_s = 2$ and max value $1/\pi$.

(h) Sampling x(t) at intervals 3/2 T_{max} means that $T_s = 3\pi$, and so $\omega_s = 2/3$ and $1/T_s = 1/3\pi$. The sketch in the Answers shows the spectra repeated with spacing $\omega_s = 2/3$ and max value $1/3\pi$. Aliasing has now taken place because we have violated the Nyquist criterion.

(i) Verifying the plots using the FFT system, with N = 900 and T = 60π. To load the pulse: Main menu, Create Values, Continue, Create X, Real, Even, 1, Continue, COS(t/2)/(pi^2 - t^2), Go, Neither, Go. You are now back on the main menu with x(t) correctly loaded. Plotting **X** confirms that. Observe the max value of $1/\pi^2$.

<u>WARNING</u>: It is impossible to evaluate the indeterminacies at t = ±π by setting t = ±π. Remember: At that value both numerator and denominator are theoretically equal to zero, and so a division by zero should be reported by the FFT system. If it were, we could then deal with it manually and specify the correct value. However most machines evaluate both the numerator and the denominator and come up with **very small but nonzero answers**, and so a division by zero is not reported. Instead they then compute some erroneous value, either close or badly off.

Thus: Examine the values in **X** at k = ±15, and be sure that they are equal to 7.9577471e-2. If not, edit them as follows: Main menu, Edit from keyboard, Continue, Change X vectors, X, Real, Specify, 15, 1/(4*pi), Specify, -15, 1/(4*pi), Skip, Continue, Go. You are now back on the main menu with those two values correctly entered.

Run Postprocessors, X, Copy X to X2, Quit. A copy of the correctly-entered pulse is now saved in **X2**.

Run ANALYSIS and plot **FRE**. We see the band-limited spectrum $X(\omega) = \cos(\pi\omega)\ \mathrm{Rect}(\omega)$ with a maximum value very close to 1.

A Dirac delta of weight 1 shows up on the FFT as a line of height $V = N/T = 900/60\pi = 15/\pi$.

To sample at steps of $T_s = T_{max} = 2\pi$: The time-domain FFT sampling interval is $T_{FFT}/N = 60\pi/900 = \pi/15$. Then 2π is 30 of these intervals. Run Postprocessors, X, Sample X, 30, Multiply X by a constant, 15/pi, Quit. Run ANALYSIS. A plot of **FRE** shows repeated copies of the band-limited spectra just touching each other, i.e. at intervals of $\omega_s = 1$. Their peak value is very close to $T_s = 1/2\pi$.

To sample at steps of $T_s = \frac{1}{2}T_{max} = \pi$: Main menu, Run Postprocessors, Copy X2 to X, Sample X, 15, Multiply X by a constant, 15/pi, Quit. Run ANALYSIS. A plot of **FRE** shows repeated copies of the band-limited spectra spaced at intervals of $\omega_s = 2$. Their peak value is very close to $1/T_s = 1/\pi$.

To sample at steps of $T_s = 3T_{max}/2 = 3\pi$: Main menu, Run Postprocessors, Copy X2 to X, Sample X, 45, Multiply X by a constant, 15/pi, Quit. Run ANALYSIS. A plot of **FRE** shows repeated copies of the band-limited spectra now badly aliased, spaced at intervals of $\omega_s = 2/3$. Their peak value is very close to $T_s = 1/3\pi$.

9.4

(a) $$x(t) = \frac{\cos(t/2)}{\pi^2 - t^2} \Longleftrightarrow X(\omega) = \cos(\pi\omega)\ \mathrm{Rect}(\omega)$$

Sketches of x(t) and X(ω) are given in the Answer section.

(b) A sketch of the shifted Dirac comb $\delta_{TS}(t) = \sum_{n=-\infty}^{\infty} \delta(t - nT_s - T_s/2)$

is given in the Answers

$$\delta_{\Omega S}(\omega) = \omega_s \sum_{n=-\infty}^{\infty} (-1)^n\, \delta(\omega - n\omega_s)$$

(c) A sketch of the sampled version of x(t), namely $x_s(t)$ is given in the Answers. It is seen that there are only two nonzero hits, all the other impulses being nulled out by the zero crossings of the waveform. Because x(t) at t = ±π is equal to 1/4π (see previous exercise) the two nonzero impulses assume weights of 1/4π each. Thus

$$x_s(t) = \frac{1}{4\pi}\left[\ \delta(t - \pi) + \delta(t + \pi)\ \right]$$

(d) Transforming $x_s(t)$ using time shift we obtain

$$X_s(\omega) = 1/4\pi\ [\exp(-j\omega\pi) + \exp(j\omega\pi)] = 1/2\pi\ \cos(\pi\omega) \qquad \text{(A)}$$

A sketch of $X_s(\omega)$ is given in the Answers.

(e) Using frequency-domain convolution to find the Fourier transform produced by the sampling in (b), namely of $x_s(t) = x(t)\ \delta_T(t)$:

$$x_s(t) = \left[\frac{\cos(t/2)}{\pi^2 - t^2}\right]\left[\sum_{n=-\infty}^{\infty} \delta(t - nT_s - T_s/2)\right]$$

Multiplication in the time domain ⟺ convolution in the frequency domain.

$$\frac{\cos(t/2)}{\ldots} \Longleftrightarrow \cos(\pi\omega)\ \mathrm{Rect}(\omega)$$

$$\sum_{n=-\infty}^{\infty} \delta(t - nT_s - T_s/2) \Longleftrightarrow \omega_s \sum_{n=-\infty}^{\infty} (-1)^n \delta(\omega - n\omega_s) \qquad (\omega_s = 2\pi/T_s = 1)$$

Thus $$x_s(t) = \frac{1}{2\pi}\int_{-\infty}^{\infty} \cos(\pi\theta)\ \mathrm{Rect}(\theta) \left[\omega_s \sum_{n=-\infty}^{\infty} (-1)^n\ \delta(\omega - n\omega_s - \theta) \right] d\theta$$

$$= \frac{\omega_s}{2\pi}\int_{-\infty}^{\infty} \sum_{n=-\infty}^{\infty} \cos(\pi\theta)\ \mathrm{Rect}(\theta)\ (-1)^n\ \delta(\omega - n\omega_s - \theta)\ d\theta$$

$$= \frac{\omega_s}{2\pi}\int_{-\infty}^{\infty} \sum_{n=-\infty}^{\infty} \cos[\pi(\omega - n\omega_s)]\ \mathrm{Rect}[(\omega - n\omega_s)]\ (-1)^n\ \delta(\omega - n\omega_s - \theta)\ d\theta$$

$$= \frac{\omega_s}{2\pi} \sum_{n=-\infty}^{\infty} (-1)^n \cos[\pi(\omega - n\omega_s)]\ \mathrm{Rect}[(\omega - n\omega_s)] \int_{-\infty}^{\infty} \delta(\omega - n\omega_s - \theta)\ d\theta$$

$$= \frac{\omega_s}{2\pi} \sum_{n=-\infty}^{\infty} (-1)^n \cos[\pi(\omega - n\omega_s)]\ \mathrm{Rect}[(\omega - n\omega_s)] \qquad \text{where } \omega_s = 1$$

$$= \frac{1}{2\pi} \sum_{n=-\infty}^{\infty} (-1)^n \cos(\pi\omega - n\pi)\ \mathrm{Rect}(\omega - n) = \frac{1}{2\pi} \sum_{n=-\infty}^{\infty} (-1)^n \cos(\pi\omega)\cos(n\pi)\,\mathrm{Rect}(\omega - n)$$

$$= \frac{1}{2\pi}\cos(\pi\omega) \sum_{n=-\infty}^{\infty} (-1)^{2n}\ \mathrm{Rect}(\omega - n) = \frac{1}{2\pi}\cos(\pi\omega) \sum_{n=-\infty}^{\infty} \mathrm{Rect}(\omega - n) \qquad \text{(B)}$$

Sketching the final sum $\sum_{n=-\infty}^{\infty} \mathrm{Rect}(\omega - n)$ we find

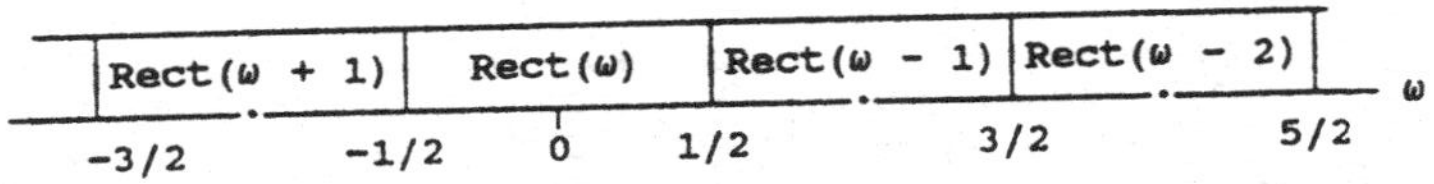

and so we continue (B), obtaining ... $= 1/2\pi \cos(\pi\omega)$ which is the same as (A).

9.5 (a) The pulse train that the student obtained is

$$f_p(t) = \sum_{n=-\infty}^{\infty} 1/\tau\ \mathrm{Rect}[(t - nT_s)/\tau]$$

A sketch of this is shown in Figure 1 below.

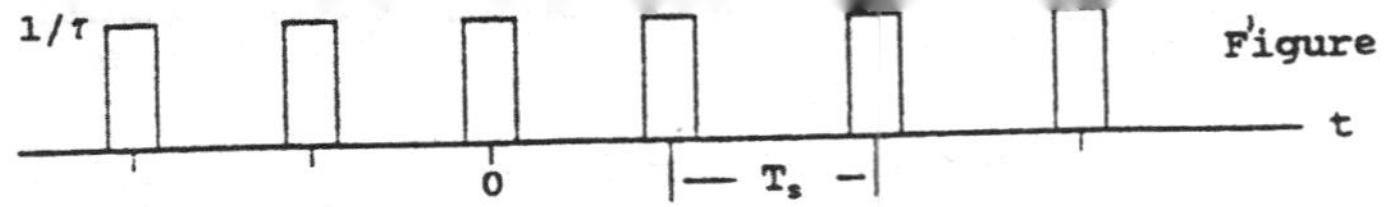

Figure 1

He then multiplied $x(t)$ by $f_p(t)$, obtaining:

$$x_s(t) = x(t)\ f_p(t) = x(t) \sum_{n=-\infty}^{\infty} 1/\tau\ \mathrm{Rect}[(t - nT_s)/\tau]$$

To see what this looks like in the frequency domain we need the Fourier series for $f_p(t)$:

(1) Transforming the central pulse in $f_p(t)$ gives

$$1/\tau\ \mathrm{Rect}(t/\tau) \Longleftrightarrow \mathrm{Sa}(\omega\tau/2)$$

(2) The period of $f_p(t)$ is T_s and so its fundamental frequency is $\omega_s = 2\pi/T_s$. Then the Fourier coefficients for $f_p(t)$ are

$$F(n) = 1/T_s\ \mathrm{Sa}(\omega\tau/2)\ \Big|_{\omega \leftarrow n\omega_s} = 1/T_s\ \mathrm{Sa}(n\omega_s\tau/2)$$

(3) The Fourier series for the pulse train is thus:

$$f_p(t) = 1/T_s \sum_{n=-\infty}^{\infty} \mathrm{Sa}(n\omega_s\tau/2)\ \exp(jn\omega_s t)$$

Then

$$x_s(t) = x(t)\ \frac{1}{T_s} \sum_{n=-\infty}^{\infty} \mathrm{Sa}(n\omega_s\tau/2)\ \exp(jn\omega_s t)$$

$$= \frac{1}{T_s} \sum_{n=-\infty}^{\infty} \mathrm{Sa}(n\omega_s\tau/2)\ x(t)\exp(jn\omega_s t)$$

We now use frequency shift to transform this, obtaining:

$$X_s(\omega) = \frac{1}{T_s} \sum_{n=-\infty}^{\infty} \mathrm{Sa}(n\omega_s\tau/2)\ X(\omega - n\omega_s)$$

This is seen to be repeated copies of the spectrum $X(\omega)$ with spacing $n\omega_s$ and with the n-th copy multiplied by $1/T_s\ \mathrm{Sa}(n\omega_s\tau/2)$. A sketch is given in Figure 2 below.

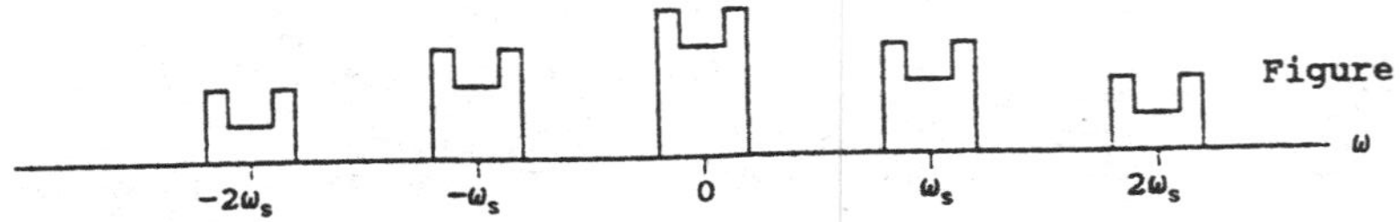

Figure 2.

For all copies to be identical we require that $Sa(n\omega_s\tau/2) = 1$ for all n, which would require that $\tau = 0$, and in that case all of the pulses in Figure 1 above would have been Dirac deltas.

(b) For $T_s = 1$ we have $\omega_s = 2\pi$. If τ = 1e-3 then:

$$Sa(n\omega_s\tau/2) = Sa(n\pi/1000)$$

For n = 78: $Sa(n\pi/1000) = 0.9900$

For n = 79: $Sa(n\pi/1000) = 0.9898$

Thus with τ = 1e-3, only the first 78 copies on each side of $\omega = 0$ are within 1% of the central one.

If τ = 71e-6 and n = 1100 then $Sa(n\pi\tau e{-6})$ = 0.989996701 and the first 1100 copies will then be within 1% of the central one.

For "pulse area = 1" their height would have to be:

$$h = 1/\tau = 1/(71e{-6}) = 14084.5$$

The envelope of the repeated spectra in Figure 2 is

$$1/T_s\ Sa(n\omega_s\tau/2) = 1/T_s\ Sa(n\pi\ \tau/T_s)$$

The smaller we make the ratio τ/T_s, the flatter will be this envelope as we increase or decrease n starting at zero. In the limit, as $\tau/T_s \longrightarrow 0$, the envelope will become eternally flat and we would then have the same spectrum as if we had used Dirac deltas. Thus the only thing that matters if we are using Rect pulses as our samplers instead of Dirac deltas is that the ratio τ/T_s be very small.

9.6 Frequency-domain impulse sampling:

(a) Loading the pulse Rect(t) with N = 1024 and T = 4: Main menu, Create Values, Continue, Create X, Real, Even, 2, 1/2, Continue, 1, 0, Go, Neither, Go. You are now back on the main menu with Rect(t) in **X**. Plotting **X** confirms that.

Run ANALYSIS and plot **FRE**. You see a single copy of $Sa(\omega/2)$ with max height 1. Run SYNTHESIS and plot **Y**. You see a single copy of Rect(t) with height 1.

(b) Main menu, Run Postprocessors, F, Sample F, 2, Multiply F by a constant, 2, Quit.

Plotting **FRE** using Zoom = 8 confirms that we have sampled F so as to retain every second value and the max value is now 2.

Run SYNTHESIS. Plot **Y**. You see repeated copies of Rect(t) with spacing 2 seconds, each with height 1.

(c) Run ANALYSIS to restore the original ...

Run Postprocessors, F, Sample F, 4, Multiply F by a constant, 4, Quit. Plotting **FRE** using Zoom = 8 confirms that we have sampled F so as to retain every fourth value and the max value is now 4. (In fact all of the values other than the central one are gone. We sampled at the zero crossings.)

Run SYNTHESIS. Plot **Y**. You see repeated copies of Rect(t) with spacing 1 seconds, each with height 1. They are now all touching each other, precisely as expected.

9.7 (a) Finding the Fourier transform of the pulse shown in Figure 9.31.

Differentiation gives $f'(t) = Rect(t - \frac{1}{2}) - \delta(t - 1)$, from which

$$f''(t) = \delta(t) - \delta(t - 1) - D\delta(t - 1)$$

$$(j\omega)^2\ F(\omega) = 1 - \exp(-j\omega) - j\omega\ \exp(-j\omega)$$

$$F(\omega) = [1 - \exp(-j\omega) - j\omega\ \exp(-j\omega)]/(j\omega)^2 \longrightarrow$$

(b) In Figure 9.32: $T_0 = 1$, $\omega_0 = 2\pi$, and so the Fourier coefficients for the periodic function shown there are

$$F(n) = 1/T_0\ F(\omega)\Big|_{\omega \leftarrow n\omega_0} = [1 - \exp(-jn2\pi) - jn2\pi\ \exp(-jn2\pi)]/(jn2\pi)^2$$

$$= j/2\pi n \qquad (n \neq 0)$$

For n = 0: F(0) = average value of waveform = ½.

Thus the Fourier series is

$$f_p(t) = \tfrac{1}{2} + j/2\pi \sum_{\substack{n=-\infty \\ n\neq 0}}^{\infty} (1/n)\ \exp(jn2\pi t)$$

(c) In Figure 9.33: $T_0 = 2$, $\omega_0 = \pi$, and so the Fourier coefficients for the periodic function shown there are

$$F(n) = 1/T_0\ F(\omega)\Big|_{\omega \leftarrow n\omega_0} = \tfrac{1}{2}[1 - \exp(-jn\pi) - jn\pi\ \exp(-jn\pi)]/(jn\pi)^2$$

$$= \tfrac{1}{2}[(-1)^n - 1 + jn\pi(-1)^n]/n^2\pi^2 \qquad (n \neq 0)$$

For n = 0: F(0) = average value of waveform = ¼.

Thus the Fourier series is

$$f_p(t) = \tfrac{1}{4} + 1/2\pi^2 \sum_{n=-\infty}^{\infty} \frac{(-1)^n - 1 + jn\pi(-1)^n}{n^2} \exp(jn\pi t)$$

9.8 (a) A sketch of $f(t) = \cos(\pi t)\,\text{Rect}(t)$ is shown in the [illegible]

(b) Finding $F(\omega)$ in four ways:

(1) By duality: From Exercise 9.3, $\dfrac{\cos(t/2)}{\pi^2 - t^2} \Longleftrightarrow \cos(\pi\omega)\,\text{Rect}(\omega)$

Then $\cos(\pi t)\,\text{Rect}(t) \Longleftrightarrow 2\pi \dfrac{\cos(\omega/2)}{\pi^2 - \omega^2}$

(2) By frequency-domain convolution:

$\cos(\pi t) \Longleftrightarrow \pi[\delta(\omega - \pi) + \delta(\omega + \pi)]$ $\qquad$ $\text{Rect}(t) \Longleftrightarrow \text{Sa}(\omega/2)$

$$\cos(\pi t)\,\text{Rect}(t) \Longleftrightarrow 1/2\pi \int_{-\infty}^{\infty} \text{Sa}(\theta/2)\, \pi[\delta(\omega - \pi - \theta) + \delta(\omega + \pi - \theta)]\, d\theta$$

$$= \tfrac{1}{2} \int_{-\infty}^{\infty} \text{Sa}[(\omega - \pi)/2]\, \delta(\omega-\pi-\theta)\, d\theta + \tfrac{1}{2} \int_{-\infty}^{\infty} \text{Sa}[(\omega + \pi)/2]\, \delta(\omega+\pi-\theta)\, d\theta$$

$$= \tfrac{1}{2}\,\text{Sa}[(\omega-\pi)/2] + \tfrac{1}{2}\,\text{Sa}[(\omega+\pi)/2] = \tfrac{1}{2} \frac{\sin[(\omega - \pi)/2]}{(\omega - \pi)/2} + \tfrac{1}{2} \frac{\sin[(\omega + \pi)/2]}{(\omega + \pi)/2}$$

$$= \frac{\cos(\omega/2)}{\pi - \omega} + \frac{\cos(\omega/2)}{\pi + \omega} = 2\pi \frac{\cos(\omega/2)}{\pi^2 - \omega^2} \longrightarrow$$

(3) By successive differentiation (using the product rule):

$$f(t) = \cos(\pi t)\,\text{Rect}(t)$$

$$f'(t) = -\pi \sin(\pi t)\,\text{Rect}(t) + \cos(\pi t)\,[\delta(t + \tfrac{1}{2}) - \delta(t - \tfrac{1}{2})]$$

$$= -\pi \sin(\pi t)\,\text{Rect}(t)$$

$$f''(t) = -\pi^2 \cos(\pi t)\,\text{Rect}(t) - \pi \sin(\pi t)\,[\delta(t + \tfrac{1}{2}) - \delta(t - \tfrac{1}{2})]$$

$$= -\pi^2 f(t) \qquad + \pi\,[\delta(t + \tfrac{1}{2}) + \delta(t - \tfrac{1}{2})]$$

$$(j\omega)^2 F(\omega) = -\pi^2 F(\omega) \qquad + \pi\,[\exp(j\omega/2) + \exp(-j\omega/2)]$$

$$(\pi^2 - \omega^2)\, F(\omega) = 2\pi \cos(\omega/2) \qquad F(\omega) = 2\pi \frac{\cos(\omega/2)}{\pi^2 - \omega^2} \longrightarrow$$

(4) By direct integration of the analysis equation:

$$\cos(\pi t)\,\text{Rect}(t) \Longleftrightarrow \int_{-\infty}^{\infty} \cos(\pi t)\,\text{Rect}(t)\,\exp(-j\omega t)\, dt$$

$$= \int_{-1/2}^{1/2} \cos(\pi t)\,[\cos(\omega t) - j \sin(\omega t)]\, dt = 2 \int_{0}^{1/2} \cos(\pi t) \cos(\omega t)\, dt$$

$$= \int_{0}^{1/2} \{\cos[(\pi+\omega)t] + \cos[(\pi-\omega)t]\}\, dt = \left. \frac{\sin[(\pi+\omega)t]}{\pi + \omega} + \frac{\sin[(\pi-\omega)t]}{\pi - \omega} \right|_{0}^{1/2}$$

$$F(\omega) = 2\pi \frac{\cos(\omega/2)}{\pi^2 - \omega^2} \longrightarrow$$

(c) Finding the Fourier transforms of the periodic functions:

(1) In Figure 9.34: $T_0 = 1$, $\omega_0 = 2\pi$, Fourier series coefficients are

$$F_p(n) = 1/T_0\, F(\omega) \Big|_{\omega \leftarrow n\omega_0} = 2\pi \frac{\cos(n\pi)}{\pi^2 - (n2\pi)^2} = \frac{2}{\pi} \frac{(-1)^n}{1 - 4n^2}$$

$$f_1(t) = \frac{2}{\pi} \sum_{n=-\infty}^{\infty} \frac{(-1)^n}{1 - 4n^2} \exp(jn2\pi t)$$

$$F_1(\omega) = \frac{2}{\pi} \sum_{n=-\infty}^{\infty} \frac{(-1)^n}{1 - 4n^2}\, 2\pi\, \delta(\omega - n2\pi) = 4 \sum_{n=-\infty}^{\infty} \frac{(-1)^n}{1 - 4n^2}\, \delta(\omega - n2\pi) \qquad \text{(A)}$$

(2) In Figure 9.35: $T_0 = 2$, $\omega_0 = \pi$, Fourier series coefficients are

$$F_p(n) = 1/T_0\, F(\omega) \Big|_{\omega \leftarrow n\omega_0} = \pi \frac{\cos(n\pi/2)}{\pi^2 - (n\pi)^2}$$

$$f_2(t) = \frac{1}{\pi} \sum_{n=-\infty}^{\infty} \frac{\cos(n\pi/2)}{1 - n^2} \exp(jn\pi t)$$

$$F_2(\omega) = \frac{1}{\pi} \sum_{n=-\infty}^{\infty} \frac{\cos(n\pi/2)}{1 - n^2}\, 2\pi\, \delta(\omega - n\pi) = 2 \sum_{n=-\infty}^{\infty} \frac{\cos(n\pi/2)}{1 - n^2}\, \delta(\omega - n\pi) \qquad \text{(B)}$$

(d) Using the FFT system to verify the weights of the frequency-domain Dirac deltas that were obtained in (c)(1) and (c)(2) above:

Steps for loading the waveform with N = 256.

- We use T = 1 because that is the value of T_0 and we intend to use the periodicity of the FFT.
- We use PULSE because we intend to find a Fourier TRANSFORM.

Create Values, Continue, Create X, Real, Even, 1, Continue, COS(pi*t), Go, Neither, Go. You are now back on the main menu with

the waveform in **X**. Plotting **X** confirms that.

Run ANALYSIS, Show Numbers, F, Eternal. The numbers that appear are the <u>weights of the Dirac deltas</u> in the Fourier transform of an eternal pulse train made up from the pulse that has been loaded. In the following two tables we show the FFT values and the exact values for the two waveforms considered above. The exact values were computed from the formulas.

(A)

n	FFT	Exact	error (%)
0	3.9999969	4	0.0000775
1	1.3333365	1.3333333	0.0002375
2	-0.26666980	-0.2666666	0.0011749
3	0.11428885	0.1143857	0.0027437
4	-6.3495201e-2	-6.3492063e-2	0.0049416

(B) Steps for loading the waveform with N = 256. Now use T = 2.

Create Values, Continue, Create X, Real, Even, 2, .5, Continue, COS(pi*t), 0, Go, Neither, Go. You are now back on the main menu with the wave form in **X**. Plotting **X** confirms that.

n	FFT	Exact	error (%)
0	1.9999937	2	0.0003150
1	1.5707963	1.5707963	0 (sic)
2	-0.66667294	-0.6666666	0.0009410
3	0	0	0
4	-0.13333961	-0.1333333	0.0047075

9.9 (a) From Exercise 9.8(c):

$$f(t) = |\cos(\pi t)| \iff 4\sum_{n=-\infty}^{\infty} \frac{(-1)^n}{1-4n^2}\,\delta(\omega - n2\pi) = F_0(\omega) \qquad (A)$$

(b) $$f''(t) \iff (j\omega)^2 F_0(\omega) = (j\omega)^2\, 4\sum_{n=-\infty}^{\infty} \frac{(-1)^n}{1-4n^2}\,\delta(\omega - n2\pi)$$

$$= 4\sum_{n=-\infty}^{\infty} \frac{(-1)^n}{1-4n^2}(j\omega)^2\,\delta(\omega - n2\pi) = 4\sum_{n=-\infty}^{\infty} \frac{(-1)^n}{1-4n^2}(jn2\pi)^2\,\delta(\omega - n2\pi)$$

$$= 4\sum_{n=-\infty}^{\infty} \frac{(-1)^n 4n^2\pi^2}{4n^2-1}\,\delta(\omega - n2\pi) = 4\pi^2\sum_{n=-\infty}^{\infty}(-1)^n\left[1 + \frac{1}{4n^2-1}\right]\delta(\omega - n2\pi)$$

$$= 4\pi^2\sum_{n=-\infty}^{\infty}(-1)^n\,\delta(\omega - n2\pi) - 4\pi^2\sum_{n=-\infty}^{\infty}(-1)^n\frac{1}{1-4n^2}\,\delta(\omega - n2\pi)$$

$$= \qquad P(\omega) \qquad + \qquad Q(\omega) \quad = \quad F_2(\omega) \longrightarrow$$

(c) Inverting $F_2(\omega)$ to the time domain:

From (A): $Q(\omega) \iff -\pi^2|\cos(\pi t)|$

To invert $P(\omega)$ we recall from Exercise 4.20 that:

$$\delta_{TS}(t) = \sum_{n=-\infty}^{\infty}\delta(t - nT_0 - T_0/2) \iff \omega_0\sum_{n=-\infty}^{\infty}(-1)^n\,\delta(\omega - n\omega_0) \qquad (\omega_0 = 2\pi/T_0)$$

In this case $T_0 = 1$, and so $\omega_0 = 2\pi$. Then

$$P(\omega) = 4\pi^2\sum_{n=-\infty}^{\infty}(-1)^n\,\delta(\omega - n2\pi) \iff 2\pi\sum_{n=-\infty}^{\infty}\delta(t - n - \tfrac{1}{2})$$

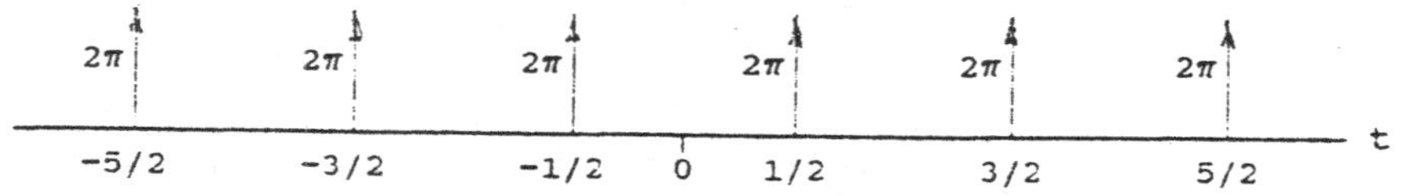

We have thus obtained:

$$F_2(\omega) \iff \left[2\pi\sum_{n=-\infty}^{\infty}\delta(t - n - \tfrac{1}{2})\right] - \pi^2|\cos(\pi t)| = f''(t) \longrightarrow$$

A sketch of $f''(t)$ is contained in the Answers.

(d) Differentiating $f(t)$ twice in the time domain: See sketch in the Answer section.

9.10 $\delta(t) \iff 1 = F(\omega)$ $\quad \delta_T(t)$ is a periodic repetition of $\delta(t)$ with period T_s. Hence by Theorem 11.3 the Fourier coefficients can be obtained from $F(\omega)$ by

$$F_p(n) = \frac{1}{T_s}F(\omega)\Big|_{\omega \leftarrow n\omega_s} \qquad \text{where } \omega_s = 2\pi/T_s$$

But $F(\omega) = 1$ and so we obtain $F_p(n) = 1/T_s$

Thus the Fourier series for $\delta_T(t)$ is

$$f_p(t) = \sum_{n=-\infty}^{\infty}F_p(n)\exp(jn\omega_s t) = \frac{1}{T_s}\sum_{n=-\infty}^{\infty}\exp(jn\omega_s t) \qquad (\omega_s = 2\pi/T_s)$$

9.11 Using successive differentiation on $f(t) = e^t$ $(0 < t < 1)$ we first rewrite it as

$$f(t) = e^t\,\mathrm{Rect}(t - \tfrac{1}{2}) \qquad (\forall\, t)$$

Differentiating using the product rule then gives

$$j\omega\, F(\omega) = F(\omega) + 1 - e\,\exp(-j\omega)$$

$$(j\omega - 1)\, F(\omega) = 1 - e\,\exp(-j\omega)$$

$$F(\omega) = \frac{e\,\exp(-j\omega) - 1}{1 - j\omega}$$

By Theorem 9.13 the expression for the required Fourier coefficients will be

$$F_p(n) = \frac{1}{T_0}\, F(\omega)\Big|_{\omega \leftarrow n\omega_0} \qquad T_0 = 2,\ \omega_0 = \pi$$

$$= \frac{1}{2}\,\frac{e\,\exp(-jn\pi) - 1}{1 - jn\pi} = \frac{1}{2}\,\frac{e(-1)^n - 1}{1 - jn\pi} \longrightarrow$$

9.12 From Exercise 2.26, $T_0 = 4$, $\omega_0 = \pi/2$, and

$$F_p(n) = \begin{cases} \dfrac{j}{4n\pi}\left[\, 2(-1)^n - 1 - \exp(-jn\pi/2) \,\right] & (n \neq 0) \\ 3/8 & (n = 0) \end{cases}$$

We now use successive differentiation starting from $f(t)$

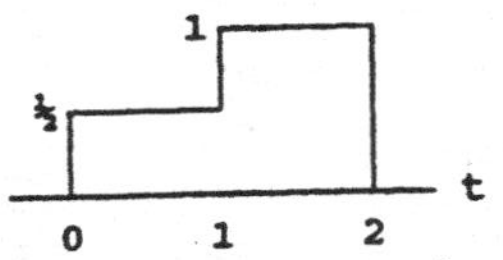

$$f'(t) = \tfrac{1}{2}\delta(t) + \tfrac{1}{2}\delta(t-1) - \delta(t-2)$$

$$j\omega\, F(\omega) = \tfrac{1}{2} + \tfrac{1}{2}\exp(-j\omega) - \exp(-j2\omega)$$

$$F(\omega) = \frac{\tfrac{1}{2} + \tfrac{1}{2}\exp(-j\omega) - \exp(-j2\omega)}{j\omega}$$

By Theorem 9.13 the expression for the required Fourier coefficients will be

$$F_p(n) = \frac{1}{T_0}\, F(\omega)\Big|_{\omega \leftarrow n\omega_0} \qquad T_0 = 4,\ \omega_0 = \pi/2$$

$$= \frac{1}{4}\,\frac{\tfrac{1}{2} + \tfrac{1}{2}\exp(-j\omega) - \exp(-j2\omega)}{j\omega}\Bigg|_{\omega \leftarrow n\pi/2}$$

$$= \frac{1}{4}\,\frac{\quad}{jn\pi/2}$$

This gives us

$$\left.\begin{aligned} F_p(n) &= \frac{j}{4n\pi}\,[2\exp(-jn\pi) - 1 - \exp(-jn\pi/2)] & (n \neq 0) \\ F_p(0) &= [\text{average value of } f_p(t)] = 3/8 & (n = 0) \end{aligned}\right] \longrightarrow$$

9.13 (a) For the gated cosine pulse shown in Figure 9.36:

$$f(t) = \cos(2\pi t/\tau)\,\mathrm{Rect}(t/\tau)$$

(1) Frequency-domain convolution:

$$\cos(2\pi t/\tau) \Longleftrightarrow \pi[\delta(\omega - 2\pi/\tau) + \delta(\omega + 2\pi/\tau)]$$

$$\mathrm{Rect}(t/\tau) \Longleftrightarrow \tau\,\mathrm{Sa}(\omega\tau/2)$$

$$F(\omega) = 1/2\pi \int_{-\infty}^{\infty} \tau\,\mathrm{Sa}(\theta\tau/2)\,\pi[\delta(\omega - 2\pi/\tau - \theta) + \delta(\omega + 2\pi/\tau - \theta)]\, d\theta$$

$$= \tau/2 \int_{-\infty}^{\infty} \mathrm{Sa}[(\omega - 2\pi/\tau)\tau/2]\,\delta(\omega - 2\pi/\tau - \theta)\, d\theta$$

$$+ \tau/2 \int_{-\infty}^{\infty} \mathrm{Sa}[(\omega + 2\pi/\tau)\tau/2]\,\delta(\omega + 2\pi/\tau - \theta)\, d\theta$$

$$= \frac{\tau}{2}\left[\, \mathrm{Sa}(\omega\tau/2 - \pi) + \mathrm{Sa}(\omega\tau/2 + \pi) \,\right] = \frac{\tau}{2}\left[\, \frac{-\sin(\omega\tau/2)}{\omega\tau/2 - \pi} + \frac{-\sin(\omega\tau/2)}{\omega\tau/2 + \pi} \,\right]$$

$$= -\frac{\tau}{2}\sin(\omega\tau/2)\,\frac{\omega\tau}{(\omega\tau/2)^2 - \pi^2} = \frac{2\omega\sin(\omega\tau/2)}{4(\pi/\tau)^2 - \omega^2} \longrightarrow$$

(2) By successive differentiation using the product rule:

$$f(t) = \cos(2\pi t/\tau)\,\mathrm{Rect}(t/\tau)$$

$$f'(t) = -2\pi/\tau\,\sin(2\pi t/\tau)\,\mathrm{Rect}(t/\tau) + \cos(2\pi t/\tau)\,[\delta(t+\tau/2) - \delta(t-\tau/2)]$$

$$= -2\pi/\tau\,\sin(2\pi t/\tau)\,\mathrm{Rect}(t/\tau) - [\delta(t + \tau/2) - \delta(t - \tau/2)]$$

$$f''(t) = -(2\pi/\tau)^2\cos(2\pi t/\tau)\,\mathrm{Rect}(t/\tau) - [D\delta(t + \tau/2) - D\delta(t - \tau/2)]$$

$$-2\pi/\tau\,\sin(2\pi t/\tau)\,[\delta(t + \tau/2) - \delta(t - \tau/2)]$$

$$= -(2\pi/\tau)^2 f(t) - [D\delta(t + \tau/2) - D\delta(t - \tau/2)]$$

$$(j\omega)^2 F(\omega) = -(2\pi/\tau)^2 F(\omega) - j\omega\,[\exp(j\omega\tau/2) - \exp(-j\omega\tau/2)]$$

$$[4(\pi/\tau)^2 - \omega^2]\,F(\omega) = -j\omega\, 2j \sin(\omega\tau/2)$$

$$F(\omega) = \frac{2\omega \sin(\omega\tau/2)}{[4(\pi/\tau)^2 - \omega^2]} \longrightarrow$$

(b) By Theorem 9.3:

$$F_p(n) = \frac{1}{T_0} F(\omega)\Big|_{\omega \leftarrow n\omega_0} = \frac{1}{T_0}\frac{2\omega \sin(\omega\tau/2)}{[4(\pi/\tau)^2 - \omega^2]}\Big|_{\omega \leftarrow n2\pi/T_0}$$

$$= \frac{1}{T_0}\frac{2(n2\pi/T_0)\sin[(n2\pi/T_0)(\tau/2)]}{[4(\pi/\tau)^2 - (n2\pi/T_0)^2]} = n\pi(\tau/T_0)^2 \frac{\sin(n\pi\tau/T_0)}{\pi^2[1 - (n\tau/T_0)^2]} \longrightarrow$$

(c)

$$F_p(n) = \lim_{T_0 \to \tau}\left[n\pi(\tau/T_0)^2 \frac{\sin(n\pi\tau/T_0)}{\pi^2[1 - (n\tau/T_0)^2]}\right] = n\pi \frac{\sin(n\pi)}{\pi^2(1 - n^2)}$$

$$= 0 \qquad (n \neq \pm 1)$$

For n = 1: $F_p(1) = \lim_{n \to 1}\left[n\pi \frac{\sin(n\pi)}{\pi^2(1 - n^2)}\right] = \lim_{n \to 1}\left[\frac{\sin(n\pi)}{\pi(1 - n^2)}\right]$

$$= \lim_{n \to 1}\left[\frac{\pi \cos(n\pi)}{\pi(-2n)}\right] = 1/2 \longrightarrow$$

Similarly $F_p(-1) = 1/2$ and so the Fourier series becomes:

$$f_p(t) = \sum_{n=-\infty}^{\infty} F_p(n) \exp(jn\omega_0 t) = \tfrac{1}{2}[\exp(j\omega_0 t) + \exp(-j\omega_0 t)]$$

(d) $f_p(t) = \cos(\omega_0 t) \longrightarrow$

(e) For a sketch of the train of pulses as they would appear when $T_0 = \tau$ see the Answer section. The sketch agrees with what was obtained in (d) ?

9.14

(a) Sampling $X(\omega)$ with spacing $\omega_0 = 4\pi/3$ in the frequency domain means replicating in the time domain with period $T_0 = 2\pi/\omega_0 = 3/2$. Inverting $X_s(t)$ to the time domain:

$$X_s(\omega) \Longleftrightarrow \sum_{n=-\infty}^{\infty} \Lambda(t - n3/2)$$

(b) Find the Fourier series of the waveform in the sketch:

$$x(t) = \Lambda(t) \Longleftrightarrow X(\omega) = Sa^2(\omega/2)$$

$$X_p(n) = \frac{1}{T_0} X(\omega)\Big|_{\omega \leftarrow n\omega_0} = 2/3\, Sa^2(n2\pi/3)$$

$$x_p(t) = \sum_{n=-\infty}^{\infty} X_p(n) \exp(jn\omega_0 t) = 2/3 \sum_{n=-\infty}^{\infty} Sa^2(n2\pi/3) \exp(jn4\pi t/3)$$

Checking the value of $X_p(0)$, i.e. the average value of $x_p(t)$:

(area in one period) = ½ + ¼ + ¼ = 1, period = 3/2, average = 2/3 ✓

9.15

(a) A sketch the pulse $f(t) = \exp(-\beta t)\,U(t)$ where $\beta = 1$ is given in the Answers.

(b) A sketch of the periodic function $f_p(t) = \sum_{n=-\infty}^{\infty} f(t - nT)$ (T = 1) is given in the Answers.

(c) Using Theorem 9.3 to find the Fourier series for $f_p(t)$:

$$f(t) = \exp(-t)\,U(t) \Longleftrightarrow F(\omega) = 1/(j\omega + 1)$$

$$F_p(n) = 1/T\, F(\omega)\Big|_{\omega \leftarrow n\omega_0} = F(\omega)\Big|_{\omega = n2\pi} = 1/(jn2\pi + 1)$$

$$f_p(t) = \sum_{n=-\infty}^{\infty} F_p(t) \exp(jn\omega_0 t) = \sum_{n=-\infty}^{\infty} \frac{1}{jn2\pi + 1}\exp(jn2\pi t)$$

(d) Finding the average value and the peak value of $f_p(t)$:

Average value = $F_p(0) = 1$

Peak value (see the sketch in the Answers)

$$= 1 + e^{-1} + e^{-2} + e^{-3} + e^{-4} + \ldots = \frac{1}{1 - e^{-1}} = 1.581976707$$

(e) Using the FFT system to verify the values obtained in (d): (N = 2048, T = 128) To load the pulse: Main menu, Create values, Continue, Create X, Real, Neither, Left, 1, Continue, EXP(-t), Go, Neither, Go. Plotting X confirms that the pulse has been correctly loaded. Run ANALYSIS.

To sample the spectrum so that $T_0 = 1$, we must use an impulse train in the frequency domain with $\omega_0 = 2\pi$. The FFT's sampling interval in the frequency domain is $\Omega = 2\pi/T = 2\pi/128$. Thus ω_0 is 128 of these steps, and so we must sample so as to retain every 128-th value.

According to (9.29), the frequency-domain Dirac deltas in the sampling train must have weights ω_0. A frequency-domain Dirac delta of weight μ on the FFT has height $V = \mu T/2\pi$, and so for $\mu = \omega_0 = 2\pi$, $V = 2\pi T/2\pi = 128$. Thus, after sampling we must multiply **F** by 128.

Run Postprocessors, F, Sample F, 128, Multiply F by a constant, 128, Quit. Run SYNTHESIS and plot **Y** with Zoom = 8. The plot shows a replication of the original pulse, with period 1 second.

Examining the numbers in **F** shows that F(0) = 128.04166. This is a value from the the FFT's estimate of a **Fourier transform** and so it is the **area** of 128 replications of the original pulse. To find the average value we must then divide by T = 128, giving 1.0003254 which compares well with the theoretical value of 1 in (d).

Plotting **Y** again we see that the peak value is shown as 1.48613. From Show Numbers we see that the peak is 1.4861296. However this value is located at one FFT sampling step beyond where the true peak would have been had it not been for the half-value. Thus to move the peak back to its true location we must multiply it by exp(Ts) where Ts is the FFT's sampling interval and is equal to T/N = 128/2048 = 0.0625. Then exp(0.0625) = 1.064494459. Multiplying 1.4861296 by this number gives 1.581976724, which agrees with the peak value obtained in (d) to 8 digit accuracy!

9.16
(a) Using time-shift to write the Fourier transform of the pulse x(t) shown in Figure 9.37: Starting from Rect(t) <==> Sa(ω/2)

$$x(t) \Longleftrightarrow X(\omega) = Sa(\omega/2)\ \exp(-j\omega/2)$$

(b) Finding the Fourier series coefficients of $x_p(t)$:

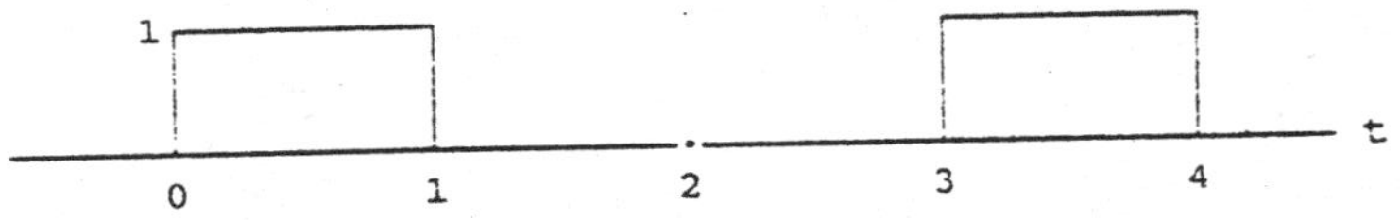

$$T_0 = 3, \quad \omega_0 = 2\pi/T_0 = 2\pi/3$$

$$X_p(n) = \frac{1}{T_0} X(\omega)\Big|_{\omega \leftarrow n\omega_0} = 1/3\ Sa(\omega/2)\ \exp(-j\omega/2)\Big|_{\omega = n2\pi/3}$$

$$= 1/3\ Sa(n\pi/3)\ \exp(-jn\pi/3) \longrightarrow$$

(c) Loading the line spectrum obtained in (b) into **F**. (N = 864, T = 3, PERIODIC): Main menu, Create Values, Continue, Create F, Complex, SIN(n*pi/3)*COS(n*pi/3)/(n*pi), Alpha = 10, Divide by zero, B, 0.3333333, Go, -SIN(n*pi/3)^2/(n*pi), Continue aliasing, Go, Go. You are now back on the main menu with the line spectrum $X_p(n)$ loaded into **F**. Run SYNTHESIS and plot **Y** with Zoom = 0.5 The plot shows a periodic repetition of x(t) with period 3.

Design project for a power supply

9.17 (a) Regarding the input voltage:

$$\omega_s = 2\pi f_0 = 2\pi 60 = 120\pi \qquad T_s = 2\pi/\omega_s = 2\pi/120\pi = 1/60$$

The input voltage is thus a cosine with period $T_s = 1/60$. Its peak value is 230√2.

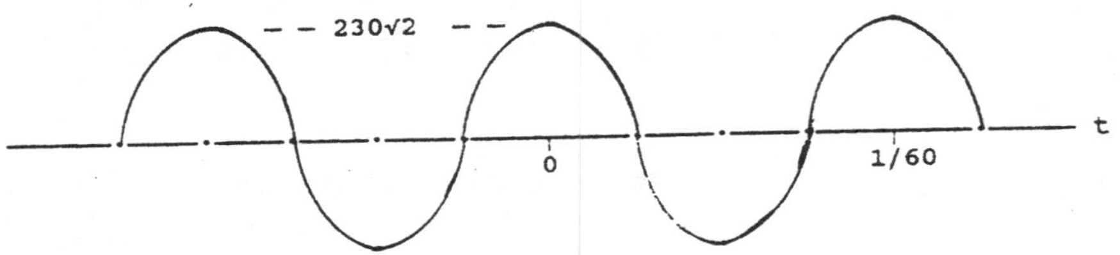

After the full-wave rectification we have the following waveform whose peak value is still 230√2 but whose period is now $T_0 = 1/120$.

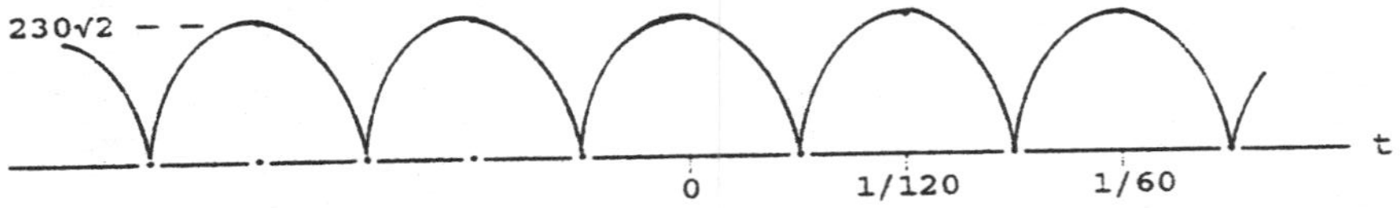

To find the Fourier series for this waveform: A single copy of the pulse making up the waveform is

$$f(t) = 230\sqrt{2}\ \cos(\pi t/T_0)\ Rect(t/T_0)$$

To find $F(\omega)$ we have $\cos(\pi t/T_0) \Longleftrightarrow \pi[\delta(\omega-\pi/T_0) + \delta(\omega+\pi/T_0)]$ and $Rect(t/T_0) \Longleftrightarrow T_0\ Sa(\omega T_0/2)$. Then, using frequency-domain convolution:

$$f(t) \Longleftrightarrow 230\sqrt{2}/2\pi \int_{-\infty}^{\infty} T_0\ Sa(\theta T_0/2)\ \pi[\delta(\omega-\pi/T_0-\theta) + \delta(\omega+\pi/T_0-\theta)]\ d\theta$$

$$= 230\ T_0/\sqrt{2}\ \{Sa[(\omega - \pi/T_0)T_0/2] + Sa[(\omega + \pi/T_0)T_0/2]\}$$

$$= 230\ T_0/\sqrt{2} \left[\frac{\sin(\omega T_0/2 - \pi/2)}{\omega T_0/2 - \pi/2} + \frac{\sin(\omega T_0/2 + \pi/2)}{\omega T_0/2 + \pi/2} \right]$$

$$= 230\ T_0/\sqrt{2} \left[\frac{-\cos(\omega T_0/2)}{\omega T_0/2 - \pi/2} + \frac{\cos(\omega T_0/2)}{\omega T_0/2 + \pi/2} \right]$$

$$= 230\ T_0/\sqrt{2}\ \cos(\omega T_0/2)\left[\frac{4\pi}{\pi^2 - (\omega T_0)^2}\right] = \frac{230\ 2\sqrt{2}\ T_0\ \pi\ \cos(\omega T_0/2)}{\pi^2 - (\omega T_0)^2}$$

Then the Fourier coefficients of the rectified waveform will be:

$$F_p(n) = 1/T_0\ F(\omega)\ \Big|_{\omega \leftarrow n\omega_0}$$

$$= \frac{230\ 2\sqrt{2}\ \pi\ \cos(n\omega_0 T_0/2)}{\pi^2 - (n\omega_0 T_0)^2} = \frac{230\ 2\sqrt{2}\ \pi\ \cos(n\pi)}{\pi^2 - (n2\pi)^2} = \frac{460\sqrt{2}}{\pi}\left[\frac{(-1)^n}{1 - 4n^2}\right] \quad \text{(A)}$$

FFT verification: Assuming that $N_1 = N_2$, the voltage at the output from the diode bridge will be $230\sqrt{2}\ |\cos(\pi t/T_0)|$ where $T_0 = 1/120$. Using N = 256, T = 1/120, PERIODIC, we load this waveform into **X** as follows:

Main menu, Create Values, Continue, Create X, Real, Even, 1, Continue, 230*SQR(2)*COS(pi*t/T), Go, Neither, Go. Plotting **X** confirms that the waveform has been entered correctly. Run ANALYSIS.

Show Numbers, F, Complex. We obtain the values shown below where we also show the formula values computed from (A).

n	FFT	Formula	error (%)
0	207.07015	207.0727526	0.001256853
1	69.026850	69.02425087	0.003765532
2	-13.807449	-13.80485017	0.018825485
3	5.9189638	5.916364360	0.043936442
4	-3.2894691	-3.286869089	0.079102968

We now have the following situation:

$\frac{N_2}{N_1} f_p(t) \longrightarrow$ 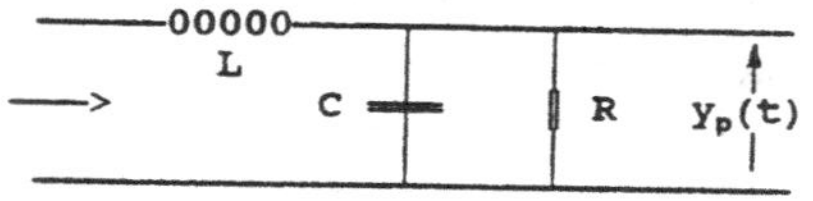

Finding $H(j\omega)$: The shunt impedance is $Z = \dfrac{1}{1/R + j\omega C} = \dfrac{R}{j\omega RC + 1}$

$$\text{Then}\quad H(j\omega) = \frac{Z}{Z + j\omega L} = \frac{\dfrac{R}{j\omega RC + 1}}{\dfrac{R}{j\omega RC + 1} + j\omega L} = \frac{R}{(j\omega)^2 RLC + j\omega L + R} \quad \text{(B)}$$

For R = 100, C = 10000e-6: RC = 1 and $H(j\omega) = \dfrac{100}{(j\omega)^2 L + j\omega L + 100}$

Then $Y_p(t) = \sum_{n=-\infty}^{\infty} H(jn\omega_0)\ \frac{N_2}{N_1} F_p(n)\ \exp(jn\omega_0 t)$ where $\omega_0 = 2\pi/T_0 = 240\pi$

$$Y_p(n) = H(jn240\pi)\ \frac{N_2}{N_1}\ F_p(n)$$

$$= \frac{N_2}{N_1}\left[\frac{100}{(jn240\pi)^2 L + jn240\pi L + 100}\right]\frac{460\sqrt{2}}{\pi}\left[\frac{(-1)^n}{1 - 4n^2}\right] \quad \text{(C)}$$

For 1000^v DC across the resistor, n = 0 must give $Y_p(n) = 1000$, and so:

$$\frac{460\sqrt{2}}{\pi}\frac{N_2}{N_1} = 1000. \quad \text{Then,}\ \frac{N_2}{N_1} = \frac{1000\pi}{460\sqrt{2}} = 4.829220587 \approx 4.829 \longrightarrow$$

Check: $\dfrac{460\sqrt{2}}{\pi} \times 4.829 = 999.954^v$ which is in error by .0046%

(b) Using this value for the turns ratio we now have:

$$Y_p(n) = \left[\frac{100000}{(jn240\pi)^2 L + jn240\pi L + 100}\right]\left[\frac{(-1)^n}{1 - 4n^2}\right] \quad \text{(D)}$$

and so, for n = 1:

$$Y_p(1) = \frac{100000/3}{100 - (240\pi)^2 L + j240\pi L} = \frac{K_1}{P_1 + jQ_1} \qquad Y_p(-1) = \frac{K_1}{P_1 - jQ_1}$$

The first harmonic is

$$H_1 = Y_p(1)\ \exp(j\omega_0 t) + Y_p(-1)\ \exp(-j\omega_0 t)$$

$$= 2\ \mathrm{Re}[Y_p(1)\ \exp(j\omega_0 t)]$$

$$= 2\ \mathrm{Re}\left[\frac{K_1}{P_1 + jQ_1}\ [\cos(\omega_0 t) + j\ \sin(\omega_0 t)]\right]$$

$$= 2\,\mathrm{Re}\left[\frac{K_1\,[P_1 - jQ_1]}{P_1^2 + Q_1^2}\,[\cos(\omega_0 t) + j\sin(\omega_0 t)]\right]$$

$$= \frac{2K_1}{P_1^2 + Q_1^2}\left[P_1\cos(\omega_0 t) + Q_1\sin(\omega_0 t)\right]$$

$$= \frac{2K_1}{\sqrt{(P_1^2 + Q_1^2)}}\cos(\omega_0 t - \theta) \qquad \text{(E)}$$

where $\cos(\theta) = P_1/\sqrt{(P_1^2 + Q_1^2)}$ and $\sin(\theta) = Q_1/\sqrt{(P_1^2 + Q_1^2)}$

Thus the peak value of the first harmonic is $2K_1/\sqrt{(P_1^2 + Q_1^2)}$ and so peak-to-peak it will be $4K_1/\sqrt{(P_1^2 + Q_1^2)}$. The specifications state that this must not exceed 0.1% of $1000^v = 1^v$. We thus have:

$$\frac{4\ (100000/3)}{\sqrt{\{[100 - (240\pi)^2 L]^2 + (240\pi L)^2\}}} = 1 \qquad \text{(F)}$$

$$[100 - (240\pi)^2 L]^2 + (240\pi L)^2 = (400000/3)^2$$

$$[(240\pi)^4 + (240\pi)^2]L^2 - 200(240\pi)^2 L + 10000 = (400000/3)^2$$

$$3.2318055\mathrm{e}11\ L^2 - 113697842.3\ L - 1.7777767\mathrm{e}10 = 0$$

$$L^2 - 0.000351809 - 0.055008777 = 0$$

$$L = \frac{0.000351809 \pm \sqrt{[(0.000351809^2 + 4(0.055008777)]}}{2}$$

$$= \frac{0.000351809 \pm 0.469079131}{2} = 0.234715470,\ -0.234363661$$

(b) Thus we must use $L = 0.234715470 \approx$ 0.235 henrys ——>

Check: Using this value for L in (F) we obtain on the LHS:

$$\left.\frac{4\ (100000/3)}{\sqrt{\{[100 - (240\pi)^2 L]^2 + (240\pi L)^2\}}}\right|_{L=0.235} = 0.998788349$$

This amounts to an error of 0.12% when compared to the required value of 1.

$$Y_p(n) = \left[\frac{100000}{(jn240\pi)^2 L + jn240\pi L + 100}\right]\left[\frac{(-1)^n}{1 - 4n^2}\right]$$

Then, proceeding as we did above for (E), the n-th harmonic is thus

$$H_n = \frac{2K}{\sqrt{(P_n^2 + Q_n^2)}}\ \frac{(-1)^n}{1 - 4n^2}\cos(n\omega_0 t - \theta)$$

where $P_n = 100 - (n240\pi)^2 L$, $Q_n = n240\pi L$, $L = 0.235$, $K = 100000$ and so the maximum value of H_n will be

$$|H_n| = \frac{200000}{\sqrt{\{[100 - (n240\pi)^2 0.235]^2 + (n240\pi 0.235)^2\}}}\ \frac{1}{4n^2 - 1}$$

From this we calculate the following:

$|H_1| = 0.499394175$ which is very close to the spec. value of 0.5

$|H_2| = 0.024955705$

$|H_3| = 0.004752974$

$|H_4| = 0.001485250$

$|H_5| = 0.000604892$

(for $|H_2|$ to $|H_5|$:) converging like $1/n^3$

<u>FFT validation</u> (N = 256, T = 1/120) From (C):

$$Y_p(n) = \left[\frac{482.9}{(jn\omega_0)^2 L + jn\omega_0 L + 100}\right]\left[\frac{460\sqrt{2}\ (-1)^n}{\pi\ (1 - 4n^2)}\right]$$

We load $H(j\omega) = \dfrac{482.9}{(j\omega)^2 L + j\omega L + 100}$ into **F2** with $L = 0.235$,

as follows: Main menu, Create values, Continue, Load H(jw), Create, 0, 2, 482.9, 100, 0.235, 0.235, Load, Alpha = 0. We are now back on the main menu. Note: We can use Alpha = 0 because $H(j\omega)$ is dying out like $1/\omega^2$. Thus little or no aliasing is needed.

Into **X** we load the full-wave rectified waveform (see earlier):

$$f(t) = 230\sqrt{2}\cos(\pi t/T_0)\ \mathrm{Rect}(t/T_0)$$

Run ANALYSIS, Run Postprocessors, F, Complex multiply F and F2. Show values, F, Real.

The following table shows the FFT valus and compares them to the theoretical values that we obtained above.

n	$|FFT|$	theoretical	error (%)
1	0.499390175	0.499394175	0.000800970
2	0.024959263	0.024955705	0.014257261
3	0.004754845	0.004752974	0.039364827
4	0.001486357	0.001485250	0.074532907
5	0.000605616	0.000604892	1.50855760

Answers:

(a) Turns ratio $N_2:N_1 = 4.829$

(b) $L = 0.235$ henrys

(c) Peak values of ripple harmonics shown above, $|H_1|$ to $|H_5|$.

10.1
(a) $19/4 = 4$ remainder 3. $|19|_4 = 3$

(b) $-35/8 = -4$ remainder -3. $|-35|_8 = -3 + 8 = 5$

(c) $59/16 = 3$ remainder 11. $|59|_{16} = 11$

(d) $-99/32 = -3$ remainder -3. $|-99|_{32} = -3 + 32 = 29$

(e) $67/25 = 2$ remainder 17. $|67|_{25} = 17$

(f) $-213/13 = -16$ remainder -5. $|-213|_{13} = -5 + 13 = 8$

(g) $214/45 = 4$ remainder 34. $|214|_{45} = 34$

(h) $-956/72 = -13$ remainder -20. $|-956|_{72} = -20 + 72 = 52$

10.2 Verifying that the following four vectors

$$\mathbf{a} = \begin{bmatrix} W^0 \\ W^0 \\ W^0 \\ W^0 \end{bmatrix} \qquad \mathbf{b} = \begin{bmatrix} W^0 \\ W^{-1} \\ W^{-2} \\ W^{-3} \end{bmatrix} \qquad \mathbf{c} = \begin{bmatrix} W^0 \\ W^{-2} \\ W^{-4} \\ W^{-6} \end{bmatrix} \qquad \mathbf{d} = \begin{bmatrix} W^0 \\ W^{-3} \\ W^{-6} \\ W^{-9} \end{bmatrix}$$

form an orthogonal set in the sense of Theorem 10.1, namely:

$$\sum_{k=0}^{N-1} \exp(j2\pi qk/N)\,\exp(j2\pi rk/N)^* = \begin{cases} 0 & \text{if } q \neq r \\ N & \text{if } q = r \end{cases} \qquad \text{(A)}$$

We showed in the text that (A) is simply the <u>inner product</u> of any of the above two vectors.

Since $W = \exp(-j2\pi/N)$ and here $N = 4$, we see that $W = -j$, and so:

$W^0 = (-j)^0 = 1$ $\qquad W^1 = (-j)^1 = -j$

$W^2 = (-j)^2 = -1$ $\qquad W^3 = (-j)^3 = j$

and so on. Thus W raised to succesive integral powers steps us around the unit circle in a counter-clockwise manner, giving

0 $\quad -j \quad -1 \quad j \quad$ 0

and so on. Forming all possible inner products (remember to conjugate the second vector!) we obtain:

$$= W^0W^0 + W^0W^0 + W^0W^0 + W^0W^0$$
$$= (1)(1) + (1)(1) + (1)(1) + (1)(1) = 4$$

$$(\mathbf{a},\mathbf{b}) = W^0W^{0*} + W^0W^{-1*} + W^0W^{-2*} + W^0W^{-3*}$$
$$= W^0W^0 + W^0W^1 + W^0W^2 + W^0W^3$$
$$= (1)(1) + (1)(-j) + (1)(-1) + (1)(j) = 0$$

$$(\mathbf{a},\mathbf{c}) = W^0W^{0*} + W^0W^{-2*} + W^0W^{-4*} + W^0W^{-6*}$$
$$= W^0W^0 + W^0W^2 + W^0W^4 + W^0W^6$$
$$= (1)(1) + (1)(-1) + (1)(1) + (1)(-1) = 0$$

$$(\mathbf{a},\mathbf{d}) = W^0W^{0*} + W^0W^{-3*} + W^0W^{-6*} + W^0W^{-9*}$$
$$= W^0W^0 + W^0W^3 + W^0W^6 + W^0W^9$$
$$= (1)(1) + (1)(j) + (1)(-1) + (1)(-j) = 0$$

$$(\mathbf{b},\mathbf{b}) = W^0W^{0*} + W^{-1}W^{-1*} + W^{-2}W^{-2*} + W^{-3}W^{-3*}$$
$$= W^0W^0 + W^{-1}W^1 + W^{-2}W^2 + W^{-3}W^3$$
$$= 1 + 1 + 1 + 1 = 4$$

$$(\mathbf{b},\mathbf{c}) = W^0W^{0*} + W^{-1}W^{-2*} + W^{-2}W^{-4*} + W^{-3}W^{-6*}$$
$$= W^0W^0 + W^{-1}W^2 + W^{-2}W^4 + W^{-3}W^6$$
$$= 1 + W^1 + W^2 + W^3$$
$$= 1 - j - 1 + j = 0$$

$$(\mathbf{b},\mathbf{d}) = W^0W^{0*} + W^{-1}W^{-3*} + W^{-2}W^{-6*} + W^{-3}W^{-9*}$$
$$= W^0W^0 + W^{-1}W^3 + W^{-2}W^6 + W^{-3}W^9$$
$$= 1 + W^2 + W^4 + W^6$$
$$= 1 - 1 + 1 - 1 = 0$$

$$(\mathbf{c},\mathbf{c}) = W^0W^{0*} + W^{-2}W^{-2*} + W^{-4}W^{-4*} + W^{-6}W^{-6*}$$
$$= W^0W^0 + W^{-2}W^2 + W^{-4}W^4 + W^{-6}W^6$$
$$= 1 + 1 + 1 + 1 = 4$$

$$(c,d) = W^0 W^{0*} + W^{-2} W^{-3*} + W^{-4} W^{-6*} + W^{-6} W^{-9*}$$
$$= W^0 W^0 + W^{-2} W^3 + W^{-4} W^6 + W^{-6} W^9$$
$$= W^0 + W^1 + W^2 + W^3$$
$$= 1 - j - 1 + j = 0$$

$$(d,d) = W^0 W^{0*} + W^{-3} W^{-3*} + W^{-6} W^{-6*} + W^{-9} W^{-9*}$$
$$= W^0 W^0 + W^{-3} W^3 + W^{-6} W^6 + W^{-9} W^9$$
$$= 1 + 1 + 1 + 1 = 4$$

10.3 (a) From Theorem 10.2:

$$F_n = \sum_{n=0}^{N-1} f_k \exp(-j2\pi nk/N) \qquad (0 \le n \le N-1)$$

For $n = 0$:

$$F_0 = \sum_{n=0}^{N-1} f_k \quad \text{and so } F_0 \text{ is real iff } f_k \text{ is real.} \longrightarrow$$

For $n = N/2$:

$$F_{N/2} = \sum_{n=0}^{N-1} f_k \exp[-j2\pi(N/2)k/N]$$

$$= \sum_{n=0}^{N-1} f_k \exp(-j\pi k) \quad = \sum_{n=0}^{N-1} (-1)^k f_k$$

and so $F_{N/2}$ is also real iff f_k is real. ⟶

(b) Proving the same proposition starting from (10.59), namely:

$F_n^* = F_{-n}$ iff f_k is real

Let $n = 0$: Then $F_0^* = F_0$ iff f_k is real, and so F_0 is real. ⟶

Let $n = N/2$: Then $F_{N/2}^* = F_{-N/2}$ iff f_k is real.

But F_n is periodic, period N, and so $F_{-N/2} = F_{N/2}$

Then $F_{N/2}^* = F_{N/2}$ iff f_k is real, and so $F_{N/2}$ is real. ⟶

10.4 (a) Sketches of the discrete functions are given in the Answers.

m	$\exp(-j2\pi m/4)$	$\exp(j2\pi m/4)$
0	1	1
1	-j	j
2	-1	-1
3	j	-j

(b) and (c) Finding the DFT's of each of these discrete functions, and then inverting the results. Sketches of the DFT's are also given in the Answers.

(1) $f_k = (1, 1, 1, 1)$ $\quad F_n = \sum_{n=0}^{N-1} f_k \exp(-j2\pi nk/N) \qquad (0 \le n \le N-1)$

$$F_n = \sum_{n=0}^{N-1} \exp(-j2\pi nk/N) \qquad (0 \le n \le N-1) \quad \text{where } N = 4.$$

$n = 0$: $\quad F_0 = \sum_{k=0}^{3} f_k = (1 + 1 + 1 + 1) = \boxed{4}$

$n = 1$: $\quad F_1 = \sum_{n=0}^{3} f_k \exp(-j2\pi k/4) = 1 - j - 1 + j = \boxed{0}$

$n = 2$: $\quad F_2 = \sum_{n=0}^{3} f_k \exp(-j4\pi k/4) = 1 - 1 + 1 - 1 = \boxed{0}$

$n = 3$: $\quad F_3 = \sum_{n=0}^{3} f_k \exp(-j6\pi k/4) = 1 + j - 1 - j = \boxed{0}$

$$F_n = (4, 0, 0, 0) \longrightarrow$$

Inverting F_n to the k-domain: $\quad f_k = \frac{1}{N}\sum_{n=0}^{N-1} F_n \exp(j2\pi nk/N) \qquad (0 \le k \le N-1)$

$k = 0$: $\quad f_0 = \frac{1}{4}\sum_{k=0}^{3} F_n = \frac{1}{4}(4 + 0 + 0 + 0) = \boxed{1}$

$k = 1$: $\quad f_1 = \frac{1}{4}\sum_{n=0}^{3} F_n \exp(j2\pi k/4) = \frac{1}{4}(4 + 0 + 0 + 0) = \boxed{1}$

$k = 2$: $\quad f_2 = \frac{1}{4}\sum_{n=0}^{3} F_n \exp(j4\pi k/4) = \frac{1}{4}(4 + 0 + 0 + 0) = \boxed{1}$

$k = 3$: $\quad f_3 = \frac{1}{4}\sum_{n=0}^{3} F_n \exp(j6\pi k/4) = \frac{1}{4}(4 + 0 + 0 + 0) = \boxed{1}$

(2) $f_k = (1, 0, 0, 0)$ $\qquad F_n = \sum_{n=0}^{N-1} f_k \exp(-j2\pi nk/N) \qquad (0 \le n \le N-1)$

$$n = 0: \quad F_0 = \sum_{k=0}^{3} f_k = 1 + 0 + 0 + 0 = \boxed{1}$$

$$n = 1: \quad F_1 = \sum_{n=0}^{3} f_k \exp(-j2\pi k/4) = 1 + 0 + 0 + 0 = \boxed{1}$$

$$n = 2: \quad F_2 = \sum_{n=0}^{3} f_k \exp(-j4\pi k/4) = 1 + 0 + 0 + 0 = \boxed{1}$$

$$n = 3: \quad F_3 = \sum_{n=0}^{3} f_k \exp(-j6\pi k/4) = 1 + 0 + 0 + 0 = \boxed{1}$$

$F_n = (1, 1, 1, 1)$ ⟶

Inverting F_n to the k-domain: $\quad f_k = \frac{1}{N} \sum_{n=0}^{N-1} F_n \exp(j2\pi nk/N) \qquad (0 \le k \le N-1)$

$$k = 0: \quad f_0 = \tfrac{1}{4} \sum_{k=0}^{3} F_n = \tfrac{1}{4}(1 + 1 + 1 + 1) = \boxed{1}$$

$$k = 1: \quad f_1 = \tfrac{1}{4} \sum_{n=0}^{3} F_n \exp(j2\pi k/4) = \tfrac{1}{4}(1 + j - 1 - j) = \boxed{0}$$

$$k = 2: \quad f_2 = \tfrac{1}{4} \sum_{n=0}^{3} F_n \exp(j4\pi k/4) = \tfrac{1}{4}(1 - 1 + 1 - 1) = \boxed{0}$$

$$k = 3: \quad f_3 = \tfrac{1}{4} \sum_{n=0}^{3} F_n \exp(j6\pi k/4) = \tfrac{1}{4}(1 - j - 1 + j) = \boxed{0}$$

(3) $f_k = (1, -1, 1, -1)$ $\qquad F_n = \sum_{n=0}^{N-1} f_k \exp(-j2\pi nk/N) \qquad (0 \le n \le N-1)$

$$n = 0: \quad F_0 = \sum_{k=0}^{3} f_k = 1 - 1 + 1 - 1 = \boxed{0}$$

$$n = 1: \quad F_1 = \sum_{n=0}^{3} f_k \exp(-j2\pi k/4) = 1 + j - 1 - j = \boxed{0}$$

$$n = 2: \quad F_2 = \sum_{n=0}^{3} f_k \exp(-j4\pi k/4) = 1 + 1 + 1 + 1 = \boxed{4}$$

$$n = 3: \quad F_3 = \sum_{n=0}^{3} f_k \exp(-j6\pi k/4) = 1 - j - 1 + j = \boxed{0}$$

$F_n = (0, 0, 4, 0)$ ⟶

Inverting F_n to the k-domain: $\quad f_k = \frac{1}{N} \sum_{n=0}^{N-1} F_n \exp(j2\pi nk/N) \qquad (0 \le k \le N-1)$

$$k = 0: \quad f_0 = \tfrac{1}{4} \sum_{k=0}^{3} F_n = \tfrac{1}{4}(0 + 0 + 4 + 0) = \boxed{1}$$

$$k = 1: \quad f_1 = \tfrac{1}{4} \sum_{n=0}^{3} F_n \exp(j2\pi k/4) = \tfrac{1}{4}(0 + 0 - 4 - 0) = \boxed{-1}$$

$$k = 2: \quad f_2 = \tfrac{1}{4} \sum_{n=0}^{3} F_n \exp(j4\pi k/4) = \tfrac{1}{4}(0 + 1 + 4 + 0) = \boxed{1}$$

$$k = 3: \quad f_3 = \tfrac{1}{4} \sum_{n=0}^{3} F_n \exp(j6\pi k/4) = \tfrac{1}{4}(0 + 0 - 4 + 4) = \boxed{-1}$$

(4) $f_k = (1, j, 1, j)$ $\qquad F_n = \sum_{n=0}^{N-1} f_k \exp(-j2\pi nk/N) \qquad (0 \le n \le N-1)$

$$n = 0: \quad F_0 = \sum_{k=0}^{3} f_k = 1 + j + 1 + j = \boxed{2 + 2j}$$

$$n = 1: \quad F_1 = \sum_{n=0}^{3} f_k \exp(-j2\pi k/4) = 1 + j(-j) - 1 + j(j) = \boxed{0}$$

$$n = 2: \quad F_2 = \sum_{n=0}^{3} f_k \exp(-j4\pi k/4) = 1 + j(-1) + 1 + j(-1) = \boxed{2 - 2j}$$

$$n = 3: \quad F_3 = \sum_{n=0}^{3} f_k \exp(-j6\pi k/4) = 1 + j(j) - 1 + j(-j) = \boxed{0}$$

$F_n = (2+2j, \ 0, \ 2-2j, \ 0)$ ⟶

Inverting F_n to the k-domain: $\quad f_k = \frac{1}{N} \sum_{n=0}^{N-1} F_n \exp(j2\pi nk/N) \qquad (0 \le k \le N-1)$

$$k = 0: \quad f_0 = \tfrac{1}{4} \sum_{k=0}^{3} F_n = \tfrac{1}{4}(2+2j + 0 + 2-2j + 0) = \boxed{1}$$

$k = 1: \quad f_1 = \frac{1}{4}\sum_{n=0}^{3} F_n \exp(j2\pi k/4) = \frac{1}{4}(2+2j + 0 + (2-2j)(-1) + 0) = \boxed{j}$

$k = 2: \quad f_2 = \frac{1}{4}\sum_{n=0}^{3} F_n \exp(j4\pi k/4) = \frac{1}{4}(2+2j + 0 + 2-2j + 0) = \boxed{1}$

$k = 3: \quad f_3 = \frac{1}{4}\sum_{n=0}^{3} F_n \exp(j6\pi k/4) = \frac{1}{4}(2+2j + 0 + (2-2j)(-1) + 0) = \boxed{j}$

(5) $f_k = (1, 2, 3, 4, 5, 6) \qquad N = 6.$

$F_n = \sum_{n=0}^{N-1} f_k \exp(-j2\pi nk/N) \qquad (0 \le n \le N-1)$

m	exp(−j2πm/6)	exp(j2πm/6)
0	1	1
1	$\frac{1}{2} - j\sqrt{3}/2$	$\frac{1}{2} + j\sqrt{3}/2$
2	$-\frac{1}{2} - j\sqrt{3}/2$	$-\frac{1}{2} + j\sqrt{3}/2$
3	-1	-1
4	$-\frac{1}{2} + j\sqrt{3}/2$	$-\frac{1}{2} - j\sqrt{3}/2$
5	$\frac{1}{2} + j\sqrt{3}/2$	$\frac{1}{2} - j\sqrt{3}/2$

$n = 0: \quad F_0 = \sum_{k=0}^{5} f_k = 1 + 2 + 3 + 4 + 5 + 6 = \boxed{21}$

$n = 1: \quad F_1 = \sum_{n=0}^{3} f_k \exp(-j2\pi k/6)$

$= 1 + 2(\frac{1}{2}-j\sqrt{3}/2) + 3(-\frac{1}{2}-j\sqrt{3}/2) + 4(-1) + 5(-\frac{1}{2}+j\sqrt{3}/2) + 6(\frac{1}{2}+j\sqrt{3}/2)$

$= \boxed{-3 + j3\sqrt{3}}$

$n = 2: \quad F_1 = \sum_{n=0}^{3} f_k \exp(-j4\pi k/6)$

$= 1 + 2(-\frac{1}{2}-j\sqrt{3}/2) + 3(-\frac{1}{2}+j\sqrt{3}/2) + 4(1) + 5(-\frac{1}{2}-j\sqrt{3}/2) + 6(-\frac{1}{2}+j\sqrt{3}/2)$

$= \boxed{-3 + j\sqrt{3}}$

$n = 3: \quad F_1 = \sum_{n=0}^{3} f_k \exp(-j4\pi k/6)$

$= 1 + 2(-1) + 3(1) + 4(-1) + 5(1) + 6(-1) = \boxed{-3}$

$n = 4: \quad F_1 = \sum_{n=0}^{3} f_k \exp(-j8\pi k/6)$

$= 1 + 2(-\frac{1}{2}+j\sqrt{3}/2) + 3(-\frac{1}{2}-j\sqrt{3}/2) + 4(1) + 5(-\frac{1}{2}+j\sqrt{3}/2) + 6(-\frac{1}{2}-j\sqrt{3}/2)$

$= \boxed{-3 - j\sqrt{3}}$

$n = 5: \quad F_1 = \sum_{n=0}^{3} f_k \exp(-j10\pi k/6)$

$= 1 + 2(\frac{1}{2}+j\sqrt{3}/2) + 3(-\frac{1}{2}+j\sqrt{3}/2) + 4(-1) + 5(-\frac{1}{2}-j\sqrt{3}/2) + 6(\frac{1}{2}-j\sqrt{3}/2)$

$= \boxed{-3 - j3\sqrt{3}}$

$F_n = (21, \ -3+j3\sqrt{3}, \ -3+j\sqrt{3}, \ -3, \ -3-j\sqrt{3}, \ -3-j3\sqrt{3}) \longrightarrow$

Inverting F_n to the k-domain: $f_k = \sum_{n=0}^{N-1} F_n \exp(-j2\pi nk/N) \qquad (0 \le n \le N-1)$

$k = 0: \quad f_0 = \frac{1}{6}\sum_{k=0}^{5} F_n$

$= \frac{1}{6}\left[21 - 3 + j3\sqrt{3} - 3 + j\sqrt{3} - 3 - 3 - j\sqrt{3} - 3 - j3\sqrt{3} \right] = \boxed{1}$

$k = 1: \quad f_1 = \frac{1}{6}\sum_{n=0}^{3} F_k \exp(j2\pi k/6)$

$= \frac{1}{6}\left[21 + (-3+j3\sqrt{3})(\frac{1}{2}+j\sqrt{3}/2) + (-3+j\sqrt{3})(-\frac{1}{2}+j\sqrt{3}/2) + (-3)(-1) + (-3-j\sqrt{3})(-\frac{1}{2}-j\sqrt{3}/2) + (-3-j3\sqrt{3})(\frac{1}{2}-j\sqrt{3}/2) \right]$

$= \frac{1}{6}\left[21 + (-3/2-9/2) + (3/2-3/2) + 3 + (3/2-3/2) + (-3/2-9/2) \right] + \frac{j\sqrt{3}}{6}\left[(3/2-3/2) + (-1/2-3/2) + (1/2+3/2) + (-3/2+3/2) \right] = \boxed{2}$

$n = 2: \quad f_2 = \frac{1}{6}\sum_{n=0}^{3} F_k \exp(j4\pi k/6)$

$$= \frac{1}{6}\Big[\, 21 + (-3+j3\sqrt{3})(-\tfrac{1}{2}+j\sqrt{3}/2) + (-3+j\sqrt{3})(-\tfrac{1}{2}-j\sqrt{3}/2) + (-3)(1) + (-3-j\sqrt{3})(-\tfrac{1}{2}+j\sqrt{3}/2) + (-3-j3\sqrt{3})(-\tfrac{1}{2}-j\sqrt{3}/2) \Big]$$

$$= \frac{1}{6}\Big[\, 21 + (3/2-9/2) + (3/2+3/2) - 3 + (3/2+3/2) + (3/2-9/2) \Big] + \frac{j\sqrt{3}}{6}\Big[(-3/2-3/2) + (-1/2+3/2) + (1/2-3/2) + (3/2+3/2) \Big] = \boxed{3}$$

$$n = 3: \quad f_3 = \frac{1}{6}\sum_{n=0}^{3} F_k \exp(j6\pi k/6)$$

$$= \frac{1}{6}\Big[\, 21 + (-3+j3\sqrt{3})(-1) + (-3+j\sqrt{3})(1) + (-3)(-1) + (-3-j\sqrt{3})(1) + (-3-j3\sqrt{3})(-1) \Big]$$

$$= \frac{1}{6}\Big[\, 21 + 3 - 3 + 3 - 3 + 3 \Big] + \frac{j\sqrt{3}}{6}\Big[-3 + 1 + -1 + 3 \Big] = \boxed{4}$$

$$n = 4: \quad f_4 = \frac{1}{6}\sum_{n=0}^{3} F_k \exp(j8\pi k/6)$$

$$= \frac{1}{6}\Big[\, 21 + (-3+j3\sqrt{3})(-\tfrac{1}{2}-j\sqrt{3}/2) + (-3+j\sqrt{3})(-\tfrac{1}{2}+j\sqrt{3}/2) + (-3)(1) + (-3-j\sqrt{3})(-\tfrac{1}{2}-j\sqrt{3}/2) + (-3-j3\sqrt{3})(-\tfrac{1}{2}+j\sqrt{3}/2) \Big]$$

$$= \frac{1}{6}\Big[\, 21 + (3/2+9/2) + (3/2-3/2) - 3 + (3/2-3/2) + (3/2+9/2) \Big] + \frac{j\sqrt{3}}{6}\Big[(-3/2+3/2) + (-1/2-3/2) + (1/2+3/2) + (3/2-3/2) \Big] = \boxed{5}$$

$$n = 5: \quad f_4 = \frac{1}{6}\sum_{n=0}^{3} F_k \exp(j10\pi k/6)$$

$$= \frac{1}{6}\Big[\, 21 + (-3+j3\sqrt{3})(\tfrac{1}{2}-j\sqrt{3}/2) + (-3+j\sqrt{3})(-\tfrac{1}{2}-j\sqrt{3}/2) + (-3)(-1) + (-3-j\sqrt{3})(-\tfrac{1}{2}+j\sqrt{3}/2) + (-3-j3\sqrt{3})(\tfrac{1}{2}+j\sqrt{3}/2) \Big]$$

$$= \frac{1}{6}\Big[\, 21 + (-3/2+9/2) + (3/2+3/2) + 3 + (3/2+3/2) + (-3/2+9/2) \Big] + \frac{j\sqrt{3}}{6}\Big[(3/2+3/2) + (-1/2+3/2) + (1/2-3/2) + (-3/2-3/2) \Big] = \boxed{6}$$

problems are as follows. (N ... N = 6.)

(1) Main menu, Create Values, Continue, Create X, Real, 1 interval, Continue, 1, Go, Go. You are now back on the main menu. Plotting **X** confirms that the function has been loaded correctly. Run ANALYSIS. Show Numbers, F. The values are identical to what we obtained above.

(2) Main menu Create Values, Continue, Create X, Real, 2 intervals, 0, 1, Continue, 1, 0, Go, Go. You are now back on the main menu. Plotting **X** confirms that the function has been loaded correctly. Run ANALYSIS. Show Numbers, F. The values are identical to what we obtained above.

(3) Main menu, Create Values, Continue, Create X, Real, 4 intervals, 0, 1, 2, 3, Continue, 1, -1, 1, -1, Go, Go. You are now back on the main menu. Plotting **X** confirms that the function has been loaded correctly. Run ANALYSIS. Show Numbers, F. The values are identical to what we obtained above.

(4) Main menu, Create Values, Continue, Create X, COmplex, 4 intervals, 0, 1, 2, 3, Continue, 1, 0, 1, 0, Go, 0, 1, 0, 1, Go, Go. You are now back on the main menu. Plotting **XRE** and **XIM** confirms that the function has been loaded correctly. Run ANALYSIS. Show Numbers, F. The values are identical to what we obtained above.

(5) Main menu, Create Values, Continue, Create X, Real, 1 interval, Continue, 1 + k, Go, Go. You are now back on the main menu. Plotting **X** confirms that the function has been loaded correctly. Run ANALYSIS. Show Numbers, F. The values are identical to what we obtained above.

10.5 Sketches of all of the functions appear in the Answer section.

(a) $f_k = 1 \quad (0 \le k \le N/2 - 1) \quad$ (pulse) $\qquad$ For $n \ne 0$:

$$F_n = \sum_{k=0}^{N-1} f_k \exp(-j2\pi nk/N) = \sum_{k=0}^{N/2-1} \exp(-j2\pi nk/N) \qquad \text{Let } \exp(-j2\pi/N) = W$$

$$= \sum_{k=0}^{N/2-1} W^{nk} = \frac{1 - (W^n)^{N/2}}{1 - W^n} = \frac{1 - \exp[(-j2\pi n/N)N/2]}{1 - \exp(-j2\pi n/N)}$$

$$= \frac{1 - \exp(-j\pi n)}{1 - \exp(-j2\pi n/N)} = \frac{1 - (-1)^n}{1 - \exp(-j2\pi n/N)} \quad (n \ne 0) \qquad \text{(A)}$$

$$F_0 = \sum_{k=0}^{N/2-1} 1 = N/2 \qquad (n = 0)$$

In order to verify this result on the FFT we will need the real and imaginary parts of F_n. From (A):

$$F_n = \frac{[1 - (-1)^n][1 - \cos(2\pi n/N) - j\sin(2\pi n/N)]}{[1 - \cos(2\pi n/N)]^2 + \sin^2(2\pi n/N)}$$

$$= \frac{[1 - (-1)^n][1 - \cos(2\pi n/N)]}{2[1 - \cos(2\pi n/N)]} - j\frac{[1 - (-1)^n]\sin(2\pi n/N)]}{2[1 - \cos(2\pi n/N)]}$$

$$A_n = \tfrac{1}{2}[1 - (-1)^n] \longrightarrow \qquad B_n = \frac{[(-1)^n - 1]\sin(2\pi n/N)]}{2[1 - \cos(2\pi n/N)]} \longrightarrow$$

FFT validation (N = 16, DISCRETE): Main menu, Create Values, Continue, Create X, Real, 2 intervals, 7, Continue, 1, 0, Go, Go. You are now back on the main menu. Plotting X confirms that f_k has been correctly loaded. Run ANALYSIS, Show Numbers, F.

n	$A_{n\ FFT}$	$A_{n\ formula}$	$B_{n\ FFT}$	$B_{n\ formula}$
0	8	8	0	0
1	1	1	-5.0273395	-5.0273395
2	0	0	0	0
3	1	1	-1.4966058	-1.4966058
4	0	0	0	0
5	1	1	-0.66817864	-0.66817864

(b) $$f_k = \begin{cases} 1 & (0 \le k \le N-2) \\ 0 & (k = N-1) \end{cases} \quad \text{(pulse)} \qquad \text{For } n \ne 0:$$

$$F_n = \sum_{k=0}^{N-1} f_k \exp(-j2\pi nk/N = \sum_{k=0}^{N-2} \exp(-j2\pi nk/N) \qquad \text{Let } \exp(-j2\pi/N) = W$$

$$= \sum_{k=0}^{N-2} W^{nk} = \frac{1 - (W^n)^{N-1}}{1 - W^n} = \frac{1 - \exp[-j2\pi n(N-1)/N]}{1 - \exp(-j2\pi n/N)} \quad (n \ne 0) \quad (A)$$

$$F_0 = \sum_{k=0}^{N-2} 1 = N - 1 \qquad (n = 0) \qquad \longrightarrow$$

In order to verify this result on the FFT we will need the real and imaginary parts of F_n. From (A):

$$F_n = \frac{\{[1-\cos[2\pi n(N-1)/N]+j\sin[2\pi n(N-1)/N]\}\{1-\cos(2\pi n/N)-j\sin(2\pi n/N)\}}{[1 - \cos(2\pi n/N)]^2 + \sin^2(2\pi n/N)}$$

$$A_n = \frac{[1-\cos[2\pi n(N-1)/N][1-\cos(2\pi n/N)] + [\sin[2\pi n(N-1)/N][\sin(2\pi n/N)]}{2[1 - \cos(2\pi n/N)]} \longrightarrow$$

$$B_n = \frac{[\sin[2\pi n(N-1)/N][1-\cos(2\pi n/N)] - [1-\cos[2\pi n(N-1)/N][\sin(2\pi n/N)]}{2[1 - \cos(2\pi n/N)]} \longrightarrow$$

FFT validation (N = 16, DISCRETE): Main menu, Create Values, Continue, Create X, Real, 2 intervals, 14, Continue, 1, 0, Go, Go. You are now back on the main menu. Plotting X confirms that f_k has been correctly loaded. Run ANALYSIS, Show Numbers, F.

n	$A_{n\ FFT}$	$A_{n\ formula}$	$B_{n\ FFT}$	$B_{n\ formula}$
0	15	15	0	0
1	-0.92387953	-0.92387953	-0.38268343	-0.38268343
2	-0.70710678	-0.70710678	-0.70710678	-0.70710678
3	-0.38268343	-0.38268343	-0.92387953	-0.92387953

(c) $f_k = 1 \quad (0 \le k \le N/4 - 1), \quad f_{k+N} = f_k, \quad f_k$ is even

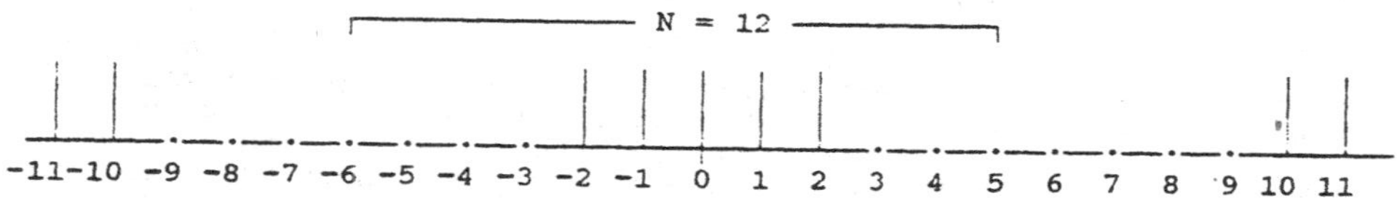

We can take the DFT over any complete period. Thus:

$$F_n = \sum_{k=-N/2}^{N/2-1} f_k \exp(-j2\pi nk/N = \sum_{k=-N/4+1}^{N/4-1} \exp(-j2\pi nk/N)$$

We now use the fact that f_k is even and we are summing over a symmetric range. Thus we continue

$$\ldots = 1 + 2\sum_{k=1}^{N/4-1} \cos(2\pi nk/N) = 1 + \sum_{k=1}^{N/4-1}\left[\exp(2\pi nk/N) + \exp(2\pi nk/N)\right]$$

$$= 1 + \sum_{k=0}^{N/4-1}\left[(W^{-n})^k + (W^n)^k\right] - 2$$

$$= \frac{1 - W^{-nN/4}}{1 - W^{-n}} + \frac{1 - W^{nN/4}}{1 - W^n} - 1 = 2\,\mathrm{Re}\left[\frac{1 - W^{nN/4}}{1 - W^n}\right] - 1$$

$$= 2\,\mathrm{Re}\left[\frac{1 - \exp(j2\pi nN/4N)}{1 - \exp(j2\pi n/N)}\right] - 1 = 2\,\mathrm{Re}\left[\frac{1 - \exp(j\pi n/2)}{1 - \exp(j2\pi n/N)}\right] - 1$$

$$= 2\,\mathrm{Re}\left[\frac{1 - \cos(\pi n/2) - j\sin(\pi n/2)}{1 - \cos(2\pi n/N) - j\sin(2\pi n/N)}\;\frac{1 - \cos(2\pi n/N) + j\sin(2\pi n/N)}{1 - \cos(2\pi n/N) + j\sin(2\pi n/N)}\right] - 1$$

$$= \left[\frac{[1 - \cos(\pi n/2)][1 - \cos(2\pi n/N)] + \sin(\pi n/2)\,\sin(2\pi n/N)}{1 - \cos(2\pi n/N)}\right] - 1$$

$(n \neq 0)$

For $n = 0$: $F_0 = \sum_{k=0}^{N-1} f_k = (N - 2)/2$ ——————>

FFT validation (N = 12, DISCRETE): Main menu, Create Values, Continue, Create X, Real, 3 intervals, 2, 9, Continue, 1, 0, 1, Go, Go. You are now back on the main menu. Plotting **X** confirms that f_k has been correctly loaded. Run ANALYSIS, Show Numbers, F.

n	$F_{n\ FFT}$	$F_{n\ formula}$
0	5	5
1	3.7320508	3.7320508
2	1.0000000	1.0000000
3	-1.0000000	-1.0000000

10.6 (a) A sketch of the function appears in the Answers.

(b) $$\sum_{k=0}^{N-1} W^k = \frac{1 - W^N}{1 - W} \quad \text{and} \quad W\frac{d}{dW}\sum_{k=0}^{N-1} W^k = W\sum_{k=0}^{N-1} kW^{k-1} = \sum_{k=0}^{N-1} kW^k$$

$$\therefore \sum_{k=0}^{N-1} kW^k = W\frac{d}{dW}\left[\frac{1 - W^N}{1 - W}\right] \qquad \text{QED}$$

(c) Finding the DFT of $f_k = k \quad (0 \le k \le N-1) \quad f_{k+N} = f_k$:

$$F_n = \sum_{k=0}^{N-1} f_k \exp(-j2\pi nk/N) = \sum_{k=0}^{N-1} k\,(W^n)^k$$

By the result derived in (b) this continues as

$$= W^n \frac{d}{dW^n}\left[\frac{1 - W^{nN}}{1 - W^n}\right] = W^n \frac{(1 - W^n)(-NW^{nN-1}) + 1 - W^{nN}}{(1 - W^n)^2}$$

$$= \frac{\quad}{(1 - W^n)^2} \qquad \frac{\quad}{(1 - W^n)^2}$$

$$= \frac{\exp(-j2\pi n/N) - N\exp(-j2\pi n) + (N - 1)\exp[-j2\pi n(N+1)/N]}{[1 - \exp(-j2\pi n/N)]^2}$$

$$= \frac{\exp(-j2\pi n/N) - N + (N - 1)\exp(-j2\pi n/N)}{[1 - \exp(-j2\pi n/N)]^2} = \frac{N\,[\exp(-j2\pi n/N) - 1]}{[1 - \exp(-j2\pi n/N)]^2}$$

$$= \frac{N}{\exp(-j2\pi n/N) - 1} = N\frac{\cos(2\pi n/N) - 1 + j\sin(2\pi n/N)}{[\cos(2\pi n/N) - 1]^2 + \sin^2(2\pi n/N)}$$

$$= N\frac{\cos(2\pi n/N) - 1 + j\sin(2\pi n/N)}{2\,[1 - \cos(2\pi n/N)]} = \frac{-N}{2} + j\frac{N\sin(2\pi n/N)}{2\,[1 - \cos(2\pi n/N)]}$$

QED.

This expression is for $n \neq 0$. For $n = 0$ we have:

$$F_0 = \sum_{k=0}^{N-1} f_k = \sum_{k=0}^{N-1} k = \tfrac{1}{2}(N - 1)N$$

(d) Verifying these results using the FFT system (N = 10, DISCRETE): Main menu, Create Values, Continue, Create X, Real, 1 interval, Continue, k, Go, Go. You are now back on the main menu. Plotting **X** confirms that f_k has been correctly loaded. Run ANALYSIS, Show Numbers, F.

n	$A_{n\ FFT}$	$A_{n\ formula}$	$B_{n\ FFT}$	$B_{n\ formula}$
0	45	45	0	0
1	-5	-5	15.388418	15.388418
2	-5	-5	6.8819096	6.8819096
3	-5	-5	3.6327126	3.6327126

10.7 (a) $f_k = \exp(\beta k) \qquad (0 \le k \le N-1) \qquad$ (pulse)

$$F_n = \sum_{k=0}^{N-1} f_k \exp(-j2\pi nk/N) = \sum_{k=0}^{N-1} \exp(\beta k)\exp(-j2\pi nk/N)$$

$$= \sum_{k=0}^{N-1} \exp[(\beta - j2\pi n/N)k] = \frac{1 - \exp[(\beta - j2\pi n/N)N]}{1 - \exp[(\beta - j2\pi n/N)]} = \frac{1 - \exp(\beta N)\exp(-j2\pi n)}{1 - \exp(\beta)\exp(-j2\pi n/N)}$$

$$= \frac{1 - \exp(\beta N)}{1 - \exp(\beta)\exp(-j2\pi n/N)}$$ ——————>

To verify this result on the FFT we will need its real and imaginary parts. Thus we continue:

$$\ldots = \frac{1 - \exp(\beta N)}{1-\exp(\beta)[\cos(2\pi n/N) - j\sin(2\pi n/N)]}\,\frac{1-\exp(\beta)[\cos(2\pi n/N)+j\sin(2\pi n/N)]}{1-\exp(\beta)[\cos(2\pi n/N)+j\sin(2\pi n/N)]}$$

$$= \frac{[1 - \exp(\beta N)]\ \{1 - \exp(\beta)[\cos(2\pi n/N) + j\ \sin(2\pi n/N)]\}}{[1 - \exp(\beta)\ \cos(2\pi n/N)]^2 + [\exp(\beta)\ \sin(2\pi n/N)]^2}$$

$$= \frac{[1 - \exp(\beta N)]\ \{1 - \exp(\beta)[\cos(2\pi n/N) + j\ \sin(2\pi n/N)]\}}{1 - 2\ \exp(\beta)\ \cos(2\pi n/N) + \exp(2\beta)}$$

$$A_n = \frac{[1 - \exp(\beta N)]\ [1 - \exp(\beta)\ \cos(2\pi n/N)]}{1 - 2\ \exp(\beta)\ \cos(2\pi n/N) + \exp(2\beta)}$$

$$B_n = \frac{[\exp(\beta N) - 1]\ \exp(\beta)\ \sin(2\pi n/N)}{1 - 2\ \exp(\beta)\ \cos(2\pi n/N) + \exp(2\beta)}$$

For $N = 20$, $\beta = 0.1$, these become:

$$A_n = \frac{[1 - \exp(2)]\ [1 - \exp(0.1)\ \cos(\pi n/10)]}{1 - 2\ \exp(0.1)\ \cos(\pi n/10) + \exp(0.2)}$$

$$B_n = \frac{[\exp(2) - 1]\ \exp(0.1)\ \sin(\pi n/10)}{1 - 2\ \exp(0.1)\ \cos(\pi n/10) + \exp(0.2)}$$

Verifying these results using the FFT system ($N = 20$, $\beta = 0.1$): Main Menu, Create Values, Continue, Create X, Real, 1 interval, Continue, EXP(0.1*k), Go, Go. You are now back on the main menu. Plotting **X** confirms that f_k has been correctly loaded. Run ANALYSIS, Show Numbers, F.

n	$A_{n\ FFT}$	$A_{n\ formula}$	$B_{n\ FFT}$	$B_{n\ formula}$
0	60.749266	60.749266	0	0
1	2.7368792	2.7368792	18.298544	18.298544
2	-1.5618422	-1.5618422	9.5807110	9.5807110
3	-2.4275794	-2.4275794	6.1944159	6.1944159

(b) $f_k = \sin(2\pi k/N) \quad (0 \le k \le N-1) \quad f_{k+N} = f_k$

$$F_n = \sum_{k=0}^{N-1} f_k \exp(-j2\pi nk/N) = \sum_{k=0}^{N-1} \sin(2\pi k/N)\ \exp(-j2\pi nk/N)$$

$$= \frac{1}{2j}\sum_{k=0}^{N-1} [\exp(j2\pi k/N) - \exp(-j2\pi k/N)]\ \exp(-j2\pi nk/N)$$

$$= \frac{1}{2j}\sum_{k=0}^{N-1} \Big[\exp[j2\pi(1 - n)k/N] - \exp[-j2\pi(1 + n)k/N] \Big]$$

$$= \frac{1}{2j}\left[\frac{1 - \exp[j2\pi(1 - n)]}{1 - \exp[j2\pi(1 - n)/N]} - \frac{1 - \exp[-j2\pi(1 + n)]}{1 - \exp[-j2\pi(1 + n)/N]}\right],$$

$= 0$ unless $n = \pm 1$

For $n = 1$:

$$\lim_{n \to 1} \frac{1 - \exp[j2\pi(1 - n)]}{1 - \exp[j2\pi(1 - n)/N]} = \lim_{n \to 1} \frac{j2\pi\ \exp[j2\pi(1 - n)]}{j2\pi/N\ \exp[j2\pi(1 - n)/N]} = N$$

For $n = -1$ we can infer that the value is $-N$ because F_n is odd. Thus:

$$F_n = \begin{cases} N/2j & (n = 1) \\ -N/2j & (n = -1) \\ 0 & (\text{otherwise}) \end{cases} \qquad (A) \qquad \longrightarrow$$

For $N = 20$ this becomes:

$$F_n = \begin{cases} 10/j & (n = 1) \\ -10/j & (n = -1) \\ 0 & (\text{otherwise}) \end{cases}$$

Verifying these results using the FFT system ($N = 20$, DISCRETE): Main menu, Create Values, Continue, Create X, Real, 1 interval, Continue, SIN(2*pi*k/N), Go, Go. You are now back on the main menu. Plotting **X** confirms that f_k has been correctly loaded. Run ANALYSIS, Show Numbers, F. The FFT values were

$$A_n = 0 \quad \text{and} \quad B_n = \begin{cases} -10 & (n = 1) \\ 10 & (n = -1) \\ 0 & (\text{otherwise}) \end{cases} \quad \text{which confirms (A).}$$

(c) $f_k = \cos(10\pi k/N) \quad (0 \le k \le N-1) \quad$ (pulse)

$$F_n = \sum_{k=0}^{N-1} f_k \exp(-j2\pi nk/N) = \sum_{k=0}^{N-1} \cos(10\pi k/N)\ \exp(-j2\pi nk/N)$$

$$= \frac{1}{2}\sum_{k=0}^{N-1} [\exp(j10\pi k/N) + \exp(-j10\pi k/N)]\ \exp(-j2\pi nk/N)$$

$$= \frac{1}{2}\sum_{k=0}^{N-1} \Big[\exp[j2\pi(5 - n)k/N] + \exp[-j2\pi(5 + n)k/N] \Big]$$

$$= \frac{1}{2}\left[\frac{1 - \exp[j2\pi(5 - n)]}{1 - \exp[j2\pi(5 - n)/N]} + \frac{1 - \exp[-j2\pi(5 + n)]}{1 - \exp[-j2\pi(5 + n)/N)]}\right]$$

$= 0$ unless $n = \pm 5$

For $n = 5$:

$$\lim_{n \to 5} \frac{1 - \exp[j2\pi(5 - n)]}{1 - \exp[j2\pi(1 - n)/N]} = \lim_{n \to 5} \frac{j2\pi \exp[j2\pi(5 - n)]}{j2\pi/N \exp[j2\pi(5 - n)/N]} = N$$

For $n = -5$ we can infer that the value is N because F_n is even. Thus:

$$F_n = \begin{cases} N/2 & (n = 1) \\ N/2 & (n = -1) \\ 0 & (\text{otherwise}) \end{cases} \qquad \text{(A)} \qquad \longrightarrow$$

For N = 20 this becomes:

$$F_n = \begin{cases} 10 & (n = 1) \\ 10 & (n = -1) \\ 0 & (\text{otherwise}) \end{cases}$$

Verifying these results using the FFT system (N = 20, DISCRETE): Main menu, Create Values, Continue, Create X, Real, 1 interval, Continue, COS(10*pi*k/N), Go, Go. You are now back on the main menu. Plotting X confirms that f_k has been correctly loaded. Run ANALYSIS, Show Numbers, F. The FFT values were

$$A_n = \begin{cases} 10 & (n = 5) \\ 10 & (n = -5) \\ 0 & (\text{otherwise}) \end{cases} \qquad \text{and} \quad B_n = 0 \quad \text{which confirms (A).}$$

10.8 (a) "$f_k \Longleftrightarrow F_n$" means "$F_n$ is the DFT of f_k", that is

$$F_n = \sum_{k=0}^{N-1} f_k \exp(-j2\pi nk/N) = \sum_{k=0}^{N-1} f_k W^{nk} \qquad [W = \exp(-j2\pi/N)]$$

Then the DFT of f_{k-m} will be

$$G_n = \sum_{k=0}^{N-1} f_{k-m} W^{nk} \qquad \text{Let } k - m = b. \text{ Then } k = b + m.$$

$$= \sum_{b=-m}^{N-1-m} f_b W^{n(b+m)}$$

In this sum, we can use <u>any complete period</u> because W is periodic with period N and the data are assumed to be extended periodically, also with period N. Thus we continue as

$$\ldots = \sum_{b=0}^{N-1} f_b W^{nb} W^{nm} = W^{nm} \sum_{k=0}^{N-1} f_k W^{nk} = W^{nm} F_n$$

Thus if $f_k \Longleftrightarrow F_n$ then $f_{k-m} \Longleftrightarrow W^{nm} F_n$ QED

That is: $f_{k-m} \Longleftrightarrow \exp(-j2\pi nm/N)\, F_n$

Compare this to $f(t-\tau) \Longleftrightarrow \exp(-j\omega\tau)\, F(\omega)$ which is its dual.

$(k = t, \quad m = \tau, \quad n = \omega)$

(b) "$f_k \Longleftrightarrow F_n$" means "$f_k$ is the IDFT of F_n", that is

$$f_k = \frac{1}{N}\sum_{n=0}^{N-1} F_n \exp(j2\pi nk/N) = \frac{1}{N}\sum_{n=0}^{N-1} F_n W^{-nk} \qquad [W = \exp(-j2\pi/N)]$$

Then the IDFT of F_{n+m} will be

$$g_k = \frac{1}{N}\sum_{n=0}^{N-1} F_{n+m} W^{-nk} \qquad \text{Let } n + m = b. \text{ Then } n = b - m.$$

$$= \frac{1}{N}\sum_{b=m}^{N-1+m} F_b W^{-(b-m)k}$$

In this sum, we can use <u>any complete period</u> because W is periodic with period N and F_n is always periodic also with period N. Thus we continue as

$$\ldots = \frac{1}{N}\sum_{b=0}^{N-1} F_b W^{-bk} W^{mk} = W^{mk} \frac{1}{N}\sum_{n=0}^{N-1} F_n W^{-nk} = W^{mk} f_k$$

Thus if $f_k \Longleftrightarrow F_n$ then $W^{mk} f_k \Longleftrightarrow F_{n+m}$ QED

That is: $\exp(-j2\pi mk/N)\, f_k \Longleftrightarrow F_{n+m}$

Compare this to $\exp(-j\omega_0 t)\, f(t) \Longleftrightarrow F(\omega + \omega_0)$ which is its dual.

$(k = t, \quad m = \omega_0, \quad n = \omega)$

(c) $$f_k = \frac{1}{N}\sum_{n=0}^{N-1} F_n \exp(j2\pi nk/N)$$

Interchanging the symbols k and n:

$$f_n = \frac{1}{N}\sum_{k=0}^{N-1} F_k \exp(j2\pi nk/N)$$

and so $f_{-n} = \frac{1}{N}\sum_{k=0}^{N-1} F_k \exp(-j2\pi nk/N)$ which states that $F_k \Longleftrightarrow N f_{-n}$

Thus, if $f_k \Longleftrightarrow F_n$ then $F_k \Longleftrightarrow N f_{-n}$ QED

10.9(a)

$$g_k = \frac{f_{k+1} - f_{k-1}}{2T_s}$$

and so, by the time-shift property (see Ex. 10.8a)

$f_{k+1} \Longleftrightarrow F_n W^{-n}$ and $f_{k-1} \Longleftrightarrow F_n W^n$ giving us

$$g_k \Longleftrightarrow \frac{F_n W^{-n} - F_n W^n}{2T_s} = F_n \frac{W^{-n} - W^n}{2T_s} \longrightarrow$$

(b) Using the fact that $W = \exp(-j2\pi/N)$ this now continues

$$\ldots = F_n \frac{\exp(j2\pi n/N) - \exp(-j2\pi n/N)}{2T_s} = F_n \frac{j \sin(2\pi n/N)}{T_s}$$

10.10 (a) $Bf_k \equiv f_{k-1}$ and so, by time-shift:

$$Bf_k = f_{k-1} \Longleftrightarrow F_n W^n$$

That is: $Bf_k \Longleftrightarrow W_n F^n$ QED

from which we see that

"Operating by B in the k-domain corresponds to multiplication by W^n in the n-domain".

Compare this to

"Operating by D in the time domain corresponds to multiplication by $j\omega$ in the frequency domain"

which is its dual.

(b) Applying the B operator twice we obtain

$$B^2 f_k = B(Bf_k) = B(f_{k-1}) = f_{k-2} \Longleftrightarrow W^{2n}F_n$$

and so, by mathematical induction:

$$B^m f_k = f_{k-m} \Longleftrightarrow W^{mn}F_n \qquad \text{QED}$$

from which we see that

"Introduction to Fourier Analysis", Instructor's Manual, Chapter 10

"Operating by B^m in the k-domain corresponds to multiplication by W^{mn} in the n-domain".

Compare this to

"Operating by D^m in the time domain corresponds to multiplication by $(j\omega)^m$ in the frequency domain"

which is its dual.

10.11 $y_k - \alpha y_{k-1} = x_k - \beta x_{k-1} \quad (\forall k)$

Rewriting this by the use of the B-operator of Exercise 10.10:

$$y_k - \alpha B y_k = x_k - \beta B x_k$$

$$[1 - \alpha B]\, y_k = [1 - \beta B]\, x_k$$

DFT-transforming, and using $Bf_k \Longleftrightarrow W^n F_n$ (from Exercise 10.10):

$$[1 - \alpha W^n]\, Y_n = [1 - \beta W^n]\, X_n$$

Consider now the difference equation

$$a_0 y_k + a_1 y_{k-1} + a_2 y_{k-2} = b_0 x_k + b_1 x_{k-1} + b_2 x_{k-2}$$

Using the B operator, this factors to

$$[a_0 + a_1 B + a_2 B^2]\, y_k = [b_0 + b_1 B + b_2 B^2]\, x_k$$

which transforms to

$$[a_0 + a_1 W^n + a_2 W^{2n}]\, y_k = [b_0 + b_1 W^n + b_2 W^{2n}]\, x_k$$

In general then, the difference equation

$$P_1(B)\, Y_n = P_2(B)\, X_n$$

transforms to

$$P_1(W^n)\, Y_n = P_2(W^n)\, X_n$$

10.12 From Exercise 10.11:

$$P_1(B)\, Y_n = P_2(B)\, X_n \Longleftrightarrow P_1(W^n)\, Y_n = P_2(W^n)\, X_n$$

from which

$$Y_n = \frac{P_2(W^n)}{P_1(W^n)} X_n$$

$$P_1(D)\ y(t) = P_2(D)\ x(t) \quad \Longleftrightarrow \quad P_1(j\omega)\ Y(\omega) = P_2(j\omega)\ X(\omega)$$

from which

$$Y(\omega) = \frac{P_2(j\omega)}{P_1(j\omega)} X(\omega)$$

which is its dual

Chapter 12 The Discrete Fourier Transform as an Estimator

12.1 Using the rectangular rule to find the area under

$y = t^2$ from $t = 0$ to $t = 1$

We divide the range of integration $b - a$ into N intervals, each of length $T_s = (b - a)/N$. In this case $b = 1$ and $a = 0$ and so $T_s = 1/N$. Then:

$$A_N \approx \sum_{k=0}^{N-1} y(t_k)\, T_s = \sum_{k=0}^{N-1} (kT_s)^2\, T_s = T_s^3 \sum_{k=0}^{N-1} k^2$$

where $T_s^3 = 1/N^3$ and $\displaystyle\sum_{k=0}^{N-1} k^2 = \frac{(N - 1)\, N\, (2N - 1)}{6}$

Thus $\displaystyle A_n = \frac{(N - 1)\, N\, (2N - 1)}{6N^3} = \frac{1}{3}(1 - 1/N)(1 - 1/2N)$

Letting $N \longrightarrow \infty$ gives $A_\infty = 1/3$ QED

12.2 Using the rectangular rule to find the area under

$y = e^t$ from $t = 0$ to $t = 1$

We divide the range of integration $b - a$ into N intervals, each of length $T_s = (b - a)/N$. In this case $b = 1$ and $a = 0$ and so $T_s = 1/N$. Then:

$$A_N \approx \sum_{k=0}^{N-1} y(t_k)\, T_s = \sum_{k=0}^{N-1} \exp(kT_s)\, T_s = T_s \sum_{k=0}^{N-1} \exp(kT_s)$$

$$= \frac{1}{N}\sum_{k=0}^{N-1} \exp(k/N) = \frac{1}{N}\sum_{k=0}^{N-1} [\exp(1/N)]^k = \frac{1}{N}\,\frac{1 - [\exp(1/N)]^N}{1 - \exp(1/N)}$$

$$= \frac{1 - e}{\dfrac{1 - \exp(1/N)}{1/N}}$$

Then $\displaystyle \lim_{N\to\infty} \frac{1 - \exp(1/N)}{1/N} = \lim_{\beta\to 0} \frac{1 - \exp(\beta)}{\beta}$ $(\beta = 1/N)$

$$= \lim_{\beta\to 0} \frac{1 - (1 + \beta + \beta^2/2! + \beta^3/3! + \dots)}{\beta}$$

$$= \lim_{\beta\to 0} \frac{-\beta - \beta^2/2! - \beta^3/3! + \dots}{\beta}$$

$$= \lim_{\beta\to 0} (-1 - \beta/2! - \beta^2/3! + \dots) = -1$$

Thus $\displaystyle A_\infty = \lim_{N\to\infty} \frac{1 - e}{N[1 - \exp(1/N)]} = \frac{1 - e}{-1} = e - 1$ QED

12.3 Applying the rectangular rule to find the Fourier coefficients of $f_p(t)$:

$$F(n) = \frac{1}{T_0}\int_0^{T_0} f_p(t)\, \exp(-jn\omega_0 t)\, dt \approx \frac{1}{T_0}\sum_{k=0}^{N-1} f_p(t_k)\, \exp(-jn\omega_0 t_k)\, T_s$$

where $\omega_0 = 2\pi/T_0$ and $t_k = kT_s$. Then $T_s = T_0/N$ and so $T_s/T_0 = 1/N$.

Also $\omega_0 t_k = (2\pi/T_0)\, kT_s = 2\pi k\, (T_s/T_0) = 2\pi k/N$ and so we continue

$$\dots = \frac{1}{N}\sum_{k=0}^{N-1} f_k \exp(-j2\pi nk/N)$$ which is $1/N$ times the DFT of f_k

Thus, as stated in Theorem 12.2:

$$\left[\begin{array}{l}\text{rectangular rule}\\ \text{applied to } f_p(t)\\ \text{for CFT spectrum}\end{array}\right] = \frac{1}{N} \times \left[\begin{array}{l}\text{FFT applied}\\ \text{to } f_p(t) \text{ for}\\ \text{FFT spectrum}\end{array}\right]$$ QED

12.4

(a) Applying the rectangular rule to find the Fourier coefficients of

$f_p(t) = 1$ $(0 < t < 4)$, $f_p(t + 8) = f_p(t)$ (Use $N = 8$)

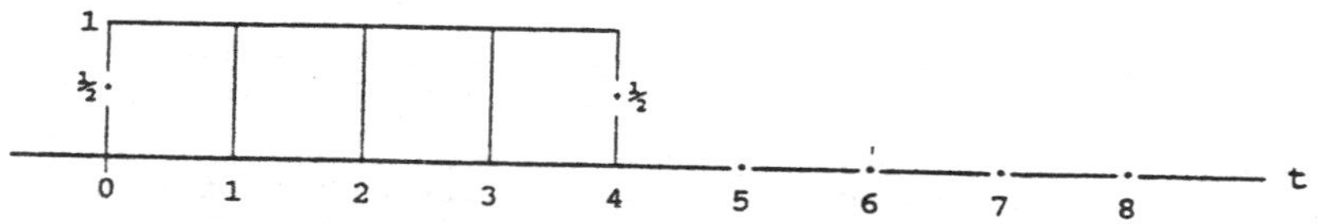

$$F(n) = \frac{1}{T_0}\int_0^{T_0} f_p(t)\, \exp(-jn\omega_0 t)\, dt \approx \frac{1}{T_0}\sum_{k=0}^{7} f_p(t_k)\, \exp(-jn\omega_0 t_k)\, T_s$$

Then $T_s/T_0 = 1/8$, and $\omega_0 t_k = k\pi/4$. Thus

$$F(n) \approx \frac{1}{8}\sum_{k=0}^{7} f_k \exp(-j\pi nk/4)$$

$$= \frac{1}{8}\left[\tfrac{1}{2} + \exp(-j\pi n/4) + \exp(-j\pi n2/4) + \exp(-j\pi n3/4) + \tfrac{1}{2}\exp(-j\pi n4/4) \right]$$

$$F(0) = \frac{1}{8}\left[\tfrac{1}{2} + 1 + 1 + 1 + \tfrac{1}{2} \right] = \tfrac{1}{2}$$

$$F(1) = \frac{1}{8}\left[\tfrac{1}{2} + \exp(-j\pi/4) + \exp(-j\pi2/4) + \exp(-j\pi3/4) + \tfrac{1}{2}\exp(-j\pi4/4) \right]$$

$$= \frac{1}{8}\left[\tfrac{1}{2} + \frac{1-j}{\sqrt{2}} - j + \frac{-1-j}{\sqrt{2}} + \tfrac{1}{2}(-1) \right] = -j(1+\sqrt{2})/8 = -j0.3017767$$

$$F(2) = \frac{1}{8}\left[\tfrac{1}{2} + \exp(-j\pi2/4) + \exp(-j\pi4/4) + \exp(-j\pi6/4) + \tfrac{1}{2}\exp(-j\pi8/4) \right]$$

$$= \frac{1}{8}\left[\tfrac{1}{2} - j + (-1) + j + \tfrac{1}{2} \right] = 0$$

$$F(3) = \frac{1}{8}\left[\tfrac{1}{2} + \exp(-j3\pi/4) + \exp(-j\pi6/4) + \exp(-j\pi9/4) + \tfrac{1}{2}\exp(-j\pi12/4) \right]$$

$$= \frac{1}{8}\left[\tfrac{1}{2} + \frac{-1-j}{\sqrt{2}} + j + \frac{1-j}{\sqrt{2}} + \tfrac{1}{2}(-1) \right] = j(1-\sqrt{2})/8 = -j5.1776695e-2$$

$$F(4) = \frac{1}{8}\left[\tfrac{1}{2} + \exp(-j4\pi/4) + \exp(-j\pi8/4) + \exp(-j\pi12/4) + \tfrac{1}{2}\exp(-j\pi16/4) \right]$$

$$= \frac{1}{8}\left[\tfrac{1}{2} - 1 + 1 - 1 + \tfrac{1}{2} \right] = 0$$

(b) By the symmetry properties of F_n for f_k real:

$$F(5) = F(3)^* = -j(1-\sqrt{2})/8 = j5.1776695e-2$$

$$F(6) = F(2)^* = 0$$

$$F(7) = F(1)^* = j(1+\sqrt{2})/8 = j0.3017767$$

(c) To verify these results using the FFT system we must run the DFT on the vector

Steps for loading (N = 8, DISCRETE): Main menu, Create Values, Continue, Create X, Real, 4 intervals, 0, 3, 4, Continue, ½, 1, ½, 0, Go, Go. You are now back on the main menu with f_k properly loaded. Plotting **X** confirms that. Run ANALYSIS, Show Numbers, F.

As the following table shows, the FFT values multiplied by T/N (see Theorem 12.1) exactly equal the values that we obtained above using the rectangular rule to find values for the Fourier transform.

n	FFT	FFT/8	Rect. Rule
0	4	½	½
1	-j2.4142136	-j0.3017767	-j0.3017767
2	0	0	0
3	-j0.41421356	-j5.1776695e-2	-j5.1776695e-2
4	0	0	0
5	j0.41421356	j5.1776695e-2	j5.1776695e-2
6	0	0	0
7	j2.4142136	j0.3017767j	j0.3017767

12.5

(a) Applying the rectangular rule to find values for the Fourier transform of

$$f(t) = t \quad (0 < t < 4) \qquad \text{(Use N = 8, T = 8)}$$

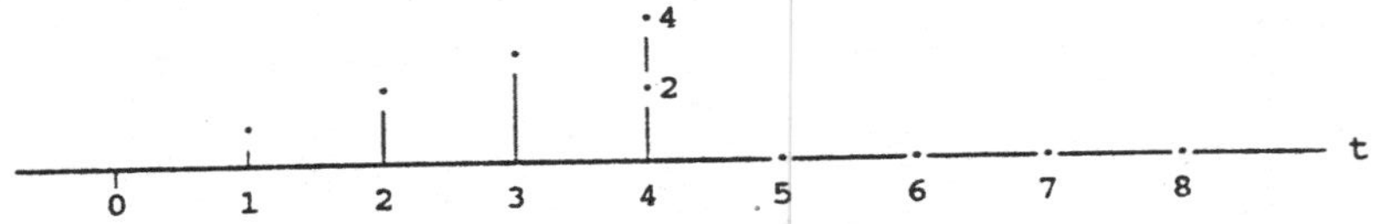

$$F(\omega) = \int_{-\infty}^{\infty} f(t)\exp(-j\omega t)\,dt = \int_0^4 t\exp(-j\omega t)\,dt$$

Sampling ω at $n\omega_0$:

$$F(n\omega_0) = \int_0^4 t\exp(-jn\omega_0 t)\,dt \approx \sum_{k=0}^{4} t_k \exp(-jn\omega_0 t_k)\,T_s$$

But T = 8 and so $\omega_0 = 2\pi/T = \pi/4$. Also, $T_s = 1$ and so $t_k = kT_s = k$

Then $\omega_0 t_k = k\pi/4$, and so we continue:

$$\ldots = \sum_{k=0}^{4} k\exp(-j\pi nk/4)$$

$= 0 + \exp(-j\pi n/4) + 2\exp(-j\pi n2/4) + 3\exp(-j\pi n3/4) + 2\exp(-j\pi n4/4)$

$F(0) = 0 + 1 + 2 + 3 + 2 = 8$

$$F(\omega_0) = 0 + \exp(-j\pi/4) + 2\exp(-j\pi 2/4) + 3\exp(-j\pi 3/4) + 2\exp(-j\pi 4/4)$$

$$= 0 + \frac{1 - j}{\sqrt{2}} - j2 + 3\frac{-1 - j}{\sqrt{2}} + 2(-1) = -(2 + \sqrt{2}) - j(2 + 2\sqrt{2})$$

$$= -3.4142136 - j4.8284271$$

$$F(2\omega_0) = 0 + \exp(-j\pi 2/4) + 2\exp(-j\pi 4/4) + 3\exp(-j\pi 6/4) + 2\exp(-j\pi 8\pi/4)$$

$$= 0 - j - 2 + j3 + 2 = j2$$

$$F(3\omega_0) = 0 + \exp(-j3\pi/4) + 2\exp(-j\pi 6/4) + 3\exp(-j\pi 9/4) + 2\exp(-j\pi 12/4)$$

$$= 0 + \frac{-1 - j}{\sqrt{2}} + j2 + 3\frac{1 - j}{\sqrt{2}} - 2 = (\sqrt{2} - 2) + j(2 - 2\sqrt{2})$$

$$= -0.5857864 - j0.8284271$$

$$F(4\omega_0) = 0 + \exp(-j4\pi/4) + 2\exp(-j\pi 8/4) + 3\exp(-j\pi 12/4) + 2\exp(-j\pi 16/4)$$

$$= 0 - 1 + 2 - 3 + 2 = 0$$

(b) By the symmetry properties of F_n for f_k real:

$$F(5\omega_0) = F(3\omega_0)^* = -0.5857864 + j0.8284271$$

$$F(6\omega_0) = F(2\omega_0)^* = -j2$$

$$F(7\omega_0) = F(1\omega_0)^* = -3.4142136 + j4.8284271$$

(c) To verify these results using the FFT system we must run the DFT on the vector $f_k = [0,\ 1,\ 2,\ 3,\ 2,\ 0,\ 0,\ 0]$

Steps for loading (N = 8, DISCRETE): Main menu, Create Values, Continue, Create X, Real, 3 intervals, 3, 4, Continue, k, ½, 0, Go, Go. You are now back on the main menu with f_k properly loaded. Plotting **X** confirms that. Run ANALYSIS, Show Numbers, F.

As the following table shows, the FFT values divided by N = 8 (see Theorem 12.2) exactly equal the values that we obtained above using the rectangular rule to find values for the Fourier coefficients.

n	A_{FFT}	A_{rect}	B_{FFT}	B_{rect}
0	8	8	0	0
1	-3.4142136	-3.4142136	-4.8284271	-4.8284271
2	0	0	0	0
3	-0.58578644	-0.58578644	-0.82842712	-0.82842712
4	0	0	0	0
5	-0.58578644	-0.58578644	0.82842712	0.82842712
6	0	0	0	0
7	-3.4142136	-3.4142136	4.8284271	4.8284271

12.6 To load the following spectrum into **F** using N = 256, T = 4, PULSE.

$$F(\omega) = \frac{\omega\sin(\omega) + 1 - \cos(\omega)}{\omega^2} + j\frac{\omega\cos(\omega) - \sin(\omega)}{\omega^2}$$

Main menu, Create Values, Continue, Create F, Complex, (w*SIN*w) + 1 - COS(w))/w^2, 0, Division by zero, B, 1.4999995, Go, Go, (w*COS(w) - SIN(w))/w^2, 0, Go, Neither, Go.

You are now back on the main menu with $F(\omega)$ in **F**. Run SYNTHESIS and plot **Y**.

You will observe the following:

- With $\alpha = 0$, ringing is present at the discontinuity but not where the pulse is continuous

Now load the spectrum again as follows: Main menu, Create Values, Continue, Use the Old Problem, Go, 20, Division by zero, B, 1.4999995, Go, Go, Go, Continue aliasing, Go, Go.

You are now back on the main menu with $F(\omega)$ in **F**. Run SYNTHESIS and plot **Y**.

You will observe the following:

- With $\alpha = 20$ most of the ringing has been removed
- With higher values of α the ringing can be removed entirely.

12.7 To load the following spectrum using N = 256, T = 2, $\alpha = 0$:

$F(\omega) = 1/(1 + j\omega)$

We note that $F(\omega) \Longleftrightarrow \exp(-t)\,U(t)$

Separating $F(\omega)$ into its real and imaginary parts:

$$F(\omega) = \frac{1}{1 + j\omega}\,\frac{1 - j\omega}{1 - j\omega} = \frac{1}{1 + \omega^2} - j\frac{\omega}{1 + \omega^2} = A(\omega) + jB(\omega)$$

Steps for loading: Main menu, Create Values, Continue, Create F, Complex, 1/(1 + w^2), 0, Go, Go, -w/(1 + w^2), 0, Go, Neither, Go.

You are now back on the main menu with $F(\omega)$ in **F**. Run SYNTHESIS and plot **Y**. (Position the origin on the left as follows: Plot menu, Reset parameters, Toggle the X and Y origin to the left, Exit)

You will observe the following:

- There is considerable ringing before and after the discontinuity caused by the fact that

to invert a spectrum that came from the ...

- The decaying exponential has only died down to about 13% of its peak value at the end of the window, i.e. t = 2. Thus there must be a significant amount of time-domain aliasing (overflow into the following periods). The values obtained from the FFT are thus highly inaccurate.

  To verify this: Show Numbers, X. At t = 1 the exact value of Y is exp(-1)= 0.36787944. The FFT (at k = 128) is 0.42545904 which amounts to an error of 15.65%.

To fix these two problems: Re-run the problem with

- T = 10. Now the decaying exponential will have died out to exp(-10) = 4.5399929e-5 which is very close to zero, and so the time-domain aliasing will be essentially nonexistent.

- Alpha = 10. This will go a long way to converting the CFT spectrum to its DFT equivalent, and should largely remove the ringing

Steps for loading: Main menu, Create Values, Continue, Create F, Complex, 1/(1 + w^2), 10, Go, -w/(1 + w^2), Continue aliasing, Go, Neither, Go.

You are now back on the main menu with F(ω) in **F**. Run SYNTHESIS and plot **Y**. (Position the origin on the left as follows: Plot menu, Reset parameters, Toggle the X and Y origin to the left, Exit)

You will observe the following:

- The decaying exponential has now died down almost to zero. Thus the time-domain aliasing will be insignificant.

- The ringing is largely gone in the plot. (There will still be a vestigial amount in the numbers.)

The FFT values are now very much more accurate. The following table verifies this.

k	$f(t)_{FFT}$	$f(t)_{exact}$	error (%)
4	0.85417896	0.8553453	0.1363621
8	0.73104769	0.7316156	0.0776280
16	0.53498806	0.5352614	0.0510720
32	0.28637486	0.2865048	0.0453524
64	8.2029516e-2	8.208500e-2	0.0675917
128	6.7382529e-3	6.737947e-3	0.0045400

Note: One step on the k-scale is equal to T_s = T/N on the t-scale. Thus k = 16 corresponds to t = 16 × 10/256 = 0.625

$$f(t) = \begin{cases} 4 + 4t + t^2 & (-2 < t < -1) \\ 2 - t^2 & (-1 < t < 1) \\ 4 - 4t + t^2 & (1 < t < 2) \end{cases}$$

is given in the Answers. Sketches of its first and second derivatives are also given. From the three sketches it is clear that f(t) is countinuous everywhere up to and including its first derivative. Hence we can expect that its Fourier transform will die out like $1/\omega^3$.

(b) We find F(ω) by successive differentiation as follows. From the sketch of f″(t) we have

$$f'''(t) = 2\delta(t + 2) - 4\delta(t + 1) + 4\delta(t - 1) - 2\delta(t - 2)$$

$$\begin{aligned}(j\omega)^3 F(\omega) &= 2 \exp(j\omega 2) - 4 \exp(j\omega) + 4 \exp(-j\omega) - 2 \exp(-j\omega 2) \\ &= 2[\exp(j\omega 2) - \exp(-j\omega 2)] - 4[\exp(j\omega) + \exp(-j\omega)] \\ &= 4j[\sin(2\omega) - 2 \sin(\omega)]\end{aligned}$$

$$F(\omega) = \frac{4[2 \sin(\omega) - \sin(2\omega)]}{\omega^3} \quad \text{(A)} \quad \longrightarrow \quad \text{QED}$$

(c) Steps for loading the pulse into **X** using N = 256 and T = 8:

Main menu, Create Values, Continue, Real, Even, 3 intervals, 1, 2, Continue, 2 - t^2, 4 - 4*t + t^2, 0, Go, Neither, Go. You are now back on the main menu with the pulse correctly loaded. Plotting **X** confirms this. Run ANALYSIS, Show Numbers, F, Single.

From the line n = 23 we see that the FFT's estimate of F(ω) at ω = 23Ω is

$$F_{23} = -2.8095879\text{e-}4 \quad \text{(B)}$$

In this case Ω = 2π/T = π/4. Then (B) is the exact DFT value of F_{23} for this pulse (to the accuracy of the MAC or DOS machine that it was computed on).

(d) From (A) we obtain

$$F(23\Omega) = \left.\frac{4[2 \sin(\omega) - \sin(2\omega)]}{\omega^3}\right|_{\omega = 23\pi/4} = -2.8108077\text{e-}4 \quad \text{(C)}$$

which of course differs from (B). Thus simply sampling (A) over -N/2 ≤ n ≤ N/2 and loading the results into **F** would give us incorrect values for the DFT spectrum.

(e) We now convert the CFT value from (A) to its FFT value by the use of aliasing as follows: Starting from (12.57), namely

$$F_n = \frac{N}{T} \sum_{m=-\infty}^{\infty} F[(n - mN)\omega_0]$$

in which
- F_n is the FFT value
- $F[(n - mN)\omega_0]$ is the CFT transform, namely $F(\omega)$, evaluated at $\omega = (n - mN)\omega_0$

We now replace ∞ by α, obtaining

$$F_n = \frac{N}{T} \sum_{m=-\alpha}^{\alpha} F[(n - mN)\omega_0] \qquad (D)$$

Then (D) enables us to change CFT values into their FFT counterparts to any desired degree of accuracy, simply by making α large enough.

We now sum the RHS of (D) for various values of α, using n = 23 and N = 256. $F(\omega)$ in this case is given in (A). The result is shown in the following table.

α	Value of RHS of (D)
0	-2.8108078e-4
1	-2.8096789e-4
2	-2.8096098e-4
3	-2.8095962e-4
4	-2.8095919e-4
5	-2.8095901e-4

Observe how, at $\alpha = 0$, we have the same value as in (C), and at $\alpha = 5$ we have nearly arrived at the value in (B). Clearly if we were to continue using larger values of α we would arrive exactly at (B).

Note: We deliberately selected a pulse whose Fourier transform goes to zero very quickly in order to display the convergence of the RHS of (D). In general the rate of convergence will not be as fast as this and we shall have to use much larger values for α.

(f) Steps for loading the spectrum $F(\omega) = 4[\,2\sin(\omega) - \sin(2\omega)\,]/\omega^3$ into the FFT with N = 256 and T = 8:

Main menu, Create Values, Continue, Create F, Real, 4*(2*SIN(w)-SIN(2*w))/w^3

We now use $\alpha = 0$. Division by zero, B, 4, Accept, Go, Go, Neither, Go. Show Numbers, F, Single.

The value for n = 23 is seen to be [illegible]

Repeating the above using $\alpha = 1$: Main menu, Create Values, Continue, Use the Old Problem, Accept the formula, Alpha = 1, Divide by zero, Go, B, 4, Accept, Go, Go. Show Numbers, F, Single.

The value for n = 23 is seen to be -2.80967879e-4 (c/f table above).

Repeating the above using α = 2, 3, 4, 5 reproduces each one of the values in the table.

For the case $\alpha = 1$, we ran SYNTHESIS. Then Show Numbers, X.

The values of the inverted spectrum are now displayed. They are remarkably accurate, as the following table shows, even though we have used little or no aliasing. This is because the CFT spectrum is converging like ω^3 and so the repeated copies of the CFT spectra have minimal interaction with each other, which means that the DFT and sampled CFT spectra are almost identical. For pulses whose spectra converge more slowly, e.g. like $1/\omega$, we shall have to use much larger values of α if we want to obtain accurate values from the FFT when using it to invert CFT spectra.

k	$f(t)_{FFT}$	$f(t)_{formula}$	error (%)
0	1.9999999	2	5.00e-6
8	1.9374999	1.9375	5.16e-6
16	1.7499999	1.75	5.71e-6
32	1.0000000	1	0
48	0.25000013	0.25	5.2e-5
64	0	0	0

Note: One step on the k-scale is equal to $T_s = T/N$ on the t-scale. Thus k = 16 corresponds to t = 16 × 8/256 = 0.5

13.1(a)
$$\sum_{\substack{m=-\infty \\ m\neq 0}}^{\infty} \frac{Z}{Z-m} = \sum_{m=1}^{\infty}\left[\frac{Z}{Z+m}+\frac{Z}{Z-m}\right] = \sum_{m=1}^{\infty}\frac{2Z^2}{Z^2-m^2} \qquad \text{(I)}$$

(b)
$$\sum_{\substack{m=-\infty \\ m\neq 0}}^{\infty} \frac{Z^2}{(Z-m)^2} = \sum_{m=1}^{\infty}\left[\frac{Z^2}{(Z+m)^2}+\frac{Z^2}{(Z-m)^2}\right]$$

$$= \sum_{m=1}^{\infty} Z^2 \frac{Z^2 - 2mZ + m^2 + Z^2 + 2mZ + m^2}{(Z^2-m^2)^2} = \sum_{m=1}^{\infty}\frac{2Z^2(Z^2+m^2)}{(Z^2-m^2)^2} \qquad \text{(II)}$$

(c)
$$\sum_{\substack{m=-\infty \\ m\neq 0}}^{\infty} \frac{Z^3}{(Z-m)^3} = \sum_{m=1}^{\infty}\left[\frac{Z^3}{(Z+m)^3}+\frac{Z^3}{(Z-m)^3}\right]$$

$$= \sum_{m=1}^{\infty} Z^3 \frac{Z^3 - 3mZ^2 + 3m^2Z - m^3 + Z^3 + 3mZ^2 + 3m^2Z + m^3}{(Z^2-m^2)^3}$$

$$= \sum_{m=1}^{\infty}\frac{2Z^3(Z^3+3m^2Z)}{(Z^2-m^2)^3} \qquad \text{(III)}$$

(d)
$$\sum_{\substack{m=-\infty \\ m\neq 0}}^{\infty} \frac{Z^4}{(Z-m)^4} = \sum_{m=1}^{\infty}\left[\frac{Z^4}{(Z+m)^4}+\frac{Z^4}{(Z-m)^4}\right]$$

$$= \sum_{m=1}^{\infty} Z^4 \frac{Z^4 - 4mZ^3 + 6m^2Z^2 - 4mZ^3 + m^4 + Z^4 + 4mZ^3 + 6m^2Z^2 + 4mZ^3 + m^4}{(Z^2-m^2)^4}$$

$$= \sum_{m=1}^{\infty}\frac{2Z^4(Z^4+6m^2Z^2+m^4)}{(Z^2-m^2)^4} \qquad \text{(IV)}$$

Summarizing:

$$k = 1: \quad \sum_{m=-\infty}^{\infty}\frac{Z}{Z-m} = \sum_{m=1}^{\infty}\frac{2Z^2}{Z^2-m^2} \qquad \text{(I)}$$

$$k = 2: \quad \sum_{m=-\infty}^{\infty}\frac{Z^2}{(Z-m)^2} = \sum_{m=1}^{\infty}\frac{2Z^2(Z^2+m^2)}{(Z^2-m^2)^2}$$

$$k = 3: \quad \sum_{m=-\infty}^{\infty}\frac{Z^3}{(Z-m)^3} = \sum_{m=1}^{\infty}\frac{2Z^3(Z^3+3m^2Z)}{(Z^2-m^2)^3} \qquad \text{(III)}$$

$$k = 4: \quad \sum_{m=-\infty}^{\infty}\frac{Z^4}{(Z-m)^4} = \sum_{m=1}^{\infty}\frac{2Z^4(Z^4+6m^2Z^2+m^4)}{(Z^2-m^2)^4} \qquad \text{(IV)}$$

From (I) and (II) we see that for large m the terms in the series die out like $1/m^2$, and so

- The series for $k = 1$ and $k = 2$ converge like $1/m^2$

From (III) and (IV) we see that for large m the terms in the series die out like $1/m^4$, and so their series converge like $1/m^4$

- The series for $k = 3$ and $k = 4$ converge like $1/m^2$

We see that all of these expressions are real and so

- The errors for canonical pulses are all real

Because $Z = n/N$ we see that $n = 0$ means $Z = 0$. All of the above are zero if $Z = 0$, and so

- The errors for canonical pulses are precisely zero when $n = 0$, regardless of the value of N.

From (I) we see that the denominator is always negative and the numerator is always positive. From (III) we see the same. Thus

- The errors are all negative for k even

From (II) we see that both the denominator and the numerator are always positive. From (IV) we see the same. Thus

- The errors are all positive for k odd

From (I), (II), (III) and (IV) the sums are all even in Z. Thus

- The error expressions are all even in n.

13.2
$$f(t) = \begin{cases} 1 & (0 < t < T/8) \\ 0 & \text{otherwise} \end{cases}$$

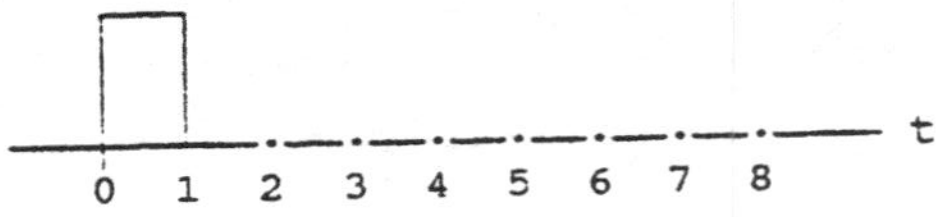

(a) Using successive differentiation to find $F(\omega)$:

$$f'(t) = \delta(t) - \delta(t-1)$$

$$F(\omega) = \frac{1 - \exp(-j\omega)}{j\omega} = \frac{1 - \cos(\omega) + j\sin(\omega)}{j\omega}$$

$$A(\omega) = \sin(\omega)/\omega \qquad B(\omega) = [\cos(\omega) - 1]/\omega \qquad \text{(I)}$$

(b) Steps for loading the pulse ($N = 256$, $T = 8$):

Main menu, Create Values, Continue, Create X, Real, Neither, Left, 2 intervals, 1, Continue, 1, 0, Go, Neither, Go. You are now back on the main menu with the pulse loaded. Plotting **X** confirms that. Run ANALYSIS, Show Numbers, F, Single.

(c) The FFT values are as follows. For n = 5:

$$A_5 = -0.17983723 \qquad B_5 = -0.43416548$$

- From (I) the exact values are calculated as follows: $\omega_0 = 2\pi/T = \pi/4$.

$$A(5\omega_0) = \sin(5\pi/4)/(5\pi/5) = -0.1800636263$$

$$B(5\omega_0) = [\cos(5\pi/4) - 1]/(5\pi/4) = -0.434711172$$

Then

$$\text{Rel. error} = \frac{[A_5 - A(5\omega_0)] + j[B_5 - B(5\omega_0)]}{A(5\omega_0) + jB(5\omega_0)]}$$

$$= \frac{2.2603300e-4 + j\,5.4569200e-4}{-0.1800636263 - j\,0.434711172}$$

$$= \frac{2.2603300e-4 + j\,5.4569200e-4}{-0.1800636263 - j\,0.434711172} \cdot \frac{-0.1800636263 + j\,0.434711172}{-0.1800636263 + j\,0.434711172}$$

$$= \frac{-2.7791864e-4 - j\,1.1770000e-11}{2.2139658e-1} = -1.2552978e-3 = -0.12552978\%$$

Thus the rel. error is real and negative.

- From Figure 13.4: $n/N = 5/256 = 0.019 \approx 0.02$

  Then the $-E(Z,0)$ curve gives: Rel error $\approx -0.125\%$

- Using the formula (13.32), namely $E(Z,0) = \pi Z \cot(\pi Z) - 1$:

  $Z = n/N = 5/256$ and so $E(Z,0) = \pi Z \cot(\pi Z) - 1$ gives

  $E(Z,0) = 5\pi/256 \cot(5\pi/256) - 1 = -1.2553003\%$

For n = 71:

$$A_{71} = -9.2965304e-3 \qquad B_{71} = -3.8507490e-3$$

- From (I) the exact values are calculated as follows:

$$\omega_0 = 2\pi/T = \pi/4.$$

$$A(71\,\omega_0) = \sin(71\pi/4)/(71\pi/5) = -1.2680511e-2$$

$$B(71\,\omega_0) = [\cos(71\pi/4) - 1]/(71\pi/4) = -5.2524397e-3$$

Then

$$\text{Rel. error} = \frac{[A_{71} - A(71\omega_0)] + j[B_{71} - B(71\omega_0)]}{A(71\omega_0) + jB(71\omega_0)]}$$

$$= \frac{3.3839809e-3 + j\,1.4016907e-3}{-1.2680511e-2 - j\,5.2524397e-3}$$

$$= \frac{3.3839809e-3 + j\,1.4016907e-3}{-1.2680511e-2 - j\,5.2524397e-3} \cdot \frac{-1.2680511e-2 + j\,5.2524397e-3}{-1.2680511e-2 + j\,5.2524397e-3}$$

$$= \frac{-5.0272905e-5 + j\,9.2e-13}{1.8838349e-4} = -2.6686470e-1 = -26.686470\%$$

Thus the rel. error is real and negative.

- From Figure 13.4: $n/N = 71/256 = 0.277 \approx 0.28$

  Then the $-E(Z,0)$ curve gives: Rel error $\approx -26\%$

- Using the formula (13.32), namely $E(Z,0) = \pi Z \cot(\pi Z) - 1$:

  $Z = n/N = 71/256$ and so $E(Z,0) = \pi Z \cot(\pi Z) - 1$ gives

  $E(Z,0) = 71\pi/256 \cot(71\pi/256) - 1 = -26.68647120\%$

13.3 (a) A sketch of the pulse

$$f(t) = \begin{cases} 0 & (t < 0) \\ t & (0 < t < 1) \\ 1 & (1 < t < 3) \\ 4 - t & (3 < t < 4) \\ 0 & (4 < t) \end{cases}$$

is given in the Answers.

(b) If each of the break points coincides with an FFT sampling instant the pulse is canonical-1 because its second derivative is only Dirac deltas.

(c) Using successsive differentiation to find the expressions for the real and imaginary parts of $F(\omega)$:

$$f''(t) = \delta(t) - \delta(t-1) - \delta(t-3) + \delta(t-4)$$

$$F(\omega) = \frac{1 - \exp(-j\omega) - \exp(-j3\omega) + \exp(-j4\omega)}{(j\omega)^2}$$

$$= \frac{1 - \cos(\omega) + j\sin(\omega) - \cos(3\omega) + j\sin(3\omega) + \cos(4\omega) - j\sin(4\omega)}{-\omega^2}$$

$$A(\omega) = [\cos(\omega) + \cos(3\omega) - \cos(4\omega) - 1]/\omega^2 \qquad \text{(I)}$$

$$B(\omega) = [\sin(4\omega) - \sin(\omega) - \sin(3\omega)]/\omega^2 \qquad \text{(II)}$$

(d) For $\omega = 0$ the value of $F(\omega)$ is equal to the area of the pulse, namely 3. This can be verified either by direct use of the analysis equation, namely

$$F(0) = \int_{-\infty}^{\infty} f(t)\,dt = \int_0^1 t\,dt + \int_1^3 dt + \int_3^4 (4-t)\,dt = 3$$

or else by using l'Hôpital's rule on $F(\omega)$.

Steps for loading the pulse ($N = 8$, $T = 8$):

Main menu, Create Values, Create X, Real, Neither, Left, 4 intervals, 1, 3, 4, Continue, t, 1, 4-t, 0, Go, Neither, Go. You are now back on the main menu with the pulse properly loaded. Plotting X confirms that. Run ANALYSIS, Show Numbers, F, Single.

We see from the FFT numbers that $F_0 = 3$ exactly.

To load the pulse using each of $N = 40$, 224 and 512 we repeat the above sequence of steps.

We see that for each of these cases, the FFT gives $F_0 = 3$ exactly. We know from the expression for the errors (see (13.22)) that $E(Z,1)$ is equal to zero for $n = 0$ regardless of the value of N, provided only that all breakpoints are at FFT sampling instatnts. The above results for F_0 confirm that.

From Chapter 12 we know that (to within the constant T/N) the FFT values are the same as those obtained from using the rectangular rule to evaluate the CFT analysis equation (see Theorem 12.1). Thus we can infer that the rectangular rule will always give the exact value for the area of a canonical-1 pulse (and, by the same argument, for the area of a canonical-k pulse, for any k) for any value of N, assuming that all of the breakpoints coincide with sampling instants.

(e) Using (13.44) plus [illegible] following spectral elements (using $T = 8$ for all three):

- $n = 137$ with $N = 800$

We load the pulse using the steps given in (d) above. From the FFT:

$A_{137} = 0 \qquad B_{137} = -1.3465017\text{e}{-4}$

From (I) and (II) above, using $\omega_0 = 2\pi/T = \pi/4$:

$A(137\omega_0) = 0 \qquad B(137\omega_0) = -1.2215018\text{e}{-4}$

$$\text{Rel error} = \frac{[A_{137} - A(137\omega_0)] + j[B_{137} - B(137\omega_0)]}{A(137\omega_0) + jB(137\omega_0)]}$$

$$= \frac{-j\,1.3465017\text{e}{-4} + j\,1.2215018\text{e}{-4}}{-j\,1.2215018\text{e}{-4}} = \frac{-j\,1.2499989\text{e}{-5}}{-j\,1.2215018\text{e}{-4}} = 10.233292624\%$$

Observe that the relative error is real and positive.

From Figure 13.4: $Z = n/N = 137/800 \approx 0.17$. Error $\approx 10\%$

From (13.33): $E(Z,1) = [(\pi Z)/\sin(\pi Z)]^2 - 1 = 10.23330190\%$

- $n = 60$ with $N = 200$

We load the pulse using the steps given in (d) above. From the FFT:

$A_{60} = -2.4445825\text{e}{-3} \qquad B_{60} = 0$

From (I) and (II) above, using $\omega_0 = 2\pi/T = \pi/4$:

$A(60\omega_0) = -1.8012654\text{e}{-3} \qquad B(60\omega_0) = 0$

$$\text{Rel error} = \frac{[A_{60} - A(60\omega_0)] + j[B_{60} - B(60\omega_0)]}{A(60\omega_0) + jB(60\omega_0)]}$$

$$= \frac{-2.4445825\text{e}{-3} + 1.8012654\text{e}{-3}}{-1.8012654\text{e}{-3}} = \frac{-6.4331701\text{e}{-4}}{-1.8012654\text{e}{-3}} = 35.7147250\%$$

Observe that the relative error is real and positive.

From Figure 13.4: $Z = n/N = 60/200 = 0.3$. Error $\approx 35\%$

From (13.33): $E(Z,1) = [(\pi Z)/\sin(\pi Z)]^2 - 1 = 35.7147234\%$

- $n = 226$ with $N = 512$

We load the pulse using the steps given in (d) above. From the FFT:

$A_{226} = -1.2630188e-4 \qquad B_{226} = 0$

From (I) and (II) above, using $\omega_0 = 2\pi/T = \pi/4$:

$A(226\omega_0) = -6.3479474e-5 \qquad B(226\omega_0) = 0$

$$\text{Rel error} = \frac{[A_{226} - A(226\omega_0)] + j[B_{226} - B(226\omega_0)]}{A(226\omega_0) + jB(226\omega_0)]}$$

$$= \frac{-1.2630188e-4 + 6.3479474e-5}{-6.3479474e-5} = \frac{-6.2822405e-5}{-6.3479474e-5} = 98.9649116\%$$

Observe that the relative error is real and positive.

From Figure 13.4: $Z = n/N = 226/512 = 0.44$. Error $\approx$ 99%

From (13.33): $E(Z,1) = [(\pi Z)/\sin(\pi Z)]^2 - 1 = 98.9648968\%$

13.4 (a) A sketch of the pulse

$$f(t) = \begin{cases} 0 & (t < 0) \\ t^2/2 & (0 < t < 1) \\ (-2t^2 + 6t - 3)/2 & (1 < t < 2) \\ (t^2 - 6t + 9)/2 & (2 < t < 3) \\ 0 & (3 < t) \end{cases}$$

is given in the Answers.

(b)

$$f'(t) = \begin{cases} 0 & (t < 0) \\ t & (0 < t < 1) \\ -2t + 3 & (1 < t < 2) \\ t - 3 & (2 < t < 3) \\ 0 & (3 < t) \end{cases}$$

$$f''(t) = \begin{cases} 0 & (t < 0) \\ 1 & (0 < t < 1) \\ -2 & (1 < t < 2) \\ 1 & (2 < t < 3) \\ 0 & (3 < t) \end{cases}$$

$$f'''(t) = \begin{cases} \delta(t) \\ -3\delta(t-1) \\ 3\delta(t-2) \\ -\delta(t-3) \end{cases}$$

From the figure for f(t) in the Answers and the following figures, we see that the pulse and its first derivative are everywhere continuous, the second derivative is discontinuous, and that the

third derivative will be only Dirac deltas. Thus the pulse is canonical-2.

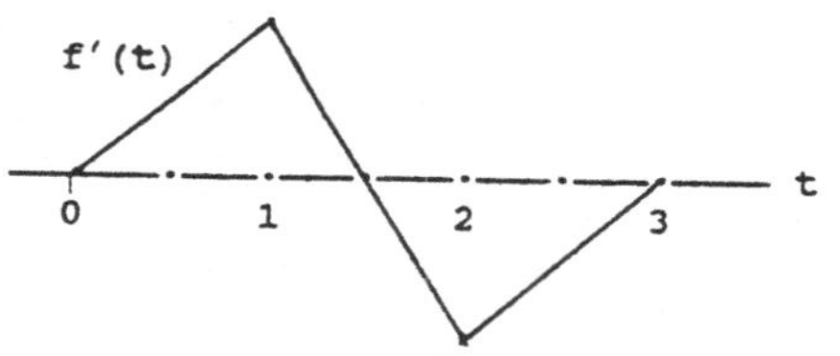

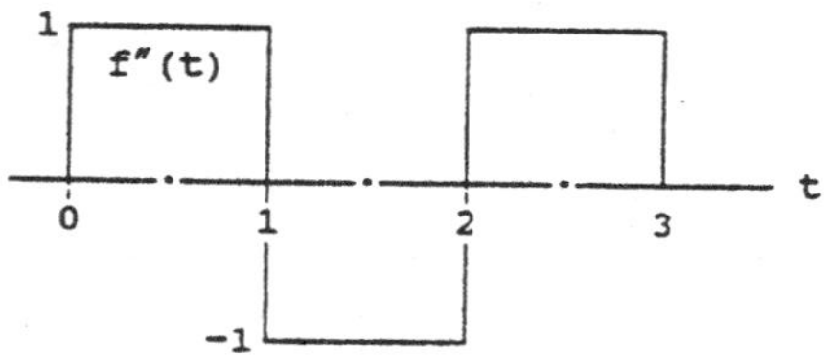

(c) Using successive differentiation, we find now the real and imaginary parts of $F(\omega)$. As shown above:

$$f'''(t) = \delta(t) - 3\delta(t-1) + 3\delta(t-2) - \delta(t-3)$$

$$F(\omega) = \frac{1 - 3\exp(-j\omega) + 3\exp(-j2\omega) - \exp(-j3\omega)}{(j\omega)^3} = \left[\frac{1 - \exp(-j\omega)}{j\omega}\right]^3$$

$$= \exp(-j\omega 3/2)\left[\frac{\exp(j\omega/2) - \exp(-j\omega/2)}{2j\omega/2}\right]^3 = \exp(-j\omega 3/2)\left[\frac{\sin(\omega/2)}{\omega/2}\right]^3$$

$$A(\omega) = \cos(3\omega/2)\ Sa^3(\omega/2) \qquad \text{(I)}$$

$$B(\omega) = -\sin(3\omega/2)\ Sa^3(\omega/2) \qquad \text{(II)}$$

(d) The value of $F(\omega)$ at $\omega = 0$ is equal to the area of the pulse. In this case it is

$$F(0) = \int_{-\infty}^{\infty} f(t)\,dt = \int_0^1 t^2/2\,dt + \int_1^2 (-2t^2+6t-3)/2\,dt + \int_2^3 (t^2-6t+9)/2\,dt$$

$$= 1/6 + \tfrac{1}{2}(-2t^3/3 + 3t^2 - 3t)\Big|_1^2 + \tfrac{1}{2}(t^3/3 - 3t^2 + 9t)\Big|_2^3$$

$$= 1/6 + 2/3 + 1/6 = 1 \longrightarrow$$

Steps for loading f(t) using N = 4, T = 4: Main menu, Create Values, Continue, Create X, Real, Neither, Left, 4 intervals, t^2/2, (-2*t^2 + 6*t - 3)/2, (t^2 - 6*t + 9)/2, Go, Neither, Go. You are now back on the main menu with the pulse properly loaded. Plotting **X** confirms that. Run **ANALYSIS**, **Show Numbers**, **F**, **Single**.

The display shows $A_0 = 1$ which is exactly the value of F(0).

Repeating the above loading procedure using N = 40, then N = 224 then N = 500, we obtain

We know from the expression for the [illegible] E(Z,2) is equal to zero for n = 0 regardless of the value of N, provided only that all breakpoints are at FFT sampling instatnts. The above results for F_0 confirm that.

From Chapter 12 we know that (to within the constant T/N) the FFT values are the same as those obtained from using the rectangular rule to evaluate the CFT analysis equation (see Theorem 12.1). Thus we can infer that the rectangular rule will always give the exact value for the area of a canonical-2 pulse (and, by the same argument, for the area of a canonical-k pulse, for any k) for any value of N, assuming that all of the breakpoints coincide with sampling instants.

(e) Using (13.44) plus FFT data to find the relative errors in the following spectral elements (using T = 4 for all three):

- n = 7 with N = 100

We load the pulse using the steps given in (d) above. From the FFT:

A_7 = 1.5042065e-3 $\quad$ B_7 = -1.5042065e-3

From (I) and (II) above, using $\omega_0 = 2\pi/T = \pi/2$:

$A(7\omega_0)$ = 1.5044447e-3 $\quad$ $B(7\omega_0)$ = -1.5044447e-3

$$\text{Rel error} = \frac{[A_7 - A(7\omega_0)] + j[B_7 - B(7\omega_0)]}{A(7\omega_0) + jB(7\omega_0)]}$$

$$= \frac{-2.3823500\text{e-}7 + j2.3823500\text{e-}7}{1.5044447\text{e-}3 - j1.5044447\text{e-}3} = \frac{-2.3823500\text{e-}7(1-j)}{1.5044447\text{e-}3(1-j)} = -.015835411\%$$

Observe that the relative error is real and negative

From Figure 13.4: Z = n/N = 7/100 ≈ 0.07. Error ≈ 0.015%

From (13.34): $E(Z,2) = [(\pi Z)^3\cot(\pi Z)]/\sin^2(\pi Z) - 1 = -0.015833700\%$

- n = 38 with N = 200

We load the pulse using the steps given in (d) above. From the FFT:

A_{38} = 0 $\quad$ B_{38} = 3.7259378e-5

From (I) and (II) above, using $\omega_0 = 2\pi/T = \pi/2$:

$A(38\omega_0)$ = 0 $\quad$ $B(38\omega_0)$ = 3.7616602e-5

$$\text{Rel error} = \frac{[A_{38} - A(38\omega_0)] + j[B_{38} - B(38\omega_0)]}{A(38\omega_0) + jB(7\omega_0)]}$$

$$= \frac{[illegible]}{j\,3.7616602\text{e-}5} = \frac{[illegible]}{j\,3.7616602\text{e-}5} = -0.949645338\%$$

Observe that the relative error is real and negative

From Figure 13.4: Z = n/N = 38/200 ≈ 0.19. Error ≈ 1%

From (13.34): $E(Z,2) = [(\pi Z)^3\cot(\pi Z)]/\sin^2(\pi Z) - 1 = -0.949649970\%$

- n = 151 with N = 500

We load the pulse using the steps given in (d) above. From the FFT:

A_{151} = 1.3895253e-7 $\quad$ B_{151} = -1.3895253e-7

From (I) and (II) above, using $\omega_0 = 2\pi/T = \pi/2$:

$A(151\omega_0)$ = 1.4987853e-7 $\quad$ $B(151\omega_0)$ = -1.4987853e-7

$$\text{Rel error} = \frac{[A_{151} - A(151\omega_0)] + j[B_{151} - B(151\omega_0)]}{A(151\omega_0) + jB(151\omega_0)]}$$

$$= \frac{-1.0926000\text{e-}8 + j1.0926000\text{e-}8}{1.4987853\text{e-}7 - j1.4987853\text{e-}7} = \frac{-1.0926000\text{e-}8(1-j)}{1.4987853\text{e-}7(1-j)} = -7.28990336\%$$

Observe that the relative error is real and negative

From Figure 13.4: Z = n/N = 151/500 ≈ 0.30. Error ≈ -7%

From (13.34): $E(Z,2) = [(\pi Z)^3\cot(\pi Z)]/\sin^2(\pi Z) - 1 = -7.289922270\%$

13.5 $$\sum_{m=-\infty}^{\infty} \frac{1}{Z - m} = \pi\cot(\pi Z) \qquad \text{(A)}$$

and so $$\sum_{m=-\infty}^{\infty} \frac{Z}{Z - m} = \pi Z\cot(\pi Z)$$

from which $$\sum_{\substack{m=-\infty \\ m\neq 0}}^{\infty} \frac{Z}{Z - m} = \pi Z\cot(\pi Z) - 1$$ which verifies (13.32).

Differentiating both sides of (A):

$$\frac{d}{dZ}\sum_{m=-\infty}^{\infty} \frac{1}{Z - m} = \sum_{m=-\infty}^{\infty} \frac{-1}{(Z - m)^2} \quad \text{and} \quad \frac{d}{dz}\pi\cot(\pi Z) = -\pi^2/\sin^2(\pi Z)$$

and so

$$\sum_{m=-\infty}^{\infty} \frac{1}{(Z - m)^2} = \pi^2/\sin^2(\pi Z) \qquad \text{(B)}$$

from which

$$\sum_{\substack{m=-\infty \\ m\neq 0}}^{\infty} \frac{Z^2}{(Z - m)^2} = (\pi Z)^2/\sin^2(\pi Z) - 1 \qquad \text{which verifies (13.33)}$$

Differentiating both sides of (B):

$$\frac{d}{dZ} \sum_{m=-\infty}^{\infty} \frac{1}{(Z - m)^2} = \sum_{m=-\infty}^{\infty} \frac{-2}{(Z - m)^3}$$

and

$$\frac{d}{dz} \pi^2/\sin^2(\pi Z) = -2\pi^3\cos(\pi Z)/\sin^3(\pi Z)$$

and so

$$\sum_{m=-\infty}^{\infty} \frac{1}{(Z - m)^3} = \pi^3\cos(\pi Z)/\sin^3(\pi Z) \qquad \text{(C)}$$

from which

$$\sum_{\substack{m=-\infty \\ m\neq 0}}^{\infty} \frac{Z^3}{(Z - m)^3} = (\pi Z)^3\cot(\pi Z)/\sin^2(\pi Z) - 1 \qquad \text{which verifies (13.33)}$$

Differentiating both sides of (C):

$$\frac{d}{dZ} \sum_{m=-\infty}^{\infty} \frac{1}{(Z - m)^3} = \sum_{m=-\infty}^{\infty} \frac{-3}{(Z - m)^4}$$

$$\frac{d}{dz} \pi^3\cos(\pi Z)/\sin^3(\pi Z) = -\pi^4 \frac{\sin^2(\pi Z) + 3\cos^2(\pi Z)}{\sin^4(\pi Z)} = -\pi^4 \frac{1 + 3\cot^2(\pi Z)}{\sin^2(\pi Z)}$$

and so

$$\sum_{m=-\infty}^{\infty} \frac{Z^4}{(Z - m)^4} = \frac{(\pi Z)^4 [1 + 3\cot^2(\pi Z)]}{3 \sin^2(\pi Z)}$$

from which

$$\sum_{\substack{m=-\infty \\ m\neq 0}}^{\infty} \frac{Z^4}{(Z - m)^4} = \frac{(\pi Z)^4 [1 + 3\cot^2(\pi Z)]}{3 \sin^2(\pi Z)} - 1 \qquad \text{which verifies (13.33)}$$

13.6 (a) $f(t) = \exp(-\beta t)\ U(t) \Longleftrightarrow F(\omega) = 1/(j\omega + 1)$

(b) From (13.8):

$$E_N(n) = \sum_{m=-\infty}^{\infty} \frac{F[(n - mN)\omega_0]}{F(n\omega_0)} = \sum_{m=-\infty}^{\infty} \frac{\beta + jn\omega_0}{\beta + j[(n - mN)\omega_0]}$$

$$= \sum_{m=-\infty}^{\infty} \frac{(\beta + jn\omega_0)/N}{\{\beta + j[(n - mN)\omega_0]\}/N} = \sum_{m=-\infty}^{\infty} \frac{\beta/N + jZ\omega_0}{\beta/N + j[Z - m)\omega_0]}$$

(c) If we now hold Z constant and let $N \longrightarrow \infty$ then we obtain

$$\lim_{N\to\infty} \sum_{m=-\infty}^{\infty} \frac{\beta/N + jZ\omega_0}{\beta/N + j[Z - m)\omega_0]} = \sum_{m=-\infty}^{\infty} \frac{Z}{Z - m} = E(Z,0) \qquad \text{QED}$$

13.7 $f(t) = \sin^2(\pi t)\ \text{Rect}(t - \tfrac{1}{2})$

(a) Finding $F(\omega)$:

$$f(t) = \sin^2(\pi t)\ \text{Rect}(t - \tfrac{1}{2}) = \tfrac{1}{2}[1 - \cos(2\pi t)]\ \text{Rect}(t - \tfrac{1}{2})$$

$$= \tfrac{1}{2}\ \text{Rect}(t - \tfrac{1}{2}) - \tfrac{1}{2}\cos(2\pi t)\ \text{Rect}(t - \tfrac{1}{2})$$

Then $\tfrac{1}{2}\ \text{Rect}(t - \tfrac{1}{2}) \Longleftrightarrow \tfrac{1}{2}\ \text{Sa}(\omega/2)\ \exp(-j\omega/2)$ (A)

and we now use successive differentiation for $-\tfrac{1}{2}\cos(2\pi t)\ \text{Rect}(t - \tfrac{1}{2})$

$$g(t) = \cos(2\pi t)\ \text{Rect}(t - \tfrac{1}{2})$$

$$g'(t) = -2\pi \sin(2\pi t)\ \text{Rect}(t - \tfrac{1}{2}) + \cos(2\pi t)\ [\delta(t) - \delta(t - 1)]$$

$$= -2\pi \sin(2\pi t)\ \text{Rect}(t - \tfrac{1}{2}) + \delta(t) - \delta(t - 1)$$

$$g''(t) = -4\pi^2 \cos(2\pi t)\ \text{Rect}(t - \tfrac{1}{2}) - 2\pi \sin(2\pi t)\ [\delta(t) - \delta(t - 1)] + D\delta(t) - D\delta(t - 1)$$

$$= -4\pi^2 \cos(2\pi t)\ \text{Rect}(t - \tfrac{1}{2}) + D\delta(t) - D\delta(t - 1)$$

$$g''(t) + 4\pi^2\ g(t) = D\delta(t) - D\delta(t - 1)$$

$$[(j\omega)^2 + 4\pi^2]\ G(\omega) = j\omega\ [1 - \exp(-j\omega) = (j\omega)^2\ \text{Sa}(\omega/2)\ \exp(-j\omega/2)$$

$$G(\omega) = \frac{(j\omega)^2\ \text{Sa}(\omega/2)\ \exp(-j\omega/2)}{(j\omega)^2 + 4\pi^2}$$ Combining this with (A) we obtain

$$F(\omega) = \tfrac{1}{2}\ \text{Sa}(\omega/2)\ \exp(-j\omega/2) - \tfrac{1}{2}\ (j\omega)^2\ \text{Sa}(\omega/2)\ \exp(-j\omega/2)/[(j\omega)^2 + 4\pi^2]$$

$$= \tfrac{1}{2}\ \text{Sa}(\omega/2)\ \exp(-j\omega/2) \left[1 - \frac{(j\omega)^2}{(j\omega)^2 + 4\pi^2} \right] = 2\pi^2 \frac{\text{Sa}(\omega/2)\ \exp(-j\omega/2)}{\ldots}$$

(b) From (13.8):

$$E_N(n) = \sum_{m=-\infty}^{\infty} \frac{F[(n - mN)\omega_0]}{F(n\omega_0)} = \sum_{m=-\infty}^{\infty} \frac{\dfrac{Sa[(n - mN)\omega_0/2],\ \exp[-j(n - mN)\omega_0/2]}{4\pi^2 - (n - mN)\omega_0^2}}{\dfrac{Sa(n\omega_0/2)\ \exp(-jn\omega_0/2)}{4\pi^2 - (n\omega_0)^2}}$$

Now $\omega_0 = 2\pi/T$ and so $mN\omega_0/2 = mN\pi/T$. Moreover, it is assumed that the end of the pulse, namely $t = 1$ is an FFT sampling instant, and for that to be true the point $t = 1$ must correspond to the sampling instant $t_k = kT_s = 1$ where k is an integer and $T_s = T/N$.

Thus $1/T_s = k$ (an integer), and so N/T is an integer.

Thus $mN\omega_0/2 = mN\pi/T$ must be an integral multiple of π. But then

$$\exp[-j(n-mN)\omega_0/2] = \exp(-jn\omega_0/2)\ \exp(jmN\omega_0/2) = \exp(-jn\omega_0/2)\ (-1)^n$$

Similarly

$$Sa[(n - mN)\omega_0/2] = \frac{\sin[(n - mN)\omega_0/2]}{(n - mN)\omega_0/2} = \frac{\sin(n\omega_0/2)\ \cos(mN\omega_0/2)}{(n - mN)\omega_0/2} = \frac{\sin(n\omega_0/2)\ (-1)^n)}{(n - mN)\omega_0/2}$$

Then

$$E_N(n) = \sum_{m=-\infty}^{\infty} \frac{\dfrac{\sin(n\omega_0/2)\ \exp(-jn\omega_0/2)\ (-1)^{2m}}{[(n - mN)\omega_0/2]\ [4\pi^2 - (n - mN)\omega_0^2]}}{\dfrac{\sin(n\omega_0/2)\ \exp(-jn\omega_0/2)}{(n\omega_0/2)\ [4\pi^2 - (n\omega_0)^2]}}$$

$$= \sum_{m=-\infty}^{\infty} \frac{(n\omega_0/2)\ [4\pi^2 - (n\omega_0)^2]}{[(n - mN)\omega_0/2]\ [4\pi^2 - (n - mN)\omega_0^2]}$$

$$= \sum_{m=-\infty}^{\infty} \frac{\dfrac{(n\omega_0/2)\ [4\pi^2 - (n\omega_0)^2]}{N^3}}{\dfrac{[(n - mN)\omega_0/2]\ [4\pi^2 - (n - mN)\omega_0^2]}{N^3}}$$

$$= \sum_{m=-\infty}^{\infty} \frac{(Z\omega_0/2)\ [4\pi^2/N^2 - (Z\omega_0)^2]}{[(Z - m)\omega_0/2]\ [4\pi^2/N^2 - (Z - m)\omega_0^2]} \quad \text{where } Z = n/N$$

$$E_N(n) = \sum_{m=-\infty}^{\infty} \frac{Z^3}{(Z - m)^3}$$

thereby confirming Theorem 13.4. QED

13.8 (a) Steps for loading $f(t) = \exp(-t)\ U(t)$ into **X** using N = 1024 and T = 16:

Main menu, Create Values, Continue, Create X, Real, Neither Left, 1 interval, Continue, EXP(-t), Go, Neither, Go. You are now back on the main menu with the pulse properly loaded. Plotting **X** confirms that. Run ANALYSIS, Show Numbers, F, Single.

n	A_n(FFT)	B_n(FFT)	error (%)	E(Z,0) (%)
0	1.00002020	0	0.002020000+j0	0
4	0.28842075	-0.45298634	-0.002997092+j0.006391469	-0.005020020
16	2.4724875e-2	-0.15509523	-0.078305442+j0.025574163	-0.080331960
64	1.6011437e-3	-0.03921320	-1.286364741+j0.102793030	-1.288419890
256	1.2206037e-4	-7.8115455e-3	-21.45757300+j0.445930127	-21.46018365
512	6.1033908e-5	0	-99.99389600+j1.227159600	-100

(b) Using (13.44) to compute the error in the FFT estimates in the table:

$$f(t) = \exp(-t)\ U(t) \Longleftrightarrow F(\omega) = 1/(j\omega + 1) = 1/(\omega^2 + 1) - j\omega/(\omega^2 + 1)$$
$$= A(\omega) + jB(\omega)$$

$$\text{Rel error} = \frac{[A_n - A(n\omega_0)] + j[B_n - B(n\omega_0)]}{A(n\omega_0) + jB(n\omega_0)]}$$

In this case $\omega_0 = 2\pi/T = \pi/8$.

- For n = 0:

$A_0 = 1.0000202$ $\quad B_0 = 0$

$A(0\omega_0) = 1$ $\quad B(0\omega_0) = 0$

$$\text{Rel error} = \frac{(1.0000202 - 1)}{1} = 0.00202\%$$ which agrees with the table.

- For n = 4:

$A_4 = 0.28842075$ $\quad B_4 = -0.45298634$

$A(4\omega_0) = 0.288400439$ $\quad B(4\omega_0) = -0.453018350$

$$\text{Rel error} = \frac{(0.28842075 - 0.28840043) + j\,(-0.45298634 + 0.453018350)}{0.288400439 - j\,0.453018350}$$

$$= \frac{2.0310900e{-5} + j3.2010400e{-5}}{0.288400439 - j\,0.453018350}\;\frac{0.288400439 + j\,0.453018350}{0.288400439 + j\,0.453018350}$$

$$= \frac{-8.6436261e{-6} + j\,1.8433023e{-5}}{(0.288400439)^2 + (0.453018350)^2} = -0.002997092 + j\,0.006391469$$

which agrees with the table.

- For n = 16:

$A_{16} = 2.4724875e{-2}$ $\quad B_{16} = -0.15509523$

$A(16\omega_0) = 2.4704523e{-2}$ $\quad B(16\omega_0) = -0.155223096$

$$\text{Rel err} = \frac{(2.4724875e{-2} - 2.4704523e{-2}) + j(-0.15509523 + 0.155223096)}{2.4704523e{-2} - j\,0.155223096}$$

$$= \frac{2.0351970e{-5} + j1.2786610e{-4}}{2.4704523e{-2} - j\,0.155223096}\;\frac{2.4704523e{-2} + j\,0.155223096}{2.4704523e{-2} + j\,0.155223096}$$

$$= \frac{-1.9344986e{-5} + j\,6.3179668e{-6}}{(2.4704523e{-2})^2 + (0.155223096)^2} = -0.078305443 + j\,0.025574130$$

which agrees with the table.

- For n = 64:

$A_{64} = 1.6011437e{-3}$ $\quad B_{64} = -0.03921320$

$A(64\omega_0) = 1.5806411e{-3}$ $\quad B(64\omega_0) = -0.039725844$

$$\text{Rel err} = \frac{(1.6011437e{-3} - 1.5806411e{-3}) + j(-0.03921320 + 0.039725844)}{1.5806411e{-3} - j\,0.039725844}$$

$$= \frac{2.0502589e{-5} + j5.1264404e{-4}}{1.5806411e{-3} - j\,0.039725844}\;\frac{1.5806411e{-3} + j\,0.039725844}{1.5806411e{-3} + j\,0.039725844}$$

$$= \frac{-2.0332809e{-5} + j\,1.6247888e{-6}}{(1.5806411e{-3})^2 + (0.039725844)^2} = -1.286364741 + j\,0.102793030$$

which agrees with the table.

- For n = 256:

$A_{256} = 1.2206037e{-4}$ $\quad B_{256} = -7.8115455e{-3}$

$A(256\omega_0) = 9.8936679e{-5}$ $\quad B(256\omega_0) = -9.9461998e{-3}$

$$\text{Rel err} = \frac{(1.2206037e{-4} - 9.8936679e{-5}) + j(-7.8115455e{-3} + 9.9461998e{-3})}{9.8936679e{-5} - j\,9.9461998e{-3}}$$

$$= \frac{2.3123690e{-5} + j\,2.1346543e{-3}}{9.8936679e{-5} - j\,9.9461998e{-3}}\;\frac{9.8936679e{-5} + j\,9.9461998e{-3}}{9.8936679e{-5} + j\,9.9461998e{-3}}$$

$$= \frac{-2.1229410e{-5} + j\,4.4118845e{-7}}{(9.8936679e{-5})^2 + (9.9461998e{-3})^2} = -21.45757335 + j\,0.445930127$$

which agrees with the table.

- For n = 512:

$A_{512} = 6.1033908e{-5}$ $\quad B_{512} = 0$

$A(512\omega_0) = 2.4736005e{-5}$ $\quad B(512\omega_0) = -4.9734689e{-3}$

$$\text{Rel error} = \frac{(6.1033908e{-5} - 2.4736005e{-5}) + j(0 + 4.9734689e{-3})}{2.4736005e{-5} - j\,4.9734689e{-3}}$$

$$= \frac{3.6297902e{-5} + j\,4.9734689e{-3}}{2.4736005e{-5} - j\,4.9734689e{-3}}\;\frac{2.4736005e{-5} + j\,4.9734689e{-3}}{2.4736005e{-5} + j\,4.9734689e{-3}}$$

$$= \frac{-2.4734495e{-5} + j\,3.0355024e{-7}}{(2.4736005e{-5})^2 + (4.9734689e{-3})^2} = -99.99389659 + j\,1.227159532$$

which agrees with the table.

(c) Using the formula $E(Z,0) = \pi Z \cot(\pi Z) - 1$ to compute the errors:

n	Z	E(Z,0) (%)
0	0	0
4	π4/1024	-0.005020020
16	π16/1024	-0.080331960
64	π64/1024	-1.288419890
256	π256/1024	-21.46018365
512	π512/1024	-99.99999988

(d) Observe how closely the complex error in the table and the error obtained in (c) agree, as predicted by Theorem 13.3.

13.9 (a) A sketch of one period of the periodic waveform

$$f_p(t) = \begin{cases} \exp(t) & (0 < t < 1) \\ 0 & (1 < t < 2) \end{cases} \qquad f_p(t+2) = f_p(t)$$

(b) Using successive [illegible] spectra of $f_{ev}(t)$ and $f_{od}(t)$:

$$f(t) = \exp(t)\ \mathrm{Rect}(t - \tfrac{1}{2})$$

$$f'(t) = \exp(t)\ \mathrm{Rect}(t - \tfrac{1}{2}) + \exp(t)\ [\delta(t) - \delta(t - 1)]$$

$$= f(t) + \delta(t) - e\ \delta(t - 1)]$$

$$j\omega\ F(\omega) = F(\omega) + 1 - e\ \exp(-j\omega)$$

$$F(\omega) = \frac{1 - e\ \exp(-j\omega)}{j\omega - 1} = \frac{e\ \cos(\omega) - 1 - je\ \sin(\omega)}{1 - j\omega}\ \frac{1 + j\omega}{1 + j\omega}$$

$$A(\omega) = \frac{e\ \cos(\omega) - 1 + e\omega\ \sin(\omega)}{1 + \omega^2} \qquad B(\omega) = \frac{\omega[e\ \cos(\omega) - 1] - e\ \sin(\omega)}{1 + \omega^2}$$

$$A_p(n) = \frac{1}{T_0}\ A(\omega)\ \Big|_{\omega \longleftarrow n\omega_0} = \tfrac{1}{2}\ \frac{e\ \cos(n\pi) - 1 + n\pi e\ \sin(n\pi)}{1 + (n\pi)^2}$$

$$= \tfrac{1}{2}\ \frac{e(-1)^n - 1}{1 + (n\pi)^2} \longrightarrow \quad (I)$$

$$jB_p(n) = \frac{1}{T_0}\ jB(\omega)\ \Big|_{\omega \longleftarrow n\omega_0} = j/2\ \frac{n\pi e\ \cos(n\pi) - n\pi - e\ \sin(n\pi)}{1 + (n\pi)^2}$$

$$= j/2\ \frac{n\pi[e(-1)^n - 1]}{1 + (n\pi)^2} \longrightarrow \quad (II)$$

(c) Sketches of the even and odd parts of $f_p(t)$ are given in the Answers, from which we see that

- $f_{ev}(t)$ is everywhere continuous
- $f_{od}(t)$ has discontinuities
- Neither $f_{ev}(t)$ nor $f_{od}(t)$ is a canonical waveform because none of their derivatives will be only Dirac deltas.

(d) According to Theorem 13.4, if we sample so that the break-points are located at sampling instants then the errors for non-canonical waveforms tend asymptotically to the associated canonical error curves as N is made larger. We would thus expect that the errors in the FFT spectrum of $f_{ev}(t)$ will tend to the canonical-1 error curve and those of $f_{od}(t)$ will tend to the canonical-0 error curve.

Steps for loading the waveform using N = 32, T = 2, PERIODIC:

Main menu, Create Values, Continue, Create X, Real, Neither, Left,

main menu with the waveform correctly loaded. Plotting X confirms that. Run ANALYSIS, Show Numbers, F, Complex.

We see that for n = N/2 - 1 = 15, the FFT values are

$$A_n = -1.8325655e\text{-}3 \qquad B_n = -5.7165302e\text{-}3$$

and repeating the above steps with N = 128, 512, 2048 we obtain the FFT values shown in the following table for A_n and B_n with n = N/2 - 1:

N	A_n	$B_p(n)$
32	-1.8325655e-3	-5.7165302e-3
128	-1.1353902e-4	-3.5653563e-4
512	-7.0923178e-6	-2.2280557e-5
2048	-4.4325478e-7	-1.3925232e-6

From the formulae for $A_p(n)$ and $B_p(n)$ derived above, namely

$$A_p(n) = \tfrac{1}{2}\ \frac{e(-1)^n - 1}{1 + (n\pi)^2} \qquad B_p(n) = \tfrac{1}{2}\ \frac{n\pi[e(-1)^n - 1]}{1 + (n\pi)^2}$$

we obtained the following values for n = N/2 - 1:

N	$A_p(n)$	$B_p(n)$
32	-8.3682475e-4	-3.9434437e-2
128	-4.7459196e-5	-9.3931400e-3
512	-2.8968868e-6	-2.3207137e-3
2048	-1.7999533e-7	-5.7847788e-4

Based on the above values for A_n and $A_p(n)$ we now obtained the errors shown in the following table, where

$$(\text{actual error in A}) = \frac{A_n - A_p(n)}{A_p(n)} \times 100$$

For the "formula error" we used $E(Z,1) = [(\pi Z)^2/\sin^2(\pi Z) - 1]\times 100$ where $Z = n/N$ and $n = N/2 - 1$, i.e.

$$Z = \frac{N/2 - 1}{N} = \tfrac{1}{2} - 1/N$$

Table 13.7

N	actual error in A	formula error	error
32	118.9903548	118.9651015	0.02 %
128	139.2350258	139.2338049	0.0009
512	144.8255068	144.8254355	0.00005
2048	146.2590356	146.2590104	0.00002
∞	—	146.7401101	—

The final column of the table is computed from

$$\text{error} = \frac{\text{column 2} - \text{column 3}}{\text{column 3}} \times 100$$

and shows clearly that as $N \longrightarrow \infty$, the actual error in A_n is tending to the formula error, as predicted by Theorem 13.3.

Based on the above values for B_n and $B_p(n)$ we now obtained the errors shown in the following table, where

$$(\text{actual error in B}) = \frac{B_n - B_p(n)}{B_p(n)} \times 100$$

Table 13.8

N	actual error in jB	formula error	error
32	85.50371083	85.49594371	0.009
128	96.20429775	96.20416281	0.001
512	99.03992654	99.03992441	0.000002
2048	99.75927802	99.75927798	0.00000004
∞	—	100.00000000	—

For the "formula error" we used $E(Z,0) = [\pi Z \cot(\pi Z) - 1]\times 100$ where $Z = n/N$ and $n = N/2 - 1$, i.e.

$$Z = \frac{N/2 - 1}{N} = \tfrac{1}{2} - 1/N$$

The final column of the table is computed from

$$\text{error} = \frac{\text{column-2} - \text{column-3}}{\text{column-3}} \times 100$$

and shows clearly that as $N \longrightarrow \infty$, the actual error in B_n is tending to the formula error, as predicted by Theorem 13.3.

13.10 From (13.11), if f(t) is canonical-0 with its discontinuities at the points g_pT_s $(p = 1, 2, 3, \ldots, r)$ then its Fourier transform will be

$$F(\omega) = 1/j\omega \sum_{p=1}^{r} \exp(-j\omega g_pT_s)$$

Since all of the discontinuities are assumed to be at the centers of FFT sampling intervals it follows that

$g_p = h_p + \tfrac{1}{2}$ where h_p is an integer. Then

$$F[(n-mN)\omega_0] = 1/j(n-mN)\omega_0 \sum_{p=1}^{r} \exp[-j(n-mN)\omega_0 g_pT_s]$$

$$= 1/j(n-mN)\omega_0 \sum_{p=1}^{r} \exp(-jn\omega_0 g_pT_s)\ \exp(jmN\omega_0 g_pT_s)$$

However,

$$mN\omega_0 g_pT_s = mN(2\pi/T)g_p(T/N) = m2\pi g_p = m2\pi(h_p + \tfrac{1}{2}) = m2\pi h_p + m\pi$$

and so

$$\exp(jmN\omega_0 g_pT_s) = \exp[j(m2\pi h_p + jm\pi)] = \exp(jm2\pi h_p)\ \exp(jm\pi) = (-1)^m$$

Thus

$$F[(n-mN)\omega_0] = 1/j[(n-mN)\omega_0] \sum_{p=1}^{r} (-1)^m \exp(-jn\omega_0 g_pT_s)$$

$$= (-1)^m/j[(n-mN)\omega_0] \sum_{p=1}^{r} \exp(-jn\omega_0 g_pT_s)$$

Also

$$F(n\omega_0) = 1/jn\omega_0 \sum_{p=1}^{r} \exp(-jn\omega_0 g_pT_s)$$

Then, using (13.8) to form the relative error:

$$E_N(n) = \sum_{\substack{m=-\infty \\ m\neq 0}}^{\infty} \frac{F[(n - mN)\omega_0]}{F(n\omega_0)}$$

we obtain

$$E_N(n) = \sum_{\substack{m=-\infty \\ m\neq 0}}^{\infty} \frac{(-1)^m/j[(n - mN)\omega_0] \sum_{p=1}^{r} \exp(-jn\omega_0 g_pT_s)}{1/jn\omega_0 \sum_{p=1}^{r} \exp(-jn\omega_0 g_pT_s)}$$

$$= \sum_{\substack{m=-\infty \\ m\neq 0}}^{\infty} \frac{(-1)^m/j[(n - mN)\omega_0]}{1/jn\omega_0} = \sum_{\substack{m=-\infty \\ m\neq 0}}^{\infty} (-1)^m \frac{n}{n - mN} \qquad \text{QED}$$

Repeating all of the above for a canonical-k pulse:

From (13.11), if f(t) is canonical-k with its breakpoints at the points g_pT_s $(p = 1, 2, 3, \ldots, r)$ then its Fourier transform

$$F(\omega) = 1/(j\omega)^{k+1} \sum_{p=1}^{r} \exp(-j\omega g_p T_s)$$

Since all of the breakpoints are assumed to be at the centers of FFT sampling intervals it follows that

$$g_p = h_p + \tfrac{1}{2}$$

where h_p is an integer. Then

$$F[(n-mN)\omega_0] = 1/[j(n-mN)\omega_0]^{k+1} \sum_{p=1}^{r} \exp[-j(n-mN)\omega_0 g_p T_s]$$

$$= 1/[j(n-mN)\omega_0]^{k+1} \sum_{p=1}^{r} \exp(-jn\omega_0 g_p T_s)\ \exp(jmN\omega_0 g_p T_s)$$

However,

$$mN\omega_0 g_p T_s = mN(2\pi/T)g_p(T/N) = m2\pi g_p = m2\pi(h_p + \tfrac{1}{2}) = m2\pi h_p + m\pi$$

and so

$$\exp(jmN\omega_0 g_p T_s) = \exp[j(m2\pi h_p + jm\pi)] = \exp(jm2\pi h_p)\ \exp(jm\pi) = (-1)^m$$

Thus

$$F[(n-mN)\omega_0] = 1/[j(n-mN)\omega_0]^{k+1} \sum_{p=1}^{r} (-1)^m \exp(-jn\omega_0 g_p T_s)$$

$$= (-1)^m/[j(n-mN)\omega_0]^{k+1} \sum_{p=1}^{r} \exp(-jn\omega_0 g_p T_s)$$

Also

$$F(n\omega_0) = 1/(jn\omega_0)^{k+1} \sum_{p=1}^{r} \exp(-jn\omega_0 g_p T_s)$$

Then, using (13.8) to form the relative error:

$$E_N(n) = \sum_{\substack{m=-\infty \\ m\neq 0}}^{\infty} \frac{F[(n - mN)\omega_0]}{F(n\omega_0)}$$

we obtain

$$E_N(n) = \sum_{\substack{m=-\infty \\ m\neq 0}}^{\infty} \frac{(-1)^m/[j(n - mN)\omega_0]^{k+1} \sum_{p=1}^{r} \exp(-jn\omega_0 g_p T_s)}{1/(jn\omega_0)^{k+1} \sum_{p=1}^{r} \exp(-jn\omega_0 g_p T_s)}$$

$$= \sum_{\substack{m=-\infty \\ m\neq 0}}^{\infty} \frac{(-1)^m/[j(n - mN)\omega_0]^{k+1}}{1/(jn\omega_0)^{k+1}} = \sum_{\substack{m=-\infty \\ m\neq 0}}^{\infty} (-1)^m \frac{n^{k+1}}{(n - mN)^{k+1}}$$

QED

13.11

(a) A sketch of the pulse $f(t) = \exp[-5(t - \tfrac{1}{2}T_s)]\ U(t - \tfrac{1}{2}T_s)$ is given in the Answers. We observe that it is a slightly shifted decaying exponential which starts at $t = \tfrac{1}{2}T_s$ and falls off exponentially as t increases.

(b) We know that $\exp(-5t)U(t) \Longleftrightarrow 1/(j\omega + 5)$ and so, using time-shift

$$f(t) = \exp[-5(t - \tfrac{1}{2}T_s)]\ U(t - \tfrac{1}{2}T_s) \Longleftrightarrow F(\omega) = \frac{\exp(-j\omega\tfrac{1}{2}T_s)}{j\omega + 5}$$

Separating $F(\omega)$ into its real and imaginary parts:

$$F(\omega) = \frac{[\cos(\omega\tfrac{1}{2}T_s) - j\ \sin(\omega\tfrac{1}{2}T_s)]\ [5 - j\omega]}{\omega^2 + 25}$$

$$= \frac{5\ \cos(\omega\tfrac{1}{2}T_s) - \omega\ \sin(\omega\tfrac{1}{2}T_s)}{\omega^2 + 25} - j\ \frac{\omega\ \cos(\omega\tfrac{1}{2}T_s) + 5\ \sin(\omega\tfrac{1}{2}T_s)}{\omega^2 + 25}$$

$$= A(\omega) + jB(\omega) \qquad \text{(I)}$$

(c) If we sample starting at $t = 0$ with sampling interval T_s then the discontinuity in $f(t)$ at $t = \tfrac{1}{2}T_s$ is not at a sampling instant and instead falls half way between two such instants. According to Section 13.8, we can then expect that the FFT-estimation errors will be less than for the pulse $\exp(-5t)\ U(t)$ whose discontinuity coincides with a sampling instant. We now verify that.

Steps for loading $f(t)$ into **X** using $N = 1024$ and $T = 4$:

Note: $T_s = T/N = 0.00390625$ and so $\tfrac{1}{2}T_s = 0.001953125$

Main menu, Create Values, Continue, Create X, Real, Neither, Left, 2 intervals, 0.001953125, Continue, 0, EXP(-5*(t - 0.001953125)), Go, Neither, Go. You are now back on the main menu with the pulse properly loaded. Plotting **X** confirms that.

Note: Draw the plot of **X** using FRACTION = 0.1 and you will see that there is no half-value, a consequence of the fact that the discontinuity did not coincide with an FFT sampling instant. Also, the max value of the pulse on the FFT should be exp(-5x0.001953125) = 0.990281904 which is confirmed in the FFT display.

Run ANALYSIS, Show Numbers, F, Single.

From the FFT we obtain the values shown in the following table where we also show the exact values of $A(n\omega_0)$ and $B(n\omega_0)$ obtained from (I), where $\omega_0 = 2\pi/T = \pi/2$

n	$A_{n(FFT)}$	$A(n\omega_0)$	$B_{n(FFT)}$	$B(n\omega_0)$
0	0.19999682	0.2	0	0
1	0.18185448	0.181857662	-5.7746849e-2	-5.7745860e-2
8	2.5637683e-2	2.5641057e-2	-6.9359161e-2	-6.9351251e-2
32	1.2910936e-5	1.9221047e-5	-1.9828185e-2	-1.9796659e-2
128	-1.8408491e-3	-1.7879372e-3	-4.7584351e-3	-4.6394621e-3
256	-1.9337755e-3	-1.7362975e-3	-1.9719158e-3	-1.7362975e-3
512	-1.9530319e-3	-1.2433499e-3	0	0

Using these values we computed the relative errors in the FFT's estimates of $F(\omega)$ from

$$\text{rel error} = \left[\frac{[A_{n(FFT)} - A(n\omega_0)]^2 + [B_{n(FFT)} - B(n\omega_0)]^2}{[A(n\omega_0)]^2 + [B(n\omega_0)]^2}\right]^{\frac{1}{2}} \times 100$$

Observe that we are using the **magnitude** of the formula in (13.44).

The results are shown in the final column of Table 13.9 on the next page.

In the first column of that table we also show the values of $-E(Z,0) = -[\pi Z\cos(\pi Z) - 1]\times 100$. In the second column of that table we show the errors in the FFT's estimates of $g(t) = \exp(-5t)\,U(t)$ computed as follows:

Steps for loading g(t) using N = 1024, T = 4: Main menu, Create Values, Continue, Create X, Real, Neither, Left, 1 interval, Continue, EXP(-5*t), Go, Neither, Go. The pulse is now correctly loaded. Plotting **X** confirms that. Run ANALYSIS, Show Numbers, F, Single.

The transform of g(t) is $G(\omega) = 1/(j\omega + 5)$ whose real and imaginary parts are

$$A(\omega) = 5/(\omega^2 + 25) \quad \text{and} \quad B(\omega) = -\omega/(\omega^2 + 25)$$

From the FFT and these formulae we obtained the following values: Note: $\omega_0 = 2\pi/T = \pi/2$

n	$A_{n(FFT)}$	$A(n\omega_0)$	$B_{n(FFT)}$	$B(n\omega_0)$
0	0.20000636	0.2	0	0
1	0.18204032	0.182033968	-5.7185660e-2	-5.7187657e-2
8	2.7341658e-2	2.7335299e-2	-6.8685122e-2	-6.8701101e-2
32	1.9659105e-3	1.9595404e-3	-1.9635493e-2	-1.9699448e-2
128	1.3016542e-4	1.2360664e-4	-4.7121921e-3	-4.9705181e-3
256	3.8142123e-5	3.0915991e-5	-1.9527525e-3	-2.4864115e-3
512	1.9072880e-5	7.7298940e-6	0	1.2433499e-[illegible]

Using these values we computed the relative errors in the FFT's estimates of $G(\omega)$ using

$$\text{rel error} = \left[\frac{[A_{n(FFT)} - A(n\omega_0)]^2 + [B_{n(FFT)} - B(n\omega_0)]^2}{[A(n\omega_0)]^2 + [B(n\omega_0)]^2}\right]^{\frac{1}{2}} \times 100$$

Observe that we are using the **magnitude** of the formula in (13.44). All of the results are shown in the final column of Table 13.9.

Table 13.9: % Errors for decaying exponential

n	E(Z,0)	exp(-5t)U(t)	f(t)
0	0.00000000	0.003180000	0.001590000
1	0.00031376	0.003489688	0.001746360
8	0.02008056	0.023259304	0.011630489
32	0.32148281	0.324657959	0.162407637
128	5.19405510	5.197232531	2.618805924
256	21.46018365	21.46332856	12.52010254
512	100.00000000	100.0022287	57.07822070

(d) Observe that

- Other than for very small values of n, the error values for exp(-5t) U(t) are almost the same as the canonical-0 values. This is to be expected from Theorem 13.3.
- The error values for the shifted exponential f(t) whose breakpoint lies **halfway through a sampling interval** are approximately one half of the canonical-0 values. This reconfirms what was discussed in Section 13.8.

13.12 Steps for loading the pulse of Exercise 13.3 are given in that exercise. Use N = 32 and T = 8 for this exercise.

After loading the pulse into the system and running ANALYSIS we were able to fill in the values in the first column of the table.

In this case the pulse is canonical-1 and so the relative error is

$$E(Z,1) = \frac{\text{FFT value} - \text{exact value}}{\text{exact value}} = \frac{\text{FFT value}}{\text{exact value}} - 1 = \frac{(\pi Z)^2}{\sin^2(\pi Z)} - 1$$

from which

$$\text{Factor} \equiv \frac{\text{FFT value}}{\quad} = \frac{(\pi Z)^2}{\quad} \qquad \text{where } Z = n/N$$

We are thus able to calculate the value of Factor once n and N are given. This enabled us to fill in the second column of the table.

It also follows that

exact value = (FFT value)/Factor

and so once the FFT value is given and Factor is known, we can calculate the exact value. This enabled us to fill in the third column of the table.

The Fourier transform of the pulse is derived in Exercise 13.3 as

$$A(\omega) = [\cos(\omega) + \cos(3\omega) - \cos(4\omega) - 1]/\omega^2 \qquad \text{(I)}$$

$$B(\omega) = [\sin(4\omega) - \sin(\omega) - \sin(3\omega)]/\omega^2 \qquad \text{(II)}$$

The FFT's frequency-domain sampling interval is $\omega_0 = 2\pi/T$ and from (I) and (II) we were able to arrive at the exact values of

Table 13.10: Error correction for a canonical-1 pulse

n	FFT value	factor	FFT/factor	$F(n\omega_0)$
0	A_0 = 3	1	3	3
1	B_1 = −2.3000166	1.003218964	−j2.292636685	−j2.292636672
2	A_2 = −0.82106695	1.012950747	−0.810569470	−0.810569469
3	B_3 = −0.26223267	1.029423491	−j0.254737406	−j0.254737409
4	A_4 = −0.42677670	1.053029288	−0.405284739	−0.405284734

$F(n\omega_0)$ for n = 1, 2, 3, 4. (For n = 0 we used l'Hôpital's rule.) These values are shown in the final column.

Observe how the final two columns are essentially identical.

In the following three exercises we demonstrate that the error expressions

$$E(Z,0) = \pi Z \cot(\pi Z) - 1 \qquad (|Z| \le \tfrac{1}{2}) \qquad (13.45)$$

and

$$E(Z,1) = \frac{(\pi Z)^2}{\sin^2(\pi Z)} - 1 \qquad (|Z| \le \tfrac{1}{2}) \qquad (13.46)$$

can be derived by direct means, without recourse to the series appearing in (13.15) or to residue theory. These exercises are somewhat more challenging than the ones above.

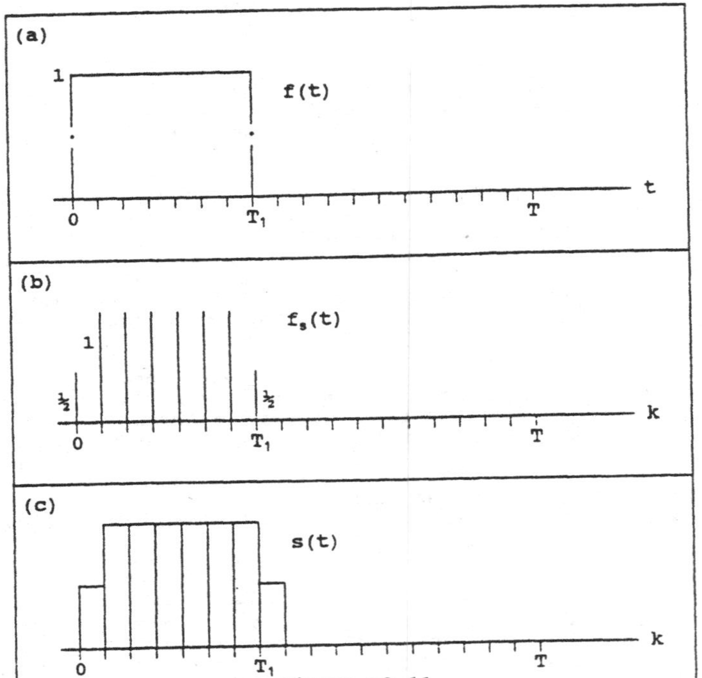

Figure 13.11

Using half-values at its discontinuities and multiplying it by the following train of unit impulses

$$\delta_T(t) = \sum_{k=-\infty}^{\infty} \delta(t - kT_s) \qquad (T_s = T/N) \qquad (13.47)$$

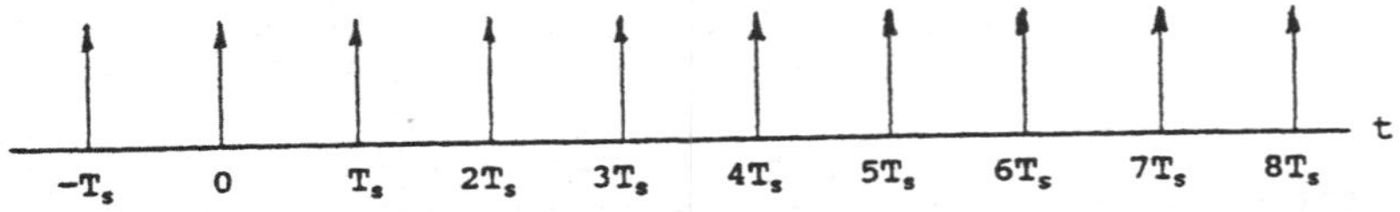

gives us the sequence of impulses shown in (b) of the figure, namely

$$f_s(t) = \sum_{k=-\infty}^{\infty} f(kT_s)\, \delta(t - kT_s) \qquad (13.48)$$

in which it is assumed that all discontinuities coincide with sampling instants.

We then use $f_s(t)$ of (13.48) as the input to a network whose impulse response $h(t)$ is shown in Figure 13.12 further down.

(a) We now show that $h(t) \Longleftrightarrow H(j\omega) = \exp(-j\omega T_s/2)\ T_s Sa(\omega T_s/2)$ (13.49)

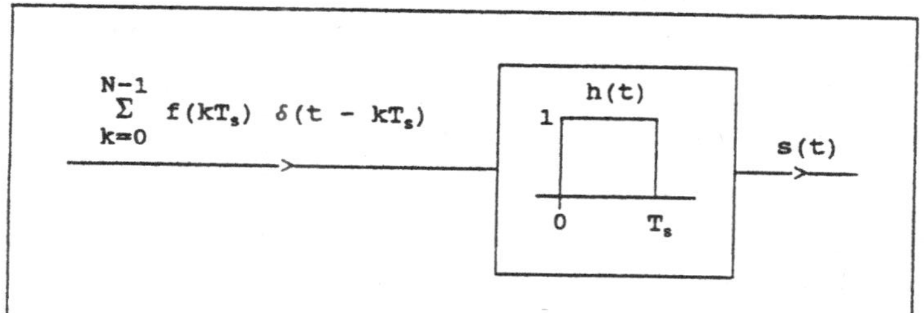

Figure 13.12

From Figure 13.12: $h(t)$ is $Rect(t/T_s)$ shifted to the right by an amount $T_s/2$. Thus

$$h(t) = Rect[(t - T_s/2)/T_s] = Rect[(t - \tau/2)/\tau] \quad \text{where } \tau = T_s$$

But $Rect(t/\tau) \Longleftrightarrow \tau\ Sa(\omega\tau/2)$, and so, using time shift, $h(t)$ transforms to

$$H(\omega) = \exp(-j\omega T_s/2)\ T_s\ Sa(\omega T_s/2) \qquad (A) \qquad \text{QED}$$

(b) To show that the response to the input $f_s(t)$ will be the pulse $s(t)$ appearing in Figure 13.11(c):

The impulse response of the filter is $h(t) = Rect[(t - T_s/2)/T_s]$

and so the k-th impulse in $f_s(t) = \sum_{k=-\infty}^{\infty} f(kT_s)\ \delta(t - kT_s)$ namely

$\delta(t - kT_s)$ must produce the response

$$h_k(t) = Rect[(t - kT_s - T_s/2)/T_s]$$

Then the entire train of impulses called $f_s(t)$ must produce the response $s(t)$ shown in Figure 11.3(c), namely

$$s(t) = \sum_{k=-\infty}^{\infty} f(kT_s)\ Rect[(t - kT_s - T_s/2)/T_s] \qquad (B) \qquad \text{QED}$$

(c) The response of the network is $s(t) \Longleftrightarrow S(\omega)$. We now verify that

$$S(n\omega_0) = \exp(-j\pi Z)\ Sa(\pi Z)\ (T/N)\ F_n \qquad (13.50)$$

where $\omega_0 = 2\pi/T$, $Z = n/N$ and F_n is the DFT of $f(t)$ sampled at the points , $t_k = kT_s$ ($k = 0, 1, 2, \ldots, N-1$).

First we need to find $S(\omega)$. By (A):

$$Rect[(t - kT_s - T_s/2)/T_s] \Longleftrightarrow T_s\ Sa(\omega T_s/2)\ \exp[-j\omega(kT_s + T_s/2)]$$

and so

$$S(\omega) = \sum_{k=-\infty}^{\infty} f(kT_s)\ T_s\ Sa(\omega T_s/2)\ \exp[-j\omega(kT_s + T_s/2)]$$

The pulse $f(t)$ was sampled at N points in the range $0 \le t < T$ and was assumed to be zero outside that range. Thus the range of summation is actually from $k = 0$ to $k = N-1$, and so

$$S(\omega) = \sum_{k=0}^{N-1} f(kT_s)\ T_s\ Sa(\omega T_s/2)\ \exp[-j\omega(kT_s + T_s/2)]$$

$$= T_s\ Sa(\omega T_s/2)\ \exp(-j\omega T_s/2) \sum_{k=0}^{N-1} f(kT_s)\ \exp(-j\omega kT_s)$$

Then, sampling ω at $n\omega_0 = = n2\pi/T$ we have

$$S(n\omega_0) = T_s\ Sa(n\pi T_s/T)\ \exp(-jn\pi T_s/T) \sum_{k=0}^{N-1} f(kT_s)\ \exp(-jn2\pi T_s/Tk)$$

But $T_s = T/N$ and so $T_s/T = 1/N$, and so

$$S(n\omega_0) = T/N\ Sa(n\pi/N)\ \exp(-jn\pi/N) \sum_{k=0}^{N-1} f(kT_s)\ \exp(-jn2\pi T_s/Tk)$$

Also $n/N = Z$ and so

$$S(n\omega_0) = T/N\ Sa(\pi Z)\ \exp(-j\pi Z) \sum_{k=0}^{N-1} f(kT_s)\ \exp(-jn2\pi T_s/Tk)$$

Finally, $\sum_{k=0}^{N-1} f(kT_s)\ \exp(-jn2\pi T_s/Tk) = F_n$, and so

$$S(n\omega_0) = \exp(-j\pi Z)\ Sa(\pi Z)\ (T/N)\ F_n \qquad \text{QED} \qquad (13.50)$$

(d) We observe that any pulse, canonical or otherwise, can be sampled in this way to give its version of (13.48) and its version of $s(t)$ in (B) which we call its **left-endpoint uniform rectangular replacement**.

(e) We also observe also that $S(n\omega_0)$ in (13.50) is a formula which can be derived for **any** pulse $f(t)$ which has a DFT.

13.14

(a) We now verify that if the canonical-0 pulse $f(t)$ of Figure 13.11 is the input to the network shown in Figure 13.13, then the response is $s(t)$ of Figure 13.11(c) i.e. its left-endpoint uniform rectangular replacement.

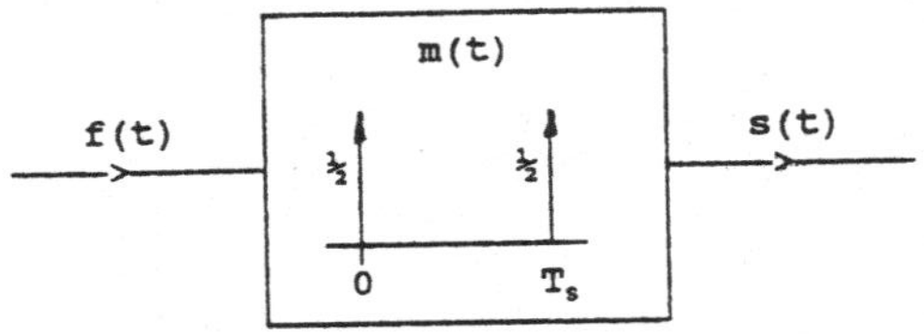

Figure 13.13

From the figure we see that $m(t) = \frac{1}{2}[\delta(t) + \delta(t - T_s)]$, and so if the input is $f(t)$ of Figure 13.11(a), then the response will be $g(t) = \frac{1}{2}[f(t) + f(t - T_s)]$. Graphically, we now have the situation shown in the following figure:

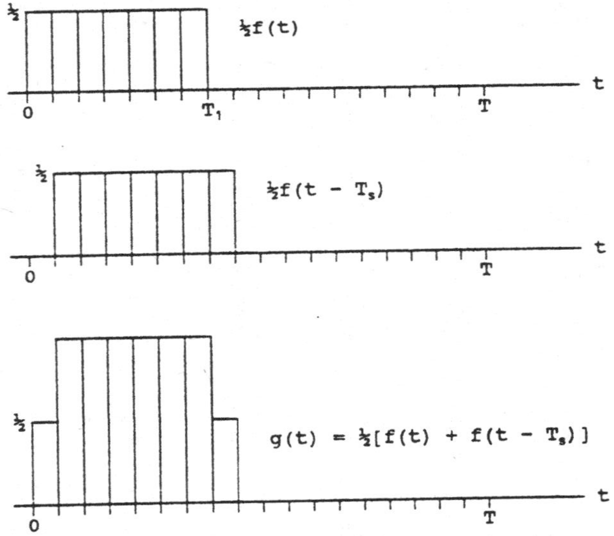

From this graphical construction we see that $g(t)$ is in fact $s(t)$, the left-endpoint uniform rectangular replacement of $f(t)$. (Note that the half-values at the discontinuities have received the proper treatment.)

Moreover, we also see that this procedure will apply to **any** canonical-0 pulse.

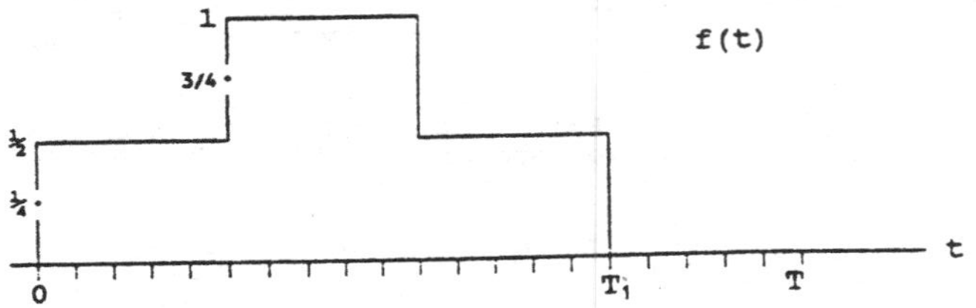

In the construction given below we show that forming the pulse $g(t) = \frac{1}{2}[f(t) + f(t - T_s)]$ results in its left-endpoint uniform replacement, complete with correct treatment of the half-values.

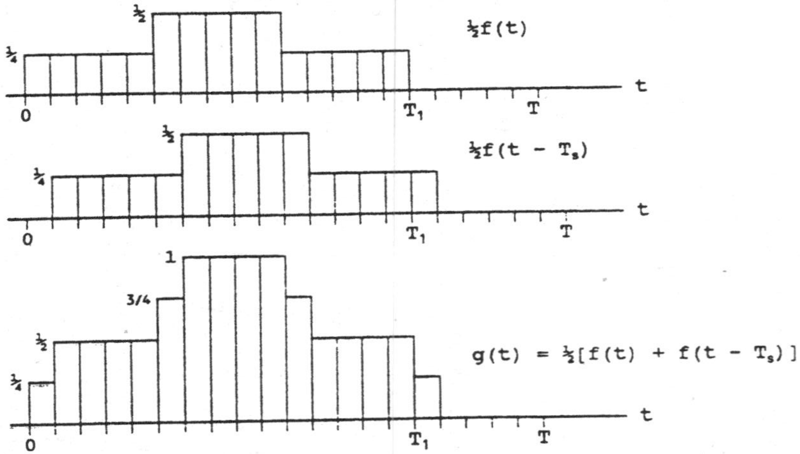

(b) From Figure 13.13:

$$m(t) = \tfrac{1}{2}[\delta(t) + \delta(t - T_s)] \Longleftrightarrow M(j\omega) = \tfrac{1}{2}[1 + \exp(-j\omega T_s)]$$

The input is $f(t) \Longleftrightarrow F(\omega)$ and so the response to $s(t)$ must have Fourier transform

$$S(\omega) = F(\omega)\, M(j\omega) \qquad (13.51)$$

where $M(j\omega) = \frac{1}{2}[1 + \exp(-j\omega T_s)]$ (13.52)

(c) We now wish to form the quantity

$$E(Z,0) = \frac{(T/N)\, F_n - F(n\omega_0)}{F(n\omega_0)}$$

which is the statement for the relative error of the FFT's estimate of $F(\omega)$. We do it as follows. From (13.50):

$$S(n\omega_0) = \exp(-j\pi Z)\ Sa(\pi Z)\ (T/N)\ F_n \qquad (P)$$

and from (13.51):

$$S(n\omega_0) = F(n\omega_0)\ M(jn\omega_0)$$

$$= F(n\omega_0)\ \tfrac{1}{2}[1 + \exp(-jn\omega_0 T_s)] \qquad \text{but } n\omega_0 T_s = n(2\pi/T)(T/N) = 2\pi n/N = 2\pi Z$$

$$= F(n\omega_0)\ \tfrac{1}{2}[1 + \exp(-j2\pi Z)]$$

$$= F(n\omega_0)\ \exp(-j\pi Z)\ \tfrac{1}{2}[\exp(j\pi Z) + \exp(-j\pi Z)]$$

$$= F(n\omega_0)\ \exp(-j\pi Z)\ \cos(j\pi Z) \qquad (Q)$$

From (P): $(T/N)\ F_n = \exp(j\pi Z)\ S(n\omega_0)/Sa(\pi Z)$

From (Q): $F(n\omega_0) = \exp(j\pi Z)\ S(n\omega_0)/\cos(\pi Z)$

Then:

$$E(Z,0) = \frac{(T/N)\ F_n - F(n\omega_0)}{F(n\omega_0)}$$

$$= \frac{\exp(j\pi Z)\ S(n\omega_0)/Sa(\pi Z) - \exp(j\pi Z)\ S(n\omega_0)/\cos(\pi Z)}{\exp(j\pi Z)\ S(n\omega_0)/\cos(\pi Z)}$$

$$= \frac{1/Sa(\pi Z) - 1/\cos(\pi Z)}{1/\cos(\pi Z)} = \cos(\pi Z)/Sa(\pi Z) - 1$$

$$= \frac{\pi Z\ \cos(\pi Z)}{\sin(\pi Z)} - 1 = \underrightarrow{\pi Z\ \cot(\pi Z) - 1} \qquad (13.53) \qquad \text{QED}$$

(d) Observe that

- We have verified (13.32) without the need for the infinite series appearing in (13.15) or residue theory
- What we have done here applies to **any** canonical-0 pulse

13.15

(a) Let $f(t)$ be the input to a network whose impulse response is a unit doublet , namely

$$d(t - T_0) = D\delta(t - T_0) \qquad (13.54)$$

Then the the response will be

$$y(t) = f(t) * D\delta(t - T_0)$$

$$\Longleftrightarrow F(\omega)\ j\omega\ \exp(-j\omega T_0) = j\omega\ F(\omega)\ \exp(-j\omega T_0) \Longleftrightarrow f'(t - T_0)$$

(b) If $f(t)$ is the input to a network which has impulse response

$$m(t) = T_s \sum_{k=1}^{\infty} d(t - kT_s) \qquad (13.55)$$

then the output will be

$$y(t) = f(t) * m(t) = f(t) * \left[T_s \sum_{k=1}^{\infty} d(t - kT_s) \right]$$

$$\Longleftrightarrow F(\omega)\ T_s \sum_{k=1}^{\infty} j\omega\ \exp(-j\omega kT_s) = T_s \sum_{k=1}^{\infty} j\omega\ F(\omega)\ \exp(-j\omega kT_s)$$

$$\Longleftrightarrow T_s \sum_{k=1}^{\infty} f'(t - kT_s) \qquad \text{QED} \qquad (13.56)$$

(c) We now show that if $f(t)$ above is any canonical-1 pulse, then the response $y(t)$ emerging from $m(t)$ will be its left-endpoint uniform rectangular replacement.

Thus, let $f(t)$ be the canonical-1 pulse of Figure 13.14(a), namely

$$f(t) = \begin{cases} 0 & (t < 0) \\ t/4T_s & (0 < t < 4T_s) \\ 2 - t/4T_s & (4T_s < t < 8T_s) \\ 0 & (8T_s < t) \end{cases}$$

Then

$$f'(t) = \begin{cases} 0 & (t < 0) \\ 1/4T_s & (0 < t < 4T_s) \\ -1/4T_s & (4T_s < t < 8T_s) \\ 0 & (8T_s < t) \end{cases}$$

from which

$$T_s f'(t) = \begin{cases} 0 & (t < 0) \\ \tfrac{1}{4} & (0 < t < 4T_s) \\ -\tfrac{1}{4} & (4T_s < t < 8T_s) \\ 0 & (8T_s < t) \end{cases}$$

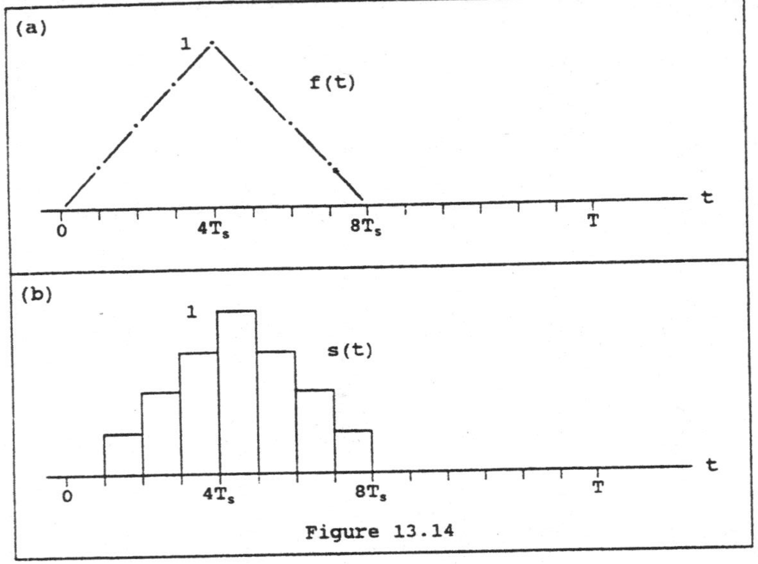

Figure 13.14

In the Answers we show the graphical construction which sketches the following:

line 1: f(t)
line 2: $T_s f'(t - 1T_s)$
line 3: $T_s f'(t - 2T_s)$
line 4: $T_s f'(t - 3T_s)$
....
line 12: $T_s f'(t - 11T_s)$
line 13: Sum of lines 2 through 12.

As shown there, line 13 is the same as s(t) of Figure 13.14(b), i.e. we have verified that the sum

$$s(t) = T_s \sum_{k=1}^{\infty} f'(t - kT_s) \qquad (13.57)$$

is the left-endpoint uniform rectangular replacement for f(t). But from (13.56) $s(t) = f(t) * m(t)$, i.e. it is the response to the pulse f(t) when it is submitted to a network with impulse response m(t). Thus we have proved our proposition at least for the pulse considered in Figure 13.4(a) above.

(d) In the following figure we show a similar construction for a second canonical-1 pulse f(t). Line-by-line the construction is as follows:

line 2: $T_s f'(t - 1T_s)$
line 3: $T_s f'(t - 2T_s)$
....
line 12: $T_s f'(t - 11T_s)$
line 13: Sum of lines 2 through 12.

Once again line 13 is seen to be the left endpoint uniform rectangular replacement of f(t).

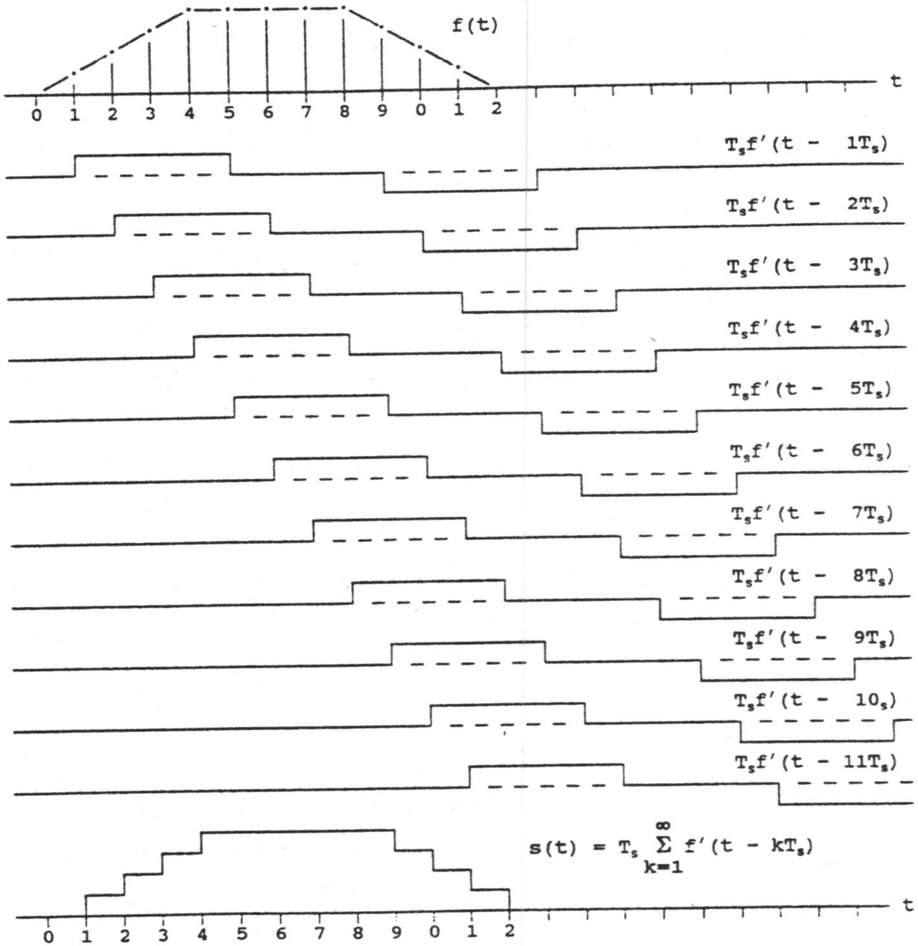

Now, any canonical-1 pulse consists of a mix of, at most, the following:

A horizontal followed by a ramp

An ramp followed by a horizontal

A ramp followed by a different ramp

and there cannot be any other possibilities. We have now examined two canonical-1 cases that cover all of these situations, either explicitly, or else by inference. In all cases the construction that we used has led correctly to the left-endpont uniform rectangular replacement of the original pulse.

Thus we can infer that m(t) will convert any canonical-1 pulse to its left-endpoint uniform rectangular replacement.

(e) From (13.55):

$$m(t) = T_s \sum_{k=1}^{\infty} d(t - kT_s)$$

$$\Longleftrightarrow M(j\omega) = T_s \sum_{k=1}^{\infty} j\omega \exp(-j\omega kT_s) = j\omega T_s \sum_{k=1}^{\infty} \exp(-j\omega kT_s) \qquad (13.58)$$

(f) Using formal summation on (13.58) we obtain

$$M(j\omega) = j\omega T_s \sum_{k=1}^{\infty} \exp(-j\omega kT_s) = j\omega T_s \exp(-j\omega T_s) \sum_{k=0}^{\infty} \exp(-j\omega kT_s)$$

$$= \frac{j\omega T_s \exp(-j\omega T_s)}{1 - \exp(-j\omega T_s)} \qquad (13.59)$$

(g) From (13.57):

$$s(t) = T_s \sum_{k=1}^{\infty} f'(t - kT_s) = f(t) * m(t) \Longleftrightarrow S(\omega) = F(\omega)\, M(\omega)$$

Thus, by (13.59)

$$S(\omega) = F(\omega) \frac{j\omega T_s \exp(-j\omega T_s)}{1 - \exp(-j\omega T_s)} \qquad (13.60)$$

(h) Sampling (13.60) at $\omega = n\omega_0$, where $\omega_0 = 2\pi/T$, we obtain

$$S(n\omega_0) = F(n\omega_0) \frac{jn\omega_0 T_s \exp(-jn\omega_0 T_s)}{1 - \exp(-jn\omega_0 T_s)}$$

But $n\omega_0 T_s = n(2\pi/T)(T/N) = 2\pi n/N = 2\pi Z$ (where $Z = n/N$) and so we continue

$$\ldots = F(n\omega_0) \frac{j2\pi Z \exp(-j2\pi Z)}{1 - \exp(-j2\pi Z)}$$

$$\ldots = F(n\omega_0) \frac{j2\pi Z \exp(-j2\pi Z)}{\exp(-j\pi Z)[\exp(j\pi Z) - \exp(-j\pi Z)]}$$

$$\ldots = F(n\omega_0) \frac{\pi Z \exp(-j\pi Z)}{\sin(\pi Z)}$$

giving us

$$F(n\omega_0) = \frac{\sin(\pi Z) \exp(j\pi Z)}{\pi Z} S(n\omega_0) \qquad (13.61)$$

(i) We now wish to form the quantity

$$E(Z,1) = \frac{(T/N)F_n - F(n\omega_0)}{F(n\omega_0)} \qquad (13.62)$$

which we know is the statement for the relative error of the FFT's estimate of $F(\omega)$. We do that as follows:

From (13.50) (which applies to any pulse that has a DFT):

$$S(n\omega_0) = \exp(-j\pi Z)\, Sa(\pi Z)\, (T/N)\, F_n$$

from which we obtain

$$(T/N)\, F_n = \frac{\exp(j\pi Z)}{Sa(\pi Z)} S(n\omega_0) \qquad (13.63)$$

From (13.61) (which applies to any canonical-1 pulse):

$$F(n\omega_0) = \frac{\sin(\pi Z) \exp(j\pi Z)}{\pi Z} S(n\omega_0) \qquad (13.64)$$

Thus from (13.63) and (13.64):

$$E(Z,1) = \frac{(T/N)F_n - F(n\omega_0)}{F(n\omega_0)}$$

$$= \frac{\dfrac{\exp(j\pi Z)}{Sa(\pi Z)} S(n\omega_0) - \dfrac{\sin(\pi Z) \exp(j\pi Z)}{\pi Z} S(n\omega_0)}{\dfrac{\sin(\pi Z) \exp(j\pi Z)}{\pi Z} S(n\omega_0)}$$

$$= \frac{\dfrac{\pi Z}{\sin(\pi Z)} - \dfrac{\sin(\pi Z)}{\pi Z}}{\dfrac{\sin(\pi Z)}{\pi Z}} = \frac{(\pi Z)^2}{\sin^2(\pi Z)} - 1 \qquad \text{QED}$$

(j) Observe that

- We have verified (13.33) without the need for the infinite series appearing in (13.15) or residue theory
- What we have done here applies to **any** canonical-1 pulse

Chapter 14 The Four Kinds of Convolution

(14.1)

k	-4	-3	-2	-1	0	1	2	3	4	5	6	7	8	9	10	11
g_k	0	0	0	0	1	0	1	0	0	2	0	0	0	0	0	0
h_k	0	0	0	0	2	1	2	3	1	0	0	0	0	0	0	0

(a) <u>Discrete linear convolution</u>: $f_k = \sum_{m=-\infty}^{\infty} g_m h_{k-m}$

First we change to the m-axis as follows:

m	-4	-3	-2	-1	0	1	2	3	4	5	6	7	8	9	10	11
g_m	0	0	0	0	1	0	1	0	0	2	0	0	0	0	0	0
h_m	0	0	0	0	2	1	2	3	1	0	0	0	0	0	0	0

Then we reverse h_m to form h_{k-m} (shown here for k = -1):

	m	-4	-3	-2	-1	0	1	2	3	4	5	6	7	8	9	10	11
	g_m	0	0	0	0	1	0	1	0	0	2	0	0	0	0	0	0
h_{k-m}	1	3	2	1	2	——————>											

Then for each value of k: Slide h_{k-m} past g_m, multiply corresponding values together and add them to obtain f_k.

k	-4	-3	-2	-1	0	1	2	3	4	5	6	7	8	9	10	11
f_k	0	0	0	0	2	1	4	4	3	7	3	4	6	2	0	0

(b) <u>Circular convolution (N = 10)</u>

The g-values are laid down on a circle which has 10 slots as shown in the following figure. The h-values are then laid down in the reverse direction.

For each value of k: The circles are laid on top of each other, corresponding values are multiplied together and added to give f_k.

The h-circle is then rotated one notch clockwise and the procedure is repeated, and so on. The resulting values for f_k are as follows:

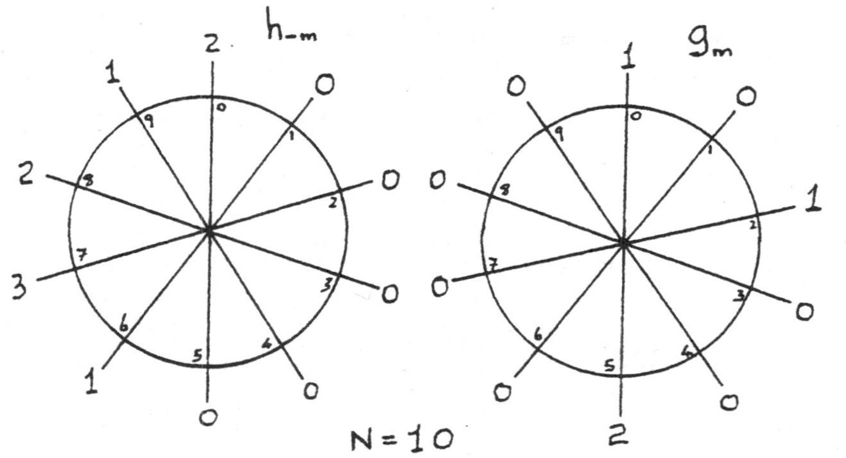

k	0	1	2	3	4	5	6	7	8	9
f_k	2	1	4	4	3	7	3	4	6	2

(c) Steps for loading the two sequences into **X** and **X2** of the FFT system (using N = 10, DISCRETE):

Main menu, Create Values, Continue, Create X and X2, Real, 10 intervals, 0, 1, 2, 3, 4, 5, 6, 7, 8, Continue, 1, 0, 1, 0, 0, 2, 0, 0, 0, 0, Go, 10 intervals, 0, 1, 2, 3, 4, 5, 6, 7, 8, Continue, 2, 1, 2, 3, 1, 0, 0, 0, 0, 0, Go, Go. You are now back on the main menu with the sequences properly loaded. Plotting **X** and **X2** or using "Show Numbers" confirms that.

Run CONVOLUTION. Show Numbers, X and Y. The values in **Y** agree exactly with those obtained in (a) by linear convolution or in (b) by circular convolution.

(d) For g_k: $C_\sigma = 6 = p$. For h_k: $C_\sigma = 5 = q$.

From (14.7) for y_k: We expect that $C_\sigma = p + q - 1 = 6 + 5 - 1 = 10$.
From the actual values obtained for y_k: $C_\sigma = 10$

The span restriction (14.10) requires that $p + q \le N + 1$.

With N = 10: For p = 6 and q = 5 we have $6 + 5 = 11 \le N + 1$ and so the pulses g and h comply with the span restriction. Thus aliasing should not take place.

In fact, the values

did not occur.

(e) Changing h_5 from 0 to 4 gives us the following sequences:

k	0	1	2	3	4	5	6	7	8	9
g_k	1	0	1	0	0	2	0	0	0	0
h_k	2	1	2	3	1	4	0	0	0	0

Now the span counts are as follows:

For g_k: $C_\sigma = 6 = p$. For h_k: $C_\sigma = 6 = q$. With N = 10 the two sequences no longer comply with the span restriction because now $p + q = 12 > N + 1$.

Thus we can expect that aliasing will take place when we perform circular convolution.

(f) Repeating the linear convolution with the new value for h_5 gives us the following:

m		-4	-3	-2	-1	0	1	2	3	4	5	6	7	8	9	10
g_m		0	0	0	0	1	0	1	0	0	2	0	0	0	0	0
h_{k-m}	4	1	3	2	1	2	———>									

k	0	1	2	3	4	5	6	7	8	9	10
f_k	2	1	4	4	3	11	3	8	6	2	8

(*)

Observe that f_k now has a span of 11. This is as expected since $p + q - 1 = 6 + 6 - 1 = 11$.

(g) Repeating the circular convolution, still with N = 10 but with the new value for h_5 gives the results shown below:

k	0	1	2	3	4	5	6	7	8	9
f_k	10	1	4	4	3	11	3	8	6	2

(**)

We observe that aliasing has in fact taken place. The final "8" in (*) has been combined with the first "2" in (*) to give us the "10" in (**). The sequences produced by linear and circular convolution no longer agree.

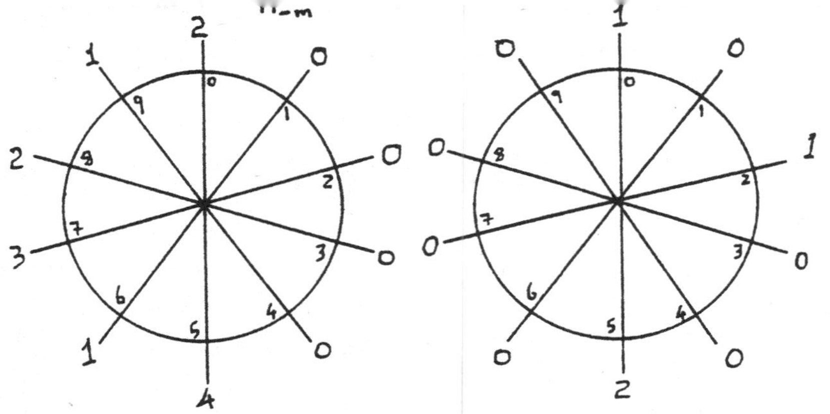

N = 10

Steps for loading the two sequences into **X** and **X2** of the FFT system (using N = 10, DISCRETE):

Main menu, Create Values, Continue, Use the Old-problem, Accept all values except h_5 which is changed from 0 to 4. You are now back on the main menu with the sequences properly loaded. PLotting **X** and **X2** or using "Show Numbers" confirms that.

Run CONVOLUTION. You will be told that the span restriction is being violated. Use "RUN CONVOLUTION REGARDLESS".

Show Numbers, X and Y. The values in **Y** agree exactly with those obtained by circular convolution above. However they do not agree with the values obtained from linear convolution in (f) because aliasing has taken place.

(h) We now change N to 12 and re-run the circular convolution as follows (see figure below). The values obtained are

k	0	1	2	3	4	5	6	7	8	9	10	11
f_k	2	1	4	4	3	11	3	8	6	2	8	0

Observe that aliasing is now no longer present. The values are identical (in the circular sense) to what we obtained in (f) using linear convolution. Observe also that even though we used N = 12, the span of f_k is still $C_\sigma = 11$, as expected.

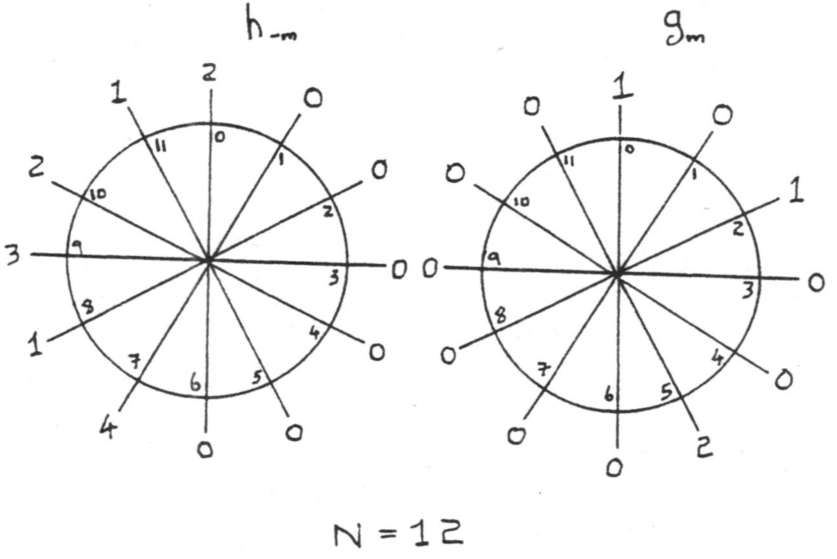

Steps for loading the two sequences into **X** and **X2** of the FFT system (using N = 12, DISCRETE):

Main menu, Create Values, Continue, Create X and X2, Real, 12 intervals, 0, 1, 2, 3, 4, 5, 6, 7, 8, 9, 10, Continue, 1, 0, 1, 0, 0, 2, 0, 0, 0, 0, 0, 0, Go, 10 intervals, 0, 1, 2, 3, 4, 5, 6, 7, 8, 9, Continue, 2, 1, 2, 3, 1, 4, 0, 0, 0, 0, 0, 0, Go, Go. You are now back on the main menu with the sequences properly loaded. PLotting **X** and **X2** or using "Show Numbers" confirms that.

Run CONVOLUTION. Show Numbers, X and Y. The values in **Y** agree exactly with those obtained in (f) by linear convolution or above by circular convolution. Aliasing is now not present.

(i) We can thus infer that by increasing N so that the span restriction is not violated, circular convolution can be successfully used to emulate linear convolution without the danger of aliasing.

(14.2)

k	-9	-8	-7	-6	-5	-4	-3	-2	-1	0	1	2	3	4	5	6
g_k	0	0	0	0	0	1	0	1	0	0	2	0	0	0	0	0
h_k	0	0	0	0	0	0	0	0	2	1	2	3	1	0	0	0

(a) <u>Discrete linear convolution:</u> $f_k = \sum_{m=-\infty}^{\infty} g_m h_{k-m}$

First we change to the m-axis as follows:

m	-9	-8	-7	-6	-5	-4	-3	-2	-1	0	1	2	3	4	5	6
g_m	0	0	0	0	0	1	0	1	0	0	2	0	0	0	0	0
h_m	0	0	0	0	0	0	0	0	2	1	2	3	1	0	0	0

Then we reverse h_m to form h_{k-m} (shown here for k = -7):

	m	-9	-8	-7	-6	-5	-4	-3	-2	-1	0	1	2	3	4	5	6
	g_m	0	0	0	0	0	1	0	1	0	0	2	0	0	0	0	0
h_{k-m}	1	3	2	1	2	———>											

Then for each value of k: Slide h_{k-m} past g_m, multiply corresponding values together and add them to obtain f_k.

k	-9	-8	-7	-6	-5	-4	-3	-2	-1	0	1	2	3	4	5	6
f_k	0	0	0	0	2	1	4	4	3	7	3	4	6	2	0	0

(b) <u>Circular convolution (N = 10)</u>

The g-values are laid down on a circle which has 10 slots as shown in the following figure. The h-values are then laid down in the reverse direction.

For each value of k: The circles are laid on top of each other, corresponding values are multiplied together and added to give f_k.

The h-circle is then rotated one notch clockwise and the procedure is repeated, and so on. The resulting values for f_k are shown further down.

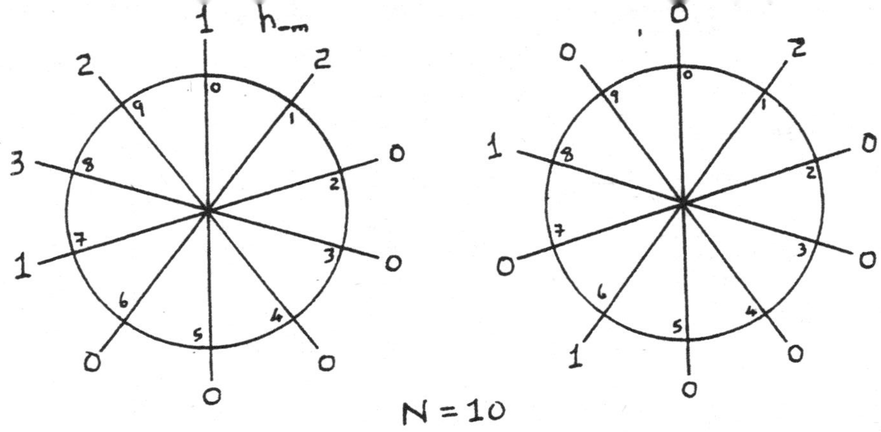

k	0	1	2	3	4	5	6	7	8	9
f_k	7	3	4	6	2	2	1	4	4	3

Observe that these are the same values in the circular sense as those obtained by linear convolution in (a)

(c) Steps for loading the two sequences into **X** and **X2** of the FFT system (using N = 10, DISCRETE):

Main menu, Create Values, Continue, Create X and X2, Real, 10 intervals, 0, 1, 2, 3, 4, 5, 6, 7, 8, Continue, 0, 2, 0, 0, 0, 0, 1, 0, 1, 0, Go, 10 intervals, 0, 1, 2, 3, 4, 5, 6, 7, 8, Continue, 1, 2, 3, 1, 0, 0, 0, 0, 0, 2, Go, Go. You are now back on the main menu with the sequences properly loaded. Plotting **X** and **X2** or using "Show Numbers" confirms that.

Run CONVOLUTION. Show Numbers, X and Y. The values in **Y** agree exactly with those obtained in (a) by linear convolution or in (b) by circular convolution.

(d) For g_k: $C_\sigma = 6 = p$. For h_k: $C_\sigma = 5 = q$.

From (14.7) for y_k: We expect that $C_\sigma = p + q - 1 = 6 + 5 - 1 = 10$.
From the actual values obtained for y_k: $C_\sigma = 10$

The span restriction (14.10) requires that $p + q \le N + 1$.

With N = 10: For p = 6 and q = 5 we have $6 + 5 = 11 \le N + 1$ and so the pulses g and h comply with the span restriction. Thus aliasing should not take place.

identical to those produced by linear convolution and so aliasing did not occur.

(e) Changing h_4 from 0 to 4 gives us the following sequences:

k	-9	-8	-7	-6	-5	-4	-3	-2	-1	0	1	2	3	4	5	6
g_k	0	0	0	0	0	1	0	1	0	0	2	0	0	0	0	0
h_k	0	0	0	0	0	0	0	0	2	1	2	3	1	4	0	0

Now the span counts are as follows:

For g_k: $C_\sigma = 6 = p$. For h_k: $C_\sigma = 6 = q$. With N = 10 the two sequences no longer comply with the span restriction because now $p + q = 12 > N + 1$.

Thus we can expect that aliasing will take place when we perform circular convolution.

(f) Repeating the linear convolution gives us the following (shown here for k = −6):

m	-9	-8	-7	-6	-5	-4	-3	-2	-1	0	1	2	3	4	5	6
g_m	0	0	0	0	0	1	0	1	0	0	2	0	0	0	0	0
h_{k-m}	4	1	3	2	1	2	——>									

Then for each value of k: Slide h_{k-m} past g_m, multiply corresponding values together and add them to obtain f_k.

k	-9	-8	-7	-6	-5	-4	-3	-2	-1	0	1	2	3	4	5	6
f_k	0	0	0	0	2	1	4	4	3	11	3	8	6	2	8	0

(*)

Observe that f_k now has a span of 11. This is as expected since $p + q - 1 = 6 + 6 - 1 = 11$.

(g) Repeating the circular convolution, still with N = 10 but with the new value for h_4 gives the following results:

k	0	1	2	3	4	5	6	7	8	9
f_k	11	3	8	6	2	10	1	4	4	3

(**)

The figure for the circular convolution is shown below.

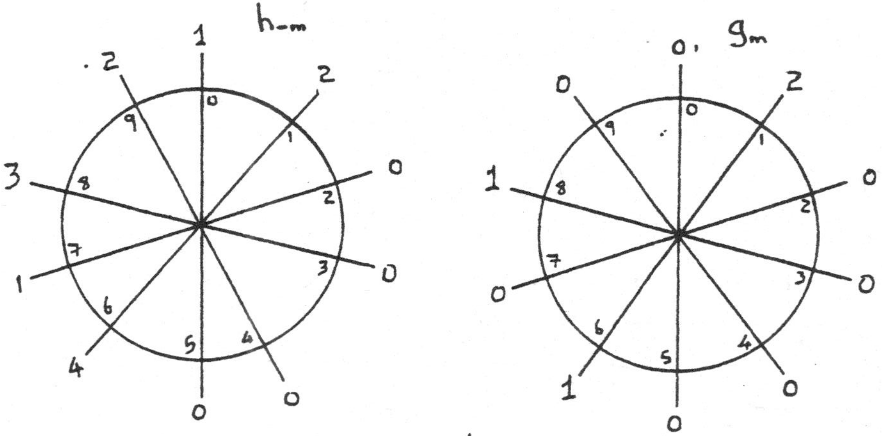

N = 10

We observe that aliasing has in fact taken place. The final "8" in (*) has been combined with the first "2" to give us the "10" in (**). The sequences produced by linear and circular convolution no longer agree.

Steps for loading the two sequences into **X** and **X2** of the FFT system (using N = 10, DISCRETE):

Main menu, Create Values, Continue, Use the Old-problem, Accept all values except h_4 which is changed from 0 to 4. You are now back on the main menu with the sequences properly loaded. PLotting **X** and **X2** or using "Show Numbers" confirms that.

Run CONVOLUTION. You will be told that the span restriction is being violated. Use "RUN CONVOLUTION REGARDLESS".

Show Numbers, X and Y. The values in **Y** agree exactly with those obtained by circular convolution above. However they do not agree with the values obtained from linear convolution in (f) because aliasing has taken place.

(h) We now change N to 12 and re-run the circular convolution as follows. The values obtained are

k	0	1	2	3	4	5	6	7	8	9	10	11
f_k	11	3	8	6	2	8	0	2	1	4	4	3

The figure for the circular convolution is as follows:

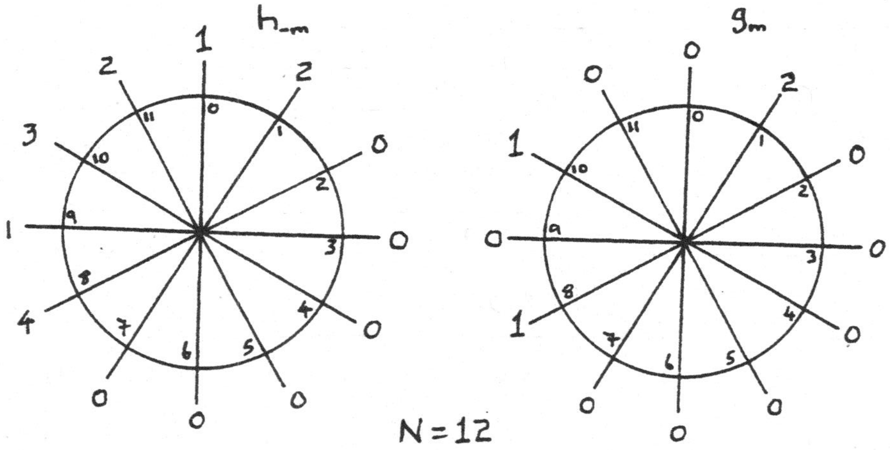

N = 12

Observe that aliasing is now no longer present. The values are identical (in the circular sense) to what we obtained in (f) using linear convolution. Observe also that even though we used N = 12, the span of f_k is still C_σ = 11, as expected.

Steps for loading the two sequences into **X** and **X2** of the FFT system (using N = 12, DISCRETE):

Main menu, Create Values, Continue, Create X and X2, Real, 12 intervals, 0, 1, 2, 3, 4, 5, 6, 7, 8, 9, 10, Continue, 0, 2, 0, 0, 0, 0, 0, 0, 1, 0, 1, 0, Go, 12 intervals, 0, 1, 2, 3, 4, 5, 6, 7, 8, 9, 10, Continue, 1, 2, 3, 1, 4, 0, 0, 0, 0, 0, 0, 2, Go, Go. You are now back on the main menu with the sequences properly loaded. PLotting **X** and **X2** or using "Show Numbers" confirms that.

Run CONVOLUTION. Show Numbers, X and Y. The values in **Y** agree exactly with those obtained in (f) by linear convolution or above by circular convolution with N = 12.

(i) We can thus infer that by increasing N so that the span restriction is not violated, circular convolution can be successfully used to emulate linear convolution.

14.3

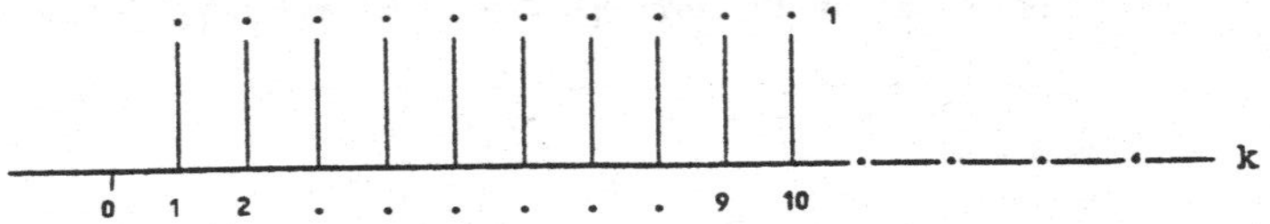

$g_k = 1 \quad (1 \le k \le 10)$

(a) Steps for loading g_k into the FFT system using N = 24, DISCRETE:

Main menu, Create Values, Continue, Create X, Real, 3 intervals, 0, 10, Continue, 0, 1, 0, Go, Go. You are now back on the main menu with the pulse correctly loaded. Plotting **X** confirms that.

Run ANALYSIS.

(b) The required values of G_n are:

n	G_n
1	0.96592583 - j7.3369351
5	0.25881905 - j0.33729955
9	-0.70710678 + j0.29289322

(c) Finding the convolution product of g_k with itself using N = 24:

m	0	1	2	3	4	5	6	7	8	9	10	11
g_m	0	1	1	1	1	1	1	1	1	1	1	0

g_{k-m}	0	1	1	1	1	1	1	1	1	1	1	0	→

k	0	1	2	3	4	5	6	7	8	9	10	11	12	13	14	15	16	17	18	19	20	21
f_k	0	0	1	2	3	4	5	6	7	8	9	10	9	8	7	6	5	4	3	2	1	0

Steps for loading f_k into the FFT system using N = 24, DISCRETE:

Main menu, Create Values, Continue, Create X, Real, 4 intervals, 0, 11, 21, Continue, 0, k-1, 21-k, 0, Go, Go. You are now back on the main menu with f_k correctly loaded. Plotting **X** confirms that.

Run ANALYSIS. The values of F_n are as follows:

n	F_n
1	-52.897603 - j14.173870
5	-4.6783687e-2 - j0.17459909
9	0.41421356 - j0.41421356

(d) We now verify that $F_n = G_n^2$:

$n = 1: G_1^2 = (0.96592583 - j7.3369351)(0.96592583 - j7.3369351)$

$= -52.897603 - j14.173870 = F_1$

$n = 5: G_5^2 = (0.25881905 - j0.33729955)(0.25881905 - j0.33729955)$

$= -4.6783686e-2 - j0.17459909 = F_5$

$n = 9: G_9^2 = (-0.70710678 + j0.29289322)(-0.70710678 + j0.29289322)$

$= 0.41421356 - j0.41421356 = F_9$

(e) Loading g_k into both **X** and **X2**, and running CONVOLUTION:

Load g_k as described in (a) above. Then: Run Postprocessors, X, Copy X to X2, Quit. You are now back on the main menu with the values in both **X** and **X2**. Run CONVOLUTION, Show Numbers, F.

The values of the spectrum of $g_k * g_k$ are as follows:

n	spectrum of $g_k * g_k$
1	-52.897603 - j14.173870
5	-4.6783687e-2 - j0.17459909
9	0.41421356 - j0.41421356

We see that they are identical to the values of F_n obtained in (c).

14.4

(a) Referring to Figures 14.16 and 14.17, we apply linear graphical convolution to the two discrete pulses g_k and h_k as follows:

We show g_m and h_{k-m} (for k = 1) in the following table:

m	0	1	2	3	4	5	6	7	8	9	10	11	12	13	14	15
g_m	0	1	1	1	1	2	2	2	2	2	0	0	0	0	0	0

h_{k-m}	2	2	2	1	1	0	0	0	0	0	0	0	→

Sliding h_{k-m} past g_m, multiplying and adding, gives the following values for f_k:

k	0	1	2	3	4	5	6	7	8	9	10	11	12	13	14	15
f_k	0	0	1	2	4	6	9	10	12	14	16	14	12	8	4	0

(b) Steps for loading g_k into **X** and h_k into **X2** of the FFT system using N = 16, DISCRETE:

Main menu, Create Values, Continue, Create X and X2, Real, 4 intervals, 0, 4, 9, Continue, 0, 1, 2, 0, Go, 4 intervals, 0, 2, 5, Continue, 0, 1, 2, 0, Go, Go. You are now back on the main menu with the pulses loaded. Plotting **X** and **X2** confirms that.

Finding the spectra of g_k and h_k: Run ANALYSIS, Show Numbers, F. The spectrum values of G_n are as follows:

n	G_n
1	−5.8614288 − j6.2756424
5	−1.0687225 + j1.3454911
7	0.44721525 − j3.3001686e-2

Run Postprocessors, X, Swop X:X2 F:F2, Quit, Run ANALYSIS, Show Numbers, F. The spectrum values of H_n are as follows:

n	H_n
1	1.6309863 − j6.7853083
5	−1.0897902 − j0.68603902
7	−0.21677275 − j1.3710948

(c) Using the values from (b), we use a hand calculator to find the values of $G_n H_n$ for each of n = 1, 5, 7, obtaining:

n = 1: G_1H_1 = (−5.8614288 −j6.2756424)(1.6309863 −j6.7853083)

= −52.14207853 + j29.53611471

n = 5: G_5H_5 = (−1.0687225 + j1.3454911)(−1.0897902 − j0.68603902)

= 2.087742703 − j0.733117678

n = 7: G_7H_7 = (0.44721525 − j3.3001686e-2)(−0.21677275 − j1.3710948)

= −0.14219252 − j0.606020646

(d) We now use the complex-multiply package in the **F** post-processor to multiply G_n and H_n together. The results for n = 1, 5, 7 are as follows:

n	G_n H_n
1	−52.142079 + j29.536115
5	2.0877427 − j0.73311773
7	−0.14219252 − j0.60602063

These are seen to essentially identical to the values obtained in (c).

(e) Run SYNTHESIS to invert the result of the multiplication in (d). Then: Show Numbers, X and Y. The values are identical to those obtined for f_k in (a).

14.5

(a) Using circular convolution by hand with N = 10 to convolve the two discrete pulses of Exercise 14.4. Checking the span restriction:

For g_k, C_σ = 9 = p. For H_k, C_σ = 5 = q. Then p + q = 14 > N + 1

The span restriction is violated and so we expect that aliasing will be present.

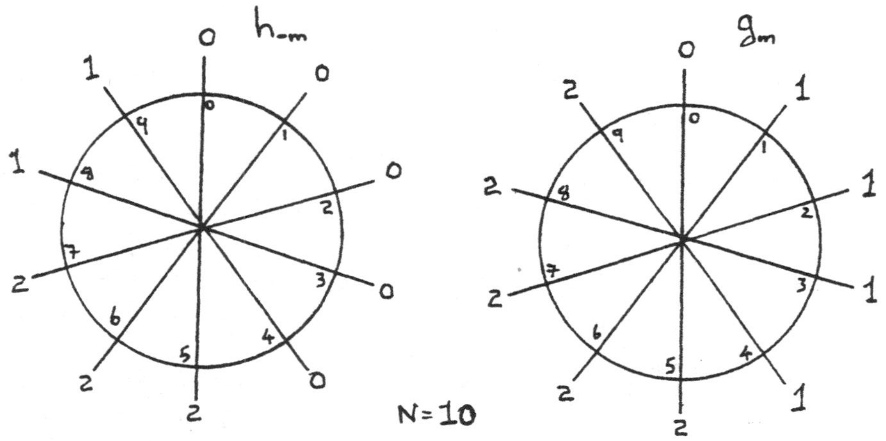

The results of the circular convolution with N = 10 are as follows:

k	0	1	2	3	4	5	6	7	8	9
f_k	16	14	13	10	8	6	9	10	12	14

(b) Carrying out the convolution of the same two pulses using the formal definition of circular convolution (14.5), i.e. computing the subscripts modulo-10:

<u>Discrete circular convolution</u>:

$$f_k = \sum_{m=0}^{N-1} g_m h_{k-m}$$ in which the subscripts are computed modulo-N

m	0	1	2	3	4	5	6	7	8	9
g_m	0	1	1	1	1	2	2	2	2	2
h_m	0	1	1	2	2	2	0	0	0	0

k=0: $f_0 = \sum_{m=0}^{9} g_m h_{0-m} = \sum_{m=0}^{9} g_m h_{10-m}$

$= (0\times0) + (1\times0) + (1\times0) + (1\times0) + (1\times0) + (2\times2) + (2\times2) + (2\times2) + (2\times1) + (2\times1) = 16$

k=1: $f_1 = \sum_{m=0}^{9} g_m h_{(1-m)\text{mod-}10}$

$= (0\times1) + (1\times0) + (1\times0) + (1\times0) + (1\times0) + (2\times0) + (2\times2) + (2\times2) + (2\times2) + (2\times1) = 14$

k=2: $f_2 = \sum_{m=0}^{9} g_m h_{(2-m)\text{mod-}10}$

$= (0\times1) + (1\times1) + (1\times0) + (1\times0) + (1\times0) + (2\times0) + (2\times0) + (2\times2) + (2\times2) + (2\times2) = 13$

k=3: $f_3 = \sum_{m=0}^{9} g_m h_{(3-m)\text{mod-}10}$

$= (0\times2) + (1\times1) + (1\times1) + (1\times0) + (1\times0) + (2\times0) + (2\times0) + (2\times0) + (2\times2) + (2\times2) = 10$

k=4: $f_4 = \sum_{m=0}^{9} g_m h_{(4-m)\text{mod-}10}$

$= (0\times2) + (1\times2) + (1\times1) + (1\times1) + (1\times0) + (2\times0) + (2\times0) + (2\times0) + (2\times0) + (2\times2) = 8$

k=5: $f_5 = \sum_{m=0}^{9} g_m h_{(5-m)\text{mod-}10}$

$= (0\times2) + (1\times2) + (1\times2) + (1\times1) + (1\times1) + (2\times0) + (2\times0) + (2\times0) + (2\times0) + (2\times0) = 6$

k=6: $f_6 = \sum_{m=0}^{9} g_m h_{(6-m)\text{mod-}10}$

$= (0\times0) + (1\times2) + (1\times2) + (1\times2) + (1\times1) + (2\times1) + (2\times0) + (2\times0) + (2\times0) + (2\times0) = 9$

k=7: $f_7 = \sum_{m=0} g_m h_{(7-m)\text{mod-}10}$

$= (0\times0) + (1\times0) + (1\times2) + (1\times2) + (1\times2) + (2\times1) + (2\times1) + (2\times0) + (2\times0) + (2\times0) = 10$

k=8: $f_8 = \sum_{m=0}^{9} g_m h_{(8-m)\text{mod-}10}$

$= (0\times0) + (1\times0) + (1\times0) + (1\times2) + (1\times2) + (2\times2) + (2\times1) + (2\times1) + (2\times0) + (2\times0) = 12$

k=9: $f_9 = \sum_{m=0}^{9} g_m h_{(9-m)\text{mod-}10}$

$= (0\times0) + (1\times0) + (1\times0) + (1\times0) + (1\times2) + (2\times2) + (2\times2) + (2\times1) + (2\times1) + (2\times0) = 14$

(c) **Steps for loading g_k into X and h_k into X2 of the FFT system using N = 10, DISCRETE:**

Main menu, Create Values, Continue, Create X and X2, Real, 3 intervals, 0, 4, Continue, 0, 1, 2, Go, 4 intervals, 0, 2, 5, Continue, 0, 1, 2, 0, Go, Go. You are now back on the main menu with the pulses loaded. Plotting X and X2 confirms that.

Run CONVOLUTION. You will be told that the span restriction is being violated. Run CONVOLUTION regardless. Show Numbers, X and Y.

The values displayed, namely

k	0	1	2	3	4	5	6	7	8	9
f_k	16	14	13	10	8	6	9	10	12	14

are identical to those obtained in (a) and (b).

14.6 Convolution of pulses with discontinuities

When a pulse f(t) has a discontinuity then we always define its value at that point to be what we have called the "half-value", and if the pulse is sampled at the discontinuity then the half-value is what is sent to the FFT. **When we are running convolution involving such pulses the half-values can cause slight errors.** There is a way in which to fix the problem and in this exercise we explore how that is done.

(a) Consider first the problem itself. Let the two inputs be

$$g(t) = h(t) = \text{Rect}(t - \tfrac{1}{2}) \qquad (14.35)$$

Using graphical convolution:

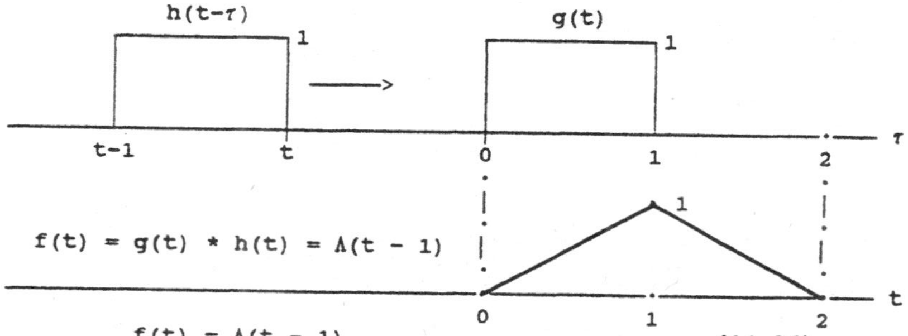

$$f(t) = \Lambda(t - 1) \qquad (14.36)$$

(b) Steps for loading g(t) and h(t) into **X** and **X2** using N = 256, T = 4:

Main menu, Create Values, Continue, Create X, Real, Neither, Left, 2 intervals, 1, Continue, 1, 0, Go, Neither, Go. You are now back on the main menu. Run Postprocessors, X, Copy X to X2, Quit.

Plotting **X** and **X2** confirms that the pulses have been correctly loaded. Observe that half-values have been created for both pulses by the system at t = 0 and t = 1, since those points are FFT sampling instants.

Run CONVOLUTION. Show Numbers. Observe that the **Y** vector agrees with a sampled version of (14.36) everywhere except at t = 0, 1 and 2 where there are slight errors.

- At t = 0 (k = 0) and t = 2 (k = 128) the exact values are 0, but the FFT system gives 3.90625e-3.
- At t = 1 (k = 64) the exact value is 1 but the system gives 0.9921875

These three errors are caused by the half-values that were present in the data that were loaded into the system.

(c) To fix the problem: **We redefine the input pulses so that their discontinuities fall midway between sampling points.** In this case we use pulses which have been shifted half a sampling interval to the right, namely:

$$g(t) = h(t) = \text{Rect}(t - \tfrac{1}{2} - T_s/2) \qquad (14.37)$$

Using time-shift we derive the theoretical result of the convolution $f(t) = g(t) * h(t)$ obtaining:

$$g(t) \Longleftrightarrow G(\omega) = \text{Sa}(\omega/2)\ \exp[-j\omega(\tfrac{1}{2} + T_s/2)]$$

$$h(t) \Longleftrightarrow H(\omega) = \text{Sa}(\omega/2)\ \exp[-j\omega(\tfrac{1}{2} + T_s/2)]$$

$$g(t) * h(t)$$

$$\Longleftrightarrow G(\omega)\ H(\omega)$$

$$= \text{Sa}(\omega/2)\ \exp[-j\omega(\tfrac{1}{2} + T_s/2)]\ \text{Sa}(\omega/2)\ \exp[-j\omega(\tfrac{1}{2} + T_s/2)]$$

$$= \text{Sa}^2(\omega/2)\ \exp[-j\omega(1 + T_s)] \Longleftrightarrow \Lambda(t - 1 - T_s)$$

Thus $f(t) = \Lambda(t - 1 - T_s)$ (14.38)

(d) Steps for loading g(t) and h(t) of (14.37) into **X** and **X2** using N = 256, T = 4:

Main menu, Create Values, Continue, Create X, Real, Neither, Left, 3 intervals, 2/256, 1 + 2/256, Continue, 0, 1, 0, Go, Neither, Go. You are now back on the main menu. Run Postprocessors, X, Copy X to X2, Quit.

Plotting **X** and **X2** confirms that the pulses have been correctly loaded. Observe that now there are no half-values, since the discontinuities fall half-way between FFT sampling instants. Run CONVOLUTION. Show Numbers. Observe that the **Y** vector now agrees with a sampled version of (14.38) **everywhere**.

14.7 We repeat Exercise 14.6 but now using as the inputs

$$g(t) = \begin{cases} \tfrac{1}{2} & (0 < t < \tfrac{1}{2}) \\ 1 & (\tfrac{1}{2} < t < 1) \\ 0 & (\text{otherwise}) \end{cases} \qquad h(t) = \text{Rect}(t - \tfrac{1}{2})$$

(a) Using graphical convolution:

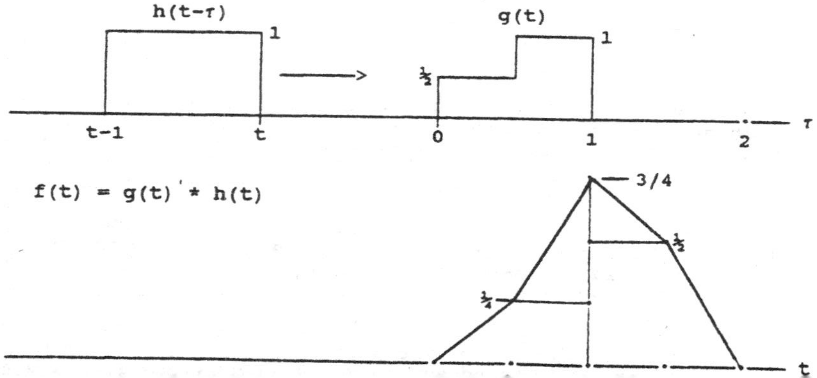

(b) Steps for loading g(t) and h(t) into **X** and **X2** using N = 256, T = 4:

Create Values, Continue, Create X and X2, Real, Neither, Left, 3 intervals, 1/2, 1, Continue, 1/2, 1, 0, Go, Neither, Left, 2 intervals, 1, Continue, 1, 0, Go, Neither, Go. You are now back on the main menu. Plotting **X** and **X2** confirms that the pulses have been correctly loaded. Observe that half-values have been created by the system at all of the discontinuities, since all of those points are FFT sampling instants.

Run CONVOLUTION. Show Numbers. The **Y** vector agrees with a sampled version of (14.36) everywhere except at:

- t = 0 (k = 0): Correct value is 0
  The system shows 1.953125e-3
- t = ¼ (k = 32): Correct value is 0.25
  The system shows 0.25195313
- t = ½ (k = 64): Correct value is 0.75
  The system shows 0.74414063
- t = 3/4 (k = 96): Correct value is ½
  The system shows 0.49804688
- t = 1 (k = 128): Correct value is 0
  The system shows 3.90625e-3

These five errors are caused by the half-values that were present in the data that were loaded into the system.

(c) To fix the problem: **We redefine the input pulses so that their discontinuities fall midway between sampling points**. In this case we use pulses which have been shifted half a sampling interval to the right, namely:

$$g(t) = \begin{cases} \tfrac{1}{2} & (T_s/2 < t < \tfrac{1}{2}+T_s/2) \\ 1 & (\tfrac{1}{2}+T_s/2 < t < 1+T_s/2) \\ 0 & (\text{otherwise}) \end{cases} \qquad h(t) = \mathrm{Rect}(t - \tfrac{1}{2})$$

Steps for loading the shifted versions of g(t) and h(t) into **X** and **X2** using N = 256, T = 4:

Main menu, Create Values, Continue, Create X and X2, Real, Neither, Left, 4 intervals, 2/256, 1/2 + 2/256, 1 + 2/256, Continue, 0, 1/2, 1, 0, Go, Neither, Left, 3 intervals, 2/256, 1 + 2/256, Continue, 0, 1, 0, Go, Neither, Go. You are now back on the main menu.

Plotting **X** and **X2** confirms that the pulses are now loaded. Observe that now there are no half-values, since the discontinuities fall half-way between FFT sampling instants.

Run CONVOLUTION. Show Numbers. Observe that the **Y** vector now agrees with a sampled version of g(t) * h(t) **everywhere**. Checking the values at k = 1, 33, 65, 97 and 129 shows that the errors that we listed earlier are no longer there.

14.8 Proving the result shown in Theorem 14.2.

Suppose that we have two **periodic functions**, $g_p(t)$ and $h_p(t)$, both with the same period T_0, and that we combine them using continuous circular convolution to obtain $f_p(t)$, where

$$f_p(t) = \int_0^{T_0} g_p(\tau)\, h_p(t-\tau)\, d\tau \qquad (14.39)$$

Letting $g_p(t)$ and $h_p(t)$ have CFT line spectra $G_p(n)$ and $H_p(n)$ respectively, we now derive the expression for the CFT line spectrum of $f_p(t)$ in terms of those two spectra.

Applying the CFT analysis equation to (14.39) gives us

$$\frac{1}{T_0}\int_0^{T_0} f_p(t)\, \exp(-jn\omega_0 t)\, dt$$

$$= \frac{1}{T_0}\int_0^{T_0}\left[\int_0^{T_0} g_p(\tau)\, h_p(t-\tau)\, d\tau\right] \exp(-jn\omega_0 t)\, dt \qquad (14.40)$$

Then, interchanging the order of integration, (14.40) continues as

$$\ldots = \frac{1}{T_0}\int_0^{T_0} g_p(\tau)\left[\int_0^{T_0} h_p(t-\tau)\, \exp(-jn\omega_0 t)\, dt\right] d\tau$$

$$= \frac{1}{T_0}\int_0^{T_0} g_p(\tau)\left[\int_0^{T_0} h_p(t-\tau)\, \exp[-jn\omega_0 (t-\tau)]\, dt\right] \exp(-jn\omega_0 \tau)\, d\tau \qquad (14.41)$$

At first sight it appears as though the final integral inside the square brackets in (14.41) depends, somehow on τ, but we shall now show that such is not the case. Define the auxiliary variable θ as follows:

$$\theta = t - \tau \qquad (14.42)$$

Then, since τ is a constant inside the square brackets, we have

$$d\theta = dt \qquad (14.43)$$

When $t = 0$ the lower limit of integration becomes $\theta = -\tau$, and when $t = T_0$ the upper limit becomes $\theta = T_0 - \tau$. Taking all of this into account that integral becomes

$$\int_0^{T_0} h_p(t-\tau)\ \exp[-jn\omega_0(t-\tau)]\ dt = \int_{-\tau}^{T_0-\tau} h_p(\theta)\ \exp(-jn\omega_0\theta)\ d\theta \qquad (14.44)$$

Now, by assumption, $h_p(t)$ is periodic with period T_0, and so, as noted in Chapter 2, when deriving its CFT coefficients it is immaterial where the limits of integration are located, **provided only that integration be over one complete period**. Thus (14.44) continues as

$$\ldots = T_0\ H_p(n) \qquad (14.45)$$

We now use this result in (14.41), and so that equation continues

$$\ldots = \frac{1}{T_0}\int_0^{T_0} g_p(\tau)\ \Big[\ T_0\ H_p(n)\ \Big]\ \exp(-jn\omega_0\tau)\ d\tau$$

$$= T_0\ H_p(n)\ \left[\ \frac{1}{T_0}\int_0^{T_0} g_p(\tau)\ \exp(-jn\omega_0\tau)\ d\tau\ \right] = T_0\ G_p(n)\ H_p(n) \qquad (14.46)$$

We have thus proved the result shown as Theorem 14.2.

14.9 Circular convolution of periodic time functions

(a) Carrying out the following convolution operation graphically:

$$f_p(t) = g_p(t) * h_p(t) \quad \text{where}$$

$$g_p(t) = h_p(t) = \sum_{k=-\infty}^{\infty} \text{Rect}(t - \tfrac{1}{2} - kT_0) \qquad (T_0 = 4)$$

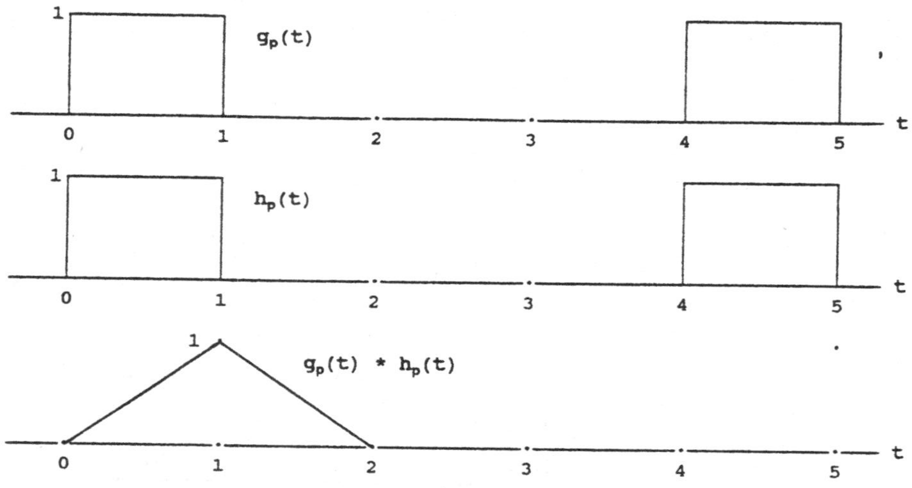

$$f_p(t) = g_p(t) * h_p(t) = \begin{cases} t & (0 < t < 1) \\ 2 - t & (1 < t < 2) \end{cases}$$

(b) Finding the expressions for the CFT coefficients of $g_p(t)$, $h_p(t)$ and $f_p(t)$:

$$g'(t) = \delta(t) - \delta(t-1) \qquad G(\omega) = \frac{1 - \exp(-j\omega)}{j\omega} \qquad \omega_0 = \pi/2$$

$$G_p(n) = \frac{1}{T_0}G(\omega)\Big|_{\omega \leftarrow n\omega_0} = \tfrac{1}{4}\ \frac{1 - \exp(-jn\pi/2)}{jn\pi/2} \longrightarrow$$

$$h'(t) = \delta(t) - \delta(t-1) \qquad H(\omega) = \frac{1 - \exp(-j\omega)}{j\omega} \qquad \omega_0 = \pi/2$$

$$H_p(n) = \frac{1}{T_0}G(\omega)\Big|_{\omega \leftarrow n\omega_0} = \tfrac{1}{4}\ \frac{1 - \exp(-jn\pi/2)}{jn\pi/2} \longrightarrow$$

$$f''(t) = 1 - 2\delta(t-1) + \delta(t-2) \qquad F(\omega) = \frac{1 - 2\exp(-j\omega) + \exp(-j2\omega)}{(j\omega)^2}$$

$$F_p(n) = \frac{1}{T_0} F(\omega) \Big|_{\omega \leftarrow n\omega_0} = \tfrac{1}{4} \frac{1 - 2\exp(-jn\pi/2) + \exp(-jn2\pi/2)}{(jn\pi/2)^2}$$

$$= \tfrac{1}{4} \left[\frac{1 - \exp(-jn\pi/2)}{jn\pi/2} \right]^2 \longrightarrow$$

(c) According to Theorem 14.2: $F_p(n) = T_0\, G_p(n)\, H_p(n)$

Using the spectra from (b), we observe that this is true.

(d) Steps for loading $g_p(t)$ and $h_p(t)$ into **X** and **X2** of the FFT system using N = 256 and T = 4, PERIODIC:

Main menu, Create Values, Continue, Create X, Real, Neither, Left, 2 intervals, 1, Continue, 1, 0, Go, Neither, Go. You are now back on the main menu. Run Postprocessors, X, Copy X to X2, Quit. The waveforms are now correctly loaded. Plotting **X** and **X2** confirms that.

Run CONVOLUTION.

The spectrum $H_p(n)$ is now in **F2** and the spectrum $F_p(n)$ is now in **F**. We have filled in the tables below, thereby verifying all of the results, including the scale factor T_0/N^2 below equation (14.30) in the box.

$$H_p(n) = \tfrac{1}{4} \frac{1 - \exp(-jn\pi/2)}{jn\pi/2} = \tfrac{1}{4} \frac{1 - \cos(n\pi/2) + j\sin(n\pi/2)}{jn\pi/2}$$

$$= \tfrac{1}{4} \frac{\sin(n\pi/2)}{n\pi/2} + j\tfrac{1}{4} \frac{\cos(n\pi/2) - 1}{n\pi/2} = A(n) + jB(n)$$

Values of $H_p(n)$:

n	FFT value A(n)	FFT value B(n)	formula value A(n)	formula value B(n)
0	0.25	0	0.25	0
1	0.15914695	-0.15914695	0.159154943	-0.159154943
2	0	-0.15912298	0	-0.159154943
3	-5.3027677e-2	-5.3027677e-2	-5.3051647e-2	-5.3051647e-2

$$F_p(n) = \tfrac{1}{4} \left[\frac{1 - \exp(-jn\pi/2)}{jn\pi/2} \right]$$

$$= \tfrac{1}{4} \exp(-jn\pi/2) \left[\frac{\exp(-jn\pi/4) - \exp(-jn\pi/2)}{2jn\pi/4} \right]^2$$

$$= \tfrac{1}{4} \left[\cos(n\pi/2) - j\sin(n\pi/2) \right] \left[\frac{\sin(n\pi/4)}{n\pi/4} \right]^2$$

$$= \tfrac{1}{4} \cos(n\pi/2) \left[\frac{\sin(n\pi/4)}{n\pi/4} \right]^2 - j\tfrac{1}{4} \sin(n\pi/2) \left[\frac{\sin(n\pi/4)}{n\pi/4} \right]^2$$

$$= A(n) + jB(n)$$

Values of $F_p(n)$:

n	FFT value A(n)	FFT value B(n)	formula value A(n)	formula value B(n)
0	0.25	0	0.25	0
1	0	-0.20262202	0	-0.202642367
2	-0.10128050	0	-0.101321184	0
3	0	2.2495476e-2	0	-2.2515818e-2

Chapter 15 Emulating Dirac Deltas and Differentiation on the FFT

15.1 $V = \mu N/T$

(a) $2\delta(t - 1)$, $N = 256$, $T = 2$:

$\mu = 2$ and so $V = 2\times256/2 = 256$ ——>

$kT_s = 1$ i.e. $kT/N = k\times2/256 = 1$, and so $k = 128$ ——>

(b) $5\delta(t - 2)$, $N = 100$, $T = 10$:

$\mu = 5$ and so $V = 5\times100/10 = 50$ ——>

$kT_s = 2$ i.e. $kT/N = k\times10/100 = 2$, and so $k = 20$ ——>

(c) $3\delta(t + 5)$, $N = 1024$, $T = 20$:

$\mu = 3$ and so $V = 3\times1024/20 = 153.6$ ——>

$kT_s = -5$ i.e. $kT/N = k\times20/1024 = -5$ and so $k = -256$ or 768 ——>

15.2 $V = \mu N/T$ and so $\mu = VT/N$

(a) $V = 256$, $k = 32$, $N = 256$, $T = 2$

$\mu = 256\times2/256 = 2$ ——>

$t_a = kT_s = 32\times T/N = 1/4$ ——>

$t_b = (N - k)T_s = (256 - 32)\times2/256 = 1.75$ ——>

(b) $V = 10240$, $k = 40$, $N = 512$, $T = 0.1$

$\mu = 10240\times0.1/512 = 2$ ——>

$t_a = kT_s = 40\times T/N = 0.0078125$ ——>

$t_b = (N - k)T_s = (512 - 40)\times0.1/512 = 0.0921875$ ——>

(c) $V = 108$, $k = 56$, $N = 72$, $T = 2$

$\mu = 108\times2/72 = 3$ ——>

$t_a = kT_s = 56\times T/N = 14/9$ ——>

$t_b = (N - k)T_s = (72 - 56)\times2/72 = 4/9$ ——>

15.3 $V = \mu T/2\pi$

(a) $2\pi\delta(\omega - 4\pi)$, $N = 256$, $T = 4$

$\mu = 2\pi$ and so $V = 2\pi\times4/2\pi = 4$ ——>

$n\Omega = 4\pi$ i.e. $n\times2\pi/4 = 4\pi$ and so $n = 8$ ——>

(b) $16\delta(\omega - \pi)$, $N = 64$, $T = 10$

$\mu = 16$ and so $V = 16\times10/2\pi = 80/\pi$ ——>

$n\Omega = \pi$ i.e. $n\times2\pi/10 = \pi$ and so $n = 5$ ——>

(c) $20\pi\delta(\omega + 400\pi)$, $N = 80$, $T = .1$

$\mu = 20\pi$ and so $V = 20\pi\times0.1/2\pi = 1$ ——>

$n\Omega = -400\pi$ i.e. $n\times2\pi/0.1 = -400\pi$ and so $n = -20$ or 60 ——>

15.4 $V = \mu T/2\pi$ and so $\mu = 2\pi V/T$

(a) $V = 2$, $n = 20$, $N = 256$, $T = 2$

$\mu = 2\pi\times2/2 = 2\pi$ ——>

$\omega_a = n\Omega = 20\times2\pi/2 = 20\pi$ ——>

$\omega_b = (N - n)\Omega = (256 - 20)\times2\pi/2 = 236\pi$ ——>

(b) $V = 100$, $n = 48$, $N = 400$, $T = 4$

$\mu = 2\pi\times100/4 = 50\pi$ ——>

$\omega_a = n\Omega = 48\times2\pi/4 = 24\pi$ ——>

$\omega_b = (N - n)\Omega = (400 - 48)\times2\pi/4 = 176\pi$ ——>

(c) $V = 1$, $n = 56$, $N = 72$, $T = 0.02$

$\mu = 2\pi\times1/0.02 = 100\pi$ ——>

$\omega_a = n\Omega = 56\times2\pi/0.02 = 5600\pi$ ——>

$\omega_b = (N - n)\Omega = (72 - 56)\times2\pi/0.02 = 1600\pi$ ——>

15.5

(a) $$3\delta(t - 2) * 2\delta(t - 3) = \int_{-\infty}^{\infty} 3\delta(\tau - 2)\, 2\delta(t - 3 - \tau)\, d\tau = 6\delta(t - 5)$$

(b) Steps for loading **X** and **X2** using $N = 400$, $T = 16$:

Create Values, Continue, Create only Diracs, Time, X, 2, 3, X2, 3, 2, Neither, Go. You are now back on the main menu. Plotting **X** and **X2** shows a value of 75 in **X** at t = 2, and a value of 50 in **X2** at t = 3.

Run CONVOLUTION. Plotting **Y** shows a value of 150 at t = 5, which is the correct representation of $6\delta(t - 5)$.

15.6

(1) When regarded as a single pulse over the range $-\pi/2 < t < \pi/2$, the spectrum is derived using convolution as follows:

$$f_1(t) = \cos(2t)\ \mathrm{Rect}(t/\pi)$$

$$g(t) = \cos(2t) \Longleftrightarrow G(\omega) = \pi[\delta(\omega - 2) + \delta(\omega + 2)]$$

$$h(t) = \mathrm{Rect}(t/\pi) \Longleftrightarrow H(\omega) = \pi\ \mathrm{Sa}(\omega\pi/2)$$

$$F_1(\omega) = 1/2\pi\ G(\omega) * H(\omega) = 1/2\pi \int_{-\infty}^{\infty} G(\theta)\ H(\omega - \theta)\ d\theta$$

$$= 1/2\pi \int_{-\infty}^{\infty} \pi[\delta(\omega - 2) + \delta(\omega + 2)]\ \pi\ \mathrm{Sa}[(\omega-\theta)\pi/2]\ d\theta$$

$$= \pi/2 \int_{-\infty}^{\infty} \left[\mathrm{Sa}[(\omega-2)\pi/2]\ \delta(\omega - 2) + \mathrm{Sa}[(\omega+2)\pi/2]\ \delta(\omega + 2) \right] d\theta$$

$$= \pi/2\ \mathrm{Sa}[(\omega-2)\pi/2] + \pi/2\ \mathrm{Sa}[(\omega+2)\pi/2]$$

$$= \pi/2 \left[\frac{\sin(\omega\pi/2 - \pi)}{\omega\pi/2 - \pi} + \frac{\sin(\omega\pi/2 + \pi)}{\omega\pi/2 + \pi} \right]$$

$$= -\tfrac{1}{2}\sin(\omega\pi/2) \left[\frac{2}{\omega - 2} + \frac{2}{\omega + 2} \right] = -\tfrac{1}{2}\sin(\omega\pi/2) \left[\frac{2\omega}{\omega^2 - 4} \right]$$

and so

$$F_1(\omega) = \frac{2\omega\sin(\omega\pi/2)}{4 - \omega^2} \longrightarrow$$

(2) When regarded as an infinite pulse train (period π) its spectrum is that of an eternal cosine $f_2(t) = \cos(2t)$, namely

$$F_2(\omega) = \pi\delta(\omega - 2) + \pi\delta(\omega + 2)$$

(a) Steps for loading $f_1(t) = \cos(2t)$ $(-\pi/2 < t < \pi/2)$ using $N = 256$ and $T = 4\pi$:

Create Values, Continue, Create X, Real, Even, 2 intervals, pi/2 Continue, COS(2*t), 0, Go, Neither, Go. You are now back on the main menu with the pulse correctly loaded. Plotting X confirms it.

Run ANALYSIS. Show Numbers, F, Single.

The following table confirms the result obtained in (1) above.
Note: $\Omega = 2\pi/T = 2\pi/4\pi = 0.5$

n	FFT value A(n)	FFT value B(n)	formula value A(nΩ)	formula value B(nΩ)
0	0	0	0	0
1	0.18870386	0	0.188561808	0
2	0.66706847	0	0.666666666	0
3	1.2126093	0	1.212183053	0

(b) Steps for loading X with the same function using $N = 256$, $T = \pi$:

Create Values, Continue, Create X, Real, Even, 1 interval, Continue, COS(2*t), Go, Neither, Go. You are now back on the main menu with the pulse correctly loaded. Plotting X confirms it. Run ANALYSIS.

There are now two ways to display the spectrum:

- Show Numbers, F, Single. This displays the signal as a **pulse**. The spectral values are seen to be:

  $F(n) = \pi/2$ for $n = \pm 1$ and zero elsewhere

  To explain this: Remember that the FFT is periodic in the time domain, and because we are using $T = \pi$ the pulses which make up the FFT's periodic extension are now all touching one another. Thus they have become the eternal cosine $\cos(2t)$ whose transform is

  $$\cos(2t) \Longleftrightarrow \pi[\delta(\omega - 2) + \delta(\omega + 2)]$$

  The spectrum of that periodic extension is thus two Dirac deltas of weight π, one at $t = 2$ and the other at $t = -2$. (In fact, because the FFT is also periodic in the frequency domain, those impulses would be repeated with periodicity $N2\pi/T$, but that need not concern us here.)

  According to Section 15.2, to create a Dirac delta of weight π at $t = 2$ we must load a single value

  $$V = \mu T/2\pi$$

  at the appropriate location. In this case $\mu = \pi$ and $T = \pi$, giving

  $$V = \pi\times\pi/2\pi = \pi/2 \longrightarrow$$

  The location of the first Dirac delta must be at $\omega = 2$. In this case $\Omega = 2\pi/T = 2$, and so $\omega = 2$ corresponds to $n = 1$. The second Dirac delta must be at $n = -1$.

  Thus, in PULSE mode, we see that the two Dirac deltas are correctly implemented by the system.

- Show Numbers, F, Eternal. Now we are regarding the signal as an **eternal train of pulses**. The spectrum values are seen to be:

$$F(n) = \pi \quad \text{for} \quad n = \pm 1 \quad \text{and zero elsewhere}$$

As before, the FFT's periodic extension has become the eternal cosine cos(2t) whose transform is

$$\cos(2t) \Longleftrightarrow \pi[\delta(\omega - 2) + \delta(\omega + 2)]$$

Again we have two Dirac deltas of weight π, one at t = 2 and the other at t = -2.

However, when in "eternal" display mode the system shows the **weights** of the Dirac deltas. (This is stated on all of the associated screens.) Thus it shows the weights correctly as π, and of course the locations must be the same as before, i.e. at $n = \pm 1$.

15.7 (a) We know that $f(t) = \exp(j2\pi t) \Longleftrightarrow F(\omega) = 2\pi\delta(\omega - 2\pi)$ and so $F(\omega)$ is a frequency-domain Dirac delta of weight $\mu = 2\pi$ located at $\omega = 2\pi$.

(b) Steps for loading one period of f(t) into **X** using N = 256, T = 1:

Create Values, Continue, Create X, Complex, Neither, Center, 1 interval, Continue, COS(2*pi*t), Go, SIN(2*pi*t), Go, Neither, Go. You are now back on the main menu with the pulse correctly loaded. Plotting **XRE** and **XIM** confirms that. Run ANALYSIS.

Show Numbers, F, Single. We see a single value of 1 at n = 1. Plotting **FRE** shows the same.

According to Section 15.2, a Dirac delta in the frequency domain appears as a single value $V = \mu T/2\pi$. In this case V = 1 and T = 1 and so $\mu = 2\pi$. Thus the weight of $F(\omega) = 2\pi\delta(\omega - 2\pi)$ has been correctly emulated.

Regarding the position of the Dirac delta: It is located at n = 1. The value of Ω is $2\pi/T$ which in this case is 2π. Thus n = 1 corresponds to $\omega = 2\pi$, again, as required.

15.8
(a) Finding the Fourier transform of the infinite train of pulses

$$f_p(t) = \sum_{k=-\infty}^{\infty} \text{Rect}(t - kT_0) \qquad (T_0 = 4)$$

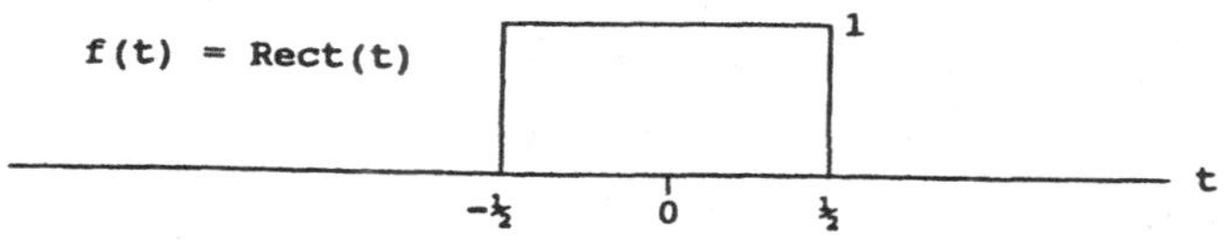

The defining pulse is f(t) = Rect(t) which transforms as follows:

$$\text{Rect}(t) \Longleftrightarrow F(\omega) = \text{Sa}(\omega/2)$$

The Fourier coefficients of $f_p(t)$ are thus

$$F_p(n) = \frac{1}{T_0} F(\omega)\Big|_{\omega \leftarrow n\omega_0} \qquad (\text{where } \omega_0 = 2\pi/T_0 = \pi/2)$$

$$= \tfrac{1}{4}\,\text{Sa}(n\pi/4)$$

The Fourier series is

$$f_p(t) = \sum_{n=-\infty}^{\infty} F_p(n)\exp(jn\omega_0 t) = \tfrac{1}{4}\sum_{n=-\infty}^{\infty} \text{Sa}(n\pi/4)\exp(jn\pi t/2)$$

The Fourier transform is

$$F_p(\omega) = \pi/2 \sum_{n=-\infty}^{\infty} \text{Sa}(n\pi/4)\,\delta(\omega - n\pi/2) \qquad \longrightarrow$$

(b) Steps for loading $f_p(t)$ into **X** using N = 1024, SAMPLED, T = 4, PULSE:

Create Values, Continue, Create X, Real, Even, 2 intervals, 1/2, Continue, 1, 0, Go, Neither, Go. You are now back on the main menu with the function correctly loaded. Plotting **X** confirms that.

Run ANALYSIS, Show Numbers, F, Eternal. The display shows the FFT's values of the weights of the Dirac deltas which make up $F_p(\omega)$.

The results and their errors when compared to the exact values are as follows:

n	FFT weight	formula weight	error (%)
0	1.5707963	1.570796327	1.7188733e-6
1	1.4142091	1.414213563	3.1558175e-4
5	-0.28282053	-0.282842713	7.8428041e-3
25	5.6457573e-2	5.6568542e-2	1.9616870e-1

15.9
(a) Steps for loading f(t) = cos(2t) as a SAMPLED PULSE with N = 256 and $T = \pi$:

Create Values, Continue, Create X, Real, Even, 1 interval, Continue, COS(2*t), Go, Neither, Go. You are now back on the main menu with the pulse correctly loaded. Plotting **X** confirms that.

Run ANALYSIS, Show Numbers, F, Eternal.

(b) The display shows $F(n) = \pi$ for $n = \pm 1$ and zero elsewhere.

The pulse transforms as follows:

$$f(t) = \cos(2t) \Longleftrightarrow F(\omega) = \pi[\delta(\omega - 2) + \delta(\omega + 2)]$$

which is two Dirac deltas of weight π located at $\omega = \pm 2$.

Thus the system has displayed the correct value of the weights. The value of Ω is $2\pi/T$ where $T = \pi$. Thus the FFT's frequency domain values appear at steps of $\omega = 2n$, and so $n = 1$ corresponds to $\omega = 2$. Thus the system values are correct in all respects.

(c) $f(t) = \cos(2t)$ differentiates to $f'(t) = -2\sin(2t)$ which transform as follows:

$$-2\sin(2t) \Longleftrightarrow j2\pi\,[\delta(\omega - 2) - \delta(\omega + 2)] \qquad (*)$$

(d) Steps for forming the derivative: Assuming that the pulse and its spectrum as created in (a) above are still in place:

Run Postprocessors, F, Central-diff 1st, Quit, Run SYNTHESIS.

Plotting Y shows that the waveform has been correctly differentiated.

Show Numbers, F, Eternal: The display shows

$F(n) = j6.2825545$ for $n = 1$ and $-j6.2825545$ for $n = -1$.

and zero elsewhere.

The weights are approximately equal to $\pm j2\pi$ (error = 0.01%)

The locations correspond to $\omega = \pm 2$.

Thus the system values are correct in all respects.

15.10
(a) Finding $F_p(\omega)$, the Fourier transform of $f_p(t) = |\sin(\pi t)|$.

(1) Considering a single period of $f_p(t)$, namely

$$f(t) = \sin(\pi t)\,\mathrm{Rect}(t - \tfrac{1}{2})$$

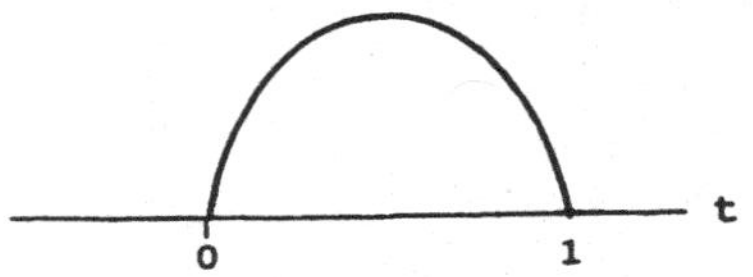

we find its Fourier transform. Thus:

$$F(\omega) = \int_{\infty}^{\infty} \sin(\pi t)\,\mathrm{Rect}(t - \tfrac{1}{2})\exp(j\omega t)\,dt$$

$$= 1/2j \int_0^1 [\exp(j\pi t) - \exp(-j\pi t)]\exp(j\omega t)\,dt$$

$$= 1/2j \int_0^1 \{\exp[j(\omega+\pi)t] - \exp[j(\omega-\pi)t]\}\,dt$$

$$= 1/2j \left[\frac{\exp[j(\omega+\pi)t]}{j(\omega+\pi)} - \frac{\exp[j(\omega-\pi)t]}{j(\omega-\omega)}\right]_0^1$$

$$= -\tfrac{1}{2}\left[\frac{\exp[j(\omega+\pi)] - 1}{\omega + \pi} - \frac{\exp[j(\omega-\pi)] - 1}{\omega - \pi}\right]$$

$$= -\tfrac{1}{2}\left[\frac{-\exp(j\omega) - 1}{\omega + \pi} - \frac{-\exp(j\omega) - 1}{\omega - \pi}\right]$$

$$= \frac{\exp(j\omega) + 1}{2}\left[\frac{1}{\omega + \pi} - \frac{1}{\omega - \pi}\right] = \frac{\exp(j\omega) + 1}{2}\left[\frac{-2\pi}{\omega^2 - \pi^2}\right]$$

$$F(\omega) = 2\pi\exp(-j\omega/2)\,\frac{\cos(\omega/2)}{\pi^2 - \omega^2} \longrightarrow$$

(2) Now we find the Fourier coefficients of an eternal train of such pulses:

$$F_p(n) = \frac{1}{T_0}\,F(\omega)\Big|_{\omega \leftarrow n\omega_0} \quad \text{where } T_0 = 1 \text{ and } \omega_0 = 2\pi$$

$$= 2\pi\exp(-j\omega/2)\,\frac{\cos(\omega/2)}{\pi^2 - \omega^2}\Big|_{\omega = n2\pi}$$

$$= 2\pi \exp(-jn\pi) \frac{\cos(n\pi)}{\pi^2 - 4\pi^2 n^2} = \frac{2}{\pi} \frac{1}{1 - 4n^2} \longrightarrow$$

(3) The Fourier series for $f_p(t)$ is thus

$$f_p(t) = \frac{2}{\pi} \sum_{n=-\infty}^{\infty} \frac{1}{1 - 4n^2} \exp(jn2\pi t)$$

(4) Then the Fourier transform of $f_p(t)$ is

$$F_p(\omega) = 4 \sum_{n=-\infty}^{\infty} \frac{1}{1 - 4n^2} \delta(\omega - n2\pi) \longrightarrow \qquad (*)$$

(b) Steps for loading the function $\sin(\pi t)$ ($0 < t < 1$) using N = 256, SAMPLED, T = 1, PULSE:

Create Values, Continue, Create X, Real, Neither. Left, 1 interval, Continue, SIN(pi*t), Go, Neither, Go. You are now back on the main menu with the waveform correctly loaded. Plotting **X** confirms that.

Run ANALYSIS. Show Numbers, F, Eternal.

The values shown are those of the weights of the Dirac Deltas which make up the Fourier transform of the eternal train $f_p(t) = |\sin(\pi t)|$. We compare the FFT values to the exact values from (*) in the following table:

n	FFT weight	formula weight	error (%)
0	3.9999498	4.000000000	0.0012625
1	-1.3333835	-1.333333333	0.0037625
5	-4.0454278e-2	-4.040404040	0.1243381
8	-1.5736571e-2	-1.5686274e-2	0.3206401

(c) Using Theorem 5.1 to find $F_{p2}(\omega)$ the Fourier transform of $f_p''(t)$. From (*):

$$f_p(t) \Longleftrightarrow F_p(\omega) = 4 \sum_{n=-\infty}^{\infty} \frac{1}{1 - 4n^2} \delta(\omega - n2\pi)$$

$$f_p''(t) \Longleftrightarrow (j\omega)^2 \quad F_p(\omega)$$

$$= 4(j\omega)^2 \sum_{n=-\infty}^{\infty} \frac{1}{1 - 4n^2} \delta(\omega - n2\pi)$$

$$= 4 \sum_{n=-\infty}^{\infty} \frac{(j\omega)^2}{1 - 4n^2} \delta(\omega - n2\pi)$$

$$= 4 \sum_{n=-\infty}^{\infty} \frac{(jn2\pi)^2}{1 - 4n^2} \delta(\omega - n2\pi)$$

$$= 4\pi^2 \sum_{n=-\infty}^{\infty} \frac{-4n^2}{1 - 4n^2} \delta(\omega - n2\pi)$$

$$= 4\pi^2 \sum_{n=-\infty}^{\infty} \left[1 + \frac{1}{4n^2 - 1} \right] \delta(\omega - n2\pi)$$

$$F_{p2}(\omega) = 4\pi^2 \sum_{n=-\infty}^{\infty} \delta(\omega - n2\pi) + 4\pi^2 \sum_{n=-\infty}^{\infty} \frac{1}{4n^2 - 1} \delta(\omega - n2\pi) \longrightarrow \quad (**)$$

(d) Inverting $F_{p2}(\omega)$ to the time domain analytically. Using the synthesis equation directly, the first part of the RHS of (**) inverts as follows :

$$4\pi^2 \sum_{n=-\infty}^{\infty} \delta(\omega - n2\pi) = \frac{4\pi^2}{2\pi} \int_{-\infty}^{\infty} \left[\sum_{n=-\infty}^{\infty} \delta(\omega - n2\pi) \right] \exp(j\omega t)\, d\omega$$

$$= 2\pi \int_{-\infty}^{\infty} \left[\sum_{n=-\infty}^{\infty} \exp(j\omega t) \right] d\omega$$

$$= 2\pi \int_{-\infty}^{\infty} \left[\sum_{n=-\infty}^{\infty} \exp(jn2\pi t)\, \delta(\omega - n2\pi) \right] d\omega$$

$$= 2\pi \sum_{n=-\infty}^{\infty} \exp(jn2\pi t)$$

This is the Fourier series of a Dirac comb with spacing

$$T_0 = 2\pi/\omega_0 = 2\pi/2\pi = 1$$

in which the impulses all have weights 2π, that is

$$4\pi^2 \sum_{n=-\infty}^{\infty} \delta(\omega - n2\pi) \Longleftrightarrow 2\pi\ \delta_T(t) \qquad (T_0 = 1)$$

$$4\pi^2 \sum_{n=-\infty}^{\infty} \frac{1}{4n^2 - 1} \delta(\omega - n2\pi) \Longleftrightarrow -\pi^2 |\sin(\pi t)|$$

Thus $\quad F_{p2}(\omega) \Longleftrightarrow -\pi^2 |\sin(\pi t)| + 2\pi\, \delta_T(t) \qquad (T_0 = 1)$

For a sketch of this time-domain waveform see the third sketch below.

(e) Differentiating $f_p(t) = |\sin(\pi t)|$ twice in the time domain:

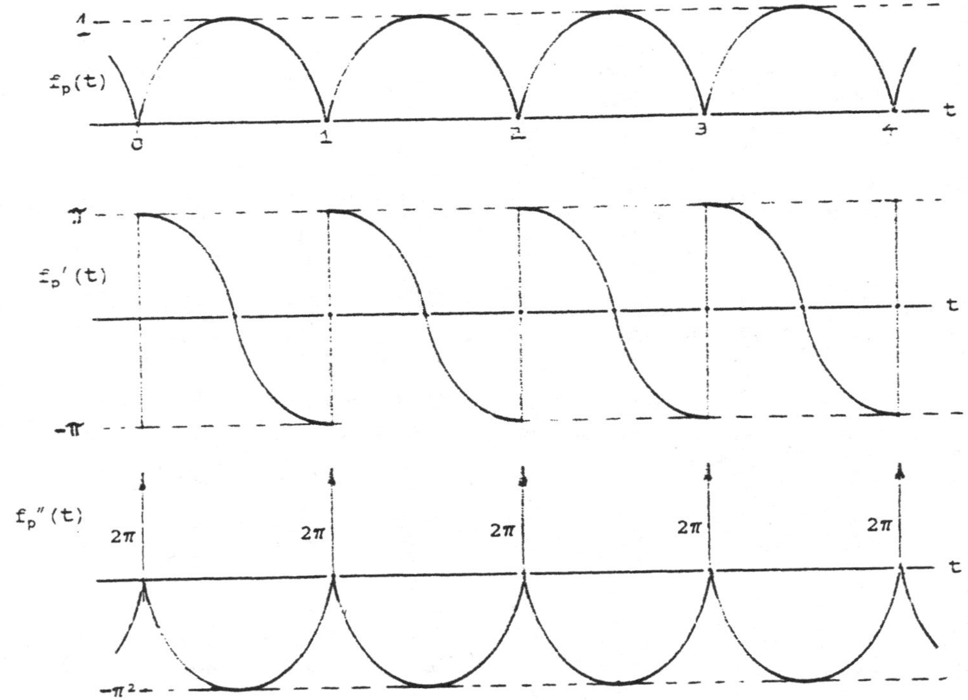

(f) Using the waveform and spectrum that were created in (b):

Run Postprocessors, F, CENTRAL-DIFF 2nd, Quit. You are now back on the main menu. Run SYNTHESIS.

Plotting **Y** shows that the original waveform has been differentiated twice.

Note: On the display of **Y** you will see the Dirac deltas at t = 0 and t = 1 with a small amount of negative-going "fuzz" in between.

If you want to inspect a plot of this, as follows: Main menu, Run Postprocessors, F, Copy F to F2, Lock F2, Run Postprocessors, X, Copy X to X2, Copy Y to X, Quit

Main menu, Edit from Keyboard, Continue, Change X vectors, X, Real, Use k = 0, Observe that the current value is 1608.46 (Actually, from Show Numbers, it is 1608.4551) Edit the value to be zero, Skip, Continue, Go. You are now back on the main menu with the value at k = 0 set to zero.

Plot **X**. You will see a negative-going sine whose peak value is -9.86948. That compares extremely well with the theoretical value of $-\pi^2$ (error = 0.00126%)

To set the system back to its state prior to this "Note": Run Postprocessors, F, Swop X:X2 F:F2, Proceed, Quit. Run SYNTHESIS.

The heights of the Dirac deltas which appear in **Y** are equal to 1608.4551. We know from Section 15.1 that a time-domain Dirac delta on the FFT appears as a line of height $V = \mu N/T$, from which

$\mu = VT/N$

In this case V = 1608.4551, T = 1 and N = 256 and so

$\mu = 6.283027734$

This compares extremely well with the theoretical value of 2π (error = 0.0025%)

15.11 Steps for loading the pulse shown in Figure 15.5 into **X** using N = 32 in two ways:

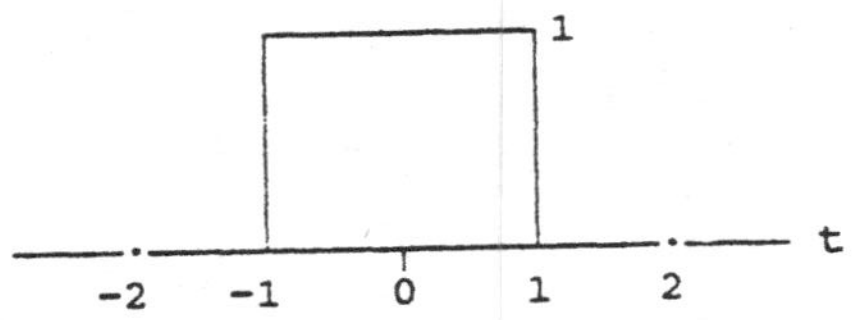

Figure 15.5:
Pulse for differentiation

(a) Using T = 2*pi, which means that the ends of the pulse will fall in the centers of sampling intervals (and so there will not be half-values present at the discontinuities):

Main menu, Create Values, Continue, Create X, Real, Even, 2 intervals, 1, Continue, 1, 0, Go, Neither, Go.

Run ANALYSIS, Run Postprocessors, F, Copy F to F2. Then:

(1) Backward-diff 1st, Quit, Run SYNTHESIS, Plot **Y**. We see two Dirac

lines, one at $k = 6$ ($t = 1.1780973$) and the other at $k = -5$ ($t = -0.9817478$). Thus, backward differencing has **delayed** the Dirac deltas slightly. The heights of the Dirac lines are $V = \pm 5.0929582$. Since $V = \mu N/T$ we have

$$\mu = VT/N = \pm 5.0929582 \times 2\pi/32 = \pm 1$$

and so the weights have been emulated correctly.

(2) Run Postprocessors, F, Copy F2 to F, Forward-diff 1st, Quit, Run SYNTHESIS, Plot **Y**. We see two Dirac lines, one at $k = 5$ ($t = 0.9817478$) and the other at $k = -6$ ($t = -1.1780973$). Thus forward differencing has **advanced** the Dirac deltas slightly. The heights of the Dirac lines are $V = \pm 5.0929582$ which leads to $\mu = 1$ as before, and so the weights have again been emulated correctly.

(3) Run Postprocessors, F, Copy F2 to F, Central-diff 1st, Quit, Run SYNTHESIS, Plot **Y**. Now we see two pairs of **double** Dirac lines, one pair at $k = 5$ ($t = 0.9817478$) and $k = 6$ ($t = 1.1780973$), and the other pair at the negatives of these values. Thus, central differencing has **doubled** the Dirac lines and **straddled** the exact locations. The heights of the double lines are 2.5464791, which, if added together give us $V = 5.0929582$ which again leads to $\mu = 1$, and so the weights have been correctly emulated in this manner.

(b) Using $T = 4$, which means that the ends of the pulse will now coincide with sampling instants, and so there will be half-values present at the discontinuities:

Main menu, Create Values, Continue, Create X, Real, Even, 2 intervals, 1, Continue, 1, 0, Go, Neither, Go.

Run ANALYSIS, Run Postprocessors, F, Copy F to F2. Then:

(1) Backward-diff 1st, Quit, Run SYNTHESIS, Plot **Y**. We see two pairs of **double** Dirac lines, one pair at $k = 8$ ($t = 1$) and $k = 9$ ($t = 1.125$), and the other pair at $k = -8$ ($t = -1$) and $k = -7$ ($t = 0.875$). Thus, when half-values are present, backward differencing has **doubled** the Dirac lines and **delayed** their locations. The heights of the double lines are each 4, which, if added together gives us $V = 8$ which again leads to $\mu = 1$, and so the weights have been correctly emulated in this manner.

(2) Run Postprocessors, F, Copy F2 to F, Forward-diff 1st, Quit, Run SYNTHESIS, Plot **Y**. We see two pairs of **double** Dirac lines, one pair at $k = 7$ ($t = 0.875$) and $k = 8$ ($t = 1$), and the other pair at $k = -9$ ($t = -1.125$) and $k = -8$ ($t = -1$). Thus, when half-values are present, forward differencing has **doubled** the Dirac lines and **advanced** the exact locations. The heights of the double lines are 4, which, if added together gives us $V = 8$ which again leads to $\mu = 1$, and so the weights have been correctly emulated in this manner.

(3) Run Postprocessors, F, Copy F2 to F, Central-diff 1st, Quit, Run SYNTHESIS, Plot **Y**. Now we see two sets of **triple** Dirac lines, one at $k = 7$ ($t = 0.875$), a larger one at $k = 8$ ($t = 1$) and a third at $k = 9$ ($t = 1.125$), with a corresponding triplet at the negatives of these values. Thus, central differencing has **tripled** the Dirac lines and **straddled** the exact locations. The heights of the triple lines are 2, 4 and 2 which, if added together gives us $V = 8$ which again leads to $\mu = 1$, and so the weights have been correctly emulated in this manner.

15.12 In the text we derived differentiation procedure (C) of Table 15.3. We now derive the remaining three shown in the table.

(a) Starting from (15.12) for **divided backward differences**:

$$f'(kT_s) \approx g_b[kT_s] \equiv \frac{f[kT_s] - f[(k-1)T_s]}{T_s} \qquad \text{(A)}$$

From the time-shift property for the DFT (see Exercise 10.8(a)) we recall that if $f_k \Longleftrightarrow F_n$ then

$$f_{k-m} \Longleftrightarrow F_n W^{nm}$$

Applying this to (A):

$$g_b[kT_s] \Longleftrightarrow \frac{F_n - F_n W^n}{T_s} = F_n \frac{1 - W^n}{T_s} = F_n \frac{1 - \exp(-j2\pi n/N)}{T_s}$$

$$= F_n \frac{1 - \cos(2\pi n/N) + j\sin(2\pi n/N)}{T_s}$$

Thus the required multiplier is

$$\frac{1 - W^n}{T_s} \quad \text{or} \quad \frac{1 - \cos(2\pi n/N) + j\sin(2\pi n/N)}{T_s} \quad \longrightarrow$$

(b) Starting from (15.14) for **divided forward differences**:

$$f'(kT_s) \approx g_f[kT_s] \equiv \frac{f[(k+1)T_s] - f[kT_s]}{T_s} \qquad \text{(B)}$$

From the time-shift property for the DFT: If $f_k \Longleftrightarrow F_n$ then $f_{k-m} \Longleftrightarrow F_n W^{nm}$. Applying this to (B):

$$g_b[kT_s] \Longleftrightarrow \frac{F_n W^{-n} - F_n}{T_s} = F_n \frac{W^{-n} - 1}{T_s} = F_n \frac{\exp(j2\pi n/N) - 1}{T_s}$$

$$= F_n \frac{\cos(2\pi n/N) - 1 + j\sin(2\pi n/N)}{}$$

Thus the required multiplier is

$$\frac{W^{-n} - 1}{T_s} \quad \text{or} \quad \frac{\cos(2\pi n/N) - 1 + j\sin(2\pi n/N)}{T_s} \quad \longrightarrow$$

(c) Starting from (15.17) for **divided central second differences:**

$$f''(kT_s) \approx h_c[kT_s] \equiv \frac{f[(k+1)T_s] - 2f[kT_s] + f[(k-1)T_s]}{T_s^2} \qquad \text{(C)}$$

From the time-shift property for the DFT: If $f_k \Longleftrightarrow F_n$ then $f_{k-m} \Longleftrightarrow F_n W^{mn}$. Applying this to (C):

$$h_c[kT_s] \Longleftrightarrow \frac{F_n W^{-n} - 2F_n + F_n W^n}{T_s} = F_n \frac{W^{-n} - 2 + W^n}{T_s}$$

$$= F_n \frac{\exp(j2\pi n/N) - 2 + \exp(-j2\pi n/N)}{T_s}$$

$$= F_n \frac{2[\cos(2\pi n/N) - 1]}{T_s}$$

Thus the required multiplier is

$$\frac{W^{-n} - 1}{T_s} \quad \text{or} \quad \frac{2[\cos(2\pi n/N) - 1]}{T_s} \quad \longrightarrow$$

15.13

(a) Steps for loading **X** with a signal plus noise as follows:

$$x(t) = 1 + 0.2\sin(16\pi t) + 0.02(\text{RND} - .5) \qquad (*)$$

where RND produces a pseuo-random number lying in the range 0 to 1. Ideally, the average value of RND is ½. We use N = 128, SAMPLED, T = 1, PULSE.

Main menu, Create Values, Continue, Create X, Real, Neither, Left, 1 interval, Continue, 1 + 0.2*SIN(16*pi*t) + 0.02*(RND - 0.5), Go, Neither, Go. You are now back on the main menu.

Plotting **X** we see the sinusoidal signal perturbed by the noise.

(b) Run ANALYSIS, Plot |**F**|. We see three strong spectral lines, one coming from the 1 in the signal, and the other two from the sinusoid.

Show Numbers, F, Single.

- Examining the values in the real part we see a value of approximately 1 in A(0) and values of approximately 1e-4 to 1e-5 in all of the other values.

approximately

0.1 in B(-8)

-0.1 in B(8)

and values of approximately 1e-4 to 1e-5 in all of the other values.

- Examining the values in the magnitude column we see values of approximately

  1 in F(0)

  0.1 in F(-8)

  0.1 in F(8)

  and values of approximately 1e-4 to 1e-5 in all of the other values.

(d) Trying to obtain the **second derivative** of the signal using the appropriate package in the **F** postprocessor:

Main menu, Run Postprocessors, F, Central-diff 2nd, Quit, Run SYNTHESIS, Plot **Y**.

We see a very noisy representation of the second derivative which is essentially useless.

(e) Run ANALYSIS, Run Postprocessors, F, Low-Pass filter F, 8.

We have now zeroed out all values of the spectrum whose spectral numbers exceed that of the sinusoid. Then

Central-diff 2nd, Quit. You are now back on the main menu.

Run SYNTHESIS, Plot **Y**.

(f) Examining **Y** we see that it is now a fairly good representation of the second derivative of the sinusoid that we started with.

It is clearly a sine wave of the form $-A\sin(16\pi t)$ where A is approximately equal to 503.254. (It will differ slightly from run to run because the random number sequence in (*) is never the same on successive occasions.)

From (*) above, the second derivative of the signal is

$$-0.2\,(16\pi)^2\sin(16\pi t)$$

and so A sould be exactly equal to 505.32375. The value that we have obtained after low-pass filtering is thus quite close to the exact value.

(g) Starting again with the raw data in **X** and filtering out **both the noise and the sine wave**, leaving only the DC component:

Run ANALYSIS, Run Postprocessors, F, Low-pass filter F, 0, Quit. Run SYNTHESIS. Plot **Y**.

The value obtained for **Y** is close to, but not exactly equal to one. This is because the raw data in **X** is contaminated by noise and so even after low-pass filtering the spectrum is not exactly equal to $2\pi\ \delta(\omega)$. In fact, on our run we obtained a single value V = 1.00067.

A frequency-domain Dirac delta appears as a line of height $V = \mu T/2\pi$ from which $\mu = V \times 2\pi/T$, and so in this case $\mu = 1.00067 \times 2\pi$.

Thus the Dirac delta's weight differed slightly from 2π because of the additive noise.

15.14

Regarding Exercise 15.13, using low-pass filtering we cannot eliminate the spectral values of the noise which lie between $n = 0$ and $n = 8$ at which the sine wave's spectrum lines are located. We can only eliminate the noise spectrum whose values of n exceed 8. However, we could examine the spectral elements and make the decision to retain **only those values whose magnitudes exceed a certain threshold.**

The **F** postprocessor package "Shave F" enables us to do precisely that.

Using the same steps for loading **X** with the function in Exercise 15.13 we obtain its spectrum by running ANALYSIS.

Then: Run Postprocessors, F, Shave F. For "CLIP" we use the following values:

CLIP = 1e-5, 1e-4, 1e-3.

After each one, QUIT and inspect the numbers in **F**.

Only after using CLIP = 1e-3 do we see that the noise values have been eliminated, with only the three strong spectral values from the signal remaining.

On our run the values in $|\mathbf{F}|$ were

$|F(\pm 8)|$ = 9.9805345e-2 and $|F(0)|$ = 1.0006698

The first two are from the $0.2\ \text{sine}(16\pi t)$ in the signal, whose transform is two Dirac deltas of weights $\pm 0.2\pi$. On the FFT a frequency-domain Dirac delta appears as a line of height $V = \mu T/2\pi$ and so $\mu = 2\pi V/T$. In this case T = 1 and so

μ = 9.9805345e-2×2π = 6.2709547e-1

This is approximately equal to the required 2π = 6.2831853, with an error of 0.19%.

We repeated the above procedure with larger values for N, the final one being N = 2048. For that case we obtained the following spectral values:

$|F(\pm 8)|$ = 9.9933869e-2 and $|F(0)|$ = 1.0000058

Observe how much closer these are to the exact values of

$|F(\pm 8)|$ = 0.1 and $|F(0)|$ = 1

The errors in $|F(\pm 8)|$ are now 0.066%. Using larger values for N will further reduce these errors.

In the following pages we offer some suggestions on how to conduct weekly computer labs based on the disk which accompanies the text. Each instructor will naturally evolve his/her own way of running the labs after a small amount of experience.

In our case the students are second-years and some of them have as yet had only limited contact with computers.

We find the weekly labs to be of immense benefit in consolidating the material that has been covered in the lectures of the previous week. Our students have attested to that fact, time after time. Seeing the lecture-concepts presented in tangible form on the computer screens has helped them significantly to grasp and absorb them.

Physically, our computer lab consists of sixty 486-50's (IBM compatibles) connected to a Novell server which holds the software. The students spend an hour to ninety minutes each week there, under the supervision of the lecturer plus three third- or fourth-year assistants who have already completed the course, and have had the benefit of using the course material in more advanced subjects such as Telecommunication Engineering, Communication Theory, Radar Signal Processing, and so on.

The students each receive a sheet containing the week's exercises.

> This worksheet should always be handed out at least a day before the lab for the students to carry out the pencil and paper portions of the exercises.

In the lab some students elect to work in pairs or threes, but the majority seem to prefer to work alone.

We have made attendance to the labs mandatory by taking roll, and then making 90% lab-attendance a prerequisite for writing a one-hour computer-lab exam at the end of the course. There are 10 marks assigned to that exam which count directly to the student's final mark out of 100. A sample of such an exam is given at the end of this section.

This is the first contact that the students have with the disk, and it is assumed that lectures have not yet proceeded to the point where meaningful problems can be run. Thus the first lab session is devoted to a simple walk-through of the two systems on the disk, namely PLOTS.EXE and FFT.EXE

- - - - - - - - - - - - - - - - - - - -

PLOTS.EXE

It is suggested that the instructor familiarize him/herself with what is contained in this system by consulting

- MAC users: README_18 in README_TWO.CC
- DOS users: README18.TXT

Therefter he/she will have little difficulty suggesting to the students what to do with the system and what facts to note.

This single walk-through is the only time that we spend on PLOTS.EXE. Many of the students are usually intrigued enough to run the demonstrations again for themselves.

- - - - - - - - - - - - - - - - - - - -

FFT.EXE

This is the main disk which accompanies the text, and is the one that is used throughout the course. The first contact with the disk can be somewhat overwhelming since there is a great deal that can be done with it. Running a few simple problems has been found to be the best way to get the students started. After the second or third lab most of them seem to find their way around it and are able to use it with confidence.

There is a complete User's Manual which covers all aspects of the system, contained in the following:

- MAC users: README_ONE.CC and README_TWO.CC
- DOS users: README1.TXT, README16.TXT and README17.TXT

The instructor is urged to make him/herself at least partially familiar with that material. Further familiarity will naturally come from repeated use of the disk, both in the labs as well as in carrying out many of the exercises at the end of the chapters.

- - - - - - - - - - - - - - - - - - - -

The only concept which the students seem to find difficult to grasp is the idea of SAMPLED vs DISCRETE.

The FFT is inherently a "discrete" system in that functions are represented by discrete sequences of numbers. In the "time-domain" they are functions of the discrete variable k and in the "frequency-domain" the discrete variable n.

Despite this, we almost always use the disk to manipulate functions of the **continuous variables** t (time) and ω (frequency). In order to do that on a discrete system such as the FFT we have to use **samples** taken from those functions. We refer to this mode of operation as **SAMPLED**.

- In **SAMPLED mode** the functions are written into the system almost exactly as they would be in the classroom, for example

  COS(pi*t) or SIN(ω/2)/(ω/2)

  Note: Math functions must always be entered in UPPERCASE.

  The system then takes **sampled values** from these expressions and places them in the appropriate vectors.

- When the functions which we are manipulating with the system are **inherently discrete** we process them using the **DISCRETE mode**. In this case the functions are written into the system as functions of the discrete variable k, for example

  1 + 2*k or k*k where k = 0, 1, 2, 3, ...

or simply as a sequence of N numbers, e.g. (1, 3, 5, 7, 9, 11, 13, 15)

- -

In the course of starting the system, the user will observe that the second screen contains the SYSTEM FLOW DIAGRAM. It is essential that this diagram be memorized since it is an invaluable roadmap with which to navigate. On it you will see that:

- The time-domain input goes into the **X** vector. Thus waveforms such as Rect(t), cos(πt), k^2, and so on are loaded into **X**.
- The function in **X** is Fourier-transformed using an operation called ANALYSIS to produce the **F** vector. We are now in the frequency-domain.
- The **F** vector can be inverse-Fourier-transformed using an operation called SYNTHESIS to bring it back to the time domain. The output of SYNTHESIS is always placed in the **Y** vector.
- There are two auxiliary vectors, **X2** and **F2** for holding additional vectors in the time and frequency domains.
- Only **X** can be operated on by ANALYSIS and only **F** can be operated on by SYNTHESIS. If the function that you wish to transform is in **X2** then it must first be moved to **X** before running ANALYSIS. Similarly, if a function that you wish to inverse-transform is in **F2** then [illegible]

In all of the plots and displays the vectors are referred to by the names **X**, **X2**, **F**, **F2**, **Y**, and for this reason the user **must** be sure in his/her mind where they lie in the flow diagram and how they are related to ANALYSIS and SYNTHESIS.

- -

We assume that the system has just been started, we have just passed through the SYSTEM FLOW DIAGRAM and we are on the main menu for the first time.

- The STATUS board on the right is a quick way to see what state the system is in.
- The SETUP option takes the user to a number of items which are discussed in detail in the User's Manual.

After covering all of the above the instructor should now step the students through the first two of the simple introductory exercises contained in

- MAC users: README_16(1) in README_ONE.CC
- DOS users: README16.TXT

Cover as many of the items discussed in those exercises as time permits.

Worksheet for COMPUTER LAB #2

(a) Using pencil and paper, find the expression for the Fourier coefficients of

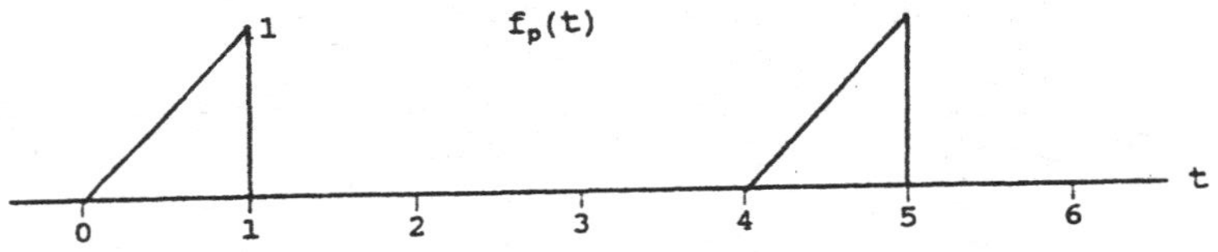

(b) Sketch the even and odd parts of $f_p(t)$ and state the expressions for their Fourier coefficients.

(c) Fill in the first of the following tables, using your hand-calculator and the above expressions:

Formula Values		
n	A(n)	B(n)
-2		
-1		
0		
1		
2		

FFT Values		
n	A(n)	B(n)
-2		
-1		
0		
1		
2		

(d) Run this problem on the FFT system using $N = 512$, and fill in the second table with the FFT values. Compare the FFT values to the formula values. Are they the same. If not, are they close?

(e) Starting from the waveform itself, use the F POSTPROCESSOR in the FFT system to create plots of the even and odd parts of the waveform.

(f) Using $N = 256$, load the expression for the Fourier coefficients that you obtained in (a) into the FFT system and invert it back to the time domain. When prompted for the ALIASING LEVEL, use 0. Observe the oscillations before and after the discontinuity.

(g) Repeat (f) by recalling the OLD PROBLEM, but this time use an ALIASING LEVEL of 5. Observe that now the oscillations are almost gone. A higher ALIASING LEVEL would reduce them further.

Steps for running COMPUTER LAB #2

(a) Finding the expression for the Fourier coefficients was done by the instructor in a problem session a few days before the computer lab, as follows:

$$F(n) = 1/T_0 \int_0^{T_0} f_p(t)\ \exp(-jn\omega_0 t)\ dt \qquad (T_0 = 4,\quad \omega_0 = 2\pi/T_0 = \pi/2)$$

$$= \tfrac{1}{4}\int_0^1 t\ \exp(-jn\pi t/2)\ dt = \tfrac{1}{4}\left[\frac{t\ \exp(-jn\pi t/2)}{-jn\pi/2} - \frac{\exp(-jn\pi t/2)}{(-jn\pi/2)^2}\right]_0^1$$

$$= \tfrac{1}{4}\left[\frac{\exp(-jn\pi/2)}{-jn\pi/2} - \frac{\exp(-jn\pi/2)] - 1}{(-jn\pi/2)^2}\right]$$

$$= \tfrac{1}{4}\left[\frac{(-jn\pi/2)\ \exp(-jn\pi/2) - \exp(-jn\pi/2)] + 1}{-n^2\pi^2/4}\right]$$

$$= \frac{[(1 + jn\pi/2)\exp(-jn\pi/2)] - 1}{n^2\pi^2} \longrightarrow \qquad (n \neq 0)$$

$$= \frac{(1 + jn\pi/2)[\cos(n\pi/2) - j\ \sin(n\pi/2)] - 1}{n^2\pi^2}$$

$$= \frac{\cos(n\pi/2) + (n\pi/2)\ \sin(n\pi/2) - 1}{n^2\pi^2} + j\ \frac{(n\pi/2)\ \cos(n\pi/2) - \sin(n\pi/2)}{n^2\pi^2}$$

$$= A(n) + j\ B(n) \qquad (n \neq 0)$$

Observe that A(n) is even and B(n) is odd.

For $n = 0$: $\quad F(0) = \tfrac{1}{4}\int_0^1 t\ dt = \tfrac{1}{4}\left.\frac{t^2}{2}\right|_0^1 = 0.125 \longrightarrow$

$= \text{average value of } f_p(t)$

(b) Sketch of even and odd parts of $f_p(t)$:

- The Fourier coefficients of $f_{ev}(t)$ are given by $A(n)$
- The Fourier coefficients of $f_{od}(t)$ are given by $j\ B(n)$

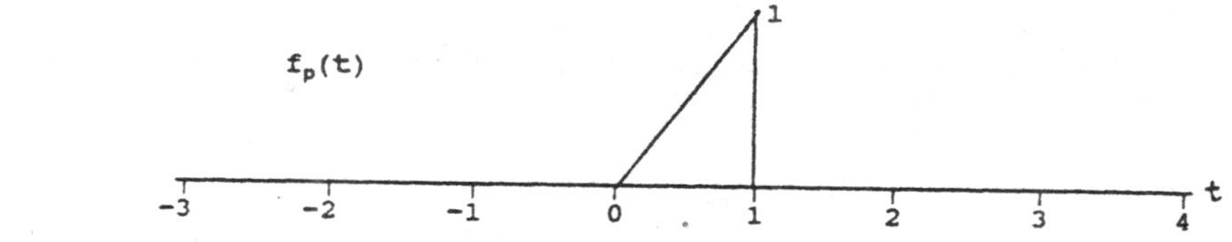

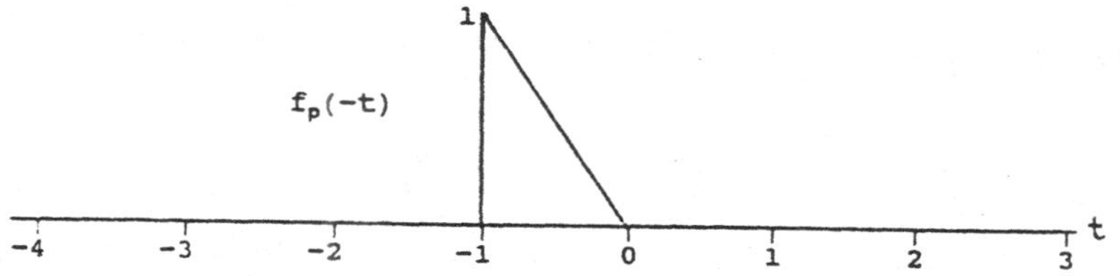

$$f_{ev}(t) = \frac{f_p(t) + f_p(-t)}{2}$$

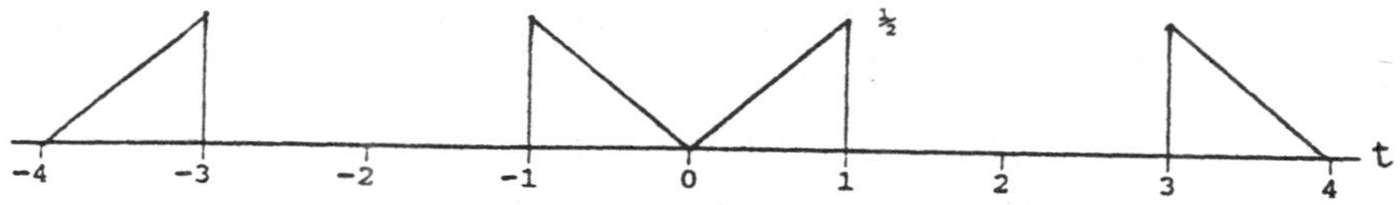

$$f_{od}(t) = \frac{f_p(t) - f_p(-t)}{2}$$

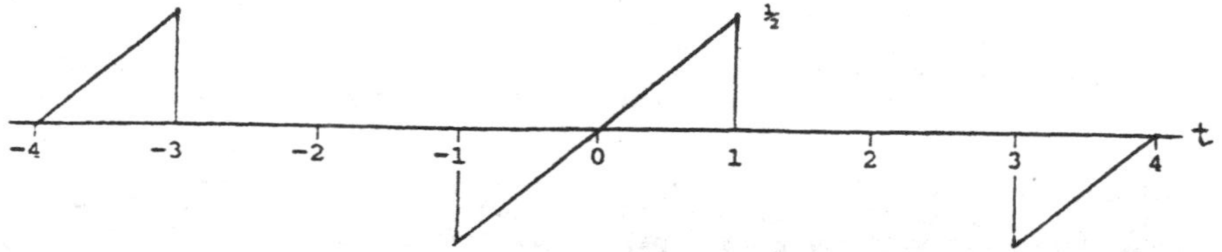

(c) Filling in the first table, using the above expressions:

n	A(n)	B(n)
-2	-5.0660591e-2	7.9577471e-2
-1	5.7833759e-2	0.10132118
0	0.125	0
1	5.7833759e-2	-0.10132118
2	-5.0660591e-2	-7.9577471e-2

(d) Steps to load the waveform, using N = 512, SAMPLED, T = 4, PERIODIC:

Main menu, Create Values, Continue, Create X, Real, Neither, Left, 2 intervals, 1, Continue, t, 0, Go, Neither, Go. You are now back on the main menu with the waveform correctly loaded. Plotting **X** confirms that.

Run ANALYSIS, Show Numbers, F, Complex. Filling in the second table.

n	$A(n)_{FFT}$	$B(n)_{FFT}$
-2	-5.0663135e-2	7.9573477e-2
-1	5.7830491e-2	0.10132246
0	0.125	0
1	5.7830491e-2	-0.10132246
2	-5.0663135e-2	-7.9573477e-2

Observe that:

- $A(n)_{FFT}$ is even and $B(n)_{FFT}$ is odd.
- The FFT values are close to those obtained from the formula but they are not identical

(e) Our strategy for producing the even part will be to

- Load the waveform into **X**
- Run ANALYSIS
- Zero out the IMAG part of $F(n)$, leaving only the REAL part

The waveform has already been loaded into X and ANALYSIS has been run. From the main menu: Run Postprocessors, F, Copy F to F2, Set FIM to zero, Quit, Run SYNTHESIS, Draw Plots, Plot Y. The even part is now on your screen.

To display the odd part: Run Postprocessors, F, Copy F2 to F, Set FRE to zero, Quit, Run SYNTHESIS, Plot Y. The odd part is now on your screen.

(f) To load the spectrum using N = 256, SAMPLED, T = 4, PERIODIC:

Main menu, Create Values, Continue, Create F, COMPLEX,

Note: Version 1.22 required that the running variable for loading the spectra of periodic waveforms should be m. All math functions should be UPPERCASE. Then

```
(COS(m*pi/2) + (m*pi/2)*SIN(m*pi/2) - 1)/((m*pi)^2)
```

Alpha = 0, Division by zero, B, 0.125, Go,

```
((m*pi/2)*COS(m*pi/2) - SIN(m*pi/2))/((m*pi)^2)
```

Continue aliasing, Go, Neither, Go. You are now back on the main menu.

Note: Version 1.30 and higher require that the running variable be n. Then the expressions are loaded as follows:

```
(COS(n*pi/2) + (n*pi/2)*SIN(n*pi/2) - 1)/((n*pi)^2)

((n*pi/2)*COS(n*pi/2) - SIN(n*pi/2))/((n*pi)^2)
```

Run SYNTHESIS and then plot the Y vector. A display of the original waveform appears with ringing before and after the discontinuity.

(g) To re-run the inversion using Alpha = 5: Main menu, Create Values, Continue, Use the Old-Problem, Accept the expression for FRE.

Alpha = 5, Division by zero, Go, B, 0.125, Go.

Accept the expression for FIM is displayed. Continue aliasing with Alpha = 5, Go, Go.

You are now back on the main menu. Run SYNTHESIS and then plot Y. A display of the original waveform appears with the ringing almost gone.

Using larger values for Alpha makes the ringing as small as we please.

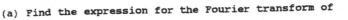

(a) Find the expression for the Fourier transform of

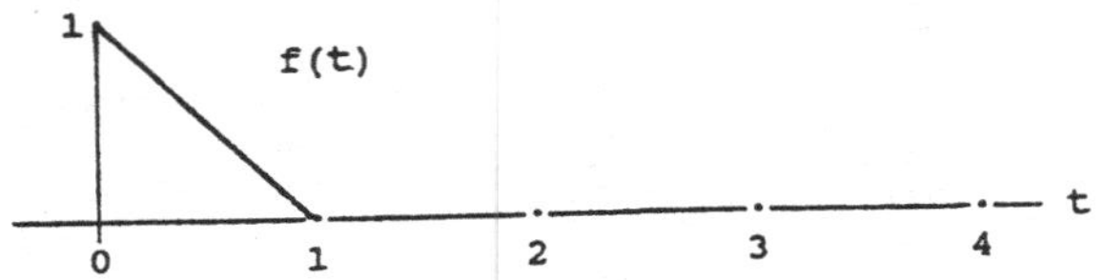

(b) Sketch the even and odd parts of f(t) and state the expressions for their Fourier coefficients.

(c) Fill in the first table below, using the above expressions:

Formula Values		
n	$A(\omega_n)$	$B(\omega_n)$
-2		
-1		
0		
1		
2		

FFT Values		
n	$A(\omega_n)$	$B(\omega_n)$
-2		
-1		
0		
1		
2		

Note: Use $\omega_n = n \times 2\pi / T$ where $T = 4$

(d) Run this problem on the FFT system using N = 512, T = 4, and fill in the same table with the FFT values. Compare the FFT values to the formula values. Are they the same. If not, are they close?

(e) Use the F Postprocessor to create plots of the even and odd parts of the waveform.

(f) Using N = 256, T = 4, load the expression for the Fourier transform that you obtained in (a) into the F vector and invert it back to the time domain. When prompted for the ALIASING LEVEL, use 0. Observe the oscillations before and after the discontinuity.

(g) Repeat (f) by recalling the OLD PROBLEM, but this time use an aliasing level of 5. Observe that now the oscillations are almost gone. A higher aliasing level would reduce them further.

(h) When we use the FFT to find the energy in a pulse we must integrate the FFT's energy spectrum using numerical integration. This involves the use of all of the FFT's energy spectral values, all of which are positive, and so the accuracy of the final answer is a good measure of the accuracy of the FFT's spectral values compared to those of the CFT (continuous Fourier transform).

Find the total energy in the pulse. Then: Using N = 1024 and T = 16, find the total energy from the FFT system and compute its relative error. Examine the plots of **FRE** and **FIM**.

(i) Repeat (h) but now use T = 1. Observe that the energy value from the FFT is very much closer to the exact answer for T = 1 than it was for T = 16, but that the plots are very much poorer.

Thus: Smaller T —> good accuracy but poor plots

Larger T —> poor accuracy but good plots

Steps for running COMPUTER LAB #3

(a) Finding the expression for the Fourier transform of f(t):

$$F(\omega) = \int_{-\infty}^{\infty} f(t)\ \exp(-j\omega t)\ dt$$

$$= \int_0^1 (1-t)\ \exp(-j\omega t)\ dt = \left[\frac{(1-t)\ \exp(-j\omega t)}{-j\omega} + \frac{\exp(-j\omega t)}{(-j\omega)^2} \right]_0^1$$

$$= \frac{-1}{-j\omega} + \frac{\exp(-j\omega) - 1}{(-j\omega)^2} = \frac{j\omega + \exp(-j\omega) - 1}{(-j\omega)^2} = \frac{1 - j\omega - \exp(-j\omega)}{\omega^2} \longrightarrow$$

$$= \frac{1 - \cos(\omega)}{\omega^2} + j\,\frac{\sin(\omega) - \omega}{\omega^2} = A(\omega) + jB(\omega) \longrightarrow$$

Observe that $A(\omega)$ is even and $B(\omega)$ is odd.

For $\omega = 0$: $$F(0) = \int_0^1 (1 - t)\ dt = -\left.\frac{(1 - t)^2}{2}\right|_0^1 = 1/2$$

= area under f(t)

(b) Sketch of even and odd parts: See next page.

- The Fourier transform of $f_{ev}(t)$ is given by $A(\omega)$
- The Fourier transform of $f_{od}(t)$ is given by $jB(\omega)$

(c) Filling in the first table, using the above expressions: T = 4 in the FFT runs and so we use $\omega_n = n\times 2\pi/T = n\times\pi/2$

Formula Values		
n	$A(\omega_n)$	$B(\omega_n)$
-2	0.202642367	0.318309886
-1	0.405284734	0.232335038
0	0.5	0
1	0.405284734	-0.232335038
2	0.202642367	-0.318309886

(d) To load the waveform, using N = 512, SAMPLED, T = 4, PULSE: Main menu, Create Values, Continue, Create X, Real, Neither, Left, 2 intervals, 1, Continue, 1 - t, 0, Go, Neither, Go. You are now back on the main menu with the waveform correctly loaded. Plotting X confirms that. Run ANALYSIS, Show Numbers, F, Single. Filling in the second table.

FFT Values		
n	$A(\omega_n)$	$B(\omega_n)$
-2	0.20265254	0.318309886
-1	0.40528902	0.232335038
0	0.5	0
1	0.40528902	-0.232335038
2	0.20265254	-0.318309886

Observe that the FFT values are close to those obtained from the formula but they are not identical.

f(t)

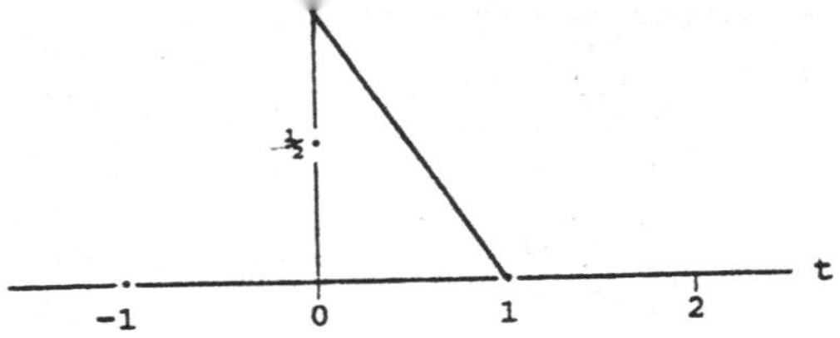

f(−t)

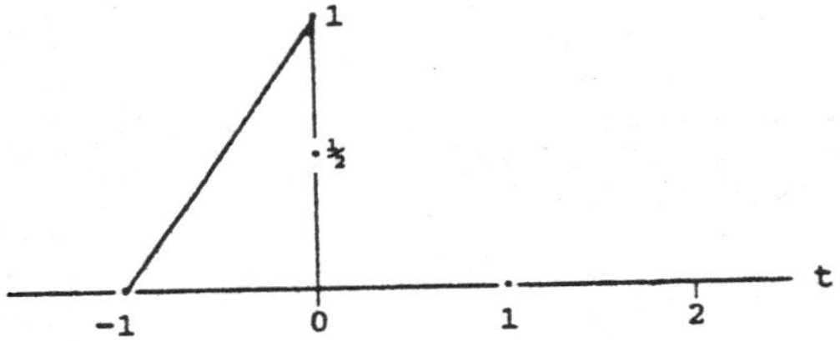

$$f_{ev}(t) = \frac{f(t) + f(-t)}{2}$$

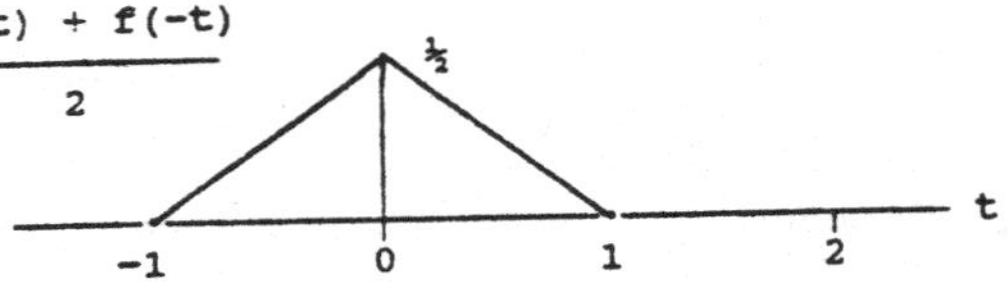

$$f_{od}(t) = \frac{f(t) - f(-t)}{2}$$

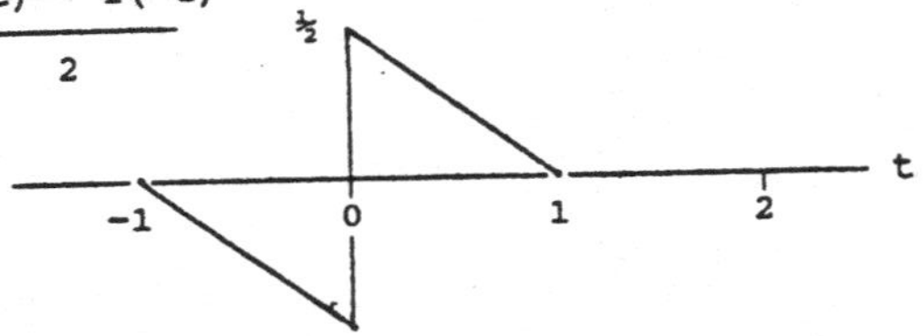

(e) Our strategy for producing the even part will be to

- Load the waveform into X
- Run ANALYSIS
- Zero out the IMAG part of $F(\omega)$, leaving only the REAL part
- Run SYNTHESIS, and display Y. It will be the even part of $f(t)$.

run. From the main menu: Run Postprocessors, F, Copy F to F2, Set FIM to zero, Quit, Run SYNTHESIS, Draw Plots, Plot Y. The even part is now on your screen.

To display the odd part: Run Postprocessors, F, Copy F2 to F, Set FRE to zero, Quit, Run SYNTHESIS, Plot Y. The odd part is now on your screen.

(f) To load the spectrum using N = 256, SAMPLED, T = 4, PULSE:

Main menu, Create Values, Continue, Create F, Complex, (1 - COS(w))/w^2

Alpha = 0, "Division by zero", Go, B, 0.5, Go, (SIN(w) - w)/w^2

Continue aliasing, Go. Neither, Go. You are now back on the main menu. Run SYNTHESIS and plot the Y vector. A display of the original waveform appears with ringing before and after the discontinuity.

(g) To re-run the inversion using Alpha = 5: Main menu, Create Values, Continue, Use the Old-Problem, Accept the expression for FRE.

Alpha = 5, "Division by zero", Go, B, 0.5, Go.

Accept the expression for FIM. Continue aliasing with Alpha = 5, Go, Go.

You are now on the main menu. Run SYNTHESIS and plot Y. A display of the original waveform appears with ringing now almost gone.

Using larger values for Alpha makes the ringing as small as we please.

(h) $$\text{Total energy} = \int_{-\infty}^{\infty} f(t)^2\,dt = \int_0^1 (1-t)^2\,dt = -\left.\frac{(1-t)^3}{3}\right|_1^1 = 1/3$$

Using N = 1024 and T = 16, load the pulse: Main menu, Create Values, Continue, Create X, Real, Neither, Left, 2, 1, Continue, 1 - t, 0, Go, Neither, Go. You are now back on the main menu. Plot X to confirm that the pulse has been loaded correctly.

Run Postprocessors, F, Energy, 512. The system shows E = 0.329468

This is a fairly good approximation for the exact value of 1/3. The relative error is 1.159%.

The inaccuracy is due mostly to the fact that the FFT spectrum is not an exact version of the CFT spectrum, plus some errors in

integrating the FFT spectrum.

Examining the plots of **FRE** and **FIM** (Zoom = 4) we see good representations of the spectra.

(i) Using N = 1024 and T = 1, load the pulse: Main menu, Create Values, Continue, Create X, Real, Neither, Left, 1 interval, Continue, 1-t, Go, Neither, Go. You are now back on the main menu. Plot **X** to confirm that the pulse has been loaded correctly.

Run Postprocessors, F, Energy, 512. The system shows E = 0.333089 The error is now only 0.073%.

We observe that the energy value from the FFT is very much closer to the exact answer for T = 1 than it was for T = 16. This is because the FFT's values are now very much closer to the CFT values.

HOWEVER: Plotting **FRE** and **FIM** shows very poor representations of those spectra when compared to (h), especially **FRE**.

Thus: Smaller T —> good accuracy but poor plots

Larger T —> poor accuracy but good plots

Worksheet for COMPUTER LAB #4

Load $f(t)' = \text{Rect}(t)$ into **X**. (Use N = 256, T = 4.) Plot it. Run ANALYSIS. Plot FRE. Store the spectrum into **F2**. Then lock **F2**.

(a) Now use LOW-PASS FILTER to zero out all elements in **F** with spectral numbers greater than 16. Draw a plot of FRE to see what you have done. Then run SYNTHESIS. Draw a plot of the resultant pulse. Explain what you see, to a demonstrator or the instructor.

(b) Restore the spectrum. Use HIGH-PASS FILTER to zero out F(0). Run "Show Numbers" and draw a plot to see what you have done. Run SYNTHESIS. Draw a plot of the resultant pulse. Explain what you see, to a demonstrator or the instructor.

(c) Restore the spectrum. Use HIGH-PASS FILTER to zero out all **F** elements with spectral numbers less than 2. Run "Show Numbers" and draw a plot to see what you have done. Run SYNTHESIS. Draw a plot of the resultant pulse. Explain to a demonstrator or the instructor.

(d) Restore the spectrum. Repeat (b) above but this time do it starting from main-menu C. Run SYNTHESIS and plot the result.

(e) Restore the spectrum. Repeat (c) above starting from main-menu C, but do it so as to keep the time-domain pulse real. Run SYNTHESIS and plot the result.

(f) Draw a hand sketch of the pulse $g(t) = \text{Rect}(t) + 0.1\delta(t)$. Then, using N = 256, T = 16, load the pulse into **X**. Draw a plot and confirm what you obtained in the sketch. Run ANALYSIS. Draw a plot of FRE. Explain what you see to a demonstrator or the instructor.

Steps for running COMPUTER LAB #4

To load f(t) using N = 256, T = 4: Main menu, Create Values, Continue, Create, X, Real, Even, 2 intervals, 0.5, Continue, 1, 0, Go, Neither, Go. You are now back on the main menu. Plot X to verify that f(t) has been correctly loaded. Run ANALYSIS and plot **FRE**. Expand it using ZOOM = 2.

To store the spectrum: Main menu, Run Postprocessors, F, Copy F to F2, Lock F2, Quit. You are now back on the main menu. "Lock F2" means that (1) it will not be zeroed whenever the system vectors are cleared, and (2) you will be warned if you accidentally try to change it.

(a) Main menu, Run Postprocessors, F, Low-pass Filter, 16, Quit. Plotting **F** shows that all spectral elements F(n) for $|n| > 16$ are now zero. Show Numbers, F, Single, shows the same.

Run SYNTHESIS and plot **Y**. We see the pulse after all spectral elements F(n) with $|n| > 16$ have been removed. The "missing parts" of the plot show what effect the high frequencies would have had.

(b) Main menu, Run Postprocessors, F, Copy F2 to F, High-pass Filter, 1, Quit. All elements F(n) with $|n| < 1$ (i.e. F(0)) have now been zeroed. Plot **F**, or Show Numbers, F, Single will confirm it.

Run SYNTHESIS and plot **Y**. We see the pulse after spectral element F(0) has been removed. Its average value has been set to zero, and so the areas above and below the horizontal axis are equal. Ideally, if we were working with the pulse Rect(t) with F(0) set to zero we would not see any change. However, the FFT is (1) discrete and (2) periodic, and so this is how it displays the effects of setting F(0) to zero.

(c) Main menu, Run Postprocessors, F, Copy F2 to F, High-pass Filter, 2, Quit. All elements F(n) with $|n| < 2$ (i.e. F(0) and F(±1)) have now been zeroed. Plot **F**, or Show Numbers, F, Single will confirm that.

Run SYNTHESIS and plot **Y**. Note: If the origin is not at the center then move it to the center as follows: From the PLOT menu, Reset Parameters, Toggle the X and Y Origins to the Center, Quit.

We see the pulse after spectral elements F(0) and F(±1) have been removed. Its average value now equal to zero, and so the areas above and below the horizontal axis are equal. **Moreover**, if you look carefully you will also see that a cosine of wavelength 4 sec. has been removed. This is the cosine formed by $F(1)\exp(j1\omega_0 t)$ and $F(-1)\exp(-j1\omega_0 t)$ where $\omega_0 = 2\pi/T = 2\pi/4 = \pi/2$ and so the cosine that has been removed is $2\ F(1)\ \cos(\pi t/2)$.

(d) Main menu, Run Postprocessors, F, Copy F2 to F, Quit. Main menu, Edit From Keyboard, Continue, Change F vectors, F, Real, Use k = 0, 0, Skip, Continue. You are now back on the main menu with spectral element F(0) set to zero. Plot **F**, or Show Numbers, F, Single confirms it.

Run SYNTHESIS and plot **Y**. We see the pulse after spectral element F(0) has been removed. Then same comments as in (b).

(e) Main menu, Run Postprocessors, F, Copy F2 to F, Quit. Main menu, Edit From Keyboard, Continue, Change F vectors, F, Real, Use k = 0, 0, Use k = 1, 0, Specify k = -1, 0, Skip, Continue. You are now back on the main menu with spectral elements F(0), F(1) and F(-1) set to zero. Plot **F**, or Show Numbers, F, Single will confirm it.

The realness of the result in the time-domain has been preserved because we zeroed F(-1) after zeroing F(1). If we had not zeroed F(-1) the spectrum would have led to a complex pulse in the time domain after running SYNTHESIS.

Run SYNTHESIS and plot **Y**. We see the pulse with spectral elements F(0), F(1) and F(-1) removed. Then same comments as in (c).

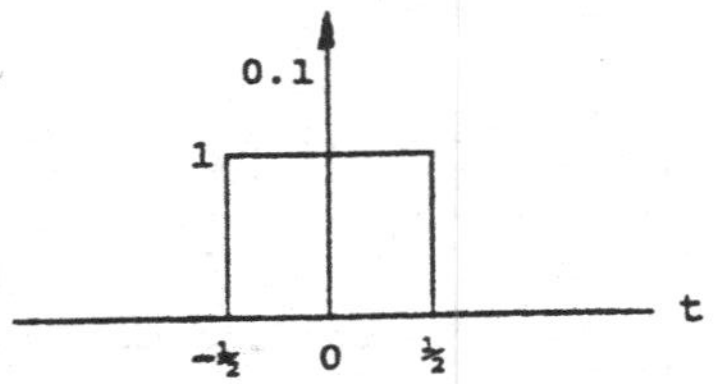

To load the pulse into **X** using N = 256, T = 16: Main menu, Continue, Create X, Real, Even, 2 intervals, 0.5, Continue, 1, 0, Go, X, 0, 0.1, Neither, Go.

You are now back on the main menu. Plot **X** to verify that f(t) has been correctly loaded. Rect(t) is clearly visible. In its center is a line of height 2.6. To see why: A Dirac delta of weight μ appears on the FFT as a line of height $V = \mu N/T$. In this case $\mu = 0.1$, N = 256 and T = 16, and so V = 1.6. Add that to the Rect and you obtain the value 2.6 that appears on the plot.

Run ANALYSIS and plot FRE. We know that

$$g(t) = \text{Rect}(t) + 0.1\delta(t) \Longleftrightarrow G(\omega) = \text{Sa}(\omega/2) + 0.1$$

The plot clearly shows the Sa to which has been added 0.1 all across its length. The max positive value is now 1.1 as expected.

Worksheet for COMPUTER LAB #5

(a) If the following is a pulse sketch its Fourier transform. If it is periodic sketch its Fourier coefficients:

(1) $3\delta(t - \frac{1}{4})$	PULSE	Sketch magnitude and phase
(2) $2\delta(t)$	PERIODIC ($T_0 = 4$)	Sketch magnitude and phase
(3) $\cos(2\pi t)$	PULSE ($\forall$ t)	Sketch $F(\omega)$
(4) $\cos(2\pi t)$	PERIODIC	Sketch $F(n)$
(5) $\sin(4\pi t)$	PULSE ($\forall$ t)	Sketch $F(\omega)$
(6) $\sin(4\pi t)$	PERIODIC	Sketch $F(n)$

(b) On the FFT

- a time-domain Dirac delta appears as a line of height $V = \mu N/T$ where μ is its weight
- a frequency-domain Dirac delta appears as a line of height $V = \mu T/2\pi$ where μ is its weight.

Starting from main-menu A:

- for the pulses in (a) use N = 256, T = 4, PULSE.
- for the periodic waveforms in (a) use N = 256, T = period, PERIODIC.

Create each of the items in (a). Then plot X and verify that the system has loaded the item correctly. Then transform and verify that each of your sketches is correct.

Steps for running COMPUTER LAB #5

(a) (1) $3\delta(t - \frac{1}{4})$ PULSE: Sketch magnitude and phase

$3\delta(t - \frac{1}{4})$ PULSE $\Longleftrightarrow F(\omega) = 3\exp(-j\omega/4)$

Magnitude: $|F(\omega)| = 3$ Phase: $\theta(\omega) = -\omega/4$

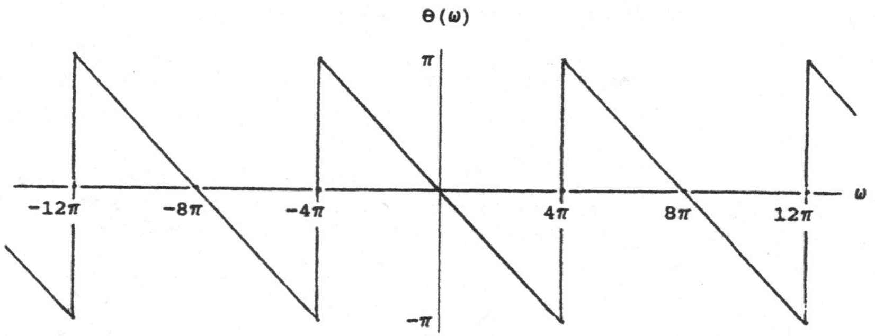

(2) $2\delta(t)$ PERIODIC ($T_0 = 4$): Sketch magnitude and phase

$f_p(t) = 2\delta(t)$ as a periodic function with $T_0 = 4$ is a Dirac comb with spacing of 4 seconds between the impulses. Its Fourier coefficients are

$$F(n) = 1/T_0 \int_{-T_0/2}^{T_0/2} f_p(t)\exp(-jn\omega_0 t)\,dt = \frac{1}{4}\int_{-\frac{1}{2}}^{\frac{1}{2}} 2\delta(t)\exp(-jn\pi/2)\,dt = \frac{1}{2}$$

Sketch of $|F(n)|$

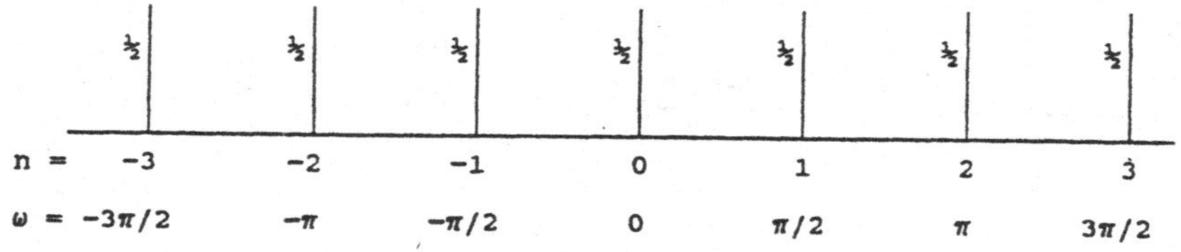

On the n-axis the lines are at every value of n. On the ω-axis the lines are at intervals of $\omega_0 = 2\pi/T_0 = 2\pi/4 = \pi/2$.

The phase plot is everywhere zero.

(3) cos(2πt) PULSE (∀ t): Sketch F(ω)

$f(t) = \cos(2\pi t) \Longleftrightarrow \pi[\delta(\omega - \omega_0) + \delta(\omega + \omega_0)]$ where $\omega_0 = 2\pi$

Thus, $F(\omega) = \pi[\delta(\omega - 2\pi) + \delta(\omega + 2\pi)]$

Sketch of F(ω)

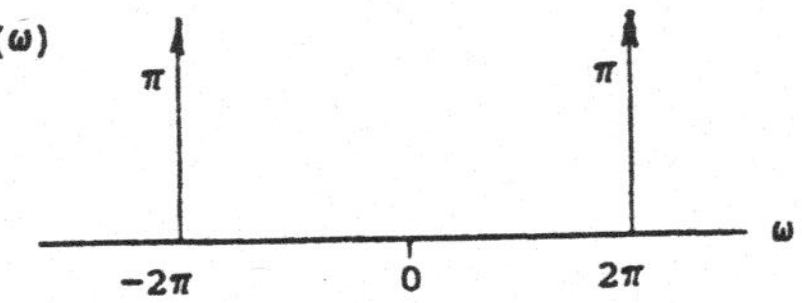

(4) cos(2πt) PERIODIC: Sketch F(n)

$f_p(t) = \cos(2\pi t) = \frac{1}{2}\exp(-j2\pi t) + \frac{1}{2}\exp(j2\pi t) \Longleftrightarrow F(-1) = \frac{1}{2} = F(1)$

Sketch of F(n)

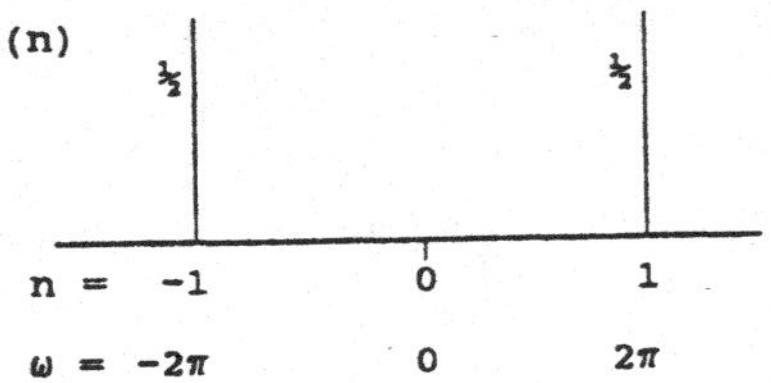

On the n-axis the lines are at ±1. On the ω-axis the lines are at intervals $\omega = \pm 2\pi$.

(5) sin(4πt) PULSE (∀ t): Sketch F(ω)

$f(t) = \sin(4\pi t) \Longleftrightarrow \pi/j[\delta(\omega - \omega_0) - \delta(\omega + \omega_0)]$ where $\omega_0 = 4\pi$

Thus, $F(\omega) = \pi/j[\delta(\omega - 4\pi) - \delta(\omega + 4\pi)]$

Sketch of F(ω)

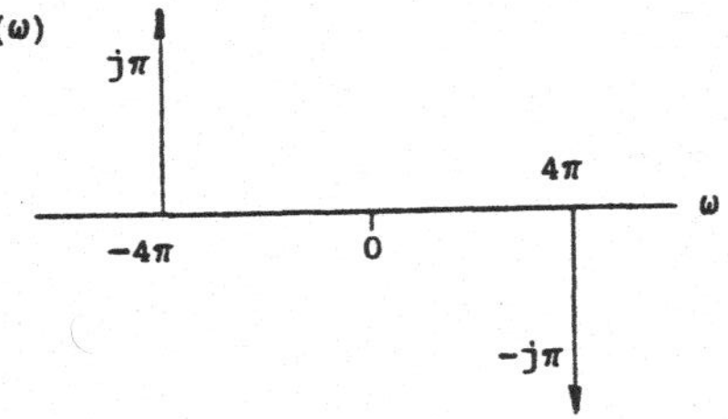

(6) sin(4πt) PERIODIC: Sketch F(n)

$f_p(t) = \sin(4\pi t) = -1/2j\ \exp(-j4\pi t) + 1/2j\ \exp(j4\pi t)$

$\Longleftrightarrow F(-1) = j/2 \quad F(1) = -j/2$

Sketches of F(n):

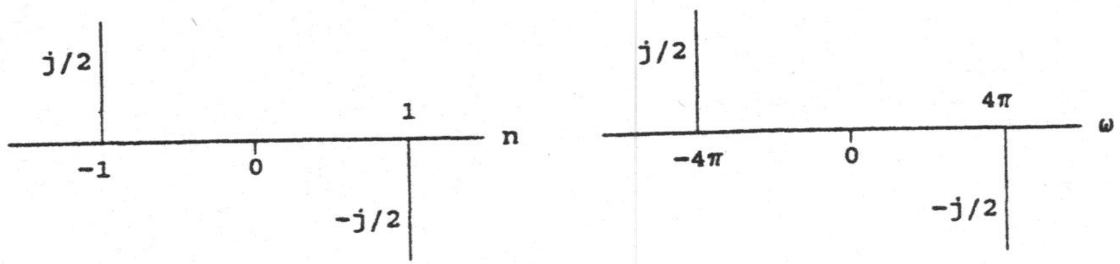

On the n-axis the lines are at ±1. On the ω-axis the lines are at $\omega = \pm 4\pi$.

(b) (1) $3\delta(t - \frac{1}{4})$ PULSE:

Main menu, T = 4, PULSE, Create Values, Continue, Create only Diracs, Time, X, 1/4, 3, Neither, Go. You are now back on the main menu. Plotting **X** shows a single line at t = 1/4 of height V = 192. This is a Dirac delta of weight

$\mu = VT/N = 192 \times 4/256 = 3$

Run ANALYSIS. Plotting |**F**| shows a value of 3 for all ω. Plotting the Phase shows falling values with the crossovers at ω = 0, 8π, 16π, etc, and the plot is odd. Thus the above sketch is correct.

(2) $2\delta(t)$ PERIODIC ($T_0 = 4$):

Main menu, T = 4, PERIODIC, Continue, Create only Diracs, Time, X, 0, 2, Neither, Go. You are now back on the main menu. Plotting **X** shows a single line at t = 0 of height V = 128. This is a Dirac delta of weight

$\mu = VT/N = 128 \times 4/256 = 2$

Run ANALYSIS. Plotting **FRE** (use ZOOM = 32) shows values of 0.5 at intervals of $\omega = \pi/2$.

To change the horizontal axis to "n": Main menu, Setup, Go into the PLOT items and toggle the horizontal axis in the frequency domain from "omega" to "n". (This will be in force for all future sessions until you toggle it back to "omega".)

Quit and return to the plot of **FRE**. We see lines of height ½ at each value of n.

(3) $\cos(2\pi t)$ PULSE:

Main menu, Create Values, T = 4, PULSE, Continue, Create X, Real, Even, 1 interval, Continue, COS(2*pi*t), Go, Neither, Go. You are now back on the main menu. Plotting **X** shows a cosine of period 1, i.e. $\cos(2\pi t)$.

Run ANALYSIS. Plotting **FRE** (use ZOOM = 16) shows lines of height $V = 2$ at $\omega = \pm 2\pi$. These are frequency-domain Dirac deltas of weight

$\mu = V2\pi/T = 2\times 2\pi/4 = \pi$

Thus the above sketch is correct.

(4) $\cos(2\pi t)$ PERIODIC:

Main menu, Create Values, T = 1, PERIODIC, Continue, Create X, Real Even, 1 interval, Continue, COS(2*pi*t), Go, Neither, Go. You are now back on the main menu. Plotting **X** shows a cosine of period 1, i.e. $\cos(2\pi t)$.

Run ANALYSIS. Plotting **FRE** (use ZOOM = 16) shows lines of height ½ at $\omega = \pm 2\pi$, or $n = \pm 1$. These are the expected spectral lines. Thus the above sketch is correct.

(5) $\sin(4\pi t)$ PULSE:

Main menu, Create Values, T = 4, PULSE, Continue, Create X, Real, Odd, 1 interval, Continue, SIN(4*pi*t), Go, Neither, Go. You are now back on the main menu. Plotting **X** shows a sine of period ½, i.e. SIN($4\pi t$).

Run ANALYSIS. Plotting **FIM** (use ZOOM = 8) shows lines of height $V = \pm 2$ at $\pm\omega = \pm 4\pi$. These are frequency-domain Dirac deltas of weight

$\mu = V2\pi/T = 2\times 2\pi/4 = \pi$

Thus the above sketch is correct.

(6) $\sin(4\pi t)$ PERIODIC:

Main menu, Create Values, T = ½, PERIODIC, Continue, Create X, Real Odd, 1 interval, Continue, SIN(4*pi*t), Go, Neither, Go. You are now back on the main menu. Plotting **X** shows a sine of period ½, i.e. SIN($4\pi t$).

Run ANALYSIS. Plotting **FIM** (use ZOOM = 32) shows lines of height ±½ at $\omega = \pm 4\pi$, or $n = \pm 1$. These are the expected spectral lines. Thus the above sketch is correct.

Worksheet for COMPUTER LAB #6

(a) Sketch each of the following and their inverse transforms:

(1) $2\pi\delta(\omega - 2\pi)$

(2) $\pi[\delta(\omega - 4\pi) + \delta(\omega + 4\pi)]$

(3) $(\pi/j)[\delta(\omega - 8\pi) - \delta(\omega + 8\pi)]$

(b) On the FFT

- a time-domain Dirac delta appears as a line of height $V = \mu N/T$ where μ is its weight
- a frequency-domain Dirac delta appears as a line of height $V = \mu T/2\pi$ where μ is its weight.

Using N = 256, T = 4, PULSE, create each of these spectra. Then verify that the system has loaded them correctly. Then inverse transform and verify that the results agree with what you obtained in the time domain.

(c) Sketch $f(t) = \text{Rect}[2(t-1)]$. Now sketch the inverse of $F(\omega)\exp(-j\omega 2)$ Using N = 256, T = 8 load Rect[2(t-1)] into **X**. Run ANALYSIS.

- Plot and verify **X**.
- Multiply **F** by exp($-j\omega 2$). Run SYNTHESIS. Plot and verify **Y**.

(d) Sketch $F(\omega)$ where $f(t) = \text{Rect}(t)$. Now sketch the transform of $f(t)\exp(j8\pi t)$.

Using N = 256, T = 8, load Rect(t) into **X**.

- Plot and verify **X**.
- Run ANALYSIS and verify the plot of **F**.
- Multiply **X** by exp($j8\pi t$). Run ANALYSIS. Plot and verify **F**.

(e) Sketch the transform of $f(t) = \text{Rect}(t/2)$. Now sketch the transform of $f(t)\cos(16\pi t)$.

Using N = 256, T = 8, load Rect(t/2) into **X** and $\cos(16\pi t)$ into **X2**.

- Plot and verify **X** and **X2**. Run ANALYSIS. Plot and verify **F**
- Multiply **X** and **X2**. Plot and verify **X**.
- Run ANALYSIS. Plot and verify **F**.

Steps for running COMPUTER LAB #6

(a) and (b)

(1) $F(\omega) = 2\pi\delta(\omega - 2\pi) \Longleftrightarrow f(t) = \exp(j2\pi t) \quad (\forall\ t)$

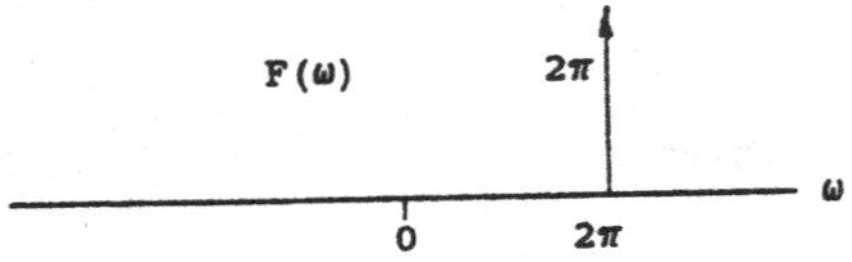

$f(t) = \exp(j2\pi t) = \cos(2\pi t) + j\sin(2\pi t) \quad (\forall\ t)$

Sketches of $\cos(2\pi t)$ and $\sin(2\pi t)$ $(\forall\ t)$ are omitted.

Steps for loading the spectrum using N = 256, T = 4, PULSE: Main menu, Create Values, Continue, Create only Diracs, Frequency, FRE, 2*pi, 2*pi, Neither, Go. You are now back on the main menu with the spectrum loaded. Plotting **FRE** (Zoom = 8) shows a single line representing the Dirac delta, of height V = 4, at $\omega = 2\pi$. Thus, the weight of the Dirac delta is $\mu = V2\pi/T = 4\times2\pi/4 = 2\pi$, which is correct.

Run SYNTHESIS. Plotting **YRE** shows a cosine of period 1, namely $\cos(2\pi t)$. Plotting **YIM** shows a sine of period 1, namely $\sin(2\pi t)$. Thus our inversion was correct.

(2) $F(\omega) = \pi[\delta(\omega - 4\pi) + \delta(\omega + 4\pi)] \Longleftrightarrow f(t) = \cos(4\pi t) \quad (\forall\ t)$

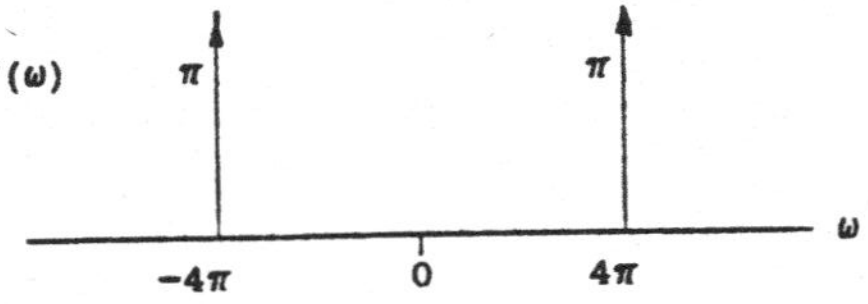

Sketch of $\cos(4\pi t)$ $(\forall\ t)$ is omitted.

Steps for loading the spectrum using N = 256, T = 4, PULSE: Main menu, Create Values, Continue, Create only Diracs, Frequency, FRE, 4*pi, pi, FRE, -4*pi, pi, Neither, Go. You are now back on the main menu with the spectrum loaded. Plotting **FRE** (Zoom = 8) shows two lines representing the Dirac deltas, of height V = 2, at $\omega = \pm4\pi$. Thus, the weights of the Dirac deltas are $\mu = V2\pi/T = 2\times2\pi/4 = \pi$, which is correct.

Run SYNTHESIS. Plotting **YRE** shows a cosine of period ½, namely $\cos(4\pi t)$. Thus our inversion was correct.

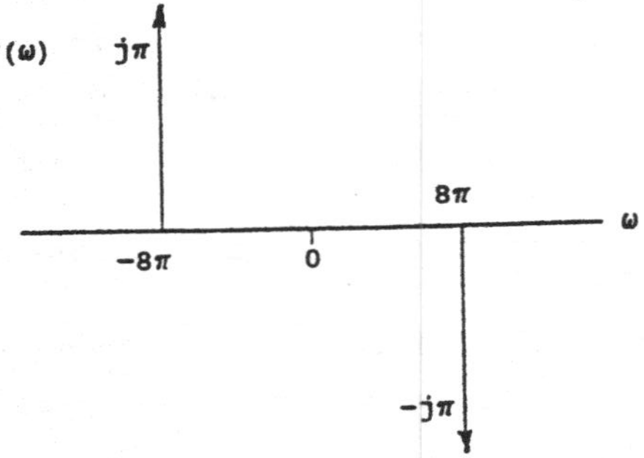

Sketch of $\sin(8\pi t)$ $(\forall\ t)$ is omitted.

Steps for loading the spectrum using N = 256, T = 4, PULSE: Main menu, Create Values, Continue, Create only Diracs, Frequency, FIM, 8*pi, -pi, FIM, -8*pi, pi, Neither, Go. You are now back on the main menu with the spectrum loaded. Plotting **FIM** (Zoom = 4) shows two lines representing the Dirac deltas, of height V = 2, at $\omega = \pm8\pi$. Thus the weights of the Dirac deltas are $\mu = V2\pi/T = 2\times2\pi/4 = \pi$, which is correct.

Run SYNTHESIS. Plotting **YRE** shows a sine of period ¼, namely $\sin(8\pi t)$. Thus our inversion was correct.

(c) Sketch of $f(t) = \text{Rect}[2(t-1)]$:

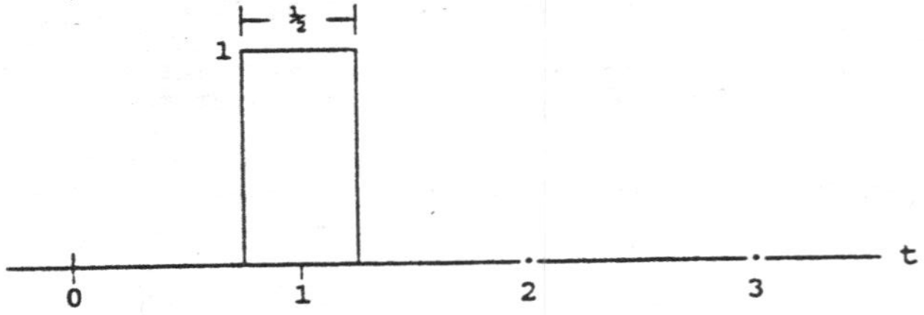

$g(t) = \text{Rect}(2t) = \text{Rect}(t/\tau)$ where $\tau = \frac{1}{2}$

$\Longleftrightarrow \tau\ \text{Sa}(\omega\tau/2) = \frac{1}{2}\ \text{Sa}(\omega/4) = G(\omega)$

$f(t) = \text{Rect}[2(t-1)] \Longleftrightarrow \frac{1}{2}\ \text{Sa}(\omega/4)\exp(-j\omega) = F(\omega)$

Then $F(\omega)\exp(-j\omega 2) \Longleftrightarrow f(t-2) = \text{Rect}[2(t-3)]$

Sketch of the inverse of $F(\omega)\ \exp(-j\omega 2)$:

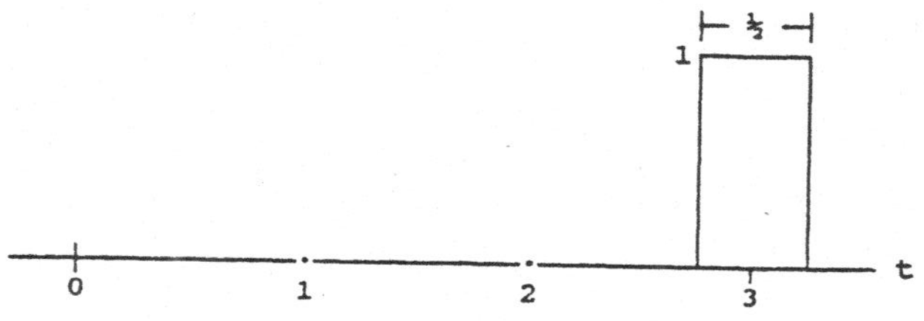

Using N = 256, T = 8: Create Values, Continue, Create X, Real, Neither, Left, 3 intervals, 3/4, 5/4, Continue, 0, 1, 0, Go, Neither, Go. You are now back on the main menu with the pulse correctly loaded. Plotting **X** (Zoom = 4) confirms that.

Run ANALYSIS. Plotting |**F**| shows |½ Sa| with the first zero at $\omega = 4\pi$. Thus it must be $|\frac{1}{2}\ \mathrm{Sa}(\omega/4)|$.

Run Postprocessors, F, Multiply F by exp(j w tau), -2, Quit, Run Synthesis. Plotting **Y** shows a Rect of width ½, centered at t = 3. This confirms our sketch above.

(d) $f(t) = \mathrm{Rect}(t) \Longleftrightarrow \mathrm{Sa}(\omega/2) = F(\omega)$

Sketch of $F(\omega)$

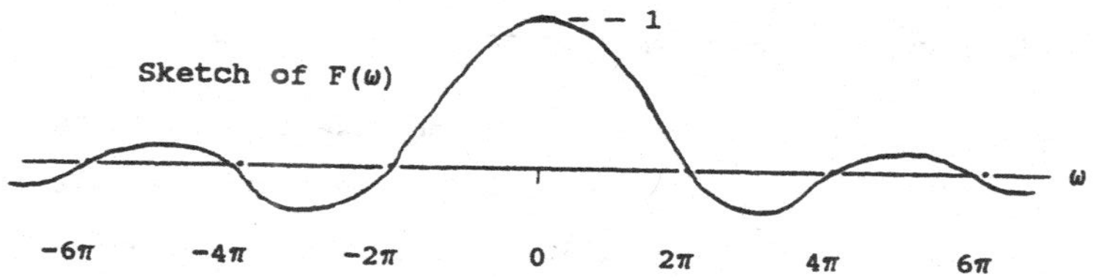

$f(t)\ \exp(j8\pi t) = \mathrm{Rect}(t)\ \exp(j8\pi t) \Longleftrightarrow F(\omega - 8\pi) = \mathrm{Sa}[(\omega - 8\pi)/2]$

This is the above Sa moved 8π to the right. Sketch is omitted.

Using N = 256, T = 8: Create Values, Continue, Create X, Real, Even, 2 intervals, ½, Continue, 1, 0,, Go, Neither, Go. You are now back on the main menu with the pulse correctly loaded. Plotting **X** confirms that.

Run ANALYSIS. Plotting **FRE** shows a Sa. Using ZOOM = 4 we see the first zero-crossing at $\omega = 2\pi$, as expected.

Run Postprocessors, X, Multiply X by exp(j wo t), 8*pi, Quit, Run ANALYSIS. Plotting **FRE** (Zoom = 2) shows the Sa moved 8π to the right, confirming our statement above.

(e) $f(t) = \mathrm{Rect}(t/2) = \mathrm{Rect}(t/\tau)$ with $\tau = 2$

$$\Longleftrightarrow 2\ \mathrm{Sa}(\omega) = F(\omega)$$

Sketch of $F(\omega)$

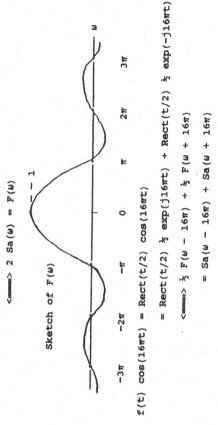

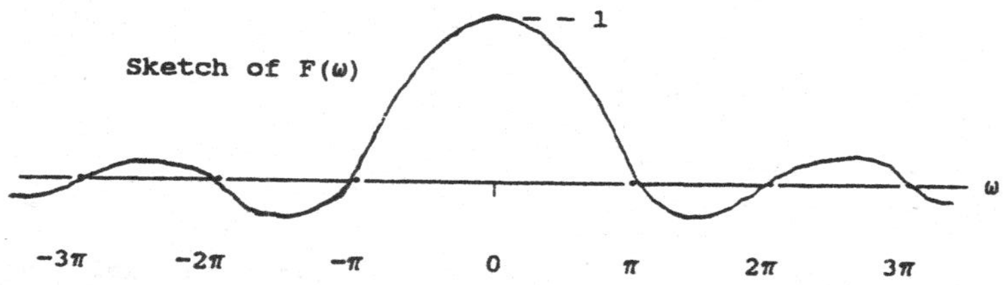

$$f(t)\ \cos(16\pi t) = \mathrm{Rect}(t/2)\ \cos(16\pi t)$$

$$= \mathrm{Rect}(t/2)\ \tfrac{1}{2}\exp(j16\pi t) + \mathrm{Rect}(t/2)\ \tfrac{1}{2}\exp(-j16\pi t)$$

$$\Longleftrightarrow \tfrac{1}{2}F(\omega - 16\pi) + \tfrac{1}{2}F(\omega + 16\pi)$$

$$= \mathrm{Sa}(\omega - 16\pi) + \mathrm{Sa}(\omega + 16\pi)$$

This is two Sa's, each of max-value 1, one centered at $\omega = 16\pi$ and the other at $\omega = -16\pi$. (Sketch ommitted)

Using N = 256, T = 8: Create Values, Continue, Create X and X2, Both Real, Even, 2 intervals, 1, Continue, 1, 0, Go, Even, 1 interval, Continue, COS(16*pi*t) Go, Neither, Go. You are now back on the main menu. Plotting **X** shows Rect(t/2) and plotting **X2** (ZOOM = 32) shows a cosine with period 0.125, namely $\cos(16\pi t)$.

Run Postprocessors, X, Multiply X by X2, Quit, Run ANALYSIS. Plotting **FRE** shows two Sa's, each of max-value 1, one centered at $\omega = 16\pi$ and the other at $\omega = -16\pi$, confirming our statement above.

Worksheet for COMPUTER LAB #7

(a) When Rect(t) is multiplied by $\cos(\omega_0 t)$ the following happens:

$$\text{Rect}(t)\ \cos(\omega_0 t) \Longleftrightarrow \tfrac{1}{2}\,\text{Sa}[(\omega - \omega_0)/2] + \tfrac{1}{2}\,\text{Sa}[(\omega + \omega_0)/2] \qquad \text{(A)}$$

Using N = 256, T = 8, create Rect(t) in **X** and $\cos(16\pi t)$ in **X2**.

(1) Plot and carefully verify **X** and **X2**. Then plot and carefully verify **F**.

(2) Multiply **X** and **X2**. Plot and carefully verify **X** once more. Then plot **FRE** and see if it agrees with (A).

(b) Sketch f(t) = Rect[(t-2)/2]. Now sketch the inverse of $F(\omega)\ \exp(-j\omega 4)$

Using N = 256, T = 8, load Rect[(t-2)/2] into **X**.

(1) Plot and verify all aspects of **X**.

(2) Multiply **F** by $\exp(-j\omega 4)$. Then plot and carefully verify **Y**.

(c) Sketch $F(\omega)$ where f(t) = Rect(2t). Now sketch the transform of f(t) $\exp(j8\pi t)$.

Using N = 256, T = 8, load Rect(2t) into **X**.

(1) Plot and carefully verify **X**. Then carefully verify the plot of **F**.

(2) Multiply **X** by $\exp(j8\pi t)$. Plot and carefully verify **X** and then **F**.

(3) Multiply **X** by $\exp(j8\pi t)$ again. Plot and verify all aspects of **X** and then **F**.

(4) If you did this again and again what would happen? Explain it to the lecturer or to an assistant.

Steps for running COMPUTER LAB #7

(a) Using N = 256, T = 8: Main menu, Create Values, Continue, Create X and X2, Real, Even, 2 intervals, 1/2, Continue, 1, 0, Go, Even, 1 interval, Continue, COS(16*pi*t), Go, Neither, Go.

(1) Plotting **X** shows a Rect that goes from $-\frac{1}{2}$ to $\frac{1}{2}$ with height 1. Clearly it is Rect(t). Observe the half-values.

Plotting **X2** (Zoom = 32) shows a sampled cosine with period 1/8 and so it must be a sampled version of cos(16*pi*t).

Run **ANALYSIS**. Plot **F**. We see a Sa of height 1. With Zoom = 8 we see the first zero crossing at $\omega = 2\pi$. Thus it must be $\text{Sa}(\omega/2)$

(2) Run Postprocessors, X, Real-multiply X and X2, Quit. Plotting **X** shows that the multiplication has now produced Rect(t) $\cos(16\pi t)$

Run **ANALYSIS** and plot **FRE**. We see two Sa's, both of height 1/2, one centered at $\omega = 16\pi$ and the other at $\omega = -16\pi$. This is in full agreement with (A).

(b) $\text{Rect}[(t-t_1)/\tau]$ is τ seconds wide and is centered at $t = t_1$.

Rect[(t-2)/2] is 2 seconds wide and is centered at t = 2.

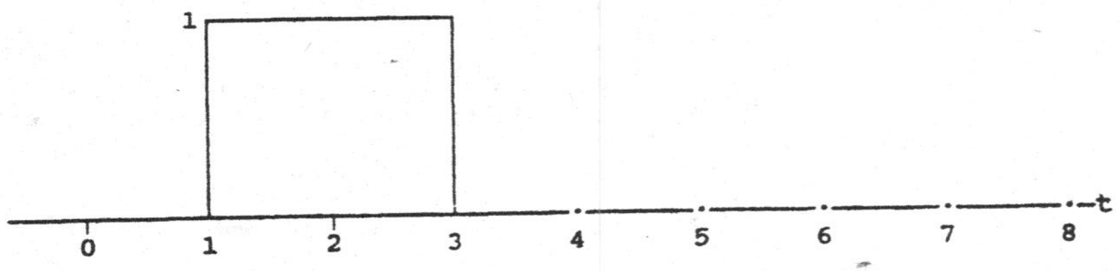

$f(t) \Longleftrightarrow F(\omega)$ Then $F(\omega)\ \exp(-j\omega 4) \Longleftrightarrow f(t - 4)$

Thus for this pulse, $F(\omega)\ \exp(-j\omega 4) \Longleftrightarrow \text{Rect}[(t - 6)/2]$.

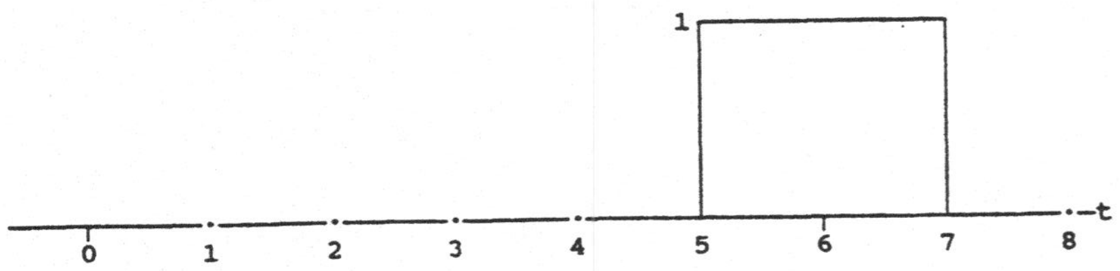

(1) Using N = 256, T = 8: Main menu, Create Values, Continue, Create X, Real, Neither, Left, 3 intervals, 1, 3, Continue, 0, 1, 0, Go Neither, Go. You are now back on the main menu. Plotting **X** confirms that the pulse f(t) = Rect[(t-2)/2] has been loaded correctly. Observe that it runs from 1 to 3 with value 1. Observe the half-values at the dicontinuities.

(2) Run ANALYSIS, Run Postprocessors, F, Multiply F by exp(j w tau), -4, Quit, Run SYNTHESIS and plot **Y**. The pulse has been shifted to the right without any other changes so that it is now centered at t = 6.

(c) $\text{Rect}(t/\tau) \Longleftrightarrow \tau\, \text{Sa}(\omega\tau/2)$. In this case $\tau = \frac{1}{2}$, and so

$f(t) = \text{Rect}(2t) \Longleftrightarrow F(\omega) = \frac{1}{2}\,\text{Sa}(\omega/4)$

Sketch of F(ω)

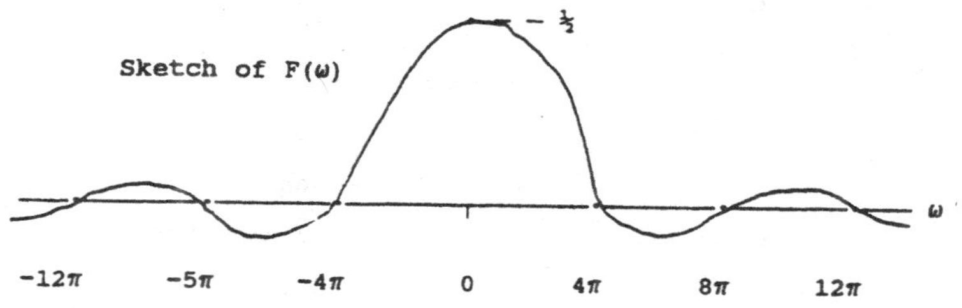

If $f(t) \Longleftrightarrow F(\omega)$ then $f(t)\exp(j8\pi t) \Longleftrightarrow F(\omega - 8\pi)$, and so

$\text{Rect}(t)\exp(j8\pi t) \Longleftrightarrow \frac{1}{2}\,\text{Sa}[(\omega-8\pi)/4]$

This spectrum is the original spectrum shifted to the right by 8π radians. (Sketch omitted.)

(1) Using N = 256, T = 8: Main menu, Create Values, Continue, Create X, Real, Even, 2, 1/4, Continue, 1, 0, Go, Neither, Go. Plotting **X** confirms that the pulse has been correctly loaded. Observe that it runs from $-\frac{1}{4}$ to $\frac{1}{4}$ with max-value 1, centered on the origin. Observe the half-values at the discontinuities.

Run ANALYSIS and plot **FRE**. Clearly it is a Sa function. Its max-value is ½. Using Zoom = 4 we see that the first zero-crossing is at 4π, and so it must be $\frac{1}{2}\,\text{Sa}(\omega/4)$ as expected.

(2) Run Postprocessors, X, Multiply X by exp(j wo t), 8*pi, Quit, Plotting **XRE** shows the original pulse now multiplied by a cosine. Using Zoom = 16 confirms that the cosine has a period of 1/4, and so it must be cos(8πt). Similarly, plotting **XIM** shows the original pulse multiplied by sin(8πt).

Run ANALYSIS and plot **FRE**. We see the same Sa function shifted to the right by 8π radians, as predicted by the frequency-shift theorem.

(3) Multiply **X** by exp(j8πt) again: Run Postprocessors, X, Multiply X by exp(j wo t), 8*pi, Quit, Plotting **XRE** shows the original pulse now multiplied by a cosine. Using Zoom = 16 confirms that the cosine has a period of 1/8, and so it must be cos(16πt). Similarly, plotting **XIM** shows the original pulse multiplied by sin(16πt).

Note: **X** has now been multiplied by exp(j8πt) twice, and so it has been multiplied by

$\exp(j16\pi t) = \cos(16\pi t) + j\sin(16\pi t)$

Run ANALYSIS and plot **FRE**. We see the Sa function shifted to the right by 16π radians, as predicted by the frequency-shift theorem. However we also see another Sa beginning to come into the display on the left. That is because the FFT is periodic and so we see the period on the left beginning to come into the display. This does not happen with the CFT which is not peiodic, but when using the FFT to emulate the CFT we have to accept the periodicity of the FFT.

(4) If we did this again and again, the periodic FFT spectrum would continue to cycle through the display window.

The period of the frequency domain on the FFT with N = 256 and T = 8 is

$N\times\text{OMEGAs} = N\times 2\pi/T = 256\times 2\pi/8 = 64\pi$

and so after 8 multiplications by exp(j8πt) we would be back at the starting point.

Worksheet for COMPUTER LAB #8

(a) What is the transfer function $H(j\omega)$ of network (A)?

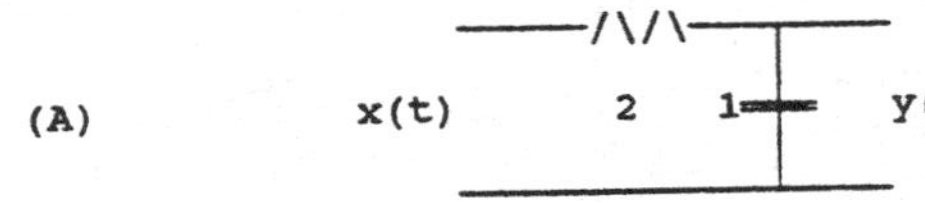

(b) What is its impulse response? Using N = 256, T = 32, Alpha = 20, load the transfer function into **F2** starting from main-menu A.

(c) Now produce its impulse response on your screen. Compare the FFT's results for t = 2 and t = 4 against the formula values from (a) and calculate the relative errors in the FFT results vs. the formula values.

Answers:
$h(2)_{FFT}$ = 0.183939721, $h(2)_{formula}$ = 0.183939721, error 0.047%.
$h(4)_{FFT}$ = 6.766667641e-2, $h(4)_{formula}$ = 6.766667641e-2, error 0.054%

(d) With Rect(t/4) as the input to the network, use the FFT to find when y(t) is at its maximum? (Ans: t = 1.875)

(e) On the FFT system, apply time-shift to the input pulse (using $\tau = -4$) by performing an appropriate operation on its spectrum before passing it through the network. At what value of t is y(t) at its maximum now? (Ans: t = 5.875)

(f) What is the transfer function $H(j\omega)$ of network (B)?

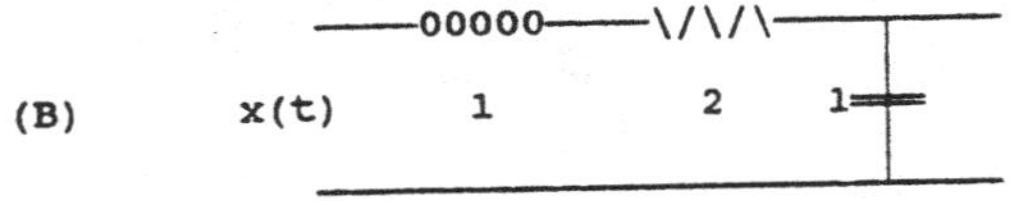

(g) What is its impulse response? Using N = 256, T = 16, Alpha = 20, load the transfer function into **F2** starting from main-menu A.

(h) Now produce its impulse response on the screen. Compare the FFT's results for t = 2 and t = 4 against the formula values and calculate the relative errors in the FFT results vs. the formula values.

Answers:
$h(1)_{FFT}$ = 0.36788014, $h(1)_{formula}$ = 0.367879441, error = 0.00019%.
$h(2)_{FFT}$ = 0.27067084, $h(2)_{formula}$ = 0.270670566, error = 0.0001%.

(i) With x(t) = Rect(t − ½) as the input , use the FFT to find when y(t) is at its maximum? (Ans: t = 1.5625)

(j) Modify the spectrum in (i) by an appropriate operation to find the response to x(t) = Rect(t − 2). At what value of t is the response now at its maximum value? (Ans: t = 3.0625)

(a) Finding the transfer function $H(j\omega)$ of network (A):

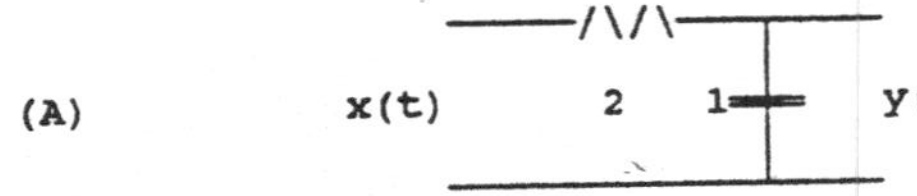

$$H(j\omega) = \frac{1/j\omega}{2 + 1/j\omega} = \frac{1}{2j\omega + 1} \longrightarrow \qquad (**)$$

(b) Impulse response: $H(j\omega) = \frac{1}{2}\,\frac{1}{j\omega + 1/2}$

$$\Longleftrightarrow h(t) = \tfrac{1}{2}\exp(-\tfrac{1}{2}t)\,U(t) \quad (*)$$

(c) Using N = 256, T = 32: Main menu, Create Values, Continue, Create H(jw), Create, 0, 1, 1, 1, 2, Load, 20. You are now back on the main menu with $H(j\omega)$ loaded into **F2**.

(d) To find the impulse response h(t) we must invert $H(j\omega)$ to the time domain. Run Postprocessors, F, Copy F2 to F, Quit, Run SYNTHESIS.

Plotting Y shows the impulse response h(t). Observe that it is a decaying exponential **and** the fact that we have used a value of T large enough so that the decay is almost to zero before the window ends. Had we used a smaller value of T which does not permit that we would have experienced time-domain aliasing caused by the overflow of the exponentials from one FFT period to the next, resulting in very poor numerical values. (Try a re-run with T = 4.)

To check the numerical values against (*): Show Numbers, X and Y.

The time steps of the FFT are T_s = T/N = 32/256 = 1/8. This is shown on the plots and number screens. Thus t = 2 corresponds to k = 16, and t = 4 corresponds to k = 32.

From (*): h(2) = ½ exp(−½×2) = 0.183939721.

From (*): h(4) = ½ exp(−½×4) = 6.766667641e-2.

For k = 16: $Y(16)_{FFT}$ = 0.18386351. Error = 0.047%

For k = 32: $Y(32)_{FFT}$ = 6.7631042e-2 Error = 0.054%

Note: We used aliasing with Alpha = 20. Had we not aliased, these errors would have been **sustantially** larger because the spectrum $H(j\omega)$ in (**) is converging **very slowly** as we increase ω. To see this, examine its magnitude spectrum without aliasing, and then examine it again after aliasing.

Indeed, re-running with Aplha = 0 we obtain the following errors:

For k = 16: $Y(16)_{FFT}$ = 0.18081545 Error = 1.7%

For k = 32: $Y(32)_{FFT}'$ = 6.6166982e-2 Error = 2.2%

(d) To load x(t) = Rect(t/4): Main menu, Create Values, Continue, Create X, Real Even, 2 intervals, 2, Continue, 1, 0, Go, Neither, Go. Plotting **X** shows that Rect(t/4) has been correctly loaded.

Run ANALYSIS. The spectrum of x(t), namely X(ω), is now in **F** and H(jω) is in **F2**. To obtain the response of the network to the Rect pulse we must form Y(ω) = H(jω) F(ω). To do that:

Run Postprocessors, F, Complex multiply F and F2, Quit. Y(ω) is now in **F**. Run SYNTHESIS and plot **Y**. The plot shows a rising exponential for 4 seconds (width of Rect(t/4)) followed by a decaying exponential after the input has returned to zero.

Show Numbers, X and Y. The maximum value is at k = 15 which corresponds to t = 1.875. That is the last instant at which the input is at 1.

(e) We now apply time-shift to the input pulse using τ = -4 before passing it through the network: Run ANALYSIS, Run Postprocessors, F, Multiply F by exp(j w tau), -4, Complex-multiply F by F2, Quit, Run SYNTHESIS. Plotting **Y** shows the same response except that it is now delayed by 4 seconds.

Show Numbers. The maximum value is now at k = 47 which corresponds to to = 5.875. This is exactly 4 seconds later than the previous maximum.

(f) To find the transfer function H(jω) of network (B):

$$H(j\omega) = \frac{1/j\omega}{j\omega + 2 + 1/j\omega} = \frac{1}{(j\omega)^2 + 2j\omega + 1} = \frac{1}{(j\omega + 1)^2} \qquad (**)$$

(g) From Theorem 4.3: If f(t) <==> F(ω) then t f(t) <==> j F′(ω)

We know that exp(-t) U(t) <==> $F(\omega) = \frac{1}{j\omega + 1}$ from which

$$j\,F'(\omega) = j\,\frac{-j}{(j\omega + 1)} = \frac{1}{(j\omega + 1)^2} \Longleftrightarrow t\,\exp(-t)\,U(t)$$

Thus the impulse response of (B) is h(t) = t exp(-t) U(t)

(h) Using N = 256, T = 16: Main menu, Create Values, Continue, Create H(jw), Create, 0, 2, 1, 1, 2, 1, Load, 20. You are now back on the main menu with H(jω) in **F2**.

To produce the impulse response we must invert H(jω) to the time domain: Run Postprocessors, F, Copy F2 to F, Quit, Run SYNTHESIS.

Plotting **Y** shows the impulse response h(t). Observe that it first rises because of the multiplier t and then decaying exponentially to zero. Observe also the fact that we have used a value of T large enough so that the decay is almost to zero before the window ends. Had we used a smaller value of T which does not permit that we would have experienced time-domain aliasing caused by the overflow of the waveforms from one FFT period to the next, resulting in very poor numerical values. (Try a re-run with T = 4.)

To check the numerical values against (*): Show Numbers, X and Y. The time steps of the FFT are T_s = T/N = 16/256 = 1/16. This is shown on the plots and number screens. Thus t = 1 corresponds to k = 16, and t = 2 corresponds to k = 32.

From (*): h(1) = 1 exp(-1) = 0.367879441

From (*): h(2) = 2 exp(-2) = 0.270670566

For k = 16: $Y(16)_{FFT}$ = 0.36788014. Error = 0.00019%

For k = 32: $Y(32)_{FFT}$ = 0.27067084. Error = 0.0001%

Note: We used aliasing with Alpha = 20. Had we not aliased, these errors would **not** have been much larger because the spectrum H(jω) in (**) is converging to zero **very rapidly** as ω is increased. To see this, examine its magnitude spectrum.

(i) Finding the response of the network using as the input x(t) = Rect(t - ½): Main menu, Create Values, Continue, Create X, Real, Neither, Left, 2 intervals, 1, Continue, 1, 0, Go, Neither, Go. Plotting **X** shows the pulse running from 0 to 1 which is Rect(t - ½), and so it has been entered correctly.

Run ANALYSIS, Run Postprocessors, F, Complex-multiply F and F2, Quit. Run SYNTHESIS and plot **Y**. The response to x(t) is shown.

To find its maximum value: Show Numbers, X and Y. We see the maximum of 0.35303767 at k = 25. This corresponds to

$t = k \times T_s = 25 \times T/N = 25 \times 16/256 = 1.5625.$

(j) Modifying the spectrum in (c) by an appropriate operation to find the response to x(t) = Rect(t - 2): The spectrum of the response is still in the **F** vector. Run Postprocessors, F, Multiply F by exp(j w tau), -3/2, Quit, Run SYNTHESIS.

Plotting **Y** shows the original response to Rect(t - ½) now delayed by a further 3/2 seconds, and so it must be the response to Rect(t - 2). Show Numbers, X and Y. The maximum is now at k = 49 which corresponds to t = 3.0625. This is exactly 1.5 seconds after the maximumm value in (i).

(a) Load Rect(2t) into **X** using N = 256, T = 4. Examine its spectrum. Now apply frequency-shift (X postprocessor) with ω_0 = 32*pi and re-examine its spectrum. Has it been correctly shifted? Make a sketch of the inverse of this shifted spectrum. Now invert to the time domain and validate your sketch.

(b) Load the same pulse as in (1). (Hint: Use the Old-problem) Examine its spectrum. Now apply time-shift (F postprocessor) with τ = 0.5 Invert to the time domain and examine the result. Has it been correctly shifted?

(c) What are the frequency transfer functions of the following networks.

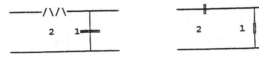

What are the expressions for their impulse responses? (A)

The transfer functions can be loaded into **X2** starting from main-menu A. How can we produce their impulse responses on the screen? Use N = 256, T = 32 and verify the results against (A).

(d) Now use Rect(t/4) as the input to the networks and examine the responses. Then apply time-shift using τ = 4, and verify the result.

We have omitted the steps for running these problems because they are so similar to ones in previous worksheets. This lab was created for students who needed some additional reinforcement.

Note: "*" means convolution in this worksheet.

Use N = 256, T = 8 for all FFT runs.

(a) Find Rect(t) * δ(t - 2) using the convolution integral. Sketch the result. Now run this on the FFT using CONVOLUTION.

(b) Sketch Rect[(t - 1)/2] * Rect(t - ½). Now run this on the FFT.

(c) Let x(t) = t (0 < t < 1). Use the convolution integral to find y(t) = x(t) * x(t) and sketch the result. Now run this on the FFT and verify your expression and sketch.

Steps for running COMPUTER LAB #10

Use N = 256, T = 8 for all FFT runs.

(a) Finding Rect(t) * δ(t - 2) using the convolution integral:

$$f(t) = \int_{-\infty}^{\infty} g(\tau)\, h(t - \tau)\, d\tau = \int_{-\infty}^{\infty} \text{Rect}(\tau)\, \delta(t - 2 - \tau)\, d\tau$$

in which Dirac-delta sampling takes place at τ = t - 2, and so we continue

$$\ldots = \int_{-\infty}^{\infty} \text{Rect}(t - 2)\, \delta(t - 2 - \tau)\, d\tau = \text{Rect}(t - 2) \longrightarrow$$

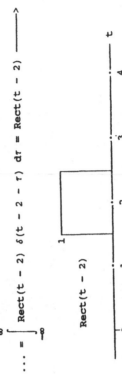

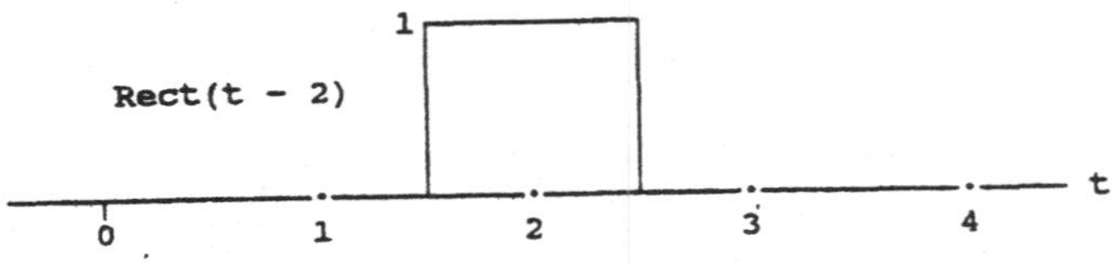

Steps for loading and running the problem (N = 256, T = 8):

Main menu, Create Values, Continue, Create X, Real, Even, 2 intervals, .5, Continue, 1, 0, Go, X2, 2, 1, Neither, Go. You are now back on the main menu. Plotting **X** shows Rect(t) and plotting **X2** shows a single line at t = 2 with height V = 32. This is a Dirac delta of weight μ = VT/N = 32x8/256 = 1.

Run CONVOLUTION. The result is now in **Y**. Plot **Y**. We see a Rect at $t = 2$ of width 1, and so is $\text{Rect}(t - 2)$ as expected.

(b) Sketch of $\text{Rect}[(t - 1)/2] * \text{Rect}(t - \frac{1}{2})$:

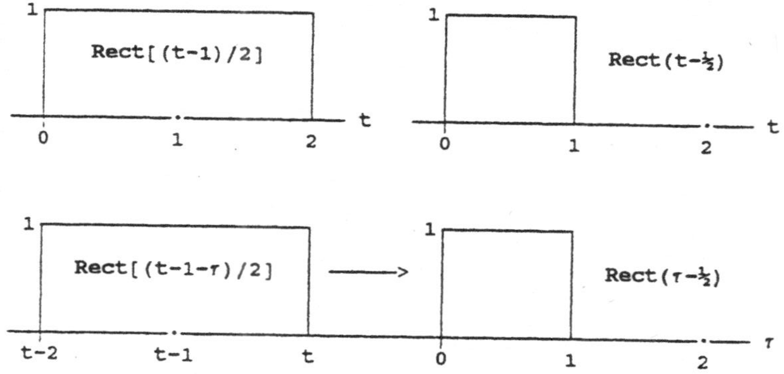

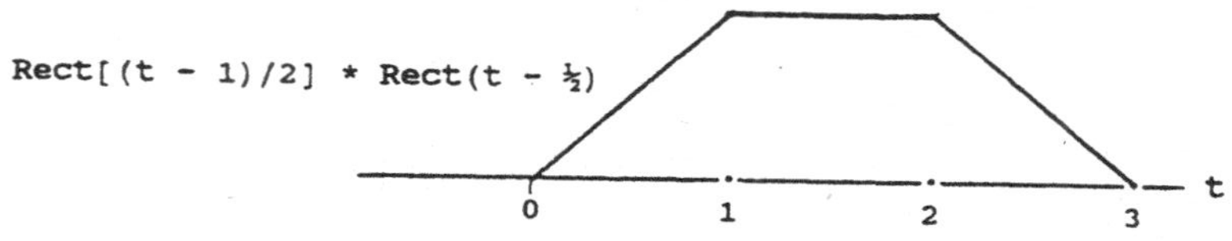

- -

Steps for loading and running the problem ($N = 256$, $T = 8$):

Main menu, Create Values, Continue, Create X and X2, Both Real, Neither, Left, 2 intervals, 2, Continue, 1, 0, Go, Left, 2 intervals, 1, Continue, 1, 0, Go, Neither, Go. You are now back on the main menu. Plotting **X** shows Rect[(t-1)/2], and **X2** shows Rect(t-½).

Run CONVOLUTION. Plotting **Y** shows the same result as in the sketch.

(c) $x(t) = t \ (0 < t < 1)$. Using the convolution integral to find $y(t) = x(t) * x(t)$:

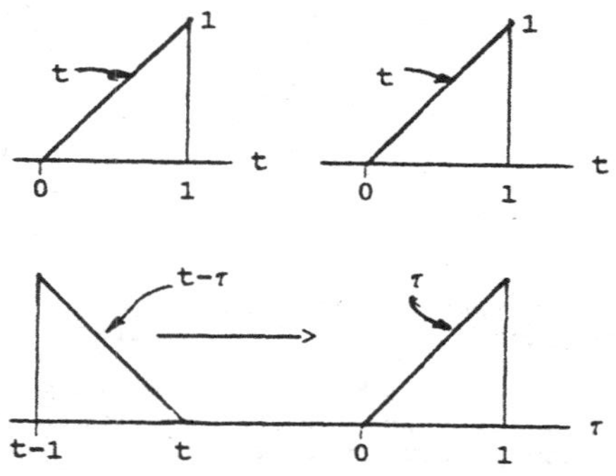

- For $t < 0$: $y(t) = 0$ ———>

- For $0 < t < 1$: $$y(t) = \int_0^t \tau(t - \tau)\, d\tau = \frac{t\tau^2}{2} - \frac{\tau^3}{6}\Big|_0^t$$
$$= \frac{t^2}{2} - \frac{t^3}{3} = t^3/6 \text{ ———>}$$

- For $1 < t < 2$: $$y(t) = \int_{t-1}^1 \tau(t - \tau)\, d\tau = \frac{t\tau^2}{2} - \frac{\tau^3}{6}\Big|_{t-1}^1$$
$$= \frac{t}{2} - \frac{1}{6} - \frac{t(t-1)^2}{2} + \frac{(t-1)^3}{3} = (-t^3 + 6t - 4)/6 \text{ ———>}$$

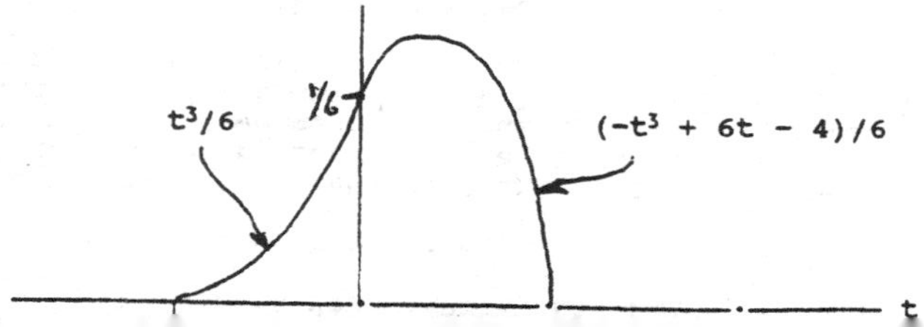

Steps for loading and running the problem (N = 256, T = 8):

Main menu, Create Values, Continue, Create X, Real, Neither, Left, 2 intervals, 2, 1, Continue, t, Go, Neither, Go. You are now back on the main menu. Run Postprocessors, X, Copy X to X2, Quit.

Plotting **X** shows t (0 < t < 1), and **X2** shows the same.

Run CONVOLUTION. Plotting **Y** shows the same result as in the sketch.

Show Numbers, X and Y. The following table verifies the results.

k	t	FFT	Formula	error (%)
16	0.5	2.0751953e-2	2.0833333e-2	0.39
32	1	0.16650391	0.16666666	0.098
48	1.5	0.27075195	0.27083333	0.03
56	1.75	0.19006348	0.19014167	0.02

Worksheet for COMPUTER LAB #11

Use N = 256, T = 8 for all FFT runs.

(a)(1) Load Rect(t) into **X**. Run ANALYSIS and verify the spectrum. Now sample **X** keeping every fourth value. To simulate impulse sampling on the FFT we must also multiply **X** by N/T which is required to make the samplers look like Dirac deltas. Plot **X** and verify that it is correct.

(2) The FFT's time-domain sampling interval is T/N. What is the value of T_s, the sampling interval which retained every fourth value?

(3) Run ANALYSIS and inspect the spectrum. According to Chapter 9 it should consist of copies of the original spectrum repeated with spacing $\omega_s = 2\pi/T_s$, each multiplied by $1/T_s$. Is that the case?

(b)(1) Load Rect(t) into **X** using N = 256, T = 8.

(2) Run ANALYSIS. Now sample **F** keeping every fourth value.

(3) The FFT's frequency-domain sampling interval is $2\pi/T$. What is the value of ω_0, the sampling interval which retained every fourth value?

To simulate impulse sampling by impulses with weights ω_0 we must also multiply **F** by $\omega_0 \times T/2\pi$ which is required to make the samplers look like the correct Dirac deltas.

(4) According to Chapter 9 if we invert a sampled spectrum to the time domain we obtain a periodic waveform, period T_0. What is the value of T_0 for this situation? Run SYNTHESIS and inspect a plot of the result. Is it correct?

(5) Start again, but now sample keeping every <u>eighth</u> value. What does the sampled spectrum look like? What is the time-domain version? Explain.

(c) Repeat (b)(5) using exp(-t) U(t). What is the peak value of the final result. Explain its value mathematically.

Steps for running COMPUTER LAB #11

(a)(1) Steps for loading and running the problem ($N = 256$, $T = 8$):

Main menu, Create Values, Continue, Create X, Real, Even, 2 intervals, 1/2, Continue, 1, 0, Go, Neither, Go. You are now back on the main menu. Plotting **X** shows Rect(t). Run ANALYSIS. Plotting **FRE** shows a Sa with max-value 1 and first zero crossing at $\omega = 2\pi$. Thus it must be $Sa(\omega/2)$.

Run Postprocessors, X, Sample X, 4, Multiply X by a Constant, 32, Quit, Run ANALYSIS.

(2) The FFT's sampling interval is $T/N = 1/32$. Four of these intervals gives us $T_s = 1/8$.

(3) According to Chapter 9 the spectrum of the unsampled pulse should now be repeated every $\omega_s = 2\pi/T_s = 16\pi$ radians, and multiplied by $1/T_s = 8$. Plotting **FRE** shows exactly that.

(b)(1) To reload Rect(t): Main menu, Create Values, Continue, Use the Old-problem.

(2) Run ANALYSIS. Run Postprocessors, F, Sample F, 4.

(3) The FFT's frequency-domain sampling interval is $2\pi/T = 2\pi/8$. Four of these gives us $\omega_0 = 4\times2\pi/8 = \pi$

To simulate impulse sampling by impulses with weights ω_0 we must also multiply **F** by $\omega_0 \times T/2\pi = \pi\times8/2\pi = 4$.

Multiply F by a constant, 4, Quit. Plotting **F** shows a Sa sampled so as to retain every fourth value with a peak value of 4.

(4) According to Chapter 9 if we invert a sampled spectrum to the time domain we obtain a periodic waveform, period T_0. In this case $T_0 = 2\pi/\omega_0 = 2$.

Run SYNTHESIS and plot **Y**. We see the original Rect(t) repeated every 2 seconds, exactly as expected.

(5) Run ANALYSIS, Run Postprocessors, F, Sample F, 8.

Now $\omega_0 = 8\times2\pi/8 = 2\pi$.

To simulate impulse sampling by impulses with weights ω_0 we must multiply **F** by $\omega_0 \times T/2\pi = 2\pi\times8/2\pi = 8$

Multiply F by a constant, 8, Quit.

Plotting **F** shows a Sa sampled so as to retain every eighth value with a peak value of 8. In fact we now have only a single nonzero value, since all of the sampling points have been zero crossings.

According to Chapter 9 if we invert this sampled spectrum to the time domain we should obtain a periodic waveform of period $T_0 = 2\pi/\omega_0 = 2\pi/2\pi = 1$

Run SYNTHESIS and plot **Y**. We see the original Rect(t) repeated every 1 second, which means that the Rects are touching one another and so the plot is 1 all along the display.

(c) Repeating (b)(5):

Main menu, Create Values, Continue, Real, Neither, Left, 1 interval, Continue, EXP(-t), Go, Neither, Go. You are now back on the main menu. Plotting **X** shows the decaying exponential.

Run ANALYSIS, Run Postprocessors, F, Sample F, 8, Multiply F by a constant, 8, Quit, Run SYNTHESIS, Plot **Y**.

We now see the decaying exponential repeated every second, as expected. The peak value is 1.53279 instead of 1. This is because of the aliasing (overlap) of the exponentials. We can account for that value almost exactly as follows:

At any one of the peaks, there are infinitely many exponentials to its the left that have contributed to that value. Thus, as a first approximation the peak should be

$$P = 1 + \exp(-1) + \exp(-2) + \ldots \qquad \text{(see figure below)}$$

$$= \sum_{n=0}^{\infty} \exp(-1)^n = \frac{1}{1 - \exp(-1)} = 1.581976707$$

The display shows **half-values** at the integer time values

$t = 0, 1, 2, \ldots$

and the peak values are one FFT sampling instant after those points. Thus the FFT is in fact showing the above value multiplied by $\exp(-Ts) = \exp(-8/256) = 0.969233235$, namely

$$P = 1.581976707 \times 0.969233235 = 1.533304401.$$

Compared to this, the FFT value of 1.53279 is in error by 0.033%

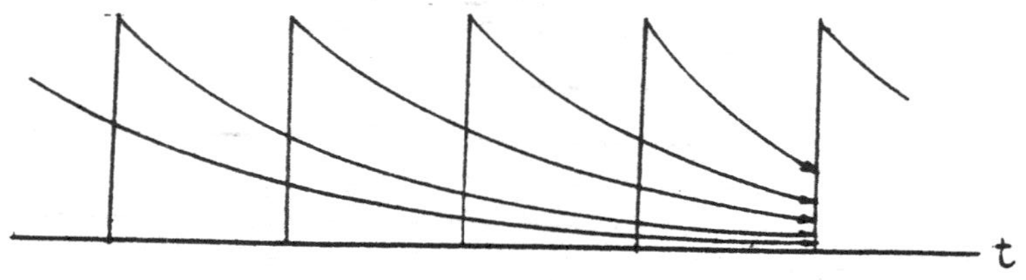

UNIVERSITY OF CAPE TOWN   SCHOOL OF MATHEMATICAL SCIENCES

SMS207-S COMPUTER LAB EXAMINATION   October 14 1994

———> Marks total to 10 <———

Calculators are not permitted, nor are they required.

You may not write any notes or answers on any paper other than this one. If you do, you will be disqualified and given a mark of zero.

Print name	
Student number	

---

(1) Using N = 128, T = 8, load the pulse x(t) shown in Figure 1.

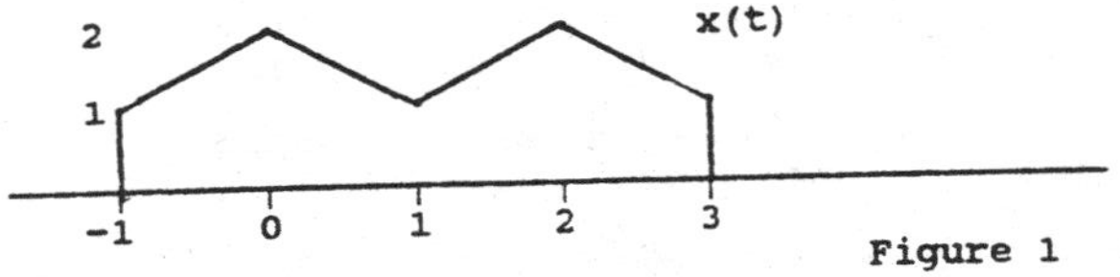

Figure 1

Now use the system to find its even part and fill in the following values:

$x_{ev}(k)$ at k = 9	
Value of t at k = 9	

[2]

- - - - - - - - - - - - - - - - - - - -

(2) Using N = 64, T = 16 load the following:

$x(t) = Rect[2(t - 3)]$, $x_2(t) = 0.1\ \delta(t + 1) + 0.2\ \delta(t - 2)$

Find $x(t) * x_2(t)$. Then fill in the following values:

F(n) at n = 8	
Value of ω at n = 8	

[2]

- - - - - - - - - - - - - - - - - - - -

(3) Using N = 128, find values for the complex Fourier coefficients for the periodic waveform shown below.

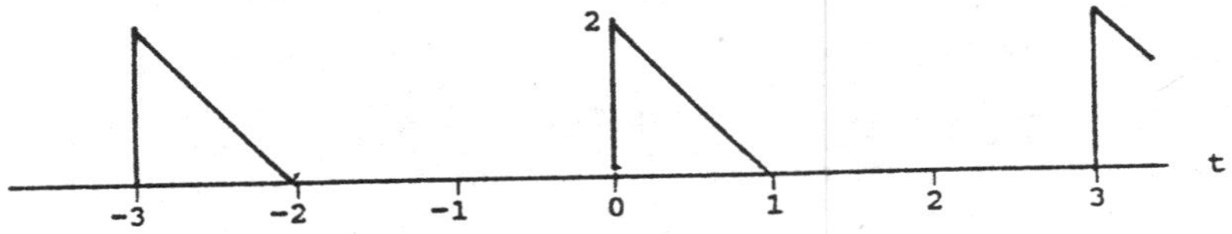

Now fill in the following values from the system's numbers:

F(1)	
Total average power in waveform	

[2]

- - - - - - - - - - - - - - - - - - - -

(4) Using N = 64, T = 128, Alpha = 2, find the impulse response of the following network:

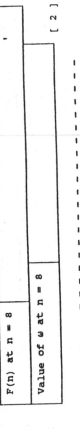

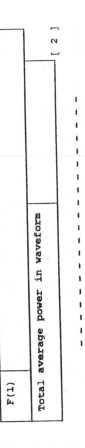

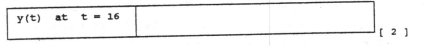

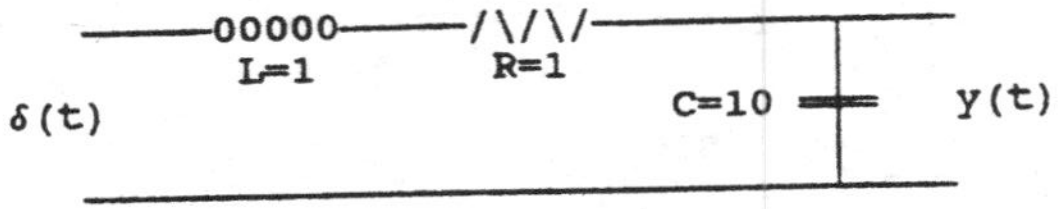

Now fill in the following value:

y(t) at t = 16	

[2]

(5) Using N = 64, T = 8, create the pulse Rect(t/2) exp(j4πt).
Then obtain and fill in the following values:

F(n) at n = 15	
Value of ω at n = 15	

[2]

UNIVERSITY OF CAPE TOWN

UNIVERSITY EXAMINATION: OCTOBER 29, 1993

School of Mathematical Sciences SMS205S
Introduction to Fourier Analysis

This is an open book examination — Time: 3 hours

Full Marks: 100 — Marks on Paper total to 122

SHOW ALL WORK THAT YOU USED IN ORDER TO OBTAIN YOUR ANSWERS

--->NO MARKS GIVEN IF REASONS FOR YOUR ANSWER ARE NOT SHOWN<---

--->NO MARKS GIVEN IF INTERMEDIATE STEPS ARE NOT SHOWN<---

---

(1) (a) For $g(t)$ and $h(t)$ as shown,

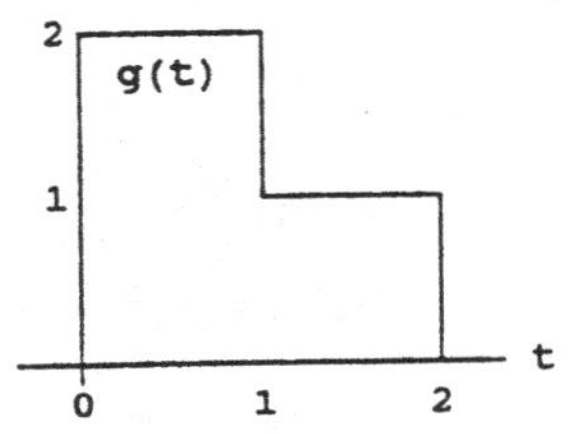

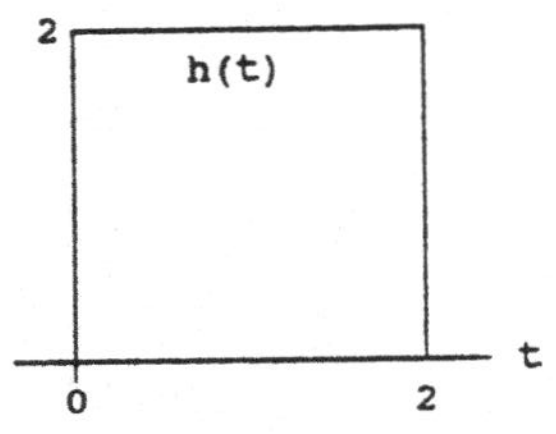

make a neat sketch of

$$f(t) = \int_{-\infty}^{\infty} g(t - \tau)\, h(\tau)\, d\tau$$

showing clearly all critical values.

(b) Using only the pulse $f(t)$ that you have sketched in (a), find its Fourier transform, $F(\omega)$.

(c) How should $F(\omega)$ in (b) be related to the transforms of the individual pulses, $G(\omega)$ and $H(\omega)$?

(d) Find $G(\omega)$ and $H(\omega)$.

(e) Verify that your statement in (c) is in fact correct. [12]

(2) (a) Use the Fourier series analysis equation for $F(n)$, the complex coefficients of

$$f_p(t) = \begin{cases} 2 & (0 < t < 1) \\ 1 & (1 < t < 2) \\ 0 & (2 < t < 4) \end{cases} \qquad f_p(t + 4) = f_p(t)$$

(b) Use the method of successive differentiation to verify your result in (a)

(c) Find the expressions for the real and imaginary parts of $F(n)$

(d) Fill in the values in the table on the last sheet of the exam paper. Print your name and student number on it and make sure that you hand it in with your paper

(e) What is the total average power in the waveform ? [12]

---

(3) (a) Let $F(\omega)$ be a fourier transform with real part $A(\omega)$ and imaginary part $B(\omega)$. Prove the following proposition:

$$\int_{-\infty}^{\infty} |F(\omega)|^2\, d\omega = \int_{-\infty}^{\infty} A(\omega)^2\, d\omega + \int_{-\infty}^{\infty} B(\omega)^2\, d\omega$$

Explain what this means in physical terms.

(b) The pulse $f(t)$ is shown here find the values of

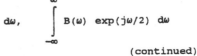

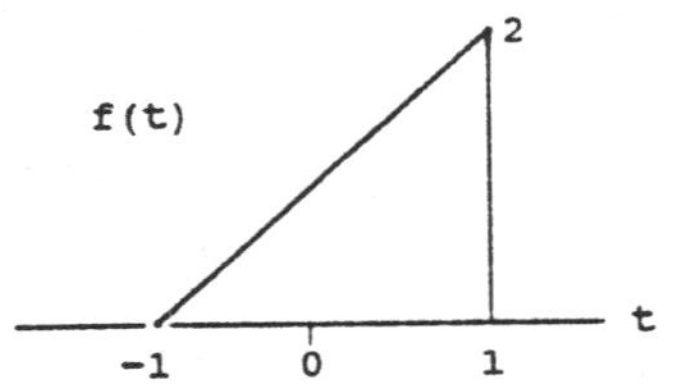

$$\int_{-\infty}^{\infty} |F(\omega)|^2\, d\omega,$$

$$\int_{-\infty}^{\infty} A(\omega)^2\, d\omega \quad \text{and} \quad \int_{-\infty}^{\infty} B(\omega)^2\, d\omega$$

(c) What are the values of

$$\int_{-\infty}^{\infty} F(\omega) \exp(j\omega)\, d\omega, \qquad \int_{-\infty}^{\infty} A(\omega) \exp(-j\omega)\, d\omega, \qquad \int_{-\infty}^{\infty} B(\omega) \exp(j\omega/2)\, d\omega$$

(continued)

(d) Make a neat sketch, showing all critical values, of

$$\int_{-\infty}^{\infty} A(\omega)\ 2\ Sa(\omega)\ \exp(j\omega t)\ d\omega$$

[12]

---

(4) (a) For the two pulses shown below, evaluate the convolution integral analytically to find $f(t) = g(t) * h(t)$.

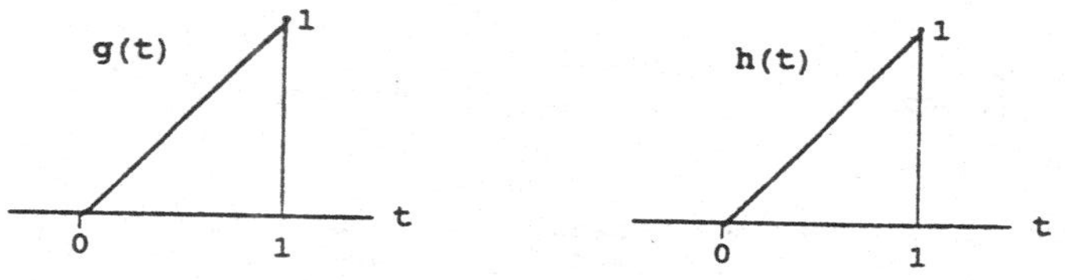

(b) Make a neat sketch of $f(t)$ showing all critical values.

(c) What is the Fourier transform of $f(t)$? [12]

---

(5) Find the inverse Fourier transform of

(a)

$$F(\omega) = \begin{cases} \omega & (0 < \omega < 1) \\ 0 & (\text{otherwise}) \end{cases}$$

(b) $F(\omega) = Rect(3\omega/2)$

Find the Fourier transform of

(c) $f(t) = Sa^2(2t)$

(d) $f(t) = Sa(t - 1)$ [12]

(6) (a) Find the frequency transfer function for the network shown.

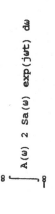

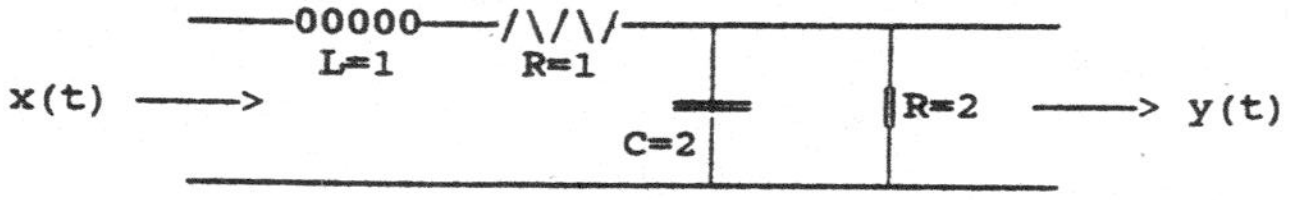

(b) Find the expression for the response of the network to

$$x_p(t) = \begin{cases} 2 & (0 < t < 1) \\ 1 & (1 < t < 2) \\ 0 & (2 < t < 8) \end{cases} \qquad x_p(t + 8) = x_p(t)$$

(c) Find the expression for the response of the network to $\delta(t - 1)$
[12]

---

(7) (a) Find the Fourier transform of $h(t) = \sin(\omega_0 t - \alpha)$

(b) Let $f(t) \Longleftrightarrow F(\omega)$. Use frequency-domain convolution to find the expression for the Fourier transform of the modulated signal

$$f_m(t) = f(t)\ \sin(\omega_0 t)$$

(c) $f_m(t)$ is demodulated by

- multiplying it by $h(t)$ above to give $g(t)$
- $g(t)$ is then passed through a low-pass filter that passes only the base-band frequencies

Find the expression for the Fourier transform of $g(t)$ and for the output from the low-pass filter.

(d) Show from (c) that if $\alpha = \pi/2$, the output of the low-pass filter is zero and if $\alpha = 0$ then it is $\frac{1}{2}f(t)$. [12]

---

(8) (a) Use the DFT analysis equation to find the discrete Fourier transform F_n of the following sequence of numbers using $N = 8$.

$$f_k = (1, 0, 1, 0, 1, 0, 1, 0)$$

(b) Now use the DFT synthesis equation to invert F_n.

(c) Make neat sketches of the magnitude and phase spectra of f_k.

(9) Find the Fourier transforms of the following two pulses using successive differentiation:

(a) $\exp(-\beta t)\, U(t) \quad (\beta > 0)$

(b) $\cos(\pi t)\, \mathrm{Rect}(t)$ [12]

---

(10) In this problem you are asked to make a number of sketches. Be sure that they are neat and that all critical values are included.

(a) We know that $\Lambda(t) \Longleftrightarrow Sa^2(\omega/2)$

Make a neat sketch of

- $\Lambda(t)$
- $Sa^2(\omega/2) \quad (-6\pi \le \omega \le 6\pi)$

(b) Prove that $Sa^2(t/2) \Longleftrightarrow 2\pi\, \Lambda(\omega)$. Sketch both functions, the first over the range $-6\pi \le t \le 6\pi$.

(c) According to the Nyquist sampling criterion, what is the largest interval with which we can impulse sample $Sa^2(t/2)$ and still be sure that we can recover it from the samples. You must give your reasons.

(d) Sketch the samples of $Sa^2(t/2)$ over $-6\pi \le t \le 6\pi$ for the situation in (c) and sketch the resulting spectrum of the samples.

(e) Using a sampling interval $T_s = 2\pi$, sketch the samples of $Sa^2(t/2)$ over $-6\pi \le t \le 6\pi$, and sketch the resulting spectrum.

(f) What are the expressions for the samples in (e) and for their spectrum? Are they a Fourier pair? [14]

SMS 205 Final Examination, October 29, 1993

Question 2(d)

Name:	Number:

n	0	1	2
A(n)			
B(n)			
|F(n)|			
θ(n)			
P(n)			

UNIVERSITY OF CAPE TOWN

UNIVERSITY EXAMINATION: NOVEMBER 7, 1994
School of Mathematical Sciences SMS207S
Introduction to Fourier Analysis

This is an open book examination Time: 3 hours

Marks total to 220. They will be divided by 2.
——> NO MARKS GIVEN IF REASONS FOR YOUR ANSWER ARE NOT SHOWN <——

(1)

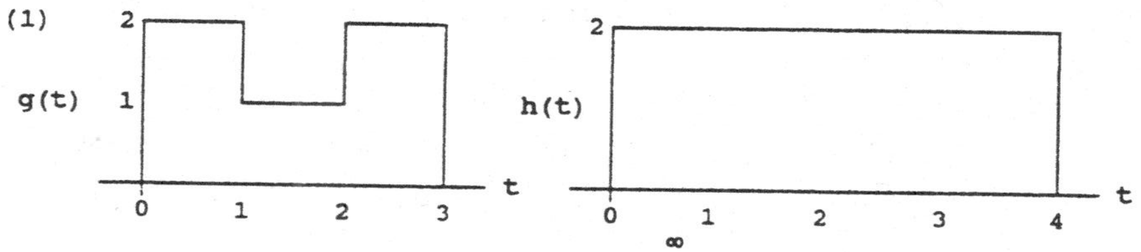

(a) Make a neat sketch of $f(t) = \int_{-\infty}^{\infty} g(t - \tau)\, h(\tau)\, d\tau$

(b) Using only the pulse $f(t)$ that you have sketched in (a), find its Fourier transform, $F(\omega)$.

(c) How should $F(\omega)$ in (b) be related to the transforms of the individual pulses, $G(\omega)$ and $H(\omega)$?

(d) Find $G(\omega)$ and $H(\omega)$.

(e) Verify that your statement in (c) is in fact correct. [20]

(2) (a) Use the analysis equation to find the expression for $F(n)$, the complex coefficients of $f_p(t)$ shown below.

(b) Use the method of successive differentiation to verify (a)

(c) What is the total average power in the waveform ?

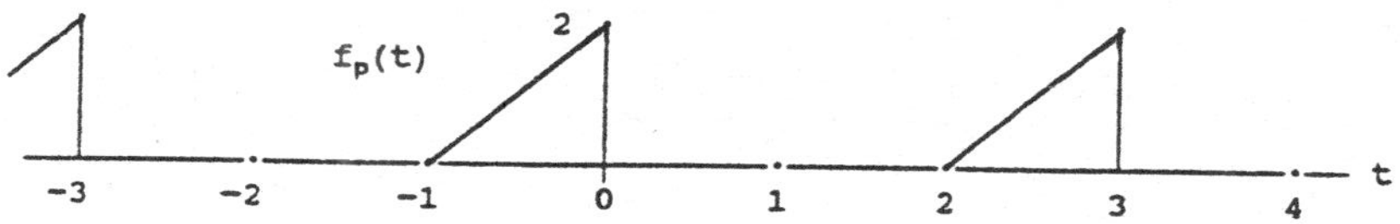

(d) Find the expressions for the real and imaginary parts of $F(n)$

(e) Fill in the values in the table on the last sheet of the exam paper. Be sure to hand it in with your script. [24]

(3) (a) Use the analysis equation to find the Fourier transform of $f(t)$.

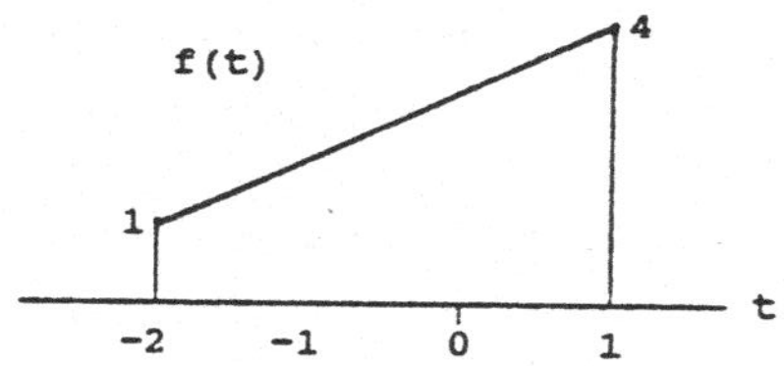

(b) Make neat sketches of

$$\frac{1}{2\pi}\int_{-\infty}^{\infty} A(\omega)\, \exp(j\omega t)\, d\omega \qquad \frac{j}{2\pi}\int_{-\infty}^{\infty} B(\omega)\, \exp(j\omega t)\, d\omega$$

(c) Find the value of $\int_{-\infty}^{\infty} |F(\omega)|^2\, d\omega$

(d) What are the values of

$$\int_{-\infty}^{\infty} F(\omega)\, \exp(j\omega)\, d\omega, \quad \int_{-\infty}^{\infty} A(\omega)\, \exp(-j\omega)\, d\omega, \quad \int_{-\infty}^{\infty} B(\omega)\, \exp(j\omega/2)\, d\omega$$

[20]

(4) For the two pulses $g(t)$ and $h(t)$ shown below, evaluate the convolution integral analytically to find $f(t) = g(t) * h(t)$.

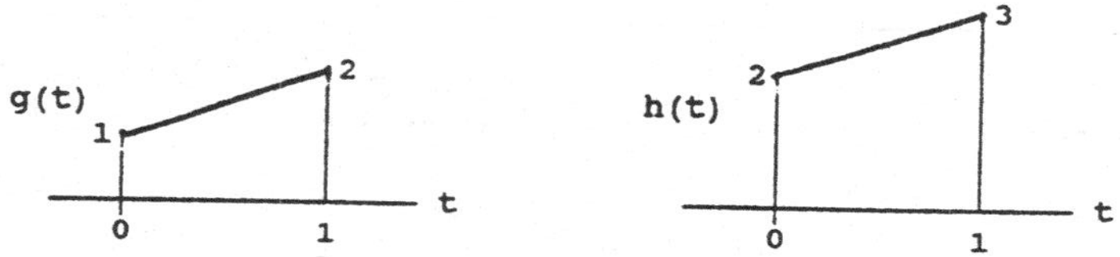

[24]

(5) Find the inverse Fourier transforms of

(a)

$$F(\omega) = \begin{cases} 1 - \omega & (0 < \omega < 1) \\ 0 & (\text{otherwise}) \end{cases}$$

(b) $F(\omega) = \text{Rect}(2\omega/3)\, \exp(-j\omega)$

Find the Fourier transforms of

(c) $f(t) = Sa^2(t/4)$

(d) $f(t) = Sa[(t - 1)/2]$

(e) $f(t) = \cos(\omega_0 t - \alpha)$ [20]

---

(6) (a) Find the frequency transfer function of the network

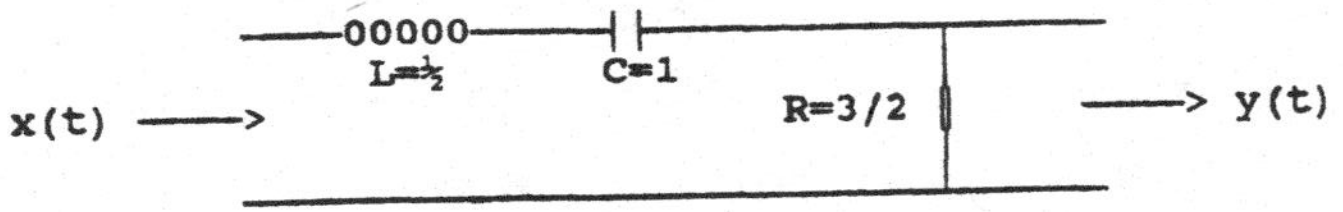

(b) Find the expression for the response to $x_p(t)$. Your answer must not refer to any items elsewhere on your paper.

$$x_p(t) = \begin{cases} 1 & (0 < t < 1) \\ -1 & (1 < t < 2) \\ 0 & (2 < t < 8) \end{cases} \qquad x_p(t + 8) = x_p(t)$$

(c) Find the impulse response. [20]

---

(7) The eternal cosine $x(t) = \cos(2\pi t)$ is sampled by

$$g(t) = \sum_{k=-\infty}^{\infty} \delta(t - kT_s) \quad (T_s = \tfrac{1}{2}) \qquad \text{to produce } y(t).$$

(a) Sketch $x(t)$, $g(t)$ and $y(t)$

(b) Use convolution to find the Fourier transform of $y(t)$. Then sketch it.

(c) $y(t)$ is passed through the recovery filter whose transform is

$H(\omega) = \text{Rect}(\omega/8\pi)$

Sketch $H(\omega)$ and state what signal emerges, giving reasons. [20]

---

(8) (a) Find the **closed expression** for the DFT, F_n, of the following:

$f_k = (1, 1, 1, 0, 0, 0, 0, 0)$ $(N = 8)$

(b) Find A_n and B_n for each value of n $(0 \le n \le N - 1)$

(c) Prove: (i) If f_k is real ... real and N is even, then $F_{N/2}$ is also real. [20]

---

(9) (a) Make a neat sketch, showing all critical items, of

$f(t) = \cos^2(\pi t)\,\text{Rect}(t)$

(b) Find the Fourier transform of $f(t)$ by successive differentiation to show that

$$F(\omega) = \frac{4\pi^2 \sin(\omega/2)}{\omega(4\pi^2 - \omega^2)}$$

(c) Now use frequency shift to transform $f(t)$ and show that the same result is obtained. [24]

---

(10) In this problem you are asked to make a number of sketches. Be sure that they are neat and that all critical values are included.

(a) Sketch $g(t) = \cos(\pi t)\,\text{Rect}(t)$

(b) Use frequency-shift to show that $g(t) \Longleftrightarrow \dfrac{2\pi \cos(\omega/2)}{\pi^2 - \omega^2}$

(c) Sketch $x(t) = \dfrac{\cos(t/2)}{\pi^2 - t^2}$ $(-5\pi \le t \le 5\pi)$

Hint: First create the following table of values

t	0	π	2π	3π	4π	5π
x(t)						

(d) Now apply duality to find the Fourier transform of $x(t)$.

(e) Sketch $X(\omega)$ obtained in (d).

(f) Is $x(t)$ strictly band-limited? If so, what is the value of ω_{max}?

(g) According to the Nyquist criterion, what is the smallest frequency with which we can impulse-sample $x(t)$ and still be sure of perfect recovery?

(h) Based on (g) what is the largest sampling interval that we can use? Call it T_{max}.

(i) Sketch $x(t)$ multiplied by the impulse train with period T_{max}.

(j) Sketch the spectrum of the sampled signal. [28]

SMS 207 Final Examination, November 7, 1994

Question 2(e)

——> PRINT

Student Name	
Number	

n	0	1
A(n)		
B(n)		
|F(n)|		
θ(n)		
P(n)		

9 780471 128489